FINE WINE AND TERROIR
The Geoscience Perspective

Edited by R.W. Macqueen and L.D. Meinert

2006

Geological Association of Canada

Geoscience Canada Reprint Series 9

Library and Archives Canada Cataloguing in Publication

Fine wine and terroir : the geoscience perspective / edited by R.W. Macqueen and L.D. Meinert.

(Geoscience Canada reprint series 9)
Includes bibliographical references.
ISBN 1-897095-21-X

1. Viticulture. 2. Wine districts. I. Macqueen, R. W. (Roger Webb), 1935- II. Meinert, L. D. (Lawrence David), 1953- III. Geological Association of Canada IV. Series.

SB388.F55 2006 634.8 C2006-905682-X

Publisher
Geological Association of Canada
c/o Department of Earth Sciences
Memorial University of Newfoundland
St. John's, Newfoundland & Labrador, Canada A1B 3X5
Phone (709) 737-7660; Fax (709) 737-2532
E-mail publications@esd.mun.ca
Website www.gac.ca

The Geological Association of Canada gratefully acknowledges the financial support of the Canadian Geological Foundation for assistance in publishing this book.

ISBN 1-897095-21-X ISSN 0821-381X

Printing: TriCo Group, Ottawa
Cover Design: Dave Sargent
Layout: Karen Dawe, with historical files from Geoscience Canada

PRINTED IN CANADA

The **GEOLOGICAL ASSOCIATION OF CANADA (GAC)** is Canada's national society for the geosciences. It was established in 1947 to advance geology and its understanding among both professionals and the general public. The GAC membership of about 2500 includes representatives of all geological disciplines from across Canada and many parts of the world. It has specialist divisions for environmental earth sciences, geophysics, marine geosciences, mineral deposits, paleontology, Precambrian, sedimentology, structural geology and tectonics, volcanology and igneous petrology, remote sensing, geomorphology, and geographic information systems. Regional sections of GAC have been set up in Victoria, Vancouver, Edmonton, Winnipeg and St. John's, and there are affiliated groups in Canada and the United States.

GAC activities include the organization and sponsorship of conferences, seminars, short courses, field trips, lecture tours and student and professional awards and grants. The Association publishes the quarterly journal Geoscience Canada and the quarterly newsmagazine GEOLOG, a Special Paper series, Short Course Notes, GEOtext series, as well as a Geoscience Canada Reprint Series and Miscellaneous Publications. GAC also maintains liaison with other earth science societies and provides advice to government and the public on geological issues. The Association was incorporated under the Canada Corporations Act in January 1984.

For information contact: Geological Association of Canada, c/o Department of Earth Sciences, Memorial University of Newfoundland, St. John's, Newfoundland & Labrador, Canada A1B 3X5; E-mail: publications@esd.mun.ca; Internet site: www.gac.ca.

L'ASSOCIATION GÉOLOGIQUE DU CANADA (AGC) est la société nationale canadienne des sciences de la terre. Elle a été fondée en1947 dans le but de contribuer à l'avancement de la géologie et de sensibiliser le grand public aux sciences de la Terre. Elle comprendenviron 2500 membres, représentant toutes les disciplines géologiques et provenant de toutes les régions du Canada et d'ailleurs dans le monde. L'AGC comprend des divisions spécialisées en géologie de l'environnement, géophysique, géologie marine, gisements minéraux, paléontologie, géologie du Précambrien, sédimentologie, géologie structurale, tectonique, volcanologie et pétrologie ignée, télédetection, géomorphologie et systèmes d'information géographique. Des sections régionales ont été établies à Victoria, Vancouver, Edmonton, Winnipeg et Saint-Jean (Terre Neuve), et il existe des groupes affiliés au Canada et aux Etats-Unis.

Les activités de l'AGC comprennent l'organisation et le parrainage de conférences, de colloques, d'ateliers, d'excursions de terrain, de tournées de conférences et l'attribution de distinctions honorifiques pour étudiants et professionnels. L'Association publie le journal trimestriel Geoscience Canada, le bulletin d'information trimestriel GEOLOG, une série de volumes spéciaux, des notes d'ateliers, une série GEOtext, une série de Geoscience Canada Reprint et des publications divers. L'AGC maintient en outre des liens avec d'autres sociétés en sciences de la Terre et agit à titre de conseillère auprès des gouvernements et du public sur des questions d'intérêt géologique. L'Association a été incorporée en janvier 1984 en vertu de la Loi sur les corporations canadiennes.

Pour de plus amples renseignements, veuillez communiquer avec: L'Association Géologique du Canada, c/o Department of Earth Sciences, Memorial University of Newfoundland, St. John's, Newfoundland et Labrador, Canada A1B 3X5; Courriel : publications@esd.mun.ca; Site internet: www.gac.ca.

GAC thanks its many Corporate Supporters in 2006:

PATRONS

Anglo American Exploration (Canada) Ltd.
Memorial University of Newfoundland

SPONSORS

Alberta Energy & Utilities Board
Falconbridge Limited
Geological Survey of Canada (Calgary)
INCO Technical Services - Exploration (Copper Cliff)
Newfoundland and Labrador Department of Natural Resources
Northwest Territories Geoscience Office
Ontario Ministry of Northern Development and Mines
Royal Tyrrell Museum of Palaeontology
Saskatchewan Industry & Resources
Yukon Energy, Mines & Resources

MEMBERS (Corporate)

Aur Resources Inc.
Barrick Gold Corporation
BLM Juneau Mineral Information Center
IBK Capital Corp.
Johnson Geo Centre
Manitoba Industry, Economic Development and Mines
SRK Consulting
Strathcona Mineral Services Limited
Suncor Energy
Voisey's Bay Nickel Company Limited

UNIVERSITIES

Acadia University
Institut national de la recherche scientifique (INRS)
McGill University
University of Calgary
University of New Brunswick
University of Toronto
University of Victoria
Utah State University

TERROIR AND FINE WINE - The Geoscience Perspective

Geoscience Canada Reprint Series 9
Edited by R.W. Macqueen and L.D. Meinert

CONTENTS

NEW PAPERS

Papers from the session "Terroir, Geology and Wine – A Tribute to Simon J. Haynes", held at the Annual Meeting of the Geological Society of America, Seattle, Washington, November, 2003

This Volume is dedicated to the Memory of

Simon John Haynes, 1944-2002

Following a long battle with cancer, Professor Simon Haynes of the Department of Earth Sciences, Brock University, St. Catharines, Ontario, passed away peacefully at home in Fonthill, Ontario, on April 6, 2002, with his family at his side. Born in wartime England on January 1, 1944, Simon completed an Honours B.Sc. in Geology at Manchester University in 1965, then emigrated to Canada, becoming a Canadian citizen and practicing geology in many parts of the world from his new home country of Canada. Field work with a number of Canadian mining companies and consultants honed his early interest in ore deposits, leading to masters studies at Carleton University in Ottawa (M.Sc. 1969, on tin-bearing sills from Tasmania), then doctoral studies at Queen's University in Kingston (Ph.D. 1975, on metallogeny and plate tectonics, Atacama, Chile). Before completing his Ph.D., Simon served as Assistant Professor at Pahlavi University in Iran, 1972-1974.

Always a wine enthusiast, particularly from his time in Iran, perhaps Simon's most satisfying preoccupation of recent years was the marriage of geology and wine. Travels in France and elsewhere enabled him to learn much about terroir, the French term to express the relationship between wine and the setting in which it is produced including climate, soil and bedrock geology. At the time of his death Simon was actively publicizing the importance of terroir to the geological community and others by serving as Founder and Editor of the *Geoscience Canada* series, Geology and Wine, on which this volume is based. As editor of *Geoscience Canada* when the series began, I too was and am enthusiastic about this topic, and we were able to get the series underway quickly in 1999. Simon's studies of terroirs of the Niagara region were the subject of the series' first full paper. Commenting on the value of these studies, Peter Gamble, former Executive Director of the Ontario Vintners Quality Alliance (VQA), stated in an e-mail message to Simon: "Just a short note to thank you for the tremendously important work you've done in identifying the different grape-growing areas in Niagara....I wanted you to know how valuable your work will be in moving forward quality wine-making in Ontario. Such seminal work is crucial to those of us choosing to make fine winemaking our future." (e-mail to Simon Haynes, December 21, 2001). Simon will be fondly remembered for his leadership of many outstanding terroir field trips in the Niagara region, the location of the majority of Ontario's, and Canada's, wine production.

Although only 58 years of age at his death, Simon packed a lot of geological living into his life, teaching at Brock for 28 years from 1974 and carrying out geological field work and consulting studies in Chile, China, Cuba, Iran and Japan as well as in Canada. Although his particular interest was in gold deposits, he had a working and teaching knowledge of a broad spectrum of metallic deposit-types, sand and gravel, gypsum, even water in the 1990s. Of recent years Simon developed competence in new areas, including land-use and environmental geology assessments of residential/industrial real estate, especially in the Niagara region. Looking over his c.v., one is impressed with how broadly-based were his interests. In this age of specialization, Simon was a kind of "renaissance geological man" who was able to apply his talent for explanation and prediction to any geological setting, often leading to new levels of insight and understanding.

Some people have knowledge, experience and an infectious enthusiasm that they are delighted to share with others. Certainly in terms of geology and wine, and likely in all other endeavours, Simon Haynes was such a person. As *Geoscience Canada* Editor it was a pleasure for me to work with and learn from Simon. His influence on the many students he taught at Brock, and his published work in many fields including geology and wine, forms a lasting legacy. Simon also was a family man and enthusiastic practitioner of life. We dedicate this volume to his memory.

Roger W. Macqueen
Calgary, Alberta, Canada
Associate Editor, *Geoscience Canada* Geology and Wine series, 2002 -
Editor, *Geoscience Canada* , 1996-2001

ACKNOWLEDGMENTS

We are sincerely grateful for financial support which has helped enable the Geological Association of Canada (GAC) to offset the cost of colour in this book, and to keep the selling price at a reasonable level. In Canada, financial supporters include Brock University Department of Earth Sciences, St. Catharines, Ontario (Simon Haynes' former Department), and the Cool Climate Oenology and Viticulture Institute of Brock University, Inniskillin Wines of Ontario, and Vintners' Quality Alliance (VQA) of Ontario; and from British Columbia, CedarCreek Estate Winery, Kettle Valley Winery, and The Wine Museum in Kelowna. Domenica and Andy Panko of St. Catharines, long time friends of Simon Haynes, also contributed. From Washington State, USA, Chateau St Michelle and Woodward Canyon wineries helped with financial support.

Authors of all papers included here are thanked, including the 11 papers that appeared between 1999 and 2006 in GAC's Geoscience Canada's Geology and Wine series, as well as the six papers included herein from the Society of Economic Geologists Session "Terroir, Geology, and Wine: A Tribute to Simon J. Haynes", held at the Geological Society of America (GSA) annual meeting in Seattle, Washington in 2003, and organized by L.D. (Larry) Meinert, co-editor of this volume.

The Geological Survey of Canada (GSC), Calgary provided a pleasant environment for Roger Macqueen to receive and process manuscripts for Geoscience Canada's Geology and Wine series and the Seattle GSA session papers. David Sargent of GSC Calgary contributed significantly to the success of the Geology and Wine series, and particularly the Seattle GSA papers, mostly by revising figures as necessary and completing the cover of this volume. Very many thanks, Dave, for myriad jobs well done! Elizabeth Macey of GSC Calgary also helped with figure revision and layout for a number of papers.

Karen Dawe, GAC Publications Director, capably set up the six Seattle GSA session papers at GAC headquarters in St. John's, Newfoundland, as well as looking after printer selection and the proofing process. We appreciate the help of GAC Books Editor Professor Sandra Barr, Geology Department, Acadia University, who helped see this book to publication. Past Geoscience Canada editor Godfrey Nowlan and current editors Sonya Dehler and Steve McCutcheon skillfully edited Geology and Wine series papers 5 through 11.

We thank the many wineries and wine industry personnel from five countries on three continents who contributed knowledge, expertise, and in many cases wine samples for analysis (not to mention a certain amount of necessary sampling), many aspects of which underlie the papers included herein. Many of these people and wineries are acknowledged by name in the individual papers.

Finally, it is a continuing pleasure for both of us to see the great progress being made in the evolving quality of new world wines, including an increasing appreciation of the role and significance of the geologic settings of vineyards everywhere. Papers in this volume contribute to and are representative of this progress, which at its core is an improved connection to the land and all that this entails.

Roger W. Macqueen
Calgary, Alberta, Canada

Lawrence D. Meinert
Northampton, Massachusetts, USA

September, 2006

INTRODUCTION

Lawrence D. Meinert
Department of Geology
Smith College
Northampton, Massachusetts, 01063
USA

This book is about the production of great wine. Rather than being focused on wine itself, however, the articles in this volume explore the physical environment that underlies and shapes the character and quality of wine. The French have a word for all aspects of the wine environment, *terroir*. Even where grape-growing and winemaking are practiced to perfection there remains an underlying mystery to great wine that is exemplified by the rather simple occurrence of two vineyards, side by side, that share most obvious aspects of climate, slope and viticulture, yet produce wines that are vastly different. Examples abound, but perhaps the most spectacular are the vineyards of Burgundy, France (Fig. 1A, B). There the Grand Cru ("Great Growth" = highest quality) white wines of Puligny-Montrachet (Loftus, 1992) and the red wines of Romanée-Conti have been highly valued for centuries, some bottles selling for thousands of dollars, whereas nearby vineyards produce only *vin ordinaire* wine selling for less than a dollar a bottle. The simple question is, "Why?", and the answer lies at least partly in terroir. Thus, a related question is "What is terroir?"

Although there is no precise translation of terroir, Johnson and Robinson (2001) offer "growing environment", and understanding of terroir has become increasingly important in the siting of vineyards and the production of great wine (e.g., Pomerol, 1989). This in no way minimizes the all-important skills of the viticulturalist in growing the grapes and the oenologist in making the wine. Without these, great wine cannot be produced.

Simply put, terroir (pronounced tehr-wahr) involves the complex interplay of climate, soil, geology and other physical factors that influence the character and quality of wine (Wilson, 1998, 2001; Fanet, 2004). It is common, although usually incorrect, to point to a single factor as the explanation: "It's the soil." "It's the water." "It's the limestone." "It's the hours of sunshine". Terroir is the *integration* of all individual factors that contribute to wine quality. But to make matters even more complicated, there is the variable of time. What may be good terroir in one year may be less so in another. For example, in years that are relatively warm and dry, vineyard "X" with a particular slope, elevation, sun angle, soil type, and bedrock may produce better wine than vineyard "Y", whereas the reverse may be true in years that are cooler and wetter (Gladstones, 1992). Terroir in the French sense does involve the skills of viticulturists to maximize the benefits of the terroir, but here we focus on the environment in which the grapes grow and with which viticulturists must work for oenologists to produce fine wine.

The 17 papers in this volume explore the concept of terroir from a variety of viewpoints and from a number of scientific disciplines. The geographic range of vineyards studied is equally wide, encompassing British Columbia (Bowen *et al.*, Eggers *et al.*, Taylor *et al.*) and Ontario (Haynes) as well as other regions of Canada (Greenough *et al.*); California (Swinchatt *et al.*), Idaho (Gillerman *et al.*), Oregon (Jones *et al.*), and Washington (Meinert and Busacca) in the United States; France (Suthren *et al.*, Wilson) and Italy (Cita and Fiore) in Europe; and South Africa (Bargmann) in the southern hemisphere. What ties all of these studies together is the application of basic science to better understand the physical environment that affects grape and wine quality. The papers appearing here are not surveys of opinion or rankings of wine on an undefined 100 point scale, rather they are carefully controlled observational, experimental, and theoretical studies of real vineyards and the wines that are produced from them. Although scientific rigour is not the same as truth or understanding, an underlying premise of this book is that a scientific approach can contribute to this understanding, and is essential to put the study of the physical environment of terroir on solid footing.

The first 11 papers in this book were published over a seven year period in the Geology and Wine series of the Canadian quarterly journal, *Geoscience Canada*. Fittingly, the first two papers (Haynes, 1999b, 2000; also see Haynes, 1999a, Macqueen, 2002, and Ziraldo, 2000) were written by Simon Haynes, Founder and first Editor of the Geology and Wine series, to whom this volume is dedicated. In these papers Haynes lays out the scientific basis for studying terroir, noting that there are five main groups of factors in terroir: meteorological, physiographic, pedological, geological, and viticultural (Haynes, 1999b, p. 191). Most unfortunately, Simon Haynes died of cancer on April 6, 2002; at his request, Roger Macqueen took over as Editor of the *Geoscience Canada* Geology and Wine series. Subsequent papers in the series focus on the terroir of specific viticultural regions (Meinert and Busacca, 2000, 2002; Bargmann, 2003; Gillerman *et al.*, 2006), the application of geographic information systems (GIS) to the study of terroir (Jones *et al.*, 2004; Bowen *et al.*, 2005), and the use of modern geochemical methods to identify the provenance and character of wine (Taylor *et al.*, 2002; Greenough *et al.*, 2005). These geochemical studies are particularly important in that they demonstrate a close correlation between the chemical composition of wine and the physical environment of the vineyards, even to the point of being able to, on occasion, fingerprint individual bottles of wine relative to their provenance. This does not say anything directly

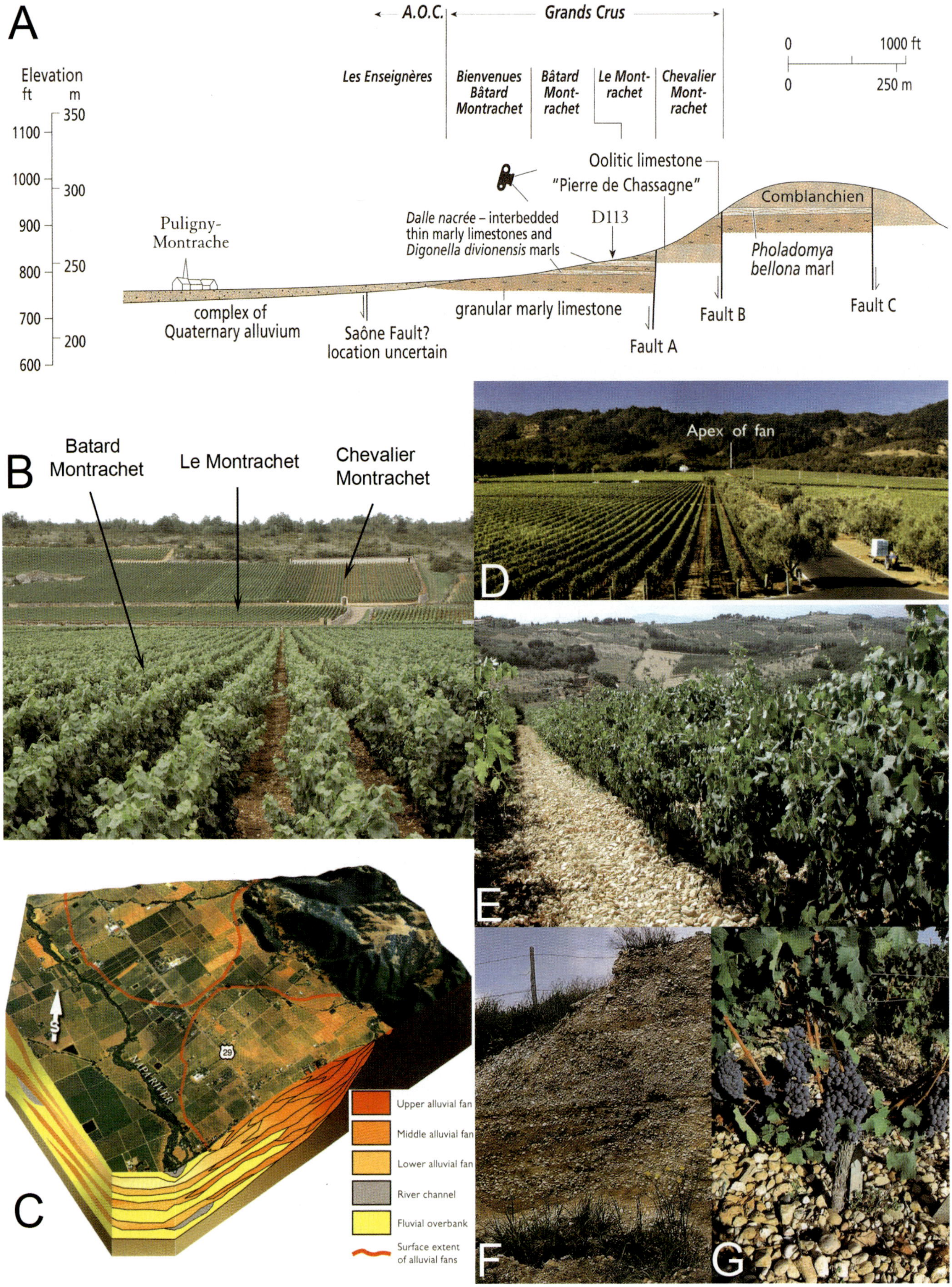
A
A.O.C.
Grands Crus
0
1000 ft
0
250 m
Elevation
ft
m
Les Enseignères
Bienvenues Bâtard Montrachet
Bâtard Mont-rachet
Le Mont-rachet
Chevalier Mont-rachet
1100
1000
900
800
700
600
350
300
250
200
Oolitic limestone
"Pierre de Chassagne"
Comblanchien
Puligny-Montrache
Dalle nacrée – interbedded thin marly limestones and Digonella divionensis marls
D113
Pholadomya bellona marl
granular marly limestone
complex of Quaternary alluvium
Saône Fault? location uncertain
Fault A
Fault B
Fault C
B
Batard Montrachet
Le Montrachet
Chevalier Montrachet
C
29
NAPA RIVER
Upper alluvial fan
Middle alluvial fan
Lower alluvial fan
River channel
Fluvial overbank
Surface extent of alluvial fans
D
Apex of fan
E
F
G

about terroir relative to wine quality, but it does establish an identifiable link between that physical environment and the resulting wine in the bottle.

The second group of six papers is published here for the first time and is an outgrowth of a terroir symposium held at the 2003 annual meeting of the Geological Society of America in Seattle, Washington. Like the first group of papers, these include terroir studies of specific viticultural regions (Cita and Fiore, 2006; Suthren *et al.*, 2006; Swinchatt *et al.*, 2006) and the use of modern geochemical and geophysical methods to identify wine provenance (Eggers *et al.*, 2006) and vineyard soil moisture characteristics (Hubbard *et al.*, 2006), respectively. In addition, there is a paper on global climate change and the potential effect that this may have on vineyards, grape production, and wine quality (Jones, 2006). Thus, the 17 papers in this book cover a wide range of topics in time - ancient to modern, and space - 3 continents, 5 countries, and more than a dozen viticultural regions.

While there is still much to learn, the collected papers of this volume provide a firm foundation for future terroir studies and a series of benchmarks for the understanding of terroir in existing vineyards. Although to some extent the terroir of each vineyard is unique, there are some common threads that are shared among the better vineyards. Probably the most important factor is the restricted availability of water – grape vines need to be stressed to a critical level for best results (the "Goldilocks" approach: just the right amount of water, not too much, not too little). Clearly, climate is the first order control of water, and it is an observational fact that there are no vineyards in rain forests or where root zones are saturated - this is too much water! Where natural precipitation is moderate to sparse, however, it is the *drainage* of that water – primarily a terroir factor - from the root zone of the vines that becomes most important. This is one connection to the predominance, in quality vineyards, of coarse clastic sediments/rocks, typically of alluvial, fluvial, or glacial origin. This is seen in the alluvial fans of Napa Valley, California (Figs. 1C, D) (Swinchatt *et al.*, 2006) and Tuscany, Italy (Fig. 1E)(Cita and Fiore, 2006); the glacial-fluvial terraces of Bordeaux (Fig. 1F, G) and New Zealand (Fig. 2A, B); and the glacial-lacustrine sediments of British Columbia (Bowen *et al.*, 2005) (Fig. 2C), Ontario (Haynes, 2000), Oregon (Jones *et al.*, 2004), and Washington (Meinert and Busacca, 2000, 2002) (Fig. 2D-F). Each of these studies documents the distribution and soil/bedrock drainage characteristics of selected vineyards: clearly, grain size and permeability in vineyard soils/bedrock are fundamental in governing the availability of water, be it natural precipitation or viticultural irrigation. Furthermore, the paper by Hubbard *et al.* (2006) demonstrates that soil moisture in the California settings of Napa and Sonoma Valleys can be mapped in detail using the advanced technique of ground-penetrating radar.

Two papers (Jones *et al.*, 2004; Bowen *et al.*, 2005) combine documentation of soil and bedrock characteristics with geographical information systems (GIS), which allows both the evaluation of existing vineyard suitabilities and the identification of new or underutilized terrains with superior vineyard potential. Bowen *et al.*'s (2005) paper takes this approach a step further by looking systematically at correlations between vineyard soil/bedrock characteristics and wine quality as measured by medals won in wine competitions. Although these authors recognize that wine medals are not a perfect indicator of wine quality, their stated assumption that better wines will tend to win more medals than worse wines allows some preliminary correlations between particular grain size and soil textures and the number of medals won. This is an important step toward extending the scientific study of terroir from strictly observational data to the elusive and highly subjective, but universally-desired, concept of wine quality.

The second important terroir factor, that of nutrient availability from soils and bedrock, is poorly understood at present, with opinions ranging from "key factor" to "of limited importance". The systematic, GIS-based mapping and observational studies of Patricia Bowen and colleagues in the Okanagan and Similkameen Valleys of British Columbia (Bowen *et al.*, 2005) offer the promise of progress in understanding how mineralogical and chemical composition, as well as soil/bedrock textural properties, may affect wine

Fig. 1 (at left) **(A)** Cross section through the Puligny-Montrachet vineyards, Côte de Beaune area, Burgundy region of France, showing subsurface faults and facies changes that underlie property boundaries between Grand Cru (Great Growth) vineyards and those of lesser quality (based on seismic, electrical resistivity, and geologic data; modified from Wilson, 1998, Fig. 4.17, p. 148). These Grand Cru vineyards are developed on pebbly, marly (= argillaceous) limestones on steep to gentle slopes. The A.O.C. (=Appellation d'Origine Contrôlée) wines, in contrast to the Grand Cru wines, are developed on a complex of Quaternary alluvium. **(B)** Position of Chevalier Montrachet, Le Montrachet, and Bâtard Montrachet Grand Cru vineyards, on a southeast-facing slope in the Côte de Beaune area. Vineyards are separated by stone walls that have existed for centuries and commonly demarcate property boundaries as well as underlying geologic discontinuities such as faults, and formation or facies boundaries. **(C)** 3-D block diagram illustrating sedimentary architecture of overlapping Oakville (top) and Rutherford (bottom) alluvial fans in Napa Valley, California (modified from Swinchatt and Howell, 2004). **(D)** Looking west from Opus One winery across vineyards on the Oakville alluvial fan, Napa, California, location as in (C) (modified from Swinchatt and Howell, 2004). **(E)** Sangiovese *vinifera* vines growing on alluvial fan gravels at Polvereto in Comune di Montespertoli, Tuscany, Italy. **(F)** Günz Pleistocene-aged gravel terrace in Bordeaux, France (from color plate 21, p. 180, Wilson, 1998). **(G)** Merlot vines growing on Günz terrace gravels at St. Julien, Bordeaux, France.

A
B
C
Person
D
loess
slackwater
sediments
E
Loess-colluvium
Gravel
lenses
Slackwater
sediments
F
G
H

quality, for better or for worse. At a broader scale, depositional processes such as floods and young glacial outwash are likely to restrict the amount of certain nutrients, such as those resulting from the slow decomposition of organic material, relative to residual soils developed over long periods of time. This, in addition to drainage, will have an impact on vine vigour such that the overall physical environment becomes inseparable from the ultimate characteristics and quality of wine. In writing about Burgundy, France, one of the world's great wine regions, Loftus (1992, p. 59), says this: "…the great wines of Burgundy are as much the products of minute variations (of soils and subsoil, aspect and drainage) which are individually almost imperceptible but collectively critical. It is the accumulation of such subtle particularities which distinguishes so clearly the character of each appellation."

Although climate is universally acknowledged as a first order control on possible vineyard locations, the reverse is also true, that terroir can influence climate, at least on the meso- and micro-scale. This ranges from the heat-absorbing and reflecting properties of rocks in vineyard soils (e.g., Figs. 1E-G, 2A, B) to the significant thermal mass of large bodies of water (Fig. 2C, G, H). The location and orientation of the water bodies shown in Figure 2C, G, and H reflect the structural grain and erosion-resistant properties of the bedrock. This is clearly seen in the north-south orientation of Okanagan Lake and similar glacial lakes in British Columbia (Bowen *et al.*, 2005). In climates marginal for wine production, like upstate New York and Ontario, the temperature-buffering effects of the lakes are quite marked, such that *vinifera* grapes can only be grown within a few kms or less of the lakes. This is dramatically illustrated by the snowfall distribution in a satellite photograph of northern New York State in early winter (Fig. 2H). Understanding the links between terroir and climate will become even more important in the future as changing climate on a global scale has the potential to alter how and where grapes are grown (Jones, 2006).

Interest in terroir shows no signs of abating. Many major international scientific meetings now include sessions on some aspect of wine and the physical factors that affect its production, characteristics, and medicinal properties. It is an increasingly interdisciplinary subject with geologists interacting with scientists in many other fields, even in the classic wine disciplines of enology and viticulture (for examples, see http://terroir.ucdavis.edu/). It is a fitting tribute to the *Geoscience Canada* Geology and Wine series and its founding editor, Simon Haynes, that terroir is taking its place among more traditional geological subjects in scientific meetings and publications, with the added benefit that the subject matter can be a very tasty beverage indeed.

REFERENCES

Bargmann, C.J., 2003: Geology and Wine 7. Geology and Wine Production in the Coastal Region, Western Cape Province, South Africa: *Geoscience Canada*, v. 30, p. 161-182.

Bowen, P.A., Bogdanoff, C.P., Estergaard, B.F., Marsh, S.G., Usher, K.B., Smith, C.A.S., and Frank, G., 2005, Geology and Wine 10. Use of Geographic Information System Technology to Assess Viticultural Performance, Okanagan and Similkameen Valleys, British Columbia: *Geoscience Canada*, v. 32, p. 161-176.

Cita, Maria B., and Fiore, Antonello, 2006, Grape Growing and Wine Making in Two Unique Italian Terroirs: Teroldego Rotaliano and Aglianico del Vulture: (p.165-181, this volume).

Eggers, N.J., Greenough, J.D., and Cernak, T., 2006, Classification of Okanagan Chardonnay Wines by Origin Using Volatile Components: (p.183-192, this volume).

Fig. 2 (at left) **(A)** The Ngaruroro River brings abundant gravel into the Hawkes Bay region of New Zealand, forming the Gimblett Gravel terrace. This is the world's first appellation defined on the basis of a stratigraphic unit, the Gimblett Gravel. **(B)** Streamcut bank exposes gravel substrate of vineyards in the Gimblett Gravel appellation. **(C)** View east across southern part of the north-south trending Okanagan Lake, British Columbia, showing vineyards on paleodeltas of glacio-lacustrine sediments deposited during previous high stand of glacial Lake Okanagan. **(D)** Exposure of multiple slackwater flood sediment layers in walls of Burlingame Canyon, Walla Walla, Washington. These units form the substrate for many vineyards in the Walla Walla region. Note person in upper right hand corner for scale. **(E)** Roadcut exposure near Touchet, Washington, of wind-blown loess overlying bedded slackwater sediments. Except in thickest loess accumulations, grapevine roots penetrate the loess into underlying slackwater sediments. **(F)** Cut bank in Tapteil vineyard, Red Mountain, Washington, exposing 3-4 metre cross section of Quaternary loess, slackwater deposits, gravel lenses, and loess-colluvium. **(G)** The mesoclimate of Wallula Vineyard, Washington, in foreground is tempered by the thermal mass of the Columbia River. The state of Oregon is on the far bank. **(H)** Satellite photograph in early winter demonstrating that the mesoclimates of most vineyards in Ontario, Canada, and New York state are tempered by large bodies of water including Lake Ontario (upper left) and the glacially carved, north-south trending Finger Lakes (lower part of photo) of New York state, as evidenced by lack of snow near Lake Ontario and the larger of the Finger Lakes.

Fanet, J., 2004, Great Wine Terroirs: University of California Press, Berkeley, California, 240 p.

Gladstones, J., 1992, Viticulture and Environment: Winetitles, Adelaide, Australia, 310 p.

Gillerman, V.S., Wilkins, D., Shellie, K., and Bitner, R., 2006, Terroir of the Western Snake River Plain, Idaho, USA: *Geoscience Canada*, v. 33, p. 37-48.

Greenough, J.D., Mallory-Greenough, L.M., and Fryer, B.J., 2005, Geology and Wine 9: Regional Trace Element Fingerprinting of Canadian Wines: *Geoscience Canada*, v. 32, p. 129-137.

Haynes, S.J., 1999a, Geology and Wine – New Series (Introduction): *Geoscience Canada*, v. 26, p. 189.

Haynes, S.J., 1999b, Geology and Wine 1. Concept of terroir and the role of geology: *Geoscience Canada*, v. 26, p. 190-194.

Haynes, S.J., 2000, Geology and Wine 2. A geological foundation for terroirs and potential sub-appellations of Niagara Peninsula wines, Ontario, Canada: *Geoscience Canada*, v. 27, p. 67-87.

Hubbard, S., Lunt, I., Grote, K., and Rubin, Y., 2006, Vineyard Soil Water Content: Mapping Small-scale Variability using Ground-Penetrating Radar: (p.193-202, this volume).

Johnson, Hugh, and Robinson, Jancis, 2001, The World Atlas of Wine: Octopus Publishing Group Ltd., London, U.K., 352 p.

Jones, G., 2006, Climate and Terroir: Impacts of Climate Variability and Change on Wine: (p.203-216, this volume).

Jones, G.V., Snead, N., and Nelson, P., 2004, Geology and Wine 8. Modeling Viticultural Landscapes: A GIS Analysis of the Terroir Potential in the Umpqua Valley of Oregon: *Geoscience Canada*, v. 31, p. 167-178.

Loftus, Simon, 1992, Puligny-Montrachet: Journal of a Village in Burgundy: Henry Holt and Co., New York, N.Y., 309 p.

Macqueen, R.W., 2002, Geology and Wine – Retrospect and Prospect: *Geoscience Canada*, v. 29, p. 147, 148.

Meinert, L.D., and Busacca, A.J., 2000, Geology and Wine 3: Terroirs of the Walla Walla Valley Appellation, Southeastern Washington State, USA: *Geoscience Canada*, v. 27, p. 149-171.

Meinert, L.D., and Busacca, A.J., 2002: Geology and Wine 6. Terroir of the Red Mountain Appellation, Central Washington State, USA: *Geoscience Canada*, v. 29, p. 149-168.

Pomerol, C., 1989, The Wines and Winelands of France – Geological Journeys (English edition of Terroirs et vins de France, Itinéraires Oenologiques et géologiques, 1984: Editions BRGM, Orléans, France, 350 p): Robertson McCarta Ltd., London, U.K., 370 p.

Swinchatt, J., Howell, D.G., 2004, The Winemaker's Dance: Exploring Terroir in the Napa Valley: University of California Press, Berkeley, California, 248 p.

Swinchatt, J., Howell, D.G., and Silacci, M., 2006, A Review of Napa Valley, California, Terroir: (p.217-231, this volume).

Suthren, R.J., Fowler, M.B., and Guion, P.D., 2006, Terroir, Wine and Geoscience Fieldwork Education in Languedoc-Rouissillon, Southern France: (p.233-246, this volume).

Taylor, V.F., Longerich, H.P., and Greenough, J.D., 2002: Geology and Wine 5. Provenance of Okanagan Valley Wines, British Columbia, using trace elements: Promise and Limitations: *Geoscience Canada*, v. 29, p. 110-120.

Wilson, J.E., 1998, Terroir: The role of geology, climate, and culture in the making of French wines: Mitchell Beazley, London, UK, 336 p.

Wilson, J.E., 2001: Geology and Wine 4. The Origin and Odyssey of Terroir: *Geoscience Canada*, v. 28, p. 139-141.

Ziraldo, D.J.P., 2000, Anatomy of a Winery - the Art of Wine at Inniskillin: Key Porter Books Ltd., Toronto, Canada, 56 p.

NEW SERIES

Geology and Wine

Simon J. Haynes
Department of Earth Sciences, Brock University, St. Catharines, Ontario L2S 3A1
simon@craton.geol.brocku.ca

This new series of articles will examine the relationship of geology to wine and the role of geology as an integral part of the *terroir* of a vineyard. The first article, which follows in this issue, presents the background to the subject and discusses the concept of *terroir.* The next article in this new series will examine geology and wine in terms of the world-wide perspective, while the article following will describe the geological factors that influence wines produced from vineyards of the Niagara Region. Although future articles will address specific regions inside and outside Canada, it is hoped that these first three articles will inspire other authors to submit manuscripts, particularly on the wine-growing regions of British Columbia, Nova Scotia and Quebec. Perhaps someone from Newfoundland will write the definitive paper on the relation of that province's own partridgeberry wine to geology: did the berry evolve after the breakup of Pangaea?

GUIDELINES FOR THE SERIES

Articles covering all aspects of the relation of wine to the earth sciences are encouraged, provided a direct relation is established between the earth science data and wine. Rehashing of geology field guides without evaluation of actual vineyard environments is not acceptable, whereas reworking of previous regional studies of clay mineralogies or geophysical surveys, in terms of vineyard terroir designations, will be gratefully appreciated. If you are uncertain, please contact me by E-mail. Articles may be in the form of Topical Articles (1000-4000 words) or Short Notes (about 1000 words or less) detailing or discussing the relationship of earth science data to wine. Longer articles in the form of Review Papers (2000-7000 words) will be considered. Such papers might discuss, for example, the interaction of wine and earth sciences with society (*e.g.,* effect of vineyard expansion on the Niagara Escarpment, a geologically defined United Nations biosphere preserve), or how such interactions may have implications for professional geologists.

There is no restriction of wine to *Vitis vinifera* table wines. Not only are articles sought on all types of *vinifera* wines, including ice wines of Canada and sun-dried wines of Chile and Italy, but also spirits derived from their distillation, like the *pisco* of the Andes, *grappa* of Italy, and *cognac* of France. Also included are grape wines from other species of the subgenus *Euvites* such as the *Vitis labrusca* of North America or the *Vitis amurensis* of northeast China. In keeping with the vastness of Canada and its harsh northern climate, wines made from indigenous fruits, like the partridgeberry, cranberry and blueberry, are welcome. How else can I keep the St. John's folk from complaining that their three wineries (Rodrigues, Notre Dame and Flynn) had been left out of the series, as yet another mainland scheme to exclude Newfoundland? Rodrigues Winery produces 12,000 cases of partridgeberry wine a year, and in Ontario, the hot-selling new wine is Stoney Ridge's cranberry wine. Unfortunately, it is difficult to source the *terroir* of fruit wines, as batches of fruit from different areas are often mixed together.

In wine, may I wish you welcome to a new century, a new millennium, and a new series.

Geology and Wine 1. Concept of Terroir and the Role of Geology

Simon J. Haynes
Department of Earth Sciences
Brock University
St. Catharines, Ontario L2S 3A1
simon@craton.geol.brocku.ca

SUMMARY

Although Peigi Wallace presented a paper on the geology of wine at the 1972 International Geological Congress in Montreal, until recently there have been few publications on this topic. However, a remarkable book by J.E. Wilson appeared in 1998 on *terroir* and the role of geology, climate and culture in the making of French wines. Wilson emphasizes vineyard geology as a precursor to the soil, and the subsoil through which the grape vine grows, and that geology controls the topography and landform type present, thus influencing drainage and microclimate. This paper examines the ideas of Wilson and other authors on the concept of *terroir,* and discusses the key role of geology in wine grape regions.

RÉSUMÉ

Depuis l'article de Peigi Wallace sur la géologie du vin présenté au Congrès international de géologie de Montréal en 1972, bien peu de publications sont parues sur le sujet. Cependant, en 1998, un livre remarquable de J.E. Wilson a été publié sur le terroir et le rôle de la géologie, du climat et des méthodes culturales dans la production des vins français. Wilson y fait état de l'importance déterminante de la géologie sur les couches du sol et du sous-sol où poussent les vignes; on y souligne également que c'est la géologie qui détermine la topographie et les formes du relief, d'où son importance sur le drainage et le microclimat. Le présent article analyse les idées de Wilson et d'autres auteurs sur le concept de terroir et discute du rôle clé de la géologie pour les régions vinicoles.

INTRODUCTION

In 1972, Peigi Wallace presented a paper on the relation of geology to wine entitled "Geology of Wine" at the 24th International Geological Congress, Montreal in Section 6: Stratigraphy and Sedimentology, convened by Digby McLaren and Gerry Middleton (Wallace, 1972). In true Canadian fashion, this Canadian first has been largely ignored in Canada, but oft-quoted in Europe by wine authors. At about the same time Peigi was presenting her talk, I had taken up a position in Shiraz, Iran, the type locality of one of the oldest-known wine grapes in the world, and was assisting a crew of Italian chemical engineers from the local oil refinery in drinking both their white Shiraz *vin nouveau* and the *grappa* that they were distilling from it.

During the lapse of 26 years until 1998, the only English-language publication on the specific role of geology to wine written by geologists was a French book by the BRGM (Bureau de Recherches Géologiques et Minières) edited by Pomerol, *Terroirs et Vins de France,* (1984); translated as *The Wine and Winelands of France, Geological Journeys* in the English version (1989). Although a definitive field guide to the geology of the French wine districts, full of interesting details such as the dates of wine festivals, and an essential accompaniment to a study of "wine and geology," it lacked information on climate and hydrogeology. Also, the geology presented, although excellent at the regional level of a French wine region (*e.g.,* the Côte d'Or of Burgundy) and often applicable at the meso-level of a wine district (*e.g.,* the Côte de Beaune, the southern part of the Côte d'Or), had only an indirect relation to the micro-level of a specific village area (*e.g.,* the *commune* of Aloxe Corton, the only Grand Cru area of Pinot Noir grapes in the Côte de Beaune).

By utter coincidence, 1998 proved to be an amazing *grande année* (vintage year) for geology and wine publications. In July, Lawrie Mintner (of Witwatersrand gold fame) led a field excursion, "Geology of the Cape Winelands," for the International Volcanological Congress in Cape Town; the guidebook (Mintner, 1998) is the only publication on the geology of African wine *terroirs.* The following month, I led a field trip, "Geology of Niagara Falls and Niagara's Vineyards and Wines," for the International Mineralogical Association meeting in Toronto, the guidebook of which (Haynes *et al.,* 1998) is the first geological publication on *terroir* and wine of the Americas, although there are previous geological field guides to the Niagara (Haynes, 1992, 1994a,b) and New York (Fakundiny *et al.,* 1989) wine districts.

However, the *crème de la crème* of the century was the publication in late 1998 of the 336-page singularly authoritative treatise, *TERROIR: the Role of Geology, Climate, and Culture in the Making of French Wines.* It was the culmination of about 20 years of post-retirement research by the ex-Shell Petroleum geologist, James E. Wilson (Wilson, 1998). The importance of this book cannot be overestimated, as it clearly outlines the major role of vineyard geology as both a precursor to the soil and the subsoil through which the grape vine will grow, as well as the determining factor for the topography and type of landform present, which themselves control the drainage and microclimate. It is the first publication illustrating the use of geology in defining a *terroir* not only at the micro level of a village but also, in some areas (*e.g.,* Aloxe Corton), down to the level of a specific vineyard, and the geologic reasons underlying its designation as Grand Cru, Premier Cru, or Commune wines. As far as I know Wilson (1998) is also the first person to use seismic and resistivity geophysical surveys to link vineyard designation to sub-surface bedrock geology and the hydrogeologic conditions of the overlying transported sediment.

TERROIR: MEANINGFUL OR MEANINGLESS?

The Concept in France

Most French-English dictionaries define *terroir* as "soil" and herein lies the source of a controversy that has been raging among English-speaking wine scientists and wine writers for decades. The implication is that, other than climate and the skill of the wine maker, soil controls the grape and hence the quality of the wine. The problem has been compounded in recent years by *terroir* having become the buzzword of the wine industry, mandatory in practically everything written in English about wine, including the label on the bottle, often without any understanding. This has led to a profusion of English-language definitions, and understandable confusion as to the correct meaning of the word.

No such confusion exists in France where the word originated. This is because France, as the world's largest grower of wine grapes, has had two millennia of hands-on viticultural experience, coupled with the last millennia of trial-and-error scientific study undertaken by the monastic orders. Thus in France, it is almost a matter of religious belief that the suitability of a particular grape variety and the general characteristics (smell, taste, colour, body, texture, *etc.*) of the wine produced from it are determined by the regional climatic and geologic setting of that particular wine district in any given wine region. These characteristics must be different, not only from other regions but, also, between districts in the same region. Indeed, I can personally attest that wines from red Pinot Noir and white Chardonnay grapes grown at Aloxe Corton, Côte de Beaune, Burgundy are totally different from those grown at Bouzy, Montagne de Reims, Champagne.

To me, the very essence of the concept of *terroir* is that the specific quality of the wine produced from a particular grape variety at the micro-level of a vineyard is a fundament of the complex interrelationships of <u>all</u> the factors, above and below the ground, that affect the grape during growth, excluding pests, diseases, herbicides, mutations, *etc.* Generally, these factors may be divided into five main groups: meteorological, physiographic, pedological, geological and viticultural. Examples of these factors include, but are not restricted to:

1. Meteorological: temperature maxima and minima, hours of sunlight, wind conditions, and rainfall;
2. Physiographic: type of landform, elevation, slope aspect and gradient, and slope drainage;
3. Pedological: composition and porosity of overlying soils, soil mineralogy and chemistry, soil grain size and texture, and clay mineralogy;
4. Geological: geology of the subsoil and the geochemistry, petrology and texture of individual strata, surface water and ground water flow rates and chemistry, and;
5. Viticultural: trellising method, row spacing, grape-bunch thinning and allowable production, fertilizing, mechanical addition of soils or rock material, systematic tile-draining, and irrigation.

Outside France, it is not unusual for the geological and even physiographic groups to be dropped from consideration, they somehow being included as part of soil science; as if geology and physical geography are not recognized as separate sciences. Or is it that the sub-surface, being unseen, doesn't exist? The geological part of the complex equation that expresses *terroir* is exemplified in the book *Terroirs et Vins de France* by Pomerol (1984; English version, 1989). Notwithstanding the above, *terroir* is claimed by the French to be untranslatable, as it is an experience of the wine lands gained by the *vigneron*: the "wine grower." The French stress that they are the only nation that has wine growers, everyone else has grape growers. From my personal experience in Champagne and Burgundy (August, 1999), the word *vigneron* appears to be used mainly for owner-growers of Grand or Premier Cru vineyards, the lower quality 3me classe (3rd class) *commune* and *vin du pays* vineyards being attended by viticulturists or agriculturists.

James E. Wilson in his book (Wilson, 1998) illustrates better than anyone the geological logic of the concept of *terroir*. An example of his studies, the Hill of Corton, is given in Figures 1 and 2 (see also the photograph on the cover of this issue). Here, for hundreds of years, the *vignerons* have known from trial and error where the vineyards that produce Grand Cru (Corton red and Corton Charlemagne white), Premier Cru red, and Commune wines should be located (Comité Interprofessionel des Vins de Bourgogne, undated). However, it took geologists to demonstrate that the boundaries of these vineyard designations corresponded exactly with the stratigraphic limits of: the Pernand marl (for the Grand Crus), with a layer of ferruginous oolite marking its lower boundary; the Dalle nacrée, an oyster-shell fossilferous lime flagstone (for the Premier Crus), and; the Digonella divionesis marl and limestone (for the Commune wines). The Hill of Corton is only one of two small *terroirs* in the Côte d'Or that supports Grand Cru white wine from Chardonnay grapes and, along strike of the Pernand marl, the only terroir of the Côte de Beaune supporting Grand Cru red wine from Pinot Noir grapes. Wilson (1998) suggests that the Chardonnay grapes prefer the cooler conditions of the west-southwest face of the hill more than the Pinot Noir on the south and south-southeast face (Figs. 1, 2), and that, " ...apparently the Pernand marl develops a particularly clayey facies on the slope which the Corton-Charlemagne occupies." Figure 1 shows the boundaries (from Comité Interprofessionel des Vins de Bourgogne, undated) of individual Grand and Premier Cru vineyards (the largest of which is 800 m along strike and 200 m across strike; most are less than 200 m in either direction) in the commune of Aloxe-Corton. Distribution of these vineyards is: Corton Charlemagne Grand Cru (four vineyards, in total 35 ha), Corton Grand Cru (14 vineyards, in total 85 ha), Corton white and red (Grand Cru) (three vineyards, in total 16 ha) and Premier Cru (eight vineyards, in total 29 ha). Each of these vineyards has a name (omitted in Figure 1 for the sake of clarity), which is its *terroir* and the name put on the label of all bottles of wine from this vineyard. The wineries may have land in several of these vineyards, so individual winemakers may produce wine with the same winemaking techniques from different vineyards, although bottled separately. This allows direct comparison of differences in the wine (colour, taste, bouquet) to differences in *terroir*, such as down-dip geologic changes and along-dip slope aspect changes. Such distinction of *terroir* at this micro-level is unique because the land has been split into smaller parcels over the long history of land-ownership in Burgundy.

One thing that must be stressed is that the French have never downplayed the skill of the winemaker. Like a good chef, the winemaker is restricted by the quality of the product he is handling, and that quality depends significantly on the five factors noted above. Unfortunately, in the New World this is usually forgotten and good wines are often attributed only to the skill of the winemaker and/or the viticulturist, with total disregard for the effects of Mother Nature on grape quality. In France, however, it is a matter of faith that only the best wine is possible from the best *terroir*, that good wines may/may not be produced from an intermediate *terroir*, and that a poor wine is the product of a poor winemaker and/or an unacceptable *terroir*. In countries outside France, the concept of *terroir* is best endorsed in the German and Italian vineyards and is endorsed in principle for the vineyards of South Africa's Western Cape, many of which were established in the 17th century. In the New World, *terroir* has either not yet been adequately quantified (Argentina, Canada and Chile) or has been rigorously opposed, as in California, Australia and New Zealand where the success of certain California wines in beating out French wines in several competitions has had a profound influence. Here, opposition by the growers to the concept of *terroir* has been supported by academics from: geography, such as Moran (1988) from New Zealand, who considers *terroir* a myth used as an excuse for increased land values, or; agriculture, like Gladstone and Smart (1994) from Australia, who wrote the following paragraph, highly detrimental to geology, as part of their explanation of the term "terroir" in Jancis Robinson's (1994) wine glossary, *The Oxford Companion to Wine*:

> An implication is that GEOLOGY, often cited as a basis of terroir, has in general no more than an indirect role. To varying degrees parent ROCK materials do contribute to the natures of the soils derived from them; they also shape local topography and therefore MESOCLIMATE. Occasionally the parent materials contribute directly because vine roots can penetrate them, as in the case of CHALK subsoils and the recent ALLUVIAL deposits of the Médoc. In the broad sense, however, it remains the soil itself, and its water relations, that play the decisive role. *The study of terroir is one of soil and mesoclimate, not of geology*" (capitalized words in the original; italics added by SJH).

Notwithstanding that they appear to restrict geology to petrology and lithology (in their section, they list geomorphology and hydrology separate from geology), it is instructive to compare their views with the quintessentially French opinion of Madame Bize-Leroy, the head of Maison Leroy and co-owner of Domaine de la Romanée-Conti (the Grand Cru estate in Burgundy), who in a letter to James E. Wilson wrote:

> Suppose that one day we might know the nature of all the elements which compose this nourishing "new earth" between the pebbly surface [the topsoil] and the Jurassic sub-basement [bedrock], would we then be able to determine exactly the physical and chemical influence of these elements on the intrinsic, fundamental character of the wine issued from them? Personally, I do not believe so, for all of this is too dynamic – I would say almost alive – to put in an equation. *I remain convinced, however, that the fundamental character of each wine depends on the nature of its subsoil*. (italics added by SJH) (Wilson, 1998, p.118)

ROLE OF GEOLOGY

Although geology and pedology must play a vital role in wine quality, the other factors of *terroir* — climate, drainage and viticultural practices — must play

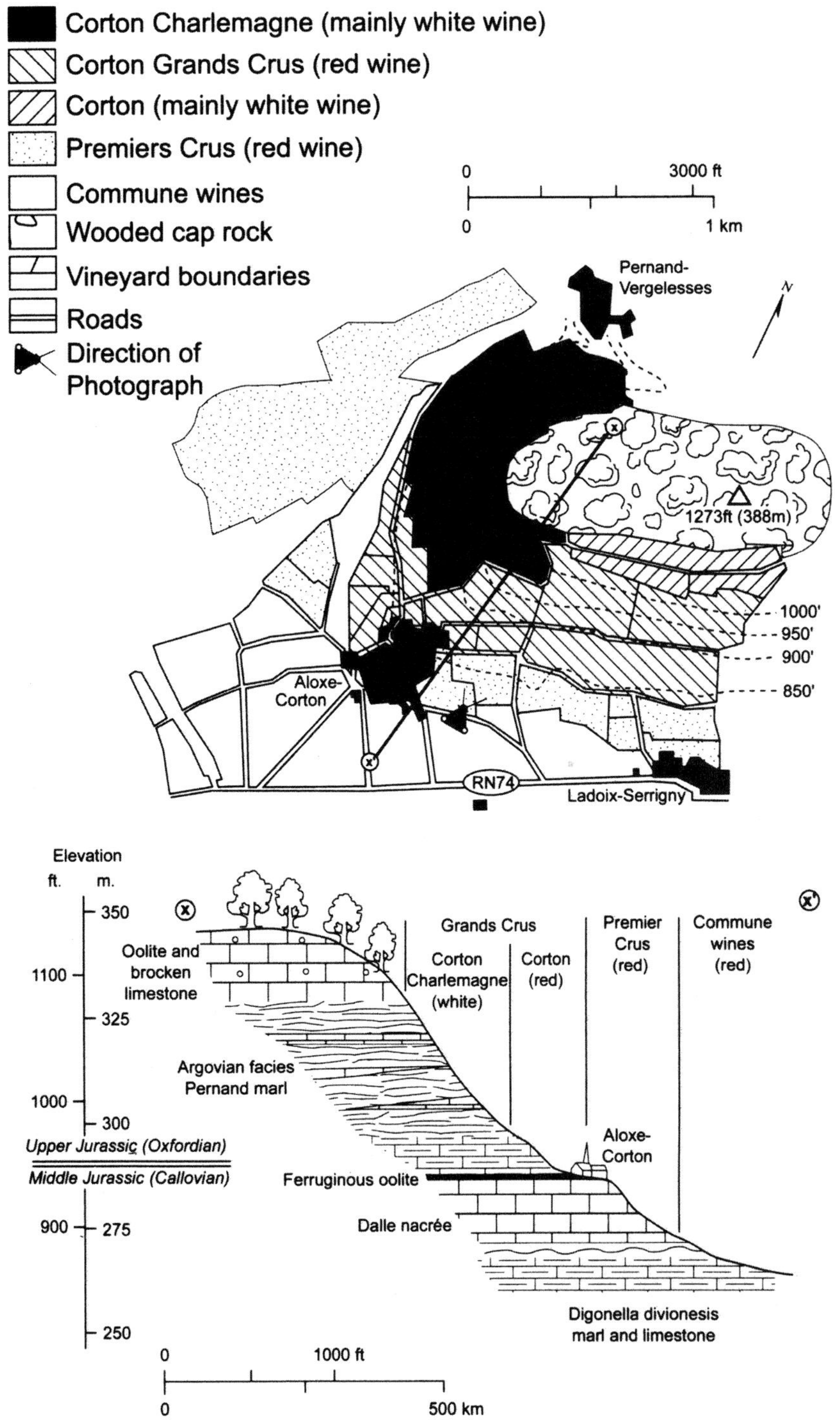

Figure 1 Plan and cross-section of the Hill of Corton, Côte d'Or wine-producing region of France, (a butte) showing the relationship of the geology to vineyard designations as Corton Charlemagne Grand Cru, Corton Grand Cru, Corton (white and red), Premier Cru (red), and Commune wines. Based on Figure 4.15 of Wilson (1998) with modification and additions of vineyard boundaries and designations from Comité Interprofessionel des Vins de Bourgogne (undated).

their part too; hence the fact that local meso- and micro-climates control the yearly difference (a good or bad "vintage"). However, given the same climate, viticulturist and winemaker, I hope Madame Bize-Leroy is right because I believe that the subsoil geology is the determining factor controlling the taste and bouquet of a wine (the very essence or "soul" of a wine). I consider that the soil primarily affects young vines in their early growth stages, but once the vine is established it is affected mainly by the geological conditions of the subsoil. Evidence for this is that it is well known from areas of old-growth vines that the wine from old vines tastes different than young vines (often better but sometimes worse). Grapes are exceptionally deep-rooting plants that dislike wetlands and grow best on well-drained soils. In older plants, the roots can penetrate more than 4 m in coarse-textured, porous sediment (Wilson, 1998, Fig. 6.3). Moreover, Bryan Coombe (1994) in his section on "root" in the *Oxford Companion to Wine* states,

> Although most vine roots occur in the top metre of soil......there are many examples where roots have penetrated to 6 m/20 ft or more. (Robinson, 1994)

Such depths are well below the A and B soil horizons, and usually below the C horizon, in either Quaternary or Tertiary unconsolidated sediments or underlying bedrock. The nature of the resultant wine encompasses the diverse facets of geology encountered by the deeply rooting grape vine as it penetrates ever downward, year after year, through the different lithologies, each with its own mineralogy, fabric and lithochemistry and specific ground water chemistry and flow rate. Thus, between different areas (communes) of a wine district, geology is highly unlikely to be a "fixed equation" as it must include, at a minimum, differences between the relevant inter-connecting factors of: physical geology (types of landform, elevation above sea level, azimuth to the sun, slope aspect and gradient.); petrology (mineralogy, texture, porosity and geochemistry of both the bedrock strata and/or the overlying unconsolidated sediments) and; hydrogeology (surface water and groundwater flow rate, direction and chemistry).

Although we are still awaiting the Ontario legislature to move on the issue of licensing geoscientists, in my opinion the professional associations of geoscientists of British Columbia and Nova Scotia (where geoscientists are licensed) should be insisting that Professional Geologists be involved in the decision-making process of the Provincial and Municipal governments and the Vintners Quality Alliance in all aspects of Canadian wine appellations and vineyard licensing. Why? Because in France, and do not forget this is the other half of Canada's cultural heritage, the refined Appellation laws of 1935 protect the quality of the winemaking regions by definition of the winery and vineyard management techniques as well as the physical environment of geology, soil and climatic conditions of the vineyards under the auspices of the INAO (Institut National des Appellations d'Origine). INAO employs a team of about six scholars, two of whom are geologists, to assess each situation. It would also appear that geophysicists may have a major input into vineyard *terroir* designation in the next century, given the evidence of Wilson (1998) that the historic distinction of several *terroirs* in Burgundy is supported by geophysical data that has revealed geological differences in the substrata.

Notwithstanding the obvious importance to geologists of the above, wine is indeed serious business, but fun to drink. And where else in the earth sciences, in the words of James E. Wilson, "can you drink on the job"? Since 1992, I have been taking time out to imbibe a large variety of wines all over the world, and ponder on their relationships to the geology of their root zones. These ruminations have led both to my offering a course, "Geology of Viticultural Regions," next year for Brock's Cool Climate Oenology and Viticulture Institute, and serving as series editor of this new *Geoscience Canada* Series.

Figure 2 Hill of Corton, Côte d'Or wine-producing region of France, showing the vineyard locations for several famous wines: Corton Charlemagne Grand Cru white, Corton white and red, Corton Grand Cru red, and Corton Premier Cru. See the cover of this issue for a colour view, and Figure 1 and text for details (photo by Simon Haynes, August 1999).

ACKNOWLEDGMENTS

I thank journal reviewers Gerry Middleton and Godfrey Nowlan, whose comments helped to improve the paper. I am indebted to Roger Macqueen for his considerable help and enthusiasm in getting this new series into print so quickly.

REFERENCES

Coombe, B.G., 1994, ROOT, *in* Robinson, J., ed., The Oxford Companion to Wine: Oxford University Press, Oxford, UK, p. 831.

Comité Interprofessionel des Vins de Bourgogne, undated, Aloxe-Corton, Bourgogne: Comité Interprofessionel des Vins de Bourgogne, rue Henri-Dunant, 21200 Beaune, France, Appellation Data Sheet and Plan, 2 p.

Fakundiny, R.H., Cadwell, D.H. and Fleisher, P.J., 1989, Geology of the Wine Country of New York: American Geophysical Union, 28th International Geological Congress, Field Trip Guidebook T388, Washington, DC, 64 p.

Gladstone, J. and Smart, R., 1994, TERROIR, *in* Robinson, J., ed., The Oxford Companion to Wine: Oxford University Press, Oxford, UK, p. 966-967

Haynes, S.J., 1992, Geology, wine and scenery of the Niagara Escarpment: Field Guidebook, 31st Annual Meeting, Ontario Petroleum Institute, Niagara Falls, 1992. Published as: Department of Geological Sciences, Brock University, Urban Geology Series 1, St. Catharines, ON, 6 p.

Haynes, S.J., 1994a, Geology, Scenery and Wines of the Niagara Escarpment: Geological Association of Canada, Mineralogical Association of Canada, Joint Annual Meeting, Waterloo 1994, Field Trip A3, Guidebook, 10 p.

Haynes, S.J., 1994b, Wine, Geology and Glaciolacustrine Soils of the Niagara Escarpment: 33rd Annual Meeting, Ontario Petroleum Institute, Niagara Falls 1994, Field Guidebook, 13 p.

Haynes, S.J., Grant, E.B. and Haynes, V.S., 1998, Geology of Niagara Falls and Niagara's Vineyards and Wines: International Mineralogical Association, General Meeting, Field Trip B7, Guidebook, 11 p.

Moran, W., 1988, The wine appellation: environmental description or economic device?: Second International Cool Climate Viticulture and Oenology Symposium, Auckland, New Zealand 1988, Proceedings, p. 356-360.

Mintner, L., ed., 1998, Geology of the Cape Winelands: International Volcanological Congress, Cape Town - July 1998, Field Excursion B3, Guide Book, 24 p.

Pomerol, C., ed., 1989, The Wine and Winelands of France; Geological Journeys: Éditions du BRGM, Orleans, France.

Robinson, J., ed., 1994, The Oxford Companion to Wine: Oxford University Press, Oxford, UK, 1088 p.

Wallace, P., 1972, Geology of wine: International Geological Congress, 24th Session, Montreal, Section 6, Stratigraphy and Sedimentology, p. 359-365.

Wilson, J.E., 1998, Terroir: the Role of Geology, Climate and Culture in the Making of French Wines: Mitchell Beazley, London, UK, 336 p.

Accepted as revised 12 November 1999

SERIES

Geology and Wine 2. A geological foundation for *terroirs* and potential sub-*appellations* of Niagara Peninsula wines, Ontario, Canada

Simon J. Haynes
Department of Earth Sciences
Brock University
St. Catharines, Ontario L2S 3A1
Simon@craton.geol.brocku.ca

> No verse can give pleasure for long, nor last, that is written by drinkers of water.
> *Horace*

SUMMARY

All of the vineyards in Niagara are planted on soils derived from a variety of Late Pleistocene to Early Holocene, subglacial and proglacial sediments. These were deposited during a series of Late Wisconsinan continental ice advances and retreats, on a landscape that had been carved into the Paleozoic bedrock by this and earlier Quaternary glaciations to form the several landforms that exist above, on, and below the cuesta of the Niagara Escarpment. This paper proposes that the existing delineation of grape viticultural areas, by climate zone, be considerably modified by the recognition that it is the significant differences in the geology that control distinction of *terroirs* and sub-*appellations* based on variation of landform physiography, soil types, and groundwater flow conditions. From considerations of these geological and climatic differences, in combination with wine profiles distinguished by J-L. Groux, winemaker at Hillebrand Estates, the Niagara Peninsula Designated Viticultural Area can be divided into at least eight sub-*appellations*.

RÉSUMÉ

Tous les vignobles de la région du Niagara croissent dans des sols issus d'une variété de sédiments sous-glaciaux et pro-glaciaux de la fin du Pléistocène et du début de l'Holocène. Ces sédiments ont été déposés lors d'avancées et de retraits glaciaires wisconsiniens sur un substratum de couches ordoviciennes sculptées par l'action de glaciations qui ont précédées et qui ont laisser les nombreuses formes de relief que l'on peut voir sur et de part et d'autre de la cuesta de l'escarpement du Niagara. Dans le présent article, nous proposons que la délimitation actuelle des diverses zones viticoles par zone climatique soit révisée en profondeur afin de tenir compte de variations significatives de nature géologique et qui déterminent les différents terroirs et sous-appellations en fonction de la physiographie du paysage, des types de sol et des paramètres de l'écoulement hydrique. En combinant de telles considérations géologiques et climatiques aux différents profils viticoles établis par J.-L. Groulx, viticulteur au domaine Hillebrand, la région viticole du Niagara pourrait être subdivisée en huit sous-divisions au moins.

INTRODUCTION

Best known internationally for the magnificent panorama of Niagara Falls, the Niagara region of Ontario is also Canada's principal producer of *Vitus vinifera* (wine grapes of European origin) and French hybrid (crosses between *Vitus vinifera* and North American species) wine grapes. Although several of the wineries are winning international medals, particularly for icewines but also for white and red table wines, it is often not appreciated that the latitude (43°N) of Niagara is the same as the famous red wine regions of Provence in France, Chianti in Italy, and Rioja in Spain. At this latitude, near the midpoint between 30° and 50°, the hours of daylight and high summer temperatures are perfectly suited to the growing of grapes (Stevenson, 1997). But because the best areas for viticulture lie between the 10°C and 20°C annual isotherms (de Blij, 1983; Unwin, 1991), the marked depression of the 10°C annual isotherm in North America (Fig. 1) creates a shorter growing season and winter severity in the Niagara Peninsula that makes the climate marginal, even as a cool-climate viticultural region. Most *Vitus vinifera* cannot endure sustained temperatures below about -20°C, whereas North American varieties such as the Concord (a *Vitus labrusca* cross) are hardy to -29°C (Smart, 1999). However, it is this same winter severity that makes Niagaran wineries the world's leading producer of icewine (see, WINES, below).

When I arrived at Brock University in September 1974, Don Ziraldo and Karl Kaiser were getting ready to pick their first vintage and open Inniskillin Wines. In what was then the Department of Geological Sciences, the late Peter Peach had started a soil mineralogy project on samples collected by the late Jaan Terasmae along a north-south transect across the northern Niagara Peninsula in order to relate glacial and post-glacial deposits to vineyard and orchard development. This led to Brock's first (and probably Canada's?) wine-related geological thesis (Gombos, 1977), 20 years before Brock opened its Cool Climate Oenology and Viticulture Institute. Unfortunately these early geological efforts were not continued and, because Niagaran viticulturists have been fully occupied in determining which grape varieties are best suited to both the growing conditions and market taste, little attention has been paid, until recently, to the effect of different geological *terroirs* in the Niagara Peninsula on the nose, palate and quality of different wine varietals. In contrast, local differences between adjacent vineyards in France, including geology, can result in one designated a *Grand Cru* and the other a simple *Vin de Pays* (Pomerol, 1989; Wilson, 1998). However, as Tim Unwin, the editor of Journal of Wine Research states in his book, *Wine and the Vine* (Unwin, 1991):

> This diversity is not only the outcome of differences in geology and climate, but it is also the result of the labour of countless generations of vine growers and wine makers, each set in their own distinctive human context. (p. 1)

Not denying the quintessential value of the viticulturists and the winemakers, there is a lack of balance in arguments that disregard vineyard geology on the grounds that: 1) the vineyard soils of Niagara have been modified by installation of sub-drains, and 2) the wine quality and flavour profile is due entirely to climate, viticultural practices (planting density, trellising techniques, canopy management, type of clone, type of rootstock, *etc.*) and the skill of the winemaker. Fortunately, I have long been encouraged in my quest to geologically define the Niagara Peninsula *terroirs*: first by David Parker, Brock geology graduate and cellarmaster at Cave Spring Cellars; next by Matthew and Paul Speck, co-owners of Henry of Pelham; and then by Paul Bosc, the Algerian-born, French-trained winemaker and owner of Château des Charmes. The results of this encouragement are five *Geology and Wine* field trip guides (Haynes, 1992a, 1994a,b, 1999a; Haynes *et al.*, 1998) and two "popular press" publications (Haynes, 1995c, 1998). However, 29 September 1999 saw both Paul-André Bosc (son of Paul Bosc) of Château des Charmes, and Jean-Laurent (J-L.) Groux, the French-born and -trained winemaker at Hillebrand Estates, release wine data in support of Niagara Peninsula *terroirs* at the 1999 Niagara Grape and Wine Festival seminar on "*Terroir*," at Brock University. Although these data are unpublished, both individuals have kindly permitted me to use their information, which has assisted considerably in the intent of this paper to quantify, in terms of geological *terroirs*, the recent statement by Tony Aspler (editor of *Winetidings* magazine) in his book, *Vintage Canada* (Aspler, 1999):

> Eventually, the geographic designations of the Niagara Peninsula and Lake Erie North Shore will be fragmented into smaller sub-zones, just as Bordeaux is broken down into communes such as St. Emillon, Pauillac, Graves, *etc.* There is a case for wines grown on the bench of the Niagara Escarpment to be differentiated from those grown on the Lakeshore plain since their taste profiles are perceptibly different. (p. 25)

It must be emphasized that this paper is concerned with the variation of wine taste/flavour profiles in relation to differences in vineyard *terroirs* and is not an assessment of different wineries. For a variety of reasons, many wineries combine grape juice from a variety of vineyards (and different growers) and many wineries blend wines prior to bottling. Thus, only those estate wineries that were known to the author to produce separate wines from specific vineyard soils are considered.

WINES OF THE NIAGARA PENINSULA

Early Developments

Historically, the first Niagaran wines are believed to have been offerings of fermented wild grape juice by aboriginal tribes to the gods, who lived at the foot of Niagara Falls, in a ceremony known as the "Wischgimi" (Aspler, 1999). Commercial wine production in Ontario until the late 1960s was based predominantly on grapes derived by American horticulturists from cross-pollination of European (*Vitus vinifera*) cuttings with native American grape varieties (mainly *Vitus labrusca*). The first crosses were the accidental results of unsuccessful attempts to grow *vinifera* cuttings in Virginia by Lord Delaware in 1619, and in Philadelphia by John Alexander (Governor John Penn's gardener) in the 1770s. Alexander's discovery resulted in the first commercial development of a North American grape hybrid (the Cape), around 1800, but this was soon eclipsed by newer American hybrids: the Isabella in 1816, and Catawba in 1823 (Aspler, 1999). Although the Catawba is still in use in Ohio today for wine (Borrello, 1995), the most successful commercial developments of *V. labrusca* crosses were the blue-skinned Concord in 1843 in Concord, Massachusetts (Schreiner, 1985) and the white Niagara in 1868 in Lockport, New York State (Aspler, 1999); both these varieties as well as several other *V. labrusca* crosses are still used in wineries around the Great Lakes in New York, Pennsylvania and Ohio and are the main source for grape juice and jam in Ontario.

20th Century Developments

Early in the 20th century, New York State's Geneva horticultural experimental station began planting hybrid vines developed by the pioneer French breeder Albert Siebel and, later, his fellow hybridizers François Baco, Eugene Kuhlmann, J-L. Vidal, and Bertille Seyve-Villard (whose names are honoured by the grape varieties Baco Noir, Vidal and Seyval Blanc). These hybrids were developed in France to overcome problems with weather and the *phylloxera* louse (a root-feeding aphid), which (together with powdery and downy mildew and other fungal diseases) had been the main reason for the failure of *Vitus vinifera* grapes in North America and which, when introduced to France on experimental North American native grape stock, had decimated the French vineyards in the latter half of the 19th century. Although tens of thousands of these "French hybrid" grape varieties were developed, most of them produced poor wines, so that by the 1950s the French began to limit their plantings, as they were concerned that they would displace traditional *V. vinifera* varieties to the detriment of their wines (Schreiner, 1985).

Beginning in about 1880, an alternative to using hybridization (cross-pollination) in order to resist infestation was the grafting of European *Vitus vinifera* cuttings onto a variety of North American native *Vitus* species rootstocks: the most effective species were *Vitus berlandieri, Vitus rupestris* and *Vitus riparia* against the *phylloxera* louse, and *Vitus aestivalis* against mildew. Today, most of the world's wine is produced from *V. vinifera* grafted onto a wide range of rootstocks that are custom selected for shallow- to deep-rooting soil conditions.

In 1946, Adhemar de Chaunac, the French-born winemaker at Bright's Wines (now part of Vincor), and J.R. van Haarlem, the grape specialist at Ontario's Vineland Horticultural Research Institute, began importing cuttings of both French hybrids and *Vitus vinifera*. This led in 1951 to the first successful commercial planting of *V. vinifera* (Chardonnay) east of the Mississippi in North America, and in 1956 the first *V. vinifera* table wines marketed by a Canadian winery (Schreiner, 1985).

Although wineries in Ontario and New York prospered with French hybrids from 1956 to the 1980s, it was the pioneering work of the Ukrainian-born Dr. Konstantin Frank, who in 1962 introduced wines from *Vitus vinifera* grapes grown at his winery in New York's Finger Lakes area (Borrello, 1995), that confounded the official skepticism of the viticulturists at both Ontario's Vineland and New York's Geneva research stations that *V. vinifera* grapes could not flourish. Frank convinced Paul Bosc, then the winemaker at Chateau-Gai, to plant *V. vinifera* red varieties such as Pinot Noir, Cabernet, Merlot and Gamay Beaujolais. Wines produced from these varieties in 1972 were impressing wine writers (Schreiner, 1985). However, it was the emergence of estate wineries, beginning with Don Ziraldo (the viticulturist) and Karl Kaiser (the winemaker) establishing Inniskillin Wines in 1975, and Paul Bosc's own winery, Château des Charmes in 1977, that set Niagara and the rest of Ontario on the track of quality table wines produced primarily from a wide range of both white and red *V. vinifera* grapes. An important aspect of progress was the work of Dr. John Paroschy on identification of winter-injury resistant *V. vinifera* clones or mutations.

French hybrids are still grown, but only the white Vidal, which is the mainstay for icewine (see below), has significant plantings, although the red Baco Noir and Maréchal Foch, and the white Seyval Blanc and Vidal (particularly as a "late-harvest" wine) have achieved some success as table wines. Today, in Niagara, the common *Vitus vinifera* grapes are: the white Chardonnay, Riesling, Gewurztraminer, Pinot Gris, and some Aligoté, Auxerrois, Muscat, Pinot Blanc, Sauvignon Blanc, Viognier; and the red Cabernet Franc, Cabernet Sauvignon, Gamay Noir, Merlot and Pinot Noir, and some Petite Syrah, Syrah and Zweigelt.

Interestingly, the Niagara wineries have become the world's largest producer of icewine. Although icewine originated in Germany, production there is sporadic because, unlike the Niagara Peninsula, consistent freezing conditions in early winter cannot be relied upon. Icewine is produced by leaving grapes to overripen on the vine until they freeze in early winter. The frozen grapes are hand-picked and pressed in the cold so that water is removed as ice crystals. The resulting "freeze-dried" liquor, which is enriched in natural sugars, is fermented slowly to produce an aromatic sweet wine of high alcohol content (10-13%) and high acidity (ideally). However, production costs are very expensive and young icewines retail at about $50 a half bottle, and older vintages may fetch double this amount. Icewines are made mainly from Vidal grapes and to a lesser extent Riesling, but also Gewurztraminer and Seyval Blanc grapes.

Canadian Wine Classification

In Canada, there are four distinct provincial wine regions that contain separate viticulture districts (Fig. 1). In descending order of size, these are Ontario (Niagara Peninsula, Lake Erie shore, Pelee Island), British Columbia (Okanagan and Similkameen valleys, Fraser Valley, Vancouver Island), Quebec (Montérégie and Estrie), and Nova Scotia. In France, the individual viticulture districts of a wine region display characteristic climatic, geologic and physiographic features and are termed an *appellation*. However, that term is defined by a large number of factors decreed by the refined Appellation laws of 1935 and each new application is evaluated by a team of about six scholars, including two geologists, under the auspices of the INAO (*Institut National des Appellations d'Origine*). The legal control is implicit, as the term *appellation* is short for *Appellation d'Origine Contrôlée* (AOC or AC). Furthermore, in France, an *appellation* can be subdivided into sub-*appellations* (communal *appellations*), and even further down to the specific vineyard (a *climat*), in terms of relative differences in its *terroir*. Outside France, the term "appellation" is used in a loose sense as a relatively large geographic area of *Vitus vinifera* vineyards that often corresponds in France to the zonal *appellations* of the separate category of French wine classification of *vin de pays*, which are areas outside the AOCs that have less strict regulations (Stevenson, 1997). In Canada, such geographic areas are officially termed Designated Viticultural Areas (DVAs) and are allocated in Ontario and British Columbia by independent alliances, the Vintners Quality Alliances (VQA), who determine the regulations governing the right of vintners to use a DVA on their labels (*e.g.*, Ziraldo, 1995). In Ontario, the largest DVA is Niagara Peninsula; the others are Lake Erie north shore and Pelee Island (Figs. 1, 2). VQA regulations governing the production and sale of wine have recently become separate provincial laws in British Columbia and Ontario, but federal legislation is needed for sale of Canadian wine in the European Economic Community. It is hoped that the separate provincial professional geoscientist associations will be pushing for representation on the VQA to give geological substance to determination of *appellations* (DVAs) and sub-*appellations*.

NIAGARA REGIONAL GEOLOGIC SETTING

The cuesta landform of the Niagara Escarpment (Fig. 2) was carved by the Quaternary glaciations (Fig. 3), and the present river valleys and waterfalls by Holocene erosion into Paleozoic bedrock. Much of the region's glaciogenic sediment was derived originally by erosion of this Paleozoic bedrock.

Paleozoic History

The Niagara Region is located on the western margin of the Appalachian Basin on the east flank of the Algonquin Arch, a prominent feature that, today forms the topographic spine of southwestern Ontario between lakes Ontario-Erie and Lake Huron, but which developed as the border between the Appalachian and Michigan basins during the Silurian and Devonian (Fig. 2). In the Ordovician (Johnson *et al.,* 1992), this area was a shallow subtropical sea that changed during the Late Ordovician–Early Silurian Taconic Orogeny to foreland deltaic, and locally to shallow marine and non-marine epicontinental environments of the Upper Ordovician red shales of the Queenston Formation (Fig. 4). During the Silurian, a shallow epeiric sea developed across the Algonquin Arch, which itself was subject to episodic uplift and subsidence. This resulted in the Early Silurian accumulation of shales and sandstones of the Medina Group, then shales and dolostones of the Clinton Group, followed later in the Middle Silurian by dolomitic reefal and algal carbonates of the Lockport, Eramosa and Guelph formations (Figs. 2, 4). This culminated in the Upper Silurian with subtidal to supratidal evaporites of the thick Salina Formation (Haynes, 1999c) and algal subtidal to intertidal dolostones of the Bertie and Bass islands formations (Haynes, 1992b). As a consequence of continental plate tectonic movements in the Michigan and Appalachian basins and base-level fluctuations, the Silurian sequences tend to thin, or pinch out, toward the Algonquin Arch, and display a number of interfingering relationships that have caused controversy in stratigraphic nomenclature and correlation (Brett *et al.,* 1994).

The Early Devonian is marked by extensive subaerial erosion that was followed by sandy and dolomitic carbonates, and then, in the Middle Devonian, with shallow tropical marine reefal carbonates and calcareous shales that intertongued with nonmarine silty and sandy clastics derived from orogenic uplift of the adjacent Appalachians by the Acadian Orogeny. The Upper Devonian saw most of eastern North America covered by the Chattanooga Sea, a broad, shallow, inland seaway, west of the Acadian Orogen, that deposited a thick accumulation of black, organic-rich shales (Johnson *et al.,* 1992).

Following filling of the Michigan and Appalachian basins with Mississippian and Pennsylvanian sediments, Permian compression in the closing stages of the collision of North America with Africa (to form Pangaea), created the thrust sheets of the Appalachian Front (Fig. 2) during the Alleghanian (Appalachian) Orogeny, and uplifted much of the Great Lakes area. This was followed by erosion and deposition of nonmarine craton-derived clastic sediments, with local marine incursions related to breakup of Pangaea and formation of the Atlantic Ocean, from the Late Triassic to the mid-Cretaceous. Following deposition of the Mesozoic, all the bedrock in Ontario appears to have undergone a period of erosion that continues to the present, but was particularly severe during the Quaternary glaciations (Johnson *et al.,* 1992).

The vineyards of the Niagara DVA are located in the northern Niagara Peninsula, where they are underlain by Upper Ordovician and Middle Silurian bedrock on, and north of, the Niagara Escarpment (Figs. 2, 4). Of the other wine regions in the Great Lakes region (Fig. 2), only the Finger Lakes is of comparable size in terms of production of *Vitus vinifera* grapes to the Niagara district. This, and most of the other Great Lakes wine regions, are located on Devonian bedrock: except the Bass Islands-Sandusky district of Ohio which is on Upper Silurian bedrock (Haynes, 1994c). Interestingly, that part of New

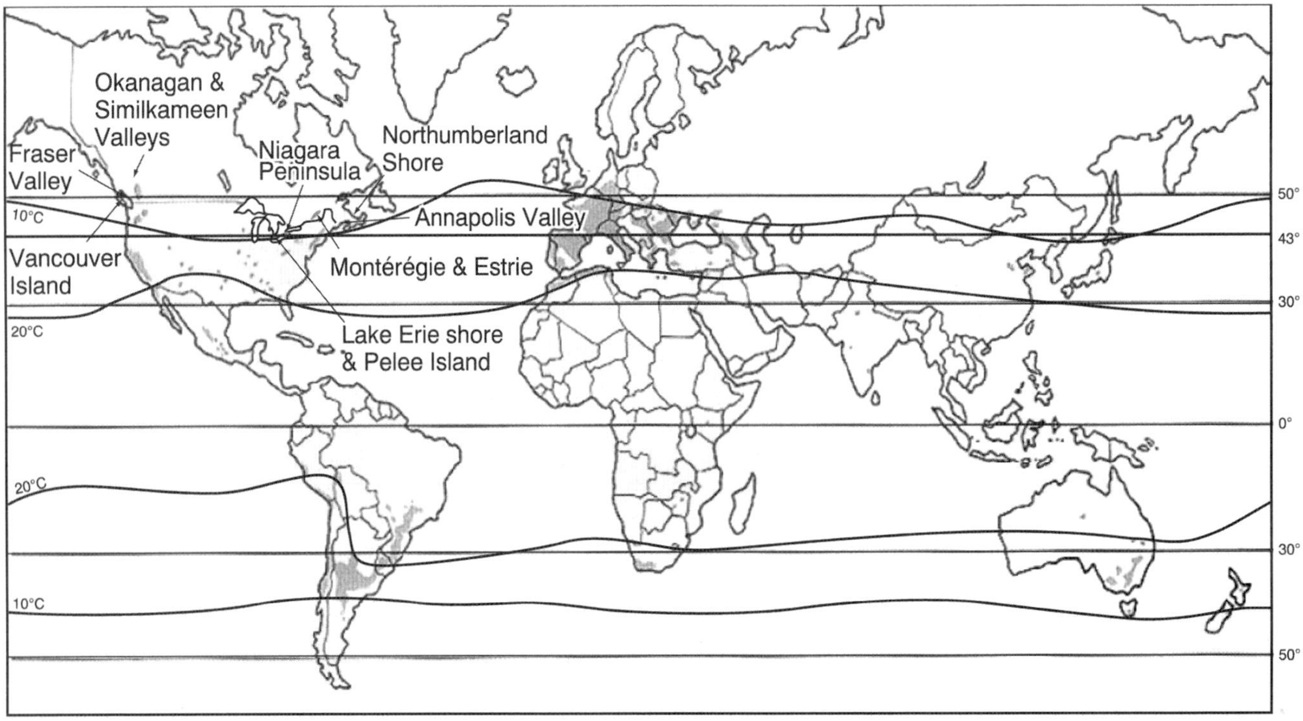

Figure 1 Location of Canadian viticulture districts in relation to the world's major areas of grape cultivation (shaded) between latitudes 30°-50°N and S (Stevenson, 1997) and the 10°C and 20°C annual isotherms (de Blij, 1983; Unwin, 1991).

York State contiguous to the Niagara DVA, north and along the Niagara Escarpment, has the same geology but has yet to be developed for vineyards.

Quaternary History

All of the vineyards in Niagara are planted on soils derived from Late Pleistocene to Early Holocene sediments. As well, the development of the present landforms of and around the vineyards, including the variable physiography of the Niagara Escarpment (a cuesta), resulted mainly from a series of Quaternary continental glaciations that carved the bedrock and laid down glacial sediment (till), consisting of rock fragments in clay or sandy silt, both during advance of the ice sheet and retreat of the ice front (Fig. 3). Quaternary deposits in southern Ontario (Barnett, 1992; Karrow, 1989) appear to represent only the last 190,000 years but include the Illinoian Glacial Stage (*circa* 135,000-190,000 years BP), the Sangamonian Interglacial Stage (*circa* 115,000-135,000 years BP), the Wisconsinan Glacial Stage (*circa* 10,000-115,000 years BP, when Ontario was covered by the Laurentide Ice Sheet), and the Holocene (*circa* 10,000 years BP to the present).

During interglacial periods between glaciations, large lakes (glaciolacustrine) and rivers (glaciofluvial) formed in front of the northward retreating ice sheets (Fig. 3). Because of the erosive power of each of these continental glaciers, usually only the results of the last glaciation (the Late Wisconsinan) are visible in most areas. Notable exceptions include: the Short Hills valley (Fig. 5), which was formed as a pre-Late Wisconsinan river gorge complete with a buried waterfall similar to the present Niagara Falls, and the wide upper terrace on the Escarpment in the Vineland area (Fig. 5) that resulted probably from a combination of ice carving by older glaciers and shore erosion by earlier lakes. As the ice front finally retreated north from the Escarpment, a series of lakes formed between the ice and the Escarpment, with deposition of glaciolacustrine clays and silt over the till on the terraces. Several smaller, local lakes were formed in the intervening period: for Niagara, Lake Warren (first-formed around 12,800 years BP) is the most important, as the large ice-marginal fan-delta at Fonthill (Fonthill Kame, Fig. 5) was constructed in a small re-entrant in the glacier margin during the early formation of Lake Warren. Late stages of Lake Warren are indicated by deposition of near-shore sediments and evidence of wave-cut platforms and shore bluffs near the crest of Fonthill (Feenstra, 1981). However, the steep northern slope of the Fonthill Kame indicates an ice-contact feature Feenstra (1981). As the ice front retreated north of Toronto, a large lake (Lake Iroquois) formed in front of it about

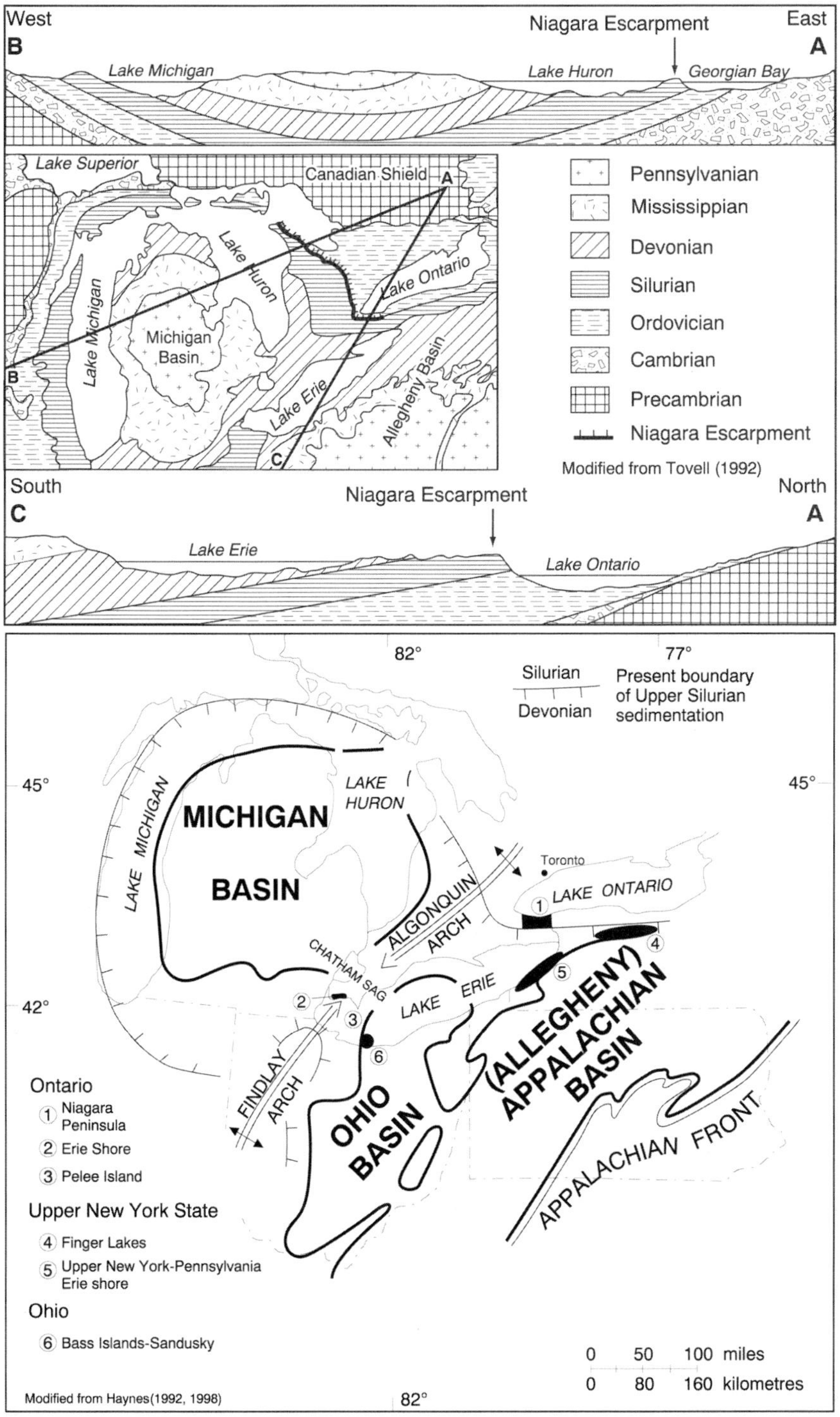

Figure 2 Geology of the Great Lakes (**top**) and relation of the wine grape regions of Ontario, Upper New York state, Pennsylvania and Ohio to major structural elements of the Appalachian, Michigan and Ohio basins, during Silurian time (**bottom**).

12,000 years ago and eroded the earlier glacigenic sediments, forming a shore bluff as a prominent ridge of beach deposits, and, below the shore bluff, lake deposits of stratified sand, silt and clay.

Glacial Lake Iroquois reached its maximum extent about 11,800 years BP; but as the ice retreated north of Brockville, Early Lake Ontario formed when water was able to drain out of Lake Iroquois to the Champlain Sea, which had also formed as melting of the ice sheet allowed the Atlantic Ocean to transgress up the valleys of the St. Lawrence and Ottawa rivers. The Champlain

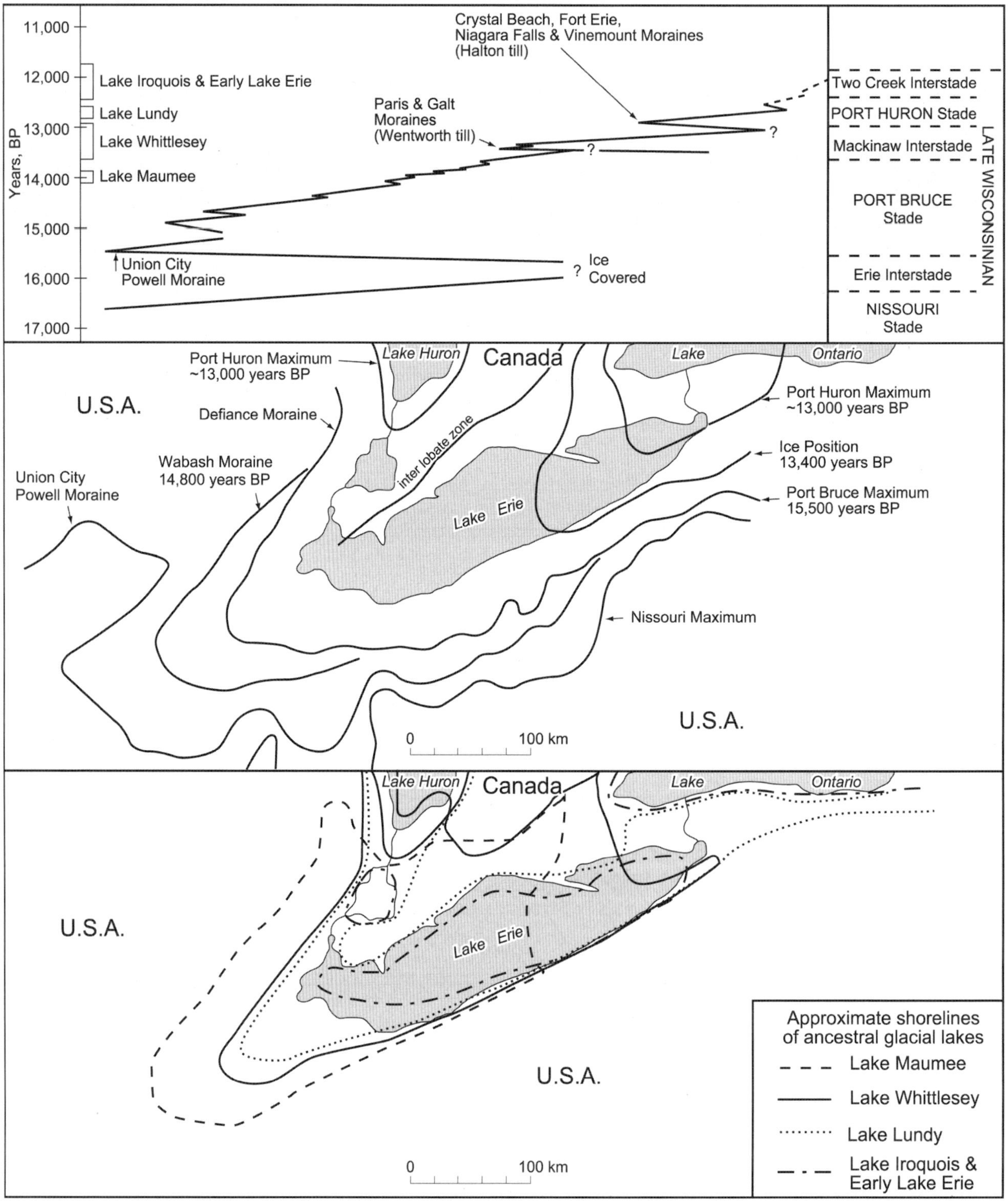

Figure 3 Generalized geochronology of glacial retreat and advance events in southern Ontario (after Barnett, 1985, 1992; Menzies, 1994.)

Sea reached its maximum extent about 11,000 years BP, but its level had fallen considerably by the beginning of the Holocene (about 10,000 years BP). At about this same time in the Niagara Peninsula, much of the area on either side of the Niagara River south of Niagara Falls, which had been inundated by northeastward-flowing waters connecting Lakes Wainfleet (that drained water from Lake Erie) and Tonawanda (whose water had drained over the escarpment at three spillways in New York and at the position of the present Niagara River) was differentially uplifted in the east, caused drainage to reverse to the west (Tesmer, 1981; Tinkler, 1994). Originally, the Niagara River had flowed through a pre-glacial gorge (the buried St. David's Gorge) that during Late Wisconsinan glaciation became filled with glacial sediment (Hobson and Terasmae, 1969; Tesmer, 1981). At the close of glaciation and the beginning of the Holocene, the Niagara River cut a new path from the pre-existing Lake Tonawanda spillway over the Escarpment back along its present path to intersect the old gorge and form a right angle bend to create the famous Whirlpool about 5000-6000 BP (Tinkler *et al.,* 1994). This was about the same time as the earliest records of cultivation of grapes for wine in the Middle East; where later, in the ancient Persian empire, official government policy dictated that (Buckley, 1994), "government ministers used to take important decisions when drunk, in the belief that the divine influence of wine would guide them."

GEOLOGY AND PHYSIOGRAPHY OF THE NIAGARA PENINSULA DVA

South of an east-west line through Niagara Falls, there is no commercial production of grapes, and the Haldimand Clay Plain (Fig. 5) is underlain, progressively southward, by the Upper Silurian Guelph, Salina and Bertie/Bass Islands formations (Fig. 4), and stops at the Onondaga Escarpment, which is capped by the Lower Devonian Bois Blanc and Middle Devonian Onondaga formations (Figs. 4, 5). North of Niagara Falls, there is commercial production of *Vitus labrusca* grapes on the Haldimand Clay Plain, the Vinemount Moraine, and the Fonthill Kame (Fig 5). These are underlain by the Middle Silurian Eramosa Member, the upper part of the Lockport Formation (Figs. 4, 5, 6), the only exposure of which is at the crest of the upper Niagara Gorge (Fig. 7a).

The principal area of *Vitus vinifera* grape wine production is located in the northern Niagara Peninsula, from the Vinemount Moraine across the Niagara Escarpment to Lake Ontario (Figs. 5, 6). Detailed descriptions of the geology, physiography and geomorphology of the Niagara Escarpment and the northern Niagara Peninsula are included in Bolton (1957), Brett *et al.,* (1994), Cheel (1991), Feenstra (1981, 1984), Haynes and Fraser (1994), Haynes (1995a,b), Hobson and Terasmae (1969), Johnson *et al.* (1992), Kingston and Presant (1989), Menzies and Taylor (1998), Tesmer (1981), Tinkler (1994), Tinkler *et al.* (1994), Tiplin and Seibel (1988), and Tovell (1992). Chemistry of ground-

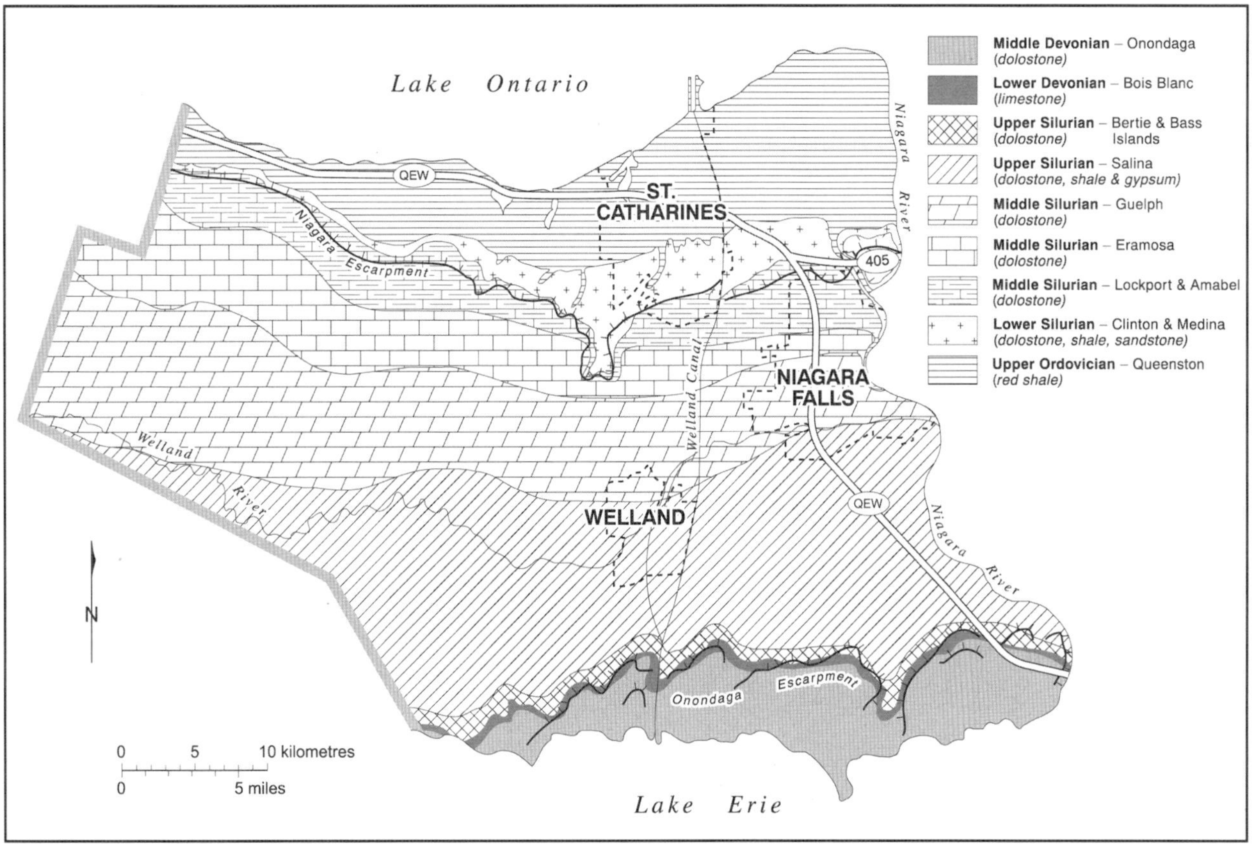

Figure 4 Bedrock geology of the Regional Municipality of Niagara (modified from Ontario Geological Survey, 1991; Haynes, 1992; Menzies and Taylor, 1998.)

waters in the Lockport Formation has been studied by Haynes and Mostaghel (1982).

The most complete vertical exposures of the Paleozoic bedrock forming the Niagara Escarpment are located at the Niagara Gorge and at the St. Catharines hydro-electric power station (Fig. 7). Here, the Escarpment forms a cliff about 90 m in height, of Lower and Middle Silurian strata (comprising, from top to bottom: Lockport and Decew dolostones; Rochester shale; Irondequoit dolostone; Reynales and Neagha limestones and shaly limestones; Thorold and Grimsby sandstones; Grimsby and Power Glen shales; the Whirlpool sandstone) and the top of the Upper Ordovician Queenston shale (Figs. 6, 7).

On the face of the Escarpment, two terraces (Bell and Irondequoit, Fig. 6) are locally wide enough to support cultivation of *Vitus vinifera*. As they are above the Lake Iroquois shorebluff, they have been termed collectively by Haynes (1994b), the Lake Iroquois Bench (Fig. 5). The plain below that extends to Lake Ontario has been been termed by Haynes (1994b) the Lake Iroquois Plain (Fig. 5). At the base of the Niagara Escarpment, the Bell Terrace (Tiplin and Seibel, 1988) lies above the prominent north-facing bluff of gravels (Fig. 6) that marks the shoreline of Lake Iroquois. Although narrow in the east and west parts of the Region Niagara, the Bell Terrace widens to about 3 km between the Welland Canal and 15-Mile Creek (Fig. 5). West of 16-Mile Creek (Fig. 5), the Escarpment is lower (about 62 m high) and, here, the Lake Iroquois Bench has been termed (Haynes, 1998; Haynes *et al.*, 1998) the Vineland Double Bench, as it comprises two broad terraces: the Bell Terrace at the base, and the higher Irondequoit Terrace (Haynes, 1994b) above the scarp of Irondequoit limestone (Fig. 6). Here, the vineyards are located on the terraces in picturesque settings reminiscent of those in northern Europe, and the major creeks (15-Mile, 16-Mile and 20-Mile creeks, Fig. 5) cascade over a series of waterfalls that vary in height from a few metres to about 24 m. These waterfalls are due to differential erosion of the three thick packages of shale (Rochester, Grimsby–Power Glen and Queenston) that underlie the two resistant sequences of dolostones (Lockport-Decew), and carbonates (Irondequoit-Reynales) overlying sandstones (Thorold-Grimsby) (*see* Fig. 6). The two terraces of the Vineland Double Bench have formed over 1) an upper, thick, ledge of Irondequoit and Reynales dolomitic and argillaceous limestones underlain by Thorold and Grimsby sandstones, and 2) a lower, thin ledge of Whirlpool sandstone, at the base of the Escarpment, that often coincides with the top of the shorebluff of proglacial Lake Iroquois (Fig. 6). The width of each rock ledge is highly variable but the lower is usually wider, being up to 2 km wide in this part of the Niagara Peninsula. Further west,

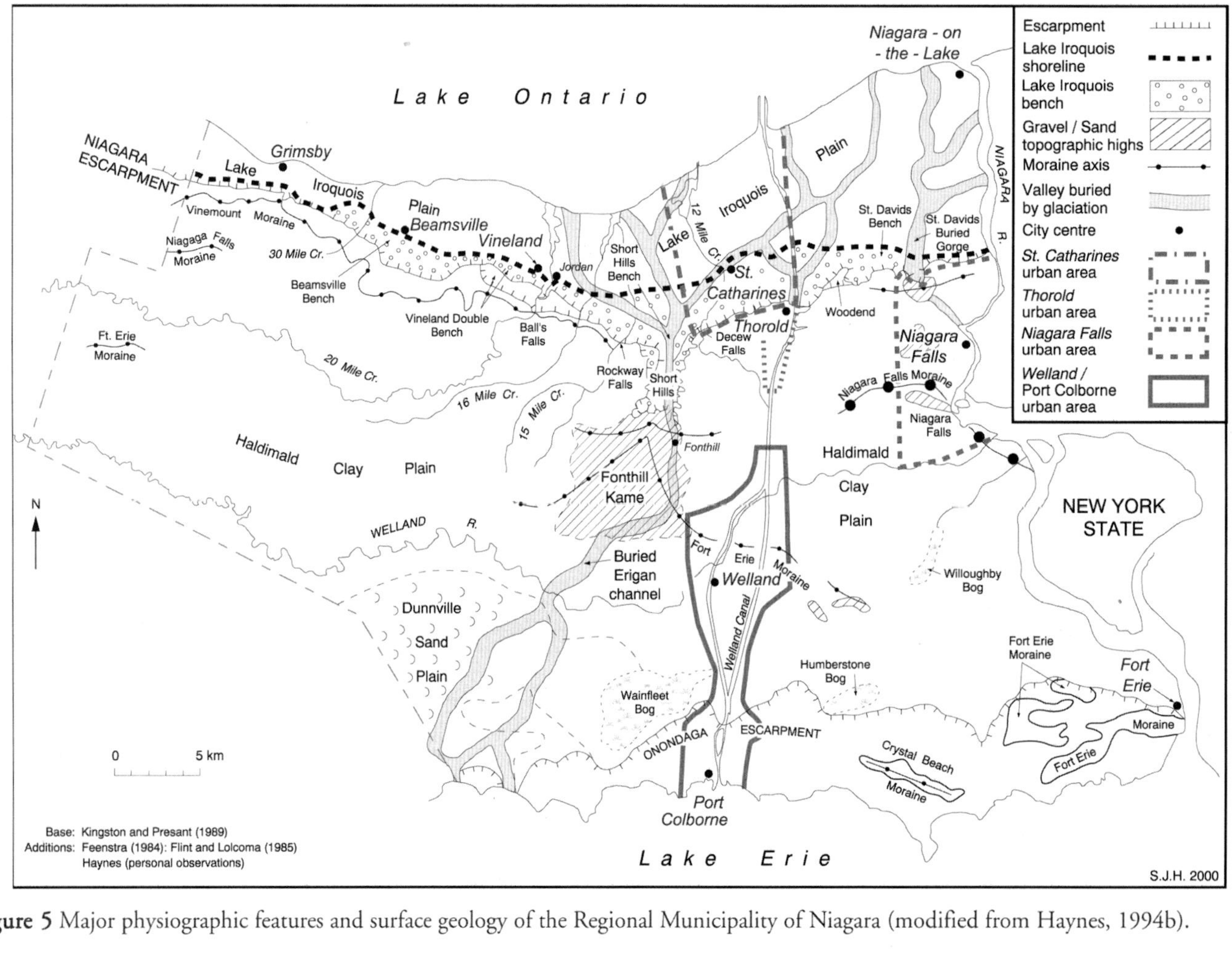

Figure 5 Major physiographic features and surface geology of the Regional Municipality of Niagara (modified from Haynes, 1994b).

toward Beamsville, the Irondequoit Terrace recedes back to the Escarpment scarp, while the Bell Terrace widens to about 1.5-2 km to form the Beamsville Bench (Fig. 5). On both Bench terraces, bedrock is overlain by a ramp of Halton till and other Quaternary glacigenic sediments (Fig. 6). Owing to limited exposure, these sediments are poorly understood and in fact may not even be Halton till but a deposit from an earlier glacial advance. The glaciolacustrine deposits on the Bell Terrace and Iroquois Plain vary from massive clay to silt to sand, with sands more common nearer Lake Ontario, suggestive to Menzies and Taylor (1998) of ice-proximal fluctuations of the ice front in the Lake Ontario basin.

The eastern end of the Vineland Double Bench is truncated by the 2-km wide, 4-km long reentrant of the Short Hills into the Escarpment (Fig. 5). This is a picturesque area, characterized by V-shaped valleys and round hills (short hills) formed during the post-glacial incision by 12-Mile Creek (and its tributaries) of glacigenic sediments deposited in the now partly buried, ancient gorge of the Erigan Channel, which used to connect Lake Erie with Lake Ontario (Flint and Lolcama, 1985). To the south, the Short Hills are terminated by the Fonthill Kame (Fig. 5), an irregular square hill, which stands 67 m above the surrounding plain of Haldimand clay, and is characterized by a steep convex-upward northern slope (ice-contact side) and shallower concave-upward southern, eastern and western slopes formed of stratified, well-washed sands and gravels.

NIAGARA PENINSULA *TERROIRS*

A French AOC *appellation* is based on the meterological (climate), physiographic (landform), viticultural (grape-growing practices), pedological (soil) and geological (subsoil) factors that constitute the *terroir* of that particular *appellation*; such that change of any factor within any one of these groups in an adjacent area could constitute reasons for designation of a separate *appellation* (see Haynes, 1999b). However, outside France, geological (and even physiographic) factors are largely ignored, and pedological factors are often restricted to soil texture.

Wine *terroirs* are, in fact, only a sub-group of the Tender Fruit category within the overall suitabilities of Niagara Peninula lands for all types of agriculture (Chapman, 1994; Fraser and Haynes, 1994). This in turn is part of a much larger framework of changing and competing land uses within the dynamics of a rapidly expanding population that is sensitive to land-use needs such as aggregates and recreational areas (Haynes and Fraser, 1994). Indeed the exact legal definition of the Niagara Escarpment and its land uses is open to debate (Haynes, 1995b), as is its position as a United Nations Biosphere Preserve (Haynes, 1995a).

In the Niagara Peninsula, no

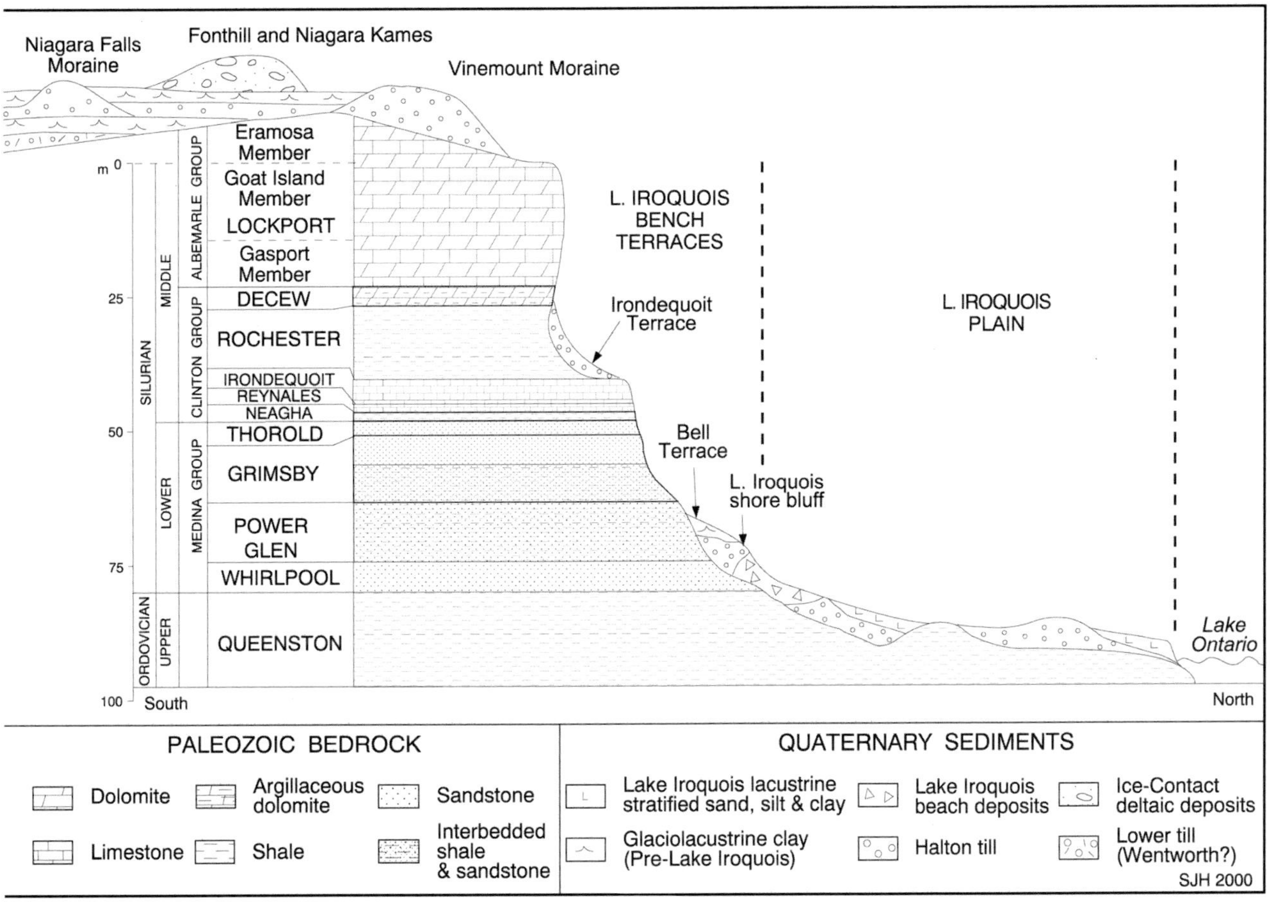

Figure 6 Composite north-south section of the geology and physiography across the Niagara Escarpment, in the Niagara Peninsula (modified from Haynes, 1994b, 1995a.)

previous comprehensive attempt to define *terroirs* has been published. Sayed (1992) presented vineyard site suitability maps that were based on a compilation of the climate zones of Wiebe and Anderson (1977) and the soil maps of Kingston and Presant (1989). Unfortunately, this study appears to have ignored both the geological reality of several of Kingston and Presant's soil types (see below) and the excellence of wines produced from several award-winning estate wineries whose vineyards were assigned poor site-suitability ratings.

Currently, the Niagara Peninsula DVA is divided into climate zones controlled by the physiography. These are discussed below and then related to physical geology and soil factors to define *terroirs*.

Niagara Peninsula Climate Zones and Physiography

Niagara's climate on and below the Niagara Escarpment is dominated by the moderating effect of its position between the large water masses of Lake Ontario and Lake Erie that, in winter, contribute to warmer-than-normal temperatures, resulting in a mean annual daily temperature of ~9°C, an average coldest monthly temperature (January) of ~ -4.7°C and a warmest (July) of 21.4°C, and the highest number of days with rainfall in spring and fall (Shaw, 1994, 1996). The relation of the average monthly temperatures at St. Catharines to those in several French wine districts (Ziraldo, 1995) is presented in Figure 8.

In addition, a study of the different effects of air flow over the eastern Niagara Escarpment (narrow terraces) and western Escarpment (wide terraces), coupled with an airborne thermal imagery study of infrared radiation along a single north-south traverse of the northern Niagara Peninsula through Vineland at an altitude of 5000 m in 1976 (Stewart *et al.*, 1977), allowed Wiebe and Anderson (1977) to propose the seven separate climate zones in the Niagara Peninsula DVA shown in Figure 8 (top). Unfortunately, this study is open to question as it was only flown in the spring months of 1974, 1975 and 1976 at different altitudes below 5000 m, and was not flown over other traverses. Notwithstanding these problems, each of Wiebe and Anderson's (1977) zones constitutes a separate possible climate *terroir*. In the zone of "level area above the Escarpment," and including the Fonthill Kame, the virtual absence of lake effects results in fewer numbers of frost-free days in spring and colder winter temperatures below -20°C, causing severe damage to the vines, and thus restricting plantings to *Vitus labrusca* varieties. North of the Escarpment, the physiography of the landforms and proximity of Lake Ontario control the air-flow pattern, such that distance north of the Escarpment is the determining climate factor in vineyard site selection in terms of the likelihood of temperatures below -20°C that cause severe damage to the vines, and hard frosts in April that cause severe bud damage (Sayed, 1992). The longest frost-free season, and therefore the least cold-injury danger to vines, is the zone of "Steep escarpment slope in western portion of Niagara," due to the escarpment-effect winds (see cross-section, Fig. 8). Moderate cold-injury danger is

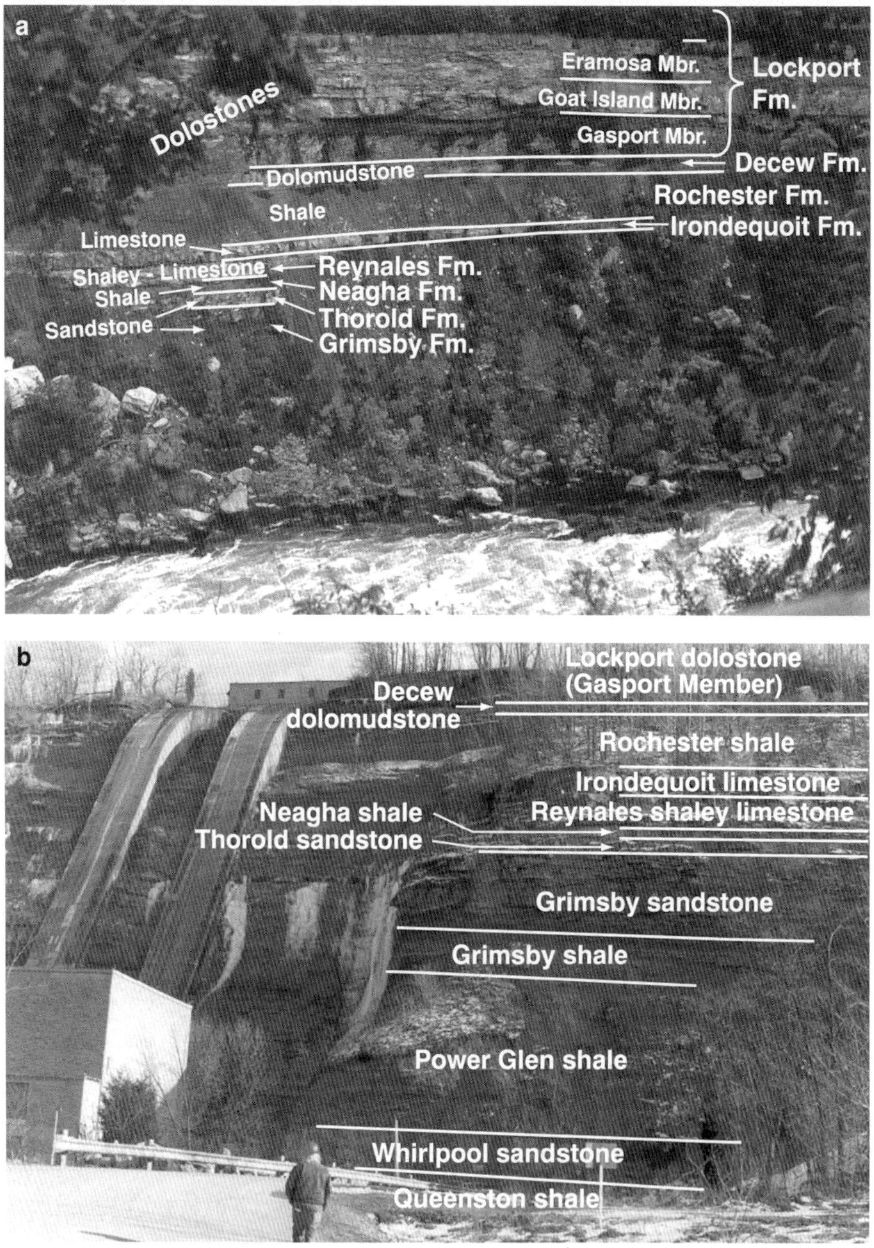

Figure 7 The sequence of Silurian strata exposed in (**a**) the Niagara River Gorge above the Whirlpool, and (**b**) the 12-mile Creek at the Power Glen hydro-electric power plant, St. Catharines.

experienced in the "Lakeshore effect zone" (which continues south along the Niagara River), and "Base of the escarpment slope (in the west) plus steep slope east of St. Catharines," zones, and, at only slightly greater risk, the zone of "Gentle slopes above escarpment edge" (Fig. 8). Interestingly, the flat "Lake plain between escarpment and lake" zone, which widens considerably in the eastern Niagara Peninsula, is at greater frost risk than the other zones below the Escarpment because it is furthest from both the Escarpment and Lake Ontario. However, four wineries and numerous vineyards are located in the northern half of this zone, east of the Welland Canal (reason enough for a rigorous climate study).

Physical Geology and Soil *Terroir* Factors

In terms of physiography, drainage, bedrock, glacial drift and soils, two distinctly different *Vitus vinifera terroirs* are present in the Niagara Peninsula (Fig. 9): 1) the Lake Iroquois Plain (the area between Lake Ontario and the Lake Iroquois shoreline bluff), and 2) the Escarpment (the terraces of the Lake Iroquois Bench and locally slopes above the Niagara Escarpment, north of the Vinemount Moraine). Topographically, the Lake Iroquois Plain has mainly gentle slopes, grading to both north and south, whereas the Lake Iroquois Bench terraces of the Escarpment grade consistently north, often with moderate to steep slopes (Figs. 6, 9). Thus, topography and climate zones indicate at least two major divisions of the Niagara Peninsula DVA.

Soils in the Niagara Peninsula are derived mainly by direct organic and weathering breakdown of the underlying Late Pleistocene-Holocene glacial, lacustrine, fluvial and alluvial sediments, which, in the vineyards, are usually thicker than 3 m (personal observations). From 1:20,000 scale mapping, Kingston and Presant (1989) demonstrated that the soils and their equivalent Pleistocene-Holocene subsoils include a large number of types that exhibit rapid lateral changes. Because they took samples to a depth of only 1 m, the soil profile penetrated by the roots of older grapes may be much more complicated, as older grape roots commonly penetrate the glacial soils by 1.5-3 m in the Niagara Peninsula (Sayed, 1992). Also, although their study has helped considerably to refine the 1:50,000 map of the Quaternary geology (Feenstra, 1984), they have imported many soil names from outside the Niagara Peninsula and in some cases have used the same soil name for soils with similar appearance but totally different Quaternary settings (*e.g.*, although the type areas of Beverly and Toledo soils are the glaciolacustrine silty clays of the Haldimand Clay Plain above of the Niagara Escarpment, they are also mapped as glaciolacustrine silty clays on the Lake Iroquois Bench; notwithstanding the *circa* 60-m vertical difference in elevation of the lake beds). However, these differences aside, the soils do exhibit a general south-to-north zonation (Fig. 9). On the Lake Iroquois Bench, Oneida soils are developed on Halton clay till, and Brantford, Beverley, Toledo (also, Cashel, Peel) soils are developed on glaciolacustrine clays and silty clays overlying Halton till. On the Lake Iroquois Plain, Chiguacousy, Peel and Jeddo soils overlie the ridges of Halton till clay, whereas fertile Grimsby, Tavistock, Vineland and Vittoria soils overlie the stratified sands silts and clays of the Lake Iroquois lakebed.

Because *Vitus vinifera* grapes hate "wet feet" good drainage is essential, so the steeper Escarpment slopes are favoured both for excellent slope drainage, as well as good groundwater drainage through the Oneida silty clay soils on the Beamsville Bench and Vineland Double

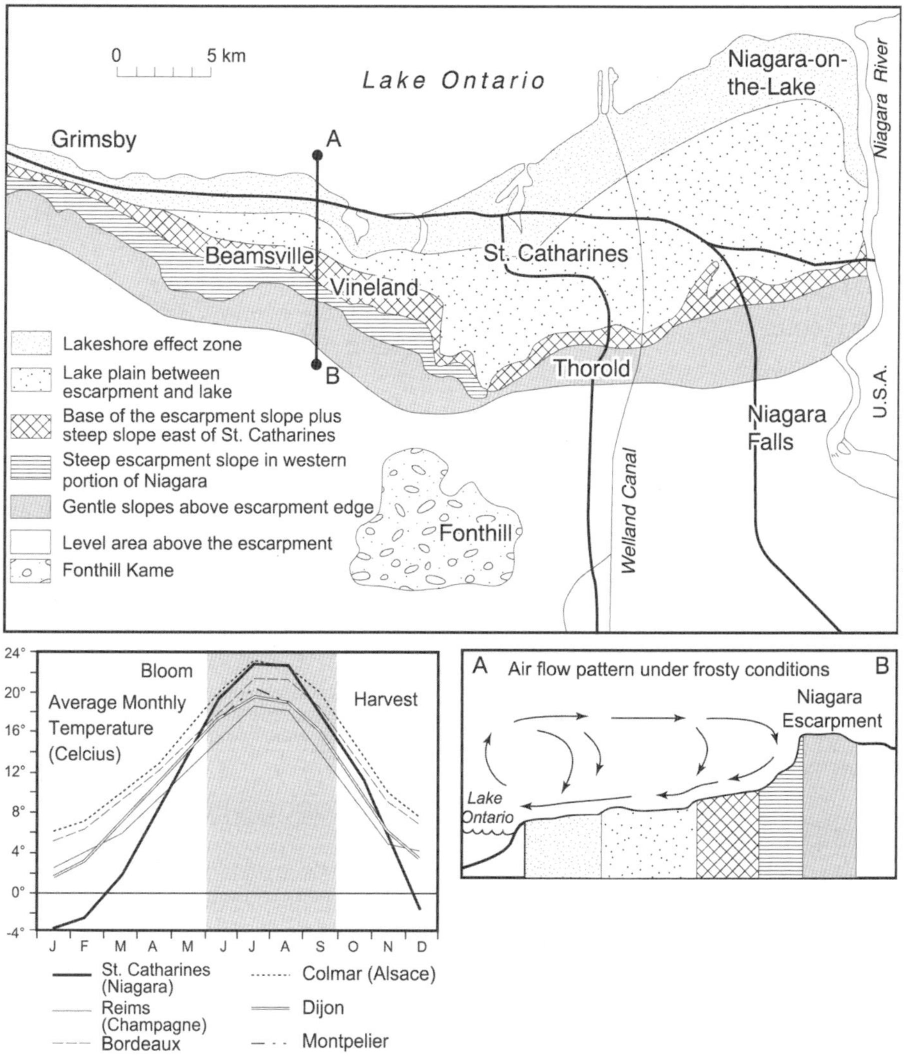

Figure 8 Plan and section of grape climatic zones in Niagara (modified from Wiebe and Anderson, 1977), and comparison of Niagara (St. Catharines) with the average monthly temperatures of French wine regions (from Ziraldo, 1995).

PROPOSED SUB-APPELLATIONS

LAKE IROQUOIS PLAIN

① Lakeshore (mixed till & lacustrine)
- a. Niagara-on-the-Lake
- b. Lakeshore West

② Central Plain (till ridge & thin lacustrine)
- a. Virgil area
- b. Niagara River

③ Queenston (clay, loam till)

④ Vineland Plain (mixed till & lacustrine)

ESCARPMENT (L. IROQUOIS BENCH)

⑤ St. Davids Bench (glaciolacustrine & till)

⑥ Short Hills Bench (glaciolacustrine & fluvial)

⑦ Vineland Double Bench (till)

⑧ Beamsville Bench (till)

Figure 9 Relation of surface geology, generalized soil types, and physiography to *Vitus vinifera* vineyard *terroirs* and proposed sub-*appellations* of the northern Niagara Peninsula (based on Feenstra, 1984; Kingston and Presant, 1989; Haynes, 1998.)

Bench, and mixed till and lacustrine silty clay soils on the St. David's Bench (Fig. 9). However, this is partly offset by the fact that today, all the vineyards employ extensive regrading of the natural topography and sub-drainage installations. Thus, successful vineyards can be located in heavy clay soils on the relatively flat Lake Iroquois Plain.

POTENTIAL *APPELLATIONS* AND SUB-*APPELLATIONS* IN REGION NIAGARA

The recognition of wine *terroirs* can be made at progressively smaller levels depending on the information available (Table 1). The largest level is a wine region (*e.g.*, Burgundy), which is the scale of southern Ontario. The next smaller is the macro-level of a French regional *appellation* (*e.g.*, the Côte de Beaune), which is a readily identifiable area of maximum contiguous viticulture. I consider that this is equivalent to the entire Niagara Peninsula viticulture area of commercial production of both *Vitus vinifera* (on and north of the Escarpment) and *Vitus labrusca* (essentially the entire area shown in Fig. 9) grapes.

At the meso-level of an *appellation*, the Côte de Beaune is divided into four separate areas on the basis of differences in physiography, geology and soil. Although these four divisions (Table 1) are today ranked separately for Chardonnay and Pinot Noir grapes in terms of quality (*Grand Cru* is highest), this is not directly applicable to the Niagara Peninsula as the Burgundian appellations have had many centuries to define the best wine grapes and the best wine-growing areas. However, these would be the *terroir* equivalent of dividing the Niagara Peninsula into four areas (Fig. 9) comprising: the roughly circular area of the Fonthill Kame; and the three longitudinal segments of the Lake Iroquois Plain (between Lake Ontario and the Lake Iroquois shoreline), the Niagara Escarpment (from the Lake Iroquois shoreline to the Vinemount Moraine), and the Haldimand Clay Plain (south of the Vinemount Moraine). Others may consider that these divisions are sub-*appellations* of a Niagara Peninsula *appellation*, there being no rules on size, nor on exact definitions.

The next smaller level (Table 1) is the mini-level of a sub-*appellation* (communal *appellation* in Burgundy) such as Aloxe-Corton or Savigny-lès-Beaune, which are village areas roughly 3-5 km along the strike of the Côte de Beaune. Previously these have not been defined in Niagara, but it is proposed that the present *Vitus vinifera* viticultural areas (Lake Iroquois Plain and Niagara Escarpment *appellations*) can be divided into a minimum of eight sub-*appellations* (see below).

The absolute smallest level of a *terroir* is the micro-level of a vineyard for a specific varietal wine. In the Côte de Beaune, this varies from tiny 1-3 hectare vineyards in the *Grand Cru appellations* to large multi-hectare vineyards in the Villages and Hautes Côtes *appellations*. In the Niagara Peninsula DVA, it is still unknown which are the best varietals, let alone which are the best vineyards: it will be many years before we can even begin to consider Niagara Peninsula *Crus* designations.

On combining the differences in the climate, landform physiography, and soil (taking into account relative stratigraphic position of the Quaternary precursor) the Lake Iroquois Plain can be provisionally subdivided into four, and

Table 1 Terroir nomenclature: comparison of Niagara Peninsula with Côte de Beaune.

Level	Name	Canada	France
Regional	Wine Region	Southern Ontario	Burgundy
Macro Level	Regional *Appellation* (DVA in Canada)	Niagara Peninsula	Côte de Beaune (south part of Côte d'Or)
Meso Level	*Appellation*	Proposed in this paper Four *appellations* 1. Lake Iroquois Plain 2. Niagara Escarpment 3. Haldimand Pland 4. Fonthill Kame	Designated by INAO as *Appellations Contrôlée* 1. Côte de Beaune Grand Cru 2. Côte de Beaune Premier Cru 3. Côte de Beaune-Villages 4. Hautes Côte de Beaune
Mini Level	Sub-*appellation* (Communal *Appellation*)	Proposed in this paper a minimum of eight for Lake Iroquois Plain and Niagara Escarpment *Appellations*	Communes in *appellations* 1 to 4 1. *e.g.*, Corton Grand Cru 2. *e.g.*, Aloxe-Corton 3. *e.g.*, Savigny-lès-Beaune 4. Not defined individually
Micro Level	Specific vineyard site (*Climat* in France)	Not yet defined	Grand Cru Corton divided into 20 *climats*: *e.g.*, Les Renardes Les Bressandes

the Escarpment into four, separate *terroirs* or sub-*appellations* (Fig. 9). In general, these correspond with those recently proposed by winemaker J-L. (Jean-Laurent) Groux of Hillebrand Estate Winery (oral presentation, September 1999, Niagara Grape and Wine Festival seminar on "*Terroir,*" at Brock University). Groux has made an excellent study of the distinctive taste, bouquet, *etc.* of wines made by him in separate batches of wine under identical conditions (thus removing the "winemaker" factor) from different vineyard *terroirs* throughout Niagara. His data base, which has been more than 10 years in the making, has allowed him to distinguish nine different *Vitus vinifera terroirs* (sub-*appellations*). However, as there are many vineyard areas that are not represented by estate wineries (and which I have not visited), and/or which have not supplied J-L. Groux with grapes, other terroirs (and new sub-*appellations*) are likely to emerge on further study.

Discussion of my eight proposed *vinifera terroirs* (sub-*appellations*) and their relation to the nine proposed sub-*appellations* of J-L. Groux follows.

Lakeshore Sub-*appellation*

This *terroir* corresponds to the "Lakeshore effect" climate zone (Fig. 8). Slopes are very gentle. It is split by the urban area of St. Catharines into Lakeshore West and Niagara-on-the-Lake (1a,b, Fig. 9). In both areas, a wide variety of lacustrine sandy, silty and clay loam soils derived from glaciolacustrine nearshore and deltaic sand and silt are present in the vicinity of the lakeshore of Lake Ontario (Tavistock, Vineland, Peel and Vittoria soils), as well as low mounds or ridges of Halton till clay soils (Chinguacousy, Peel, and Jeddo). Also, newer vineyards are planted on good, sandy loam soils in competition with tree fruits (Sayed, 1992).

J-L. Groux considers my Lakeshore subdivisions (1a,b, Fig. 9) to be two separate sub-*appellations*. Also, he extends his Lakeshore sub-*appellation* further east than my Lakeshore West (1b, Fig. 9). His Niagara-on-the-Lake sub-*appellation* expands much further south than mine (1a, Fig. 9) to include, of my Central Plain sub-*appellation* (see below), all of my Niagara River *terroir* (2b, Fig. 9) and the eastern part of my Virgil area *terroir* (2a, Fig. 9). He considers his Lakeshore sub-*appellation* to be one of the best in Niagara for Chardonnay varietal wines, as they exhibit a floral profile and a long aftertaste, and are especially suited for "Champagne-style" sparkling wine. Merlot is good here, while Gamay is good in his Niagara-on-the-Lake sub-*appellation*. Niagara-on-the-Lake is the site of Konzelman Estates, Niagara's most northerly winery and vineyards (located next to Lake Ontario), a producer of excellent German-style wines.

Although there are no clear geological or climatic reasons for separating this sub-*appellation* into separate Niagara-on-the-Lake and Lakeshore West sub-*appellations*, the different physiographic separation of these two areas from the Niagara Escarpment (Fig. 9) may, in future, govern their distinction as separate *terroirs*.

Central Plain Sub-*appellation*

This *terroir* (2, Fig. 9) corresponds to the part east of the Welland Canal of the "Lake plain between escarpment and lake" climate zone (Fig. 8). Gentle slopes grade north and south from an east-west striking low ridge-like mound of Halton Till. I have divided it into the Virgil area and Niagara River *terroirs* (2a,b, Fig. 9) along the line of the buried river channel that links south to the St. Davids Buried Gorge that marks the pre-Late Wisconsinan ancestral Niagara River. Not only is this a natural "geologic break" but also, the vineyards close to the Niagara River (2b, Fig. 9) are subject to the moderating effects of the large body of moving water such that the climate is similar to that of the Lakeshore effect zone (Fig. 8). Thus the till soils and very gentle slopes are major factors in making this *terroir* famous for the fine wines of Inniskillin, Marynissen Estates and Reif Estates.

J-L. Groux includes the eastern part of my Virgil area, and all of my Niagara River *terroirs* in his Niagara-on-the-Lake sub-*appellation* (see above). The western part of my Virgil area (2a, Fig. 9) is included in his Grantham sub-*appellation*, which he recognizes as an area of very heavy clay soils only recently used for vineyards after installation of extensive sub-drainage. J-L. Groux claims that these heavy clay soils create wines with more depth and are good for red Pinot Noir and Merlot, and white Chardonnay and Sauvignon Blanc.

Queenston Sub-*appellation*

This *terroir* (3, Fig. 9) was first distinguished by J-L. Groux. It is located in the far south-eastern part of the "Lake plain climate zone," and is the farthest *terroir* from Lake Ontario and its moderating influence, which results in the warmest summers but cold winter vine damage. It has heavy clay soil (Halton Till exposed in the lakebed of Lake Iroquois) and slopes are gentle to the north. Groux considers that the Chardonnay wines from here are superb.

Vineland Plain Sub-*appellation*

This is the western part (4, Fig. 9) of the "Lake Plain" climate zone (Fig. 8), and slopes are gentle to the north. It is underlain by Halton till and a variety of Lake Iroquois lake and deltaic sediments. Although recognized also by J-L. Groux, it is not well known as a sub-*appellation* and does not appear to produce wines of distinction.

St. Davids Bench Sub-*appellation*

This (5, Fig. 9) is the widest part of the "steep slope east of St. Catharines" climate zone (Fig. 8). Slopes are all northward and range from steep in the south to moderate in the north. Soils are clay loams derived from pre-Lake Iroquois glaciolacustrine silty clays overlying Halton till that merge to the east with nearshore and deltaic sands (Fig. 10). The Irondequoit Terrace is present but is only 100-200 m wide. The main part of the St. Davids Bench viticultural area extends from the top of the Thorold sandstone, at the base of the low scarp of Irondequoit limestone, north across Highway 81 toward the Lake Iroquois shore bluff (Fig. 10). This is the preserve of Paul Bosc and his family, who developed the St. Davids Bench viticultural area from scratch, and named it. This is the premier example in Ontario of a French-trained viticulturist and winemaker deliberately choosing a virgin *terroir*, testing the climate, soils and drainage, buying the property, then developing (with sub-drains and surface-regrading) flagship vineyards (Château des Charmes) that produce award-winning wines. Paul Bosc tells me he was struck by the similarity of the topography and limestone geology to Burgundy, and how the groundwater underflow from the Escarpment irrigates his vines in hot summer months. However, it wasn't until I visited Burgundy in 1999 that I realized how much the shape and orientation of the hill of Woodend at the end of the St Davids Bench (Fig. 10), separating this *terroir* from the other Escarpment bench *terroirs*, resembles the hill of Corton at the end of the Côte de Beaune, separating the southern from the northern part of the Côte d'Or (Haynes, 1999: Figs. 1, 2 and cover photograph).

A wide variety of good wines are produced here. Paul Bosc has pioneered commercial development of different varieties, including two little-known but classical white grape varieties from northern France: Auxerrois, from Alsace; and Aligoté from Burgundy. The latter is often referred to as "that other white wine" because, except for the village of Bouzeron where it is a fine wine with its own *appellation* (Bourgogne Aligoté-Bouzeron), it is relegated to the highest and lowest slopes of the Côte d'Or (although it preceeded Chardonnay on the Hill of Corton in times past). J-L. Groux considers the St. Davids Bench is very good for Chardonnay, Cabernet Franc and Cabernet Sauvignon.

Paul-André Bosc (son of Paul Bosc) has his own ideas on *terroir*, which he presented at the 1999 seminar on "*Terroir*" at Brock University. This included the first scientific evidence for the existence in the Niagara Peninsula of *terroirs* at the micro-level of a vineyard. This evidence consisted of differences in both the flavour profiles (from the tastings we conducted) and analytical data (Table 2) of two Chardonnay wines produced from similar grapes grown in vineyards on the north side (Toledo clay loam) and south side (Beverly silty clay loam) of Highway 81 (*see* Fig. 10). Except for different spacings of the sub-drain (5-m separation for the Beverly soils, and 2.5-m for the Toledo soils), which is a reflection of the differences in soil texture, porosity and permeability, all other viticultural and wine-making factors were the same for both vineyards. As the climate cannot be significantly different across a road, only differences in the soil parentage and/or ground water chemistry

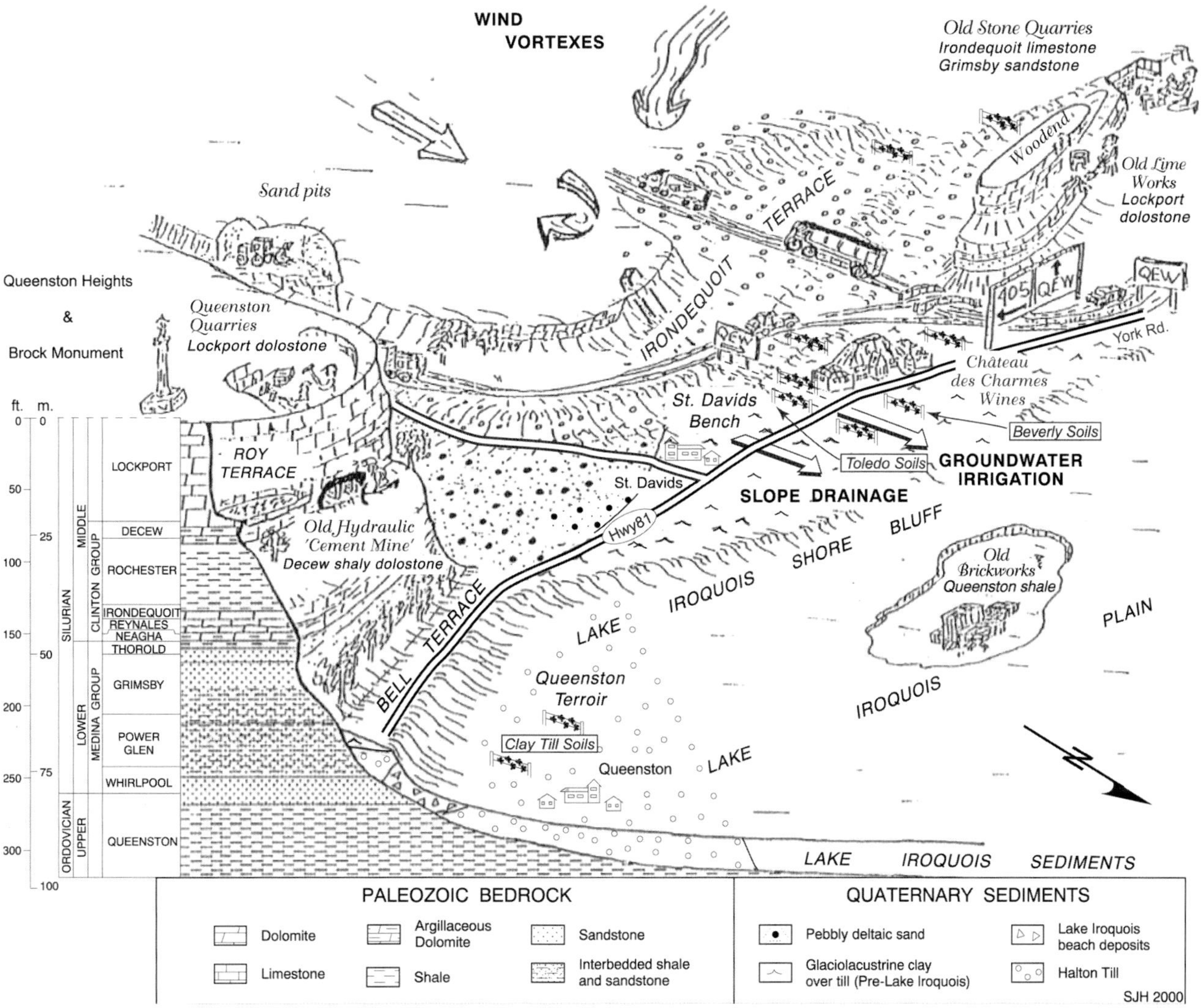

Figure 10 Simplified geological sketch of the Niagara Escarpment between Queenston Heights and Woodend, illustrating the relation of geology to the St. Davids Bench and Queenston sub-*appellations* (modified from Haynes, 1994b).

and flow rate can account for the differences presented in Table 2 between both the grape juice and the wine from the two vineyards.

Paul-André Bosc considers that the numerous different clones of specific grape varieties need to be tested for all the different soils. Interestingly, a mutant Gamay (the grape for Beaujolais) grew on the original Bosc home farm from cuttings Paul Bosc Sr. brought from Europe. Cuttings from this mutant now occupy several acres, and the first commercial quantities were produced in 1999. Paul-André Bosc informs me (oral communication, and imbibing, March 2000) that the clone is now recognized by the official body in France so this will be, in a sense, the first Canadian vinifera. Certainly, "tank-tasting" (opening the tap

Table 2 Comparison of Chardonnay juice and wine from adjacent Toledo and Beverly soil vineyards at Château des charmes.

	St. David's Bench Vineyard	Paul Bosc Vineyard
Vineyard Location	North of Highway 81	South of Highway 81
Soil Name	Toledo	Beverly
Juice		
Total acidity	8.55	9.30
PH	3.37	3.37
Brix (sugar)	23.5	23.2
Wine		
PH	3.32	3.47
Total acidity	7.13	7.58
Alcohol	13.50%	13.50%

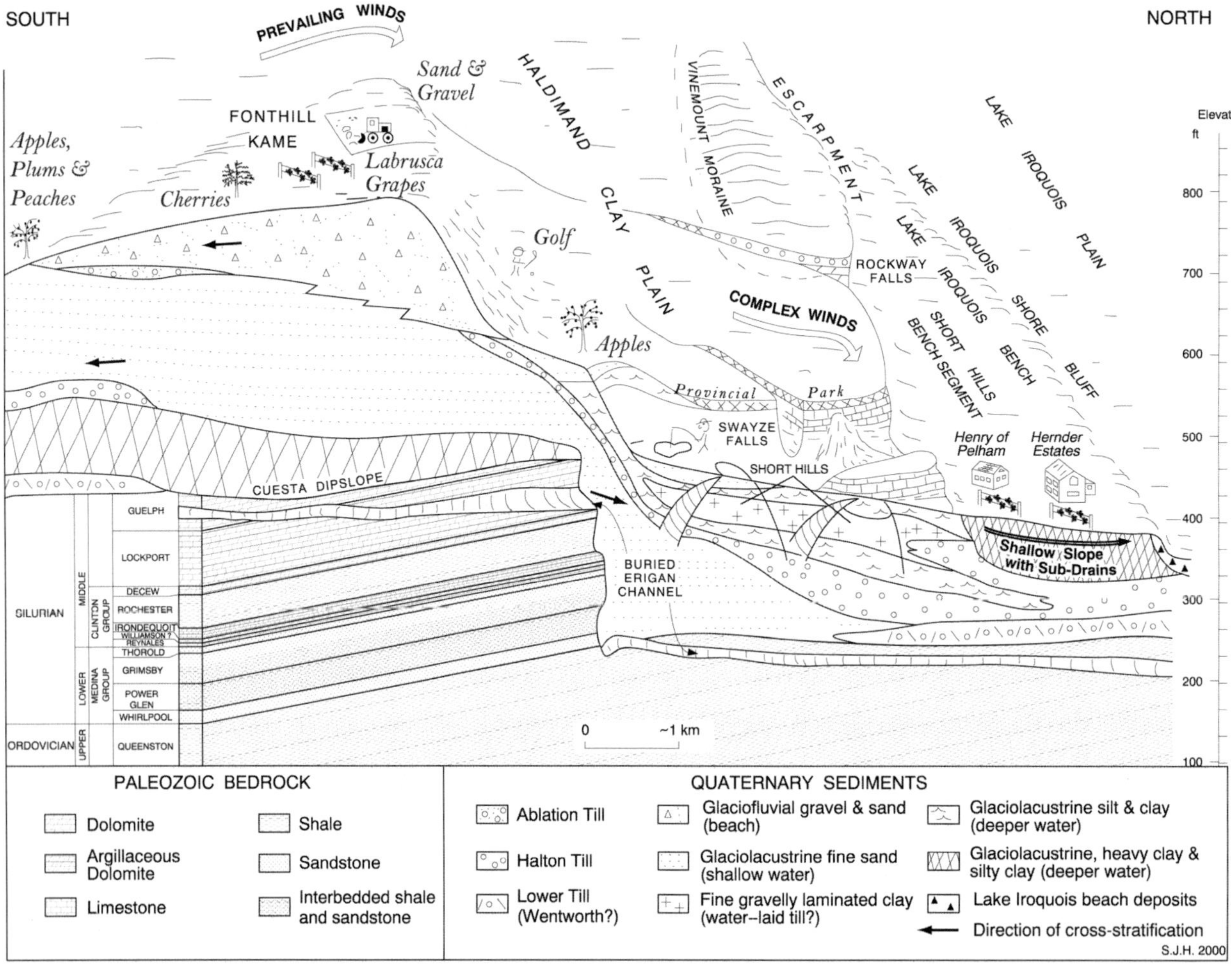

Figure 11 Simplified north-south cross-section of the geology, physiography and land uses in the Short Hills-Fonthill Kame segment of the Escarpment and their relation to wineries on the Short Hills Bench sub-*appellation* (modified after Haynes, 1994b and from data of Feenstra, 1981; Kingston and Presant, 1989.)

of the maturation tank and drawing off a quantity to be tasted) of the 1999 vintage indicates a full-bodied wine of exceptional character.

J-L. Groux recognizes a separate sub-*appellation*, Iroquois Lakeshore, along the Lake Iroquois shore bluff that is good for Cabernet Franc and Cabernet Sauvignon. I have included this as the northern part of the St. Davids Bench as the geology, and apparently the wines, are not significantly different.

Short Hills Bench Sub-*appellation*

This is situated at the mouth of the Short Hills Reentrant, on the west side of 12-Mile Creek (6, Fig. 9). This is the most unusual of the Escarpment *terroirs* as the landforms, climate and Quaternary sediments and soils are all products of the development of the Fonthill kame-delta above one of the world's largest waterfalls, that dwarfed Niagara Falls, at the head of a *circa* 2-km wide, 90-m deep, pre-Late Wisconsinan gorge now completely buried in its southern part by glacigenic sediments (Fig. 11). Within the actual Short Hills reentrant itself, the Late Wisconsinan glaciogenic sediments have been carved down about 50-60 m during the Holocene by 12-Mile Creek and its tributaries into a series of flat-topped, conical hills (hence the name, Short Hills). This flat-top surface can be traced from the top of the Lake Iroquois shore bluff south of the Escarpment back about half way to Swayze Falls (Figs. 9, 11) as a very gentle north-facing slope. North of the Short Hills reentrant, this slope is underlain by more than 43 m of very heavy, massive, blue-tinted brown, locally colour-banded lacustrine clay (personal on-site observations of Waterloo University core, summer 1999 drilling project) that represents a pre-Lake Iroquois lakebed. To the south, the slope of this north-facing surface increases, apparently in response to a change to fluvial-deltaic deposition of coarser-grained facies (fine, gravelly laminated clay, then fine sands) as it approaches the Fonthill kame-delta and its upper coarser sand and gravel units (Fig. 11). Here, the deep valleys trap the cold air from Fonthill, creating a southward bulge in the cold "Lake Plain" climate zone (Fig. 8). In front, and to the west side of the Escarpment edge of the Short Hills reentrant, a northward bulge of the warmer "Base of the escarpment slope" and "Steep escarpment slope" climate zones (Fig. 8) is contained between the Short Hills reentrant and 15-Mile Creek (Figs. 8, 9). Here, the gently north-sloping ancient lake-bed surface, now regraded and sub-drained, is the site of Henry of Pelham's Estate vineyards (Fig. 12).

This sub-*appellation* was not recognized by J-L. Groux, because he has not made wine from here, but it is adjacent to his West St. Catharines sub-*appellation*, which, although well-known as a producer of *Vitus labrusca* grapes for juice and jam, has few *Vitus vinifera* vineyards due to the effect of cold air channelling down 12-Mile Creek. The Reserve Chardonnay from Henry of Pelham's very thick, heavy glaciolacustrine clay vineyard (which I confirmed by digging a 2.5-m pit with their backhoe) is excellent and a great medal winner, the clay making the wine of depth and characteristic strong flavour. Also, this is one of only a few places in eastern North America to produce a good Baco Noir.

Figure 12 View southwest across Henry of Pelham Estate vineyards on the Short Hills Bench to the steep high scarp of the Niagara Escarpment at Rockway.

Vineland Double Bench Sub-*appellation*

This *terroir* (7, Fig. 9) is located about the town of Vineland and village of Jordan (Fig. 5) between Rockway Falls and an unnamed headland of the Escarpment west of Vineland. It is characterized by picturesque vineyards and waterfalls sited on the moderate northward slopes of both the *circa* 1-2 km wide upper (Irondequoit) and lower (Bell) terraces (Figs. 13, 14). Because the steepest slope of the Escarpment is here the scarp face of the Irondequoit Terrace, cartographers refer to this as the Niagara Escarpment, rather than the subdued scarp of Lockport dolostone to the south. This has influenced the climate zones (*i.e.*, the topographic base for the map of Wiebe and Anderson, 1977), such that the upper Irondequoit Terrace is in the moderate cold-injury danger "Gentle slopes above escarpment edge" climate zone, while the lower Bell Terrace is in the slight cold-injury danger "Steep escarpment slope in western portion of Niagara" zone (Figs. 5, 8, 9).

On both terraces, Oneida and Chinguacousy light and heavy till loam soils overlie Halton till (Fig. 13). Also, both terraces are well slope-drained and the loam soils experience summer irrigation by ground waters flowing north from the Escarpment. Wind vortexes, created by the interplay of the prevailing wind from Lake Erie with winds from Lake Ontario, break up spring frosts.

Riesling in all of its delightful styles is king in this *terroir*, not least because of the viticultural and winemaking skills of the Schmidt family, who pioneered the growing of wine grapes on the Irondequoit Terrace and brought their German wine experience to the creation of Vineland Estates winery (the new sparkling Riesling wines and icewines are a "must try"). J-L. Groux groups, as do most people, the Vineland Double Bench with the Beamsville Bench, as the soils and general setting on the Escarpment are the same (see below). He considers that this whole zone yields excellent Riesling and Gewurztraminer with the required high acidity (high acidity makes these German varietal wines "steely," which is much appreciated by wine connoisseurs), and Chardonnay with higher acid and more floral tones compared to the other *terroirs*. On the Bell Terrace, Ed and Lorraine Gurinskas of Lakeview Cellars Estate winery have pioneered Cabernet Sauvignon and a great Pinot Gris. Both terraces are sites of many new vineyards and wineries.

Beamsville Bench Sub-*appellation*

Although the Beamsville Bench (8, Fig. 9) is usually attributed to the whole of the "Steep escarpment slope in western portion of Niagara" climate zone, from Grimsby to the Short Hills reentrant (Fig. 8), I consider that it should be restricted to the bay-like reentrant of both the Escarpment and this climate zone, immediately south of the town of Beamsville (Figs. 5, 8, 9). Here, the Irondequoit terrace has merged with the face of the Escarpment, and the generally northerly Escarpment face presents a mixture of landforms, with a variety of microclimates due to differing east through west exposures. In the east, a sheer northwest-facing cliff of exposed strata from the Grimsby to the Lockport (Fig. 6) has formed by presently-active erosion as fissure caves (the location of the Cave Springs) along regionally predominant northeast-striking joints, parallel to the cliff. On the moderate slopes of the Bell Terrace below, Cave Spring Cellars vineyards, under owner Len Pennachetti and winemaker Angelo Pavan, are renowned for their Chardonnay and for pioneering efforts to establish the Beamsville Bench as one of the top quality *terroirs*. In the central and western parts of the Bench, opposite Beamsville (Fig. 5), the Escarpment face is more subdued as it is covered by a concave-upward ramp of Halton till. This part is therefore higher than the main Bell Terrace to the north and has steep-to-moderate slopes that were incised during the Holocene by numerous streams, creating an intricate network of picturesque hills and valleys now favoured by many of the newer wineries.

CONCLUSIONS

Although *Vitus labrusca* grapes are grown on the Haldimand Clay Plain and the Fonthill Kame, south of the Niagara Escarpment, the area of the Niagara Peninsula that can support commercial *Vitus vinifera* viticulture is on and north of the Escarpment. This is governed by the fact that *V. labrusca* grapes are able to withstand temperatures down to about -29°C, whereas *V. vinifera* grapes suffer vine injury if winter temperatures fall consistently below -20°C, and spring bud damage in frost-prone areas. The area of *Vitus* cultivation is termed, by the Vintners Quality Alliance (VQA), the Niagara Peninsula Designated Viticultural Area (Niagara Peninsula DVA), which corresponds to the macro-level of a regional *appellation* in France.

At the meso-level of a French *appellation*, clear differences in climate, physiography, Paleozoic bedrock lithology, and Quaternary drift geology/pedology dictate that the Niagara Peninsula DVA can be divided into two *Vitus vinifera appellations* and two *Vitus labrusca appellations*. The two *V. vinifera appellations* comprise:

1. Lake Iroquois Plain: the area between the shore of Lake Ontario and the base of the Lake Iroquois shoreline;
2. Niagara Escarpment: from the top of the Lake Iroquois shoreline, southward over the Escarpment terraces of the Lake Iroquois Bench to the Vinemount Moraine.

The two *V. labrusca appellations* comprise the Haldimand Clay Plain and the

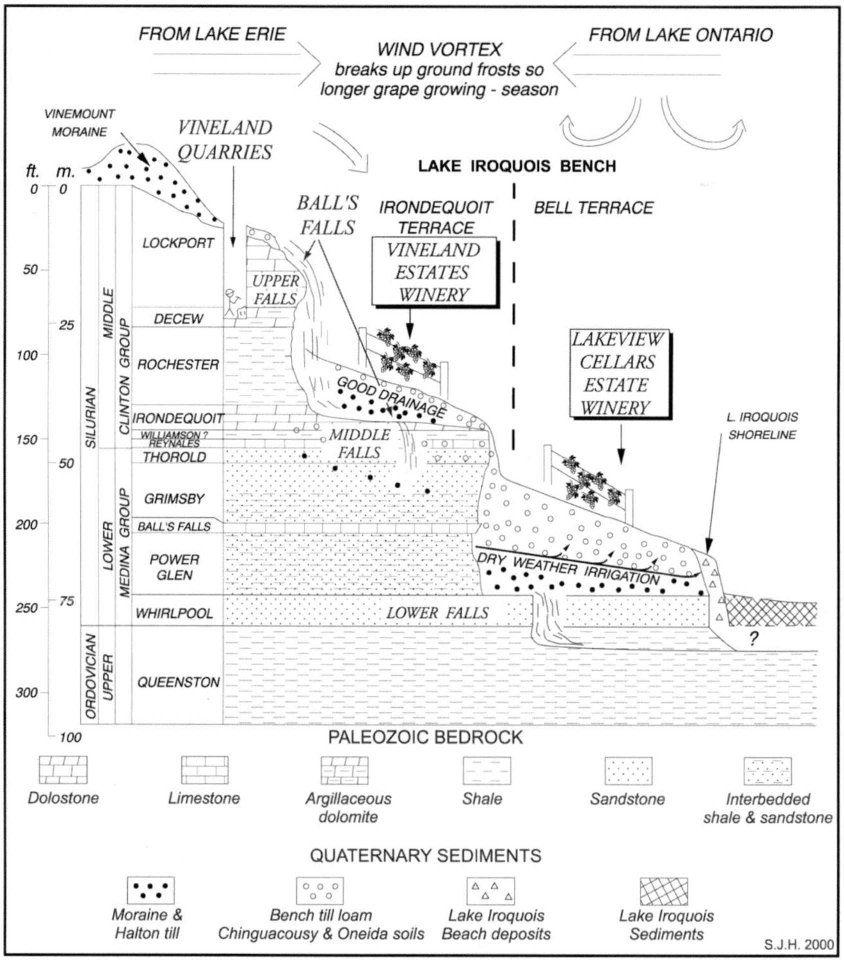

Figure 13 Simplified stratigraphic section of the Vineland Double Bench showing relation of geology to the vineyards on the Irondequoit and Bell Terraces and the waterfalls of 20-Mile Creek at Ball's Falls (modified from Haynes, 1994a).

Fonthill Kame, and are not considered in this article as they do not commercially produce wine. However, in future, it is likely that these will support the new varieties of *V. vinifera* that have been developed to be more cold-resistant, particularly should the warm winter climate-trend of the past decade continue. In spite of this, predictions of their success must be tinged with caution as temperatures below -20°C are common here, and even the Vineland research station, in the warm lakeshore climate zone about Lake Ontario, recorded -26°C on Christmas Eve, 1980.

At the mini-level, at least eight, and possibly 10, distinctive geological-pedological-physiographic *terroirs* define sub-*appellations* that are in general agreement with similar areas distinguished by J-L. Groux, winemaker at Hillebrand Estates, on the basis of grape variety suitability and differences in wine flavour profiles. These comprise:

1. Four (possibly six) sub-*appellations* of the Lake Iroquois Plain *Appellation*: a) Lakeshore (which may constitute two separate sub-*appellations* of Lakeshore West and Niagara-on-the-Lake); b) Central Plain (which may constitute two separate sub-*appellations* of Virgil and Niagara River); c) Queenston; d) Vineland Plain.
2. Four sub-*appellations* of the Niagara Escarpment *Appellation*: a) St. Davids Bench; b) Short Hills Bench; c) Vineland Double Bench; d) Beamsville Bench.

Distinction of micro- *terroirs* at the vineyard level is limited by lack of data, and is a fertile field for future research. The physical-chemical distinction of juice and wine (produced under identical viticultural and oenological conditions) from two neighbouring vineyards with different soils at Château des Charmes indicates to the owners, the Bosc family, the validity of the concept that geological *terroirs* play a major role in determining the flavour profiles of Niagaran wines.

Worldwide, many famous vineyards are on poorly consolidated silty-to-gravelly soils, so the till and clay soil vineyards of Niagara are unusual. However, it is the combination of these unusual soils and climate that give Niagara wines their distinctive character and, particularly the icewines, increasing medal-winning international recognition. Eventually, a time may come when it is possible to define the best places to plant the most suitable grape varieties to produce the very best wines. However, since there are many years of work ahead in accurately defining the mini-geology and mini-climate of Niagaran *terroirs* at the sub-*appellation* level, and even more work at the micro-level of a vineyard, it is worth heeding the following justification of such analysis, by Paul-André Bosc (oral presentation, September 1999, Seminar on "*Terroir*" Brock University),

> *Terroir*: it wouldn't be interesting unless people were interested in it; *terroir* is the future.

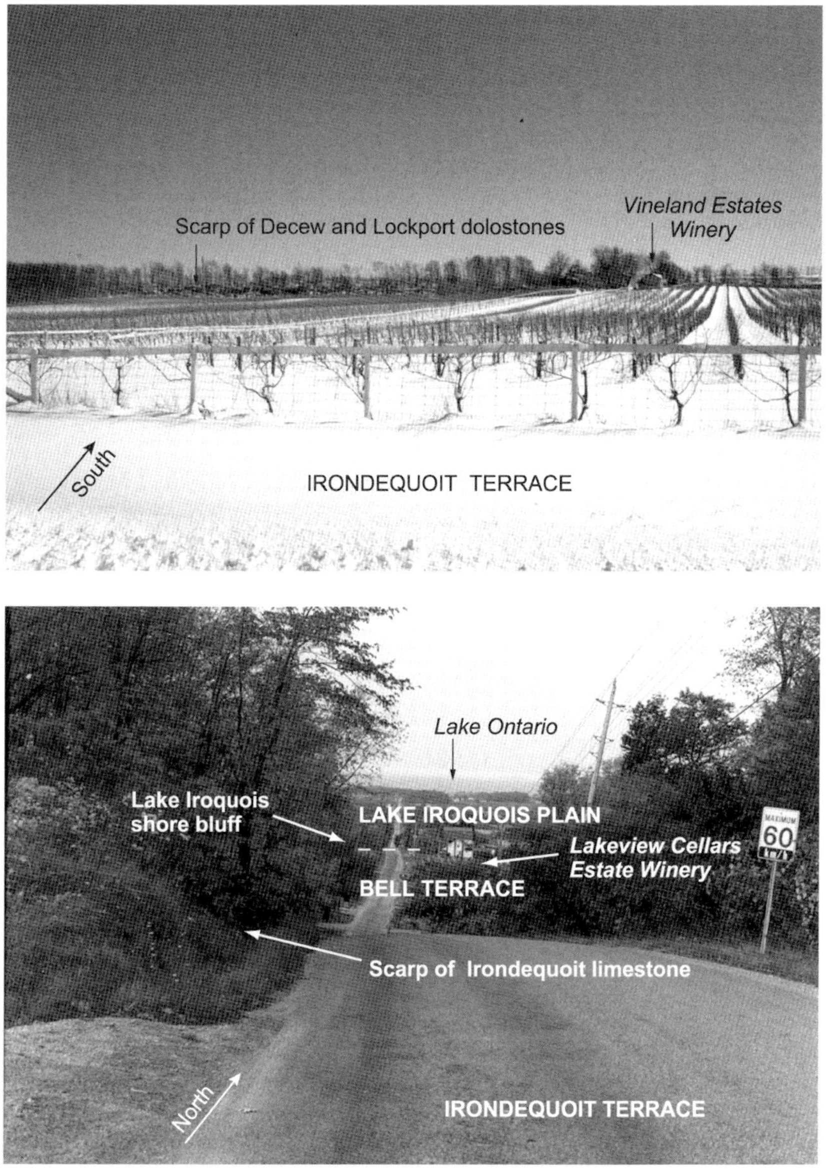

Figure 14 The Vineland Double Bench of the Lake Iroquois Bench. **(Top)** View north across Vineland Estates vineyards, on the Irondequoit terrace, to the low scarp of Decew and Lockport dolostones. **(Bottom)** view south from the Irondequoit terrace across Lakeview Cellars Estate vineyards, on the Bell Terrace, to the Lake Iroquois Plain.

ACKNOWLEDGMENTS

I am indebted to both Jean-Laurent Groux and Paul-André Bosc for unselfishly allowing me to use their data and ideas on wine *terroirs* of the Niagara Peninsula. Particular gratitude is extended to Mike Lozon, Department of Earth Sciences, Brock University for the computer artwork of the figures and plates. I also thank the vineyard managers and wine makers of the Niagara Region who, over the years, have imparted their experiences and ideas, particularly the members of the Bosc and Speck families, and Dave Parker. The manuscript was vastly improved by the editorial skills of Roger Macqueen and the critical comments of the reviewers, Terry Carter and A.D. (Sandy) McCracken.

REFERENCES

Aspler, T., 1999, Vintage Canada: the complete guide to Canadian wines: McGraw-Hill Ryerson, 3rd ed., 223 p.

Barnett, P.J., 1985, Glacial retreat and lake levels, north-central Lake Erie Basin, Ontario, *in* Karrow, P.F. and Calkin, P.E., eds., Quaternary Evolution of the Great Lakes: Geological Association of Canada, Special Paper 30, p. 185-194.

Barnett, P.J., 1992, Quaternary geology of Ontario, *in* Geology of Ontario: Ontario Geological Survey, Special Volume 4, Part 2, p.1011-1088.

Bolton, T.E., 1957, Silurian stratigraphy and paleontology of the Niagara Escarpment in Ontario: Geological Survey of Canada, Memoir 289, 145p.

Borrello, J., 1995, Wineries of the Great Lakes: a guidebook: Raptor Press, Lapeer MI, 175 p.

Brett, C.E., Goodman, W.M. and Loduca, S.T., 1994, Silurian Sequences of the Niagara Peninsula: Fourth Canadian Paleontolgy Conference Field Trip, Guidebook, 46 p.

Buckley, R., ed., 1994, The world of wine: The Grape and Civilization. Understanding Global Issues: European Schoolbooks Publishing, Cheltenham, UK.

Chapman, P., 1994, Agriculture in Niagara: an overview, *in* Gaylor, H.J., ed., Niagara's Changing Landscapes: Carleton University Press, Ottawa ON, p. 279-299.

Cheel, R.J., ed., 1991, Sedimentology and Depositional Environments of Silurian Strata of the Niagara Escarpment, Ontario and New York: Geological Association of Canada-Mineralogical Association of Canada, Joint Annual Meeting, Toronto, Field Trip B4, Guidebook, 99 p.

de Blij, H.J., 1983, Wine: A Geographic Appreciation: Rowman and Allenheld, Totowa, NJ.

Feenstra, B.H., 1981, Quaternary Geology and Industrial Minerals of the Niagara -Welland Area, Southern Ontario: Ontario Geological Survey, Open File Report 5361, 260 p.

Feenstra, B.H., 1984, Quaternary geology of the Niagara Welland Area: Ontario Geological Survey, Map 2496, scale 1:50 000.

Flint, J.J. and Lolcama, J., 1985, Buried ancestral drainage between Lakes Erie and Ontario: Geological Society of America Bulletin, v. 97, p. 75-84.

Fraser, J.Z. and Haynes, S.J., 1994, Landscape dynamics on the Niagara Escarpment, Niagara Peninsula, Part 2: quantitative analysis of land-use change on the Fonthill Kame: Ontario Ministry of Environment and Energy, Niagara Escarpment Commission, Ontario Heritage Foundation, Canadian Heritage and Parks Canada; Leading Edge '94, Conference, Hockley Valley, Proceedings, p. 475-487.

Gombos, F., 1977, The Role of Vermiculite in the Fixation of Potassium in Soils and the Availability of Potassium for Vine Nutrition in Some Niagara Vineyards: M.Sc. thesis, Brock University, St. Catharines ON, 116 p.

Haynes, S.J., 1992a, Geology, wine and scenery of the Niagara Escarpment: Ontario Petroleum Institute, 31st Annual Meeting, Niagara Falls, Field Guidebook. Published as: Department of Geological Sciences, Brock University, Urban Geology Series, n. 1, St. Catharines ON, 6 p.

Haynes, S.J., 1992b, Lithofacies and Stratigraphic Relationship of the Bass Islands Formation, Onondaga Escarpment, Niagara Peninsula: Ontario Petroleum Institute, 31st Annual Meeting, Niagara Falls, Proceedings, Technical Paper 11, 19 p.

Haynes, S.J., 1994a, Geology, Scenery and Wines of the Niagara Escarpment: Geological Association of Canada-Mineralogical Association of Canada, Joint Annual Meeting, Waterloo ON, Field Trip A3, Guidebook, 10 p.

Haynes, S.J., 1994b, Wine, Geology and Glaciolacustrine Soils of the Niagara Escarpment: Ontario Petroleum Institute, 33rd Annual Meeting, Niagara Falls ON, Field Guidebook, 13 p.

Haynes, S.J., 1994c, Breccias Ancient and Modern, Cayugan Series: Ontario Petroleum Institute, 33rd Annual Meeting, Niagara Falls, Technical Paper 12, 15 p.

Haynes, S.J., 1995a, Geology, landscape dynamics and land-use of the southern Niagara Escarpment: *PROTAMBI '95*, Moa, Cuba; Minera y Geologia, v. 12, n. 3, p. 55-64.

Haynes, S.J., 1995b, Geological definition of the Escarpment and its landforms: relevance to conflicting land-uses in the eastern Niagara Escarpment Plan area: Ontario Ministry of Environment and Energy, Niagara Escarpment Commission, Ontario Heritage Foundation, Canadian Heritage and Parks Canada; Leading Edge '95, Conference, Collingwood ON, Proceedings, p. 82-88.

Haynes, S.J., 1995c The evolution of wine and grape, *in* Welcome to Wine Country, Rannie Publications, Beamsville ON, p. 6.

Haynes, S.J., 1998, The "terroir" of Niagara's vineyards: St. Catharines Chamber of Commerce, Chamber News, fall issue, p. 1, 12.

Haynes, S.J., 1999a, The last "Geology of Niagara Vineyards," field trip of the millennium: Ontario Petroleum Institute, 38th Annual Meeting, Niagara Falls; Published by Siashadal Inc., Box 397, Fonthill ON, Field Trip Guidebook, 20 p.

Haynes, S.J., 1999b, Geology and Wine 1. Concept of terroir and the role of geology: Geoscience Canada, v. 26, n. 4, p. 190-194.

Haynes, S.J., 1999c, Vertical and Lateral Evaporitic Facies Changes Within the Salina Formation: Ontario Petroleum Institute, 38th Annual Meeting, Niagara Falls ON, Proceedings, Technical Paper, 15 p.

Haynes, S.J. and Fraser, J.Z., 1994, Landscape dynamics on the Niagara Escarpment, Niagara Peninsula, Part 1: geologic controls and implications for land-use change: Ontario Ministry of Environment and Energy, Niagara Escarpment Commission, Ontario Heritage Foundation, Canadian Heritage and Parks Canada; Leading Edge '94, Conference, Hockley Valley, Proceedings, p. 373-385.

Haynes, S.J., Grant, E.B. and Haynes, V.S., 1998, Geology of Niagara Falls and Niagara's Vineyards and Wines: International Mineralogical Association, General Meeting, Toronto, Field Trip B7, Guidebook, 11 p.

Haynes, S.J., and Mostaghel, M.A., 1982, Present-day precipitation of lead and zinc from groundwaters: Mineralium Deposita, v. 17, p. 213-228.

Hobson, G.D. and Terasmae, J., 1969, Pleistocene Geology of the Buried St. Davids Gorge, Niagara Falls, Ontario: Geophysical and Palynological Studies: Geological Survey of Canada, Paper 68-67, 16 p.

Johnson, M.D., Armstrong, D.K., Sanford, B.V., Telford, P.G. and Rutka, M.A., 1992, Paleozoic and Mesozoic geology of Ontario, *in* Geology of Ontario: Ontario Geological Survey, Special Volume 4, Part 2, p. 907-1010.

Karrow, P.F., 1989, Quaternary geology of the Great Lakes subregion, *in* Fulton, R.J., ed., Quaternary Geology of Canada and Greenland: Geological Survey of Canada, Geology of Canada, n. 1, chapter 4, p. 326-350.

Kingston, M.S. and Presant, E.W., 1989, The soils of the Regional Municipality of Niagara: Ontario Institute of Pedology, Report 60, Volumes 1 and 2, plus 7 maps scale 1:25,000.

Menzies, J., 1994, Ideas in transition: some perspectives on landscape evolution in the Niagara Peninsula, *in* Gaylor, H.J., ed., Niagara's Changing Landscapes: Carleton University Press, Ottawa ON, p. 53-79.

Menzies, J. and Taylor, E.M., 1998, Urban geology of St. Catharines-Niagara Falls, Region Niagara, *in* Karrow, P.F. and White, O.L., eds., Urban Geology of Canadian Cities: Geological Association of Canada, Special Paper 42, p. 287-321.

Ontario Geological Survey, 1991, Bedrock geology of Ontario, southern sheet: Ontario Geological Survey, Map 2544.

Pomerol, C., ed., 1989, The Wine and Winelands of France; Geological Journeys: Editions du BRGM, Orleans, France.

Sayed, H., 1992, Vineyard site suitability in Ontario: Horticultural Research Institute of Ontario, Vineland Station, Ontario, Ontario Ministry of Agriculture, 23 p.

Schreiner, J., 1985, The World of Canadian Wine: Douglas and McIntyre, Vancouver BC, 286 p.

Shaw, T.B., 1994, Climate of the Niagara region, *in* Gaylor, H.J., ed., Niagara's Changing Landscapes: Carleton University Press, Ottawa ON, p. 111-138.

Shaw, T.B., 1996, An assessment of the growing season thermal and moisture environment for timing disease control in Niagara vineyards: 4th International Symposium on Cool Climate Viticulture and Enology, Proceedings, v. 3, p. 10-15.

Smart, R.E., 1999, Winter freeze, *in* Robinson, J., ed., The Oxford Companion to Wine, 2nd edition: Oxford University Press, Oxford UK, p. 779-780.

Stevenson, T., 1997, The New Sotheby's Wine Encyclopedia: Firefly Books, Willowdale ON, 600 p.

Stewart, R.B., Mukammal, E.I. and Wiebe, J., 1977, Delineation of frost prone areas in the Niagara fruit belt: Joint study by Atmospheric Environment Service of Environment Canada and Horticultural Research Institute of Ontario Ministry of Agriculture and Food, 16 p.

Tesmer, I.H., ed., 1981, Colossal Cataract: The Geologic History of Niagara Falls: State University of New York Press, Albany NY, 219 p.

Tinkler, K.J., 1994, Entre lacs: a postglacial peninsula physiography, *in* Gaylor, H.J., ed., Niagara's Changing Landscapes: Carleton University Press, Ottawa ON, p. 13-52.

Tinkler, K.J., Pengelly, J.W., Parkins, W.G. and Asselin, G., 1994, Postglacial recession of Niagara Falls in relation to the Great Lakes: Quaternary Research, v. 42, p. 20-29.

Tiplin, A.H. and Seibel, G.A., 1988, Our Romantic Niagara: A Geological History of the River and the Falls: The Niagara Falls Heritage Foundation, 210 p.

Tovell, W.M., 1992, Guide to the geology of the Niagara Escarpment: Niagara Escarpment Commission, Ashton-Potter, Concord ON.

Unwin, T., 1991, Wine and the vine: Routledge, New York NY, p. 392.

Wiebe, J. and Anderson, E.T., 1977, Site selection for grapes in the Niagara Peninsula: Horticultural Research Institute of Ontario, Vineland Station ON, 1-page map.

Wilson, J.E., 1998, Terroir: the role of geology, climate and culture in the making of French wines: Mitchell Beazley, London UK, 336 p.

Ziraldo, D.J.P., 1995, Anatomy of a Winery, The Art of Wine at Inniskillin: Key Porter Books, Toronto ON, 48 p.

Accepted as revised 3 May 2000

SERIES

Geology and Wine 3: *Terroirs* of the Walla Walla Valley *appellation*, southeastern Washington State, USA[1]

Lawrence D. Meinert
Department of Geology
Washington State University
Pullman, WA 99164-2812
meinert@wsu.edu

Alan J. Busacca
Department of Crop and Soil Sciences
Washington State University
Pullman, WA 99164-6420
busacca@wsu.edu

SUMMARY

Terroir of the Walla Walla Valley *appellation* of Washington State is influenced by 1) the rain shadow effect and volcanic tephra of the Cascade Mountain Range, 2) soils derived from Quaternary glacial sediments and wind-blown loess overlying Miocene basalt, and 3) a warm, dry climate with abundant sunshine and cool nights due to high latitude (45-48°N) and elevation. Wine flavours of Walla Walla Valley varietals (including Cabernet Sauvignon, Chardonnay, Merlot, and Syrah) appear to be influenced by low humidity, long growing season (1500-1800°C degree days), soil type and drainage, drip irrigation, pruning to restrict vine vigour, trellising methods, and overall topographic climatic setting of the vineyards. Microclimates within the Walla Walla Valley *appellation*, combined with variations in soil and bedrock stratigraphy, yield a range of wine styles as great as many larger regions and some countries.

RÉSUMÉ

Le terroir de la vallée de Walla Walla d'appellation de l'État de Washington est caractérisé par 1) l'effet parapluie de la topographie locale et la nature téphritique des dépôts de la chaîne des monts Cascade, 2) les sols issus de dépôts glaciaires et de loess éoliens recouvrant un basalte Miocène, et 3) un climat chaud et sec très ensoleillé, avec les nuits fraîches des hautes latitudes (N45° à N48°) et de l'altitude élevée. Les goûts des vins des variétés de la vallée de Walla Walla (incluant le Cabernet Sauvignon, le Chardonnay, le Merlot, et le Syrah) semblent dépendre de la faible humidité de la région, sa saison végétative longue (1 500°C à 1 800°C degrés-jours), de la nature de son sol et de son drainage, d'une irrigation goutte à goutte, de la taille de la vigne pour en restreindre la vigueur végétative, des méthodes de treillages et, des conditions topographiques et climatiques générales où baignent ces vignobles. L'effet combiné des microclimats de la région d'appellation de la vallée de Walla Walla avec les variétés de ses sols et de la stratigraphie du substratum rocheux régional permet la production de styles de vin aussi prestigieux que bien des régions reconnues et même de certains pays.

INTRODUCTION

Terroir is a relatively simple term to describe the complex interplay of climate, soil, geology, and other physical factors that influence the character and quality of wine. Although the term has long been used in France, recent publications such as Halliday (1993, 1999), Wilson (1998), Haynes (1999, 2000), and Tesic (2000) have explored the concept of *terroir* in other parts of the world with considerable success. The purpose of the present paper is to describe the *terroir* of the Walla Walla Valley *appellation*, a relatively small and young winemaking region in southeastern Washington and overlapping into northeastern Oregon (Fig. 1) that is producing some outstanding wines, particularly stout red wines such as Cabernet Sauvignon, Merlot, and Syrah.

Washington State is a region of superlatives, both oenological and geological, with exposures of some of the world's largest and most spectacular flood basalts, dune and loess fields, and glacial outburst flood deposits. All of these play a part in the *terroirs* of Washington State wines. In addition, recent volcanic

[1] Geology and wine is an international topic, much like *Geoscience Canada's* current series on Oceanic Lithosphere, and Economic Geology Models. Accordingly we are pleased to publish this third paper in the Geology and Wine series, on the Walla Walla Valley *appellation* of Washington State. Although the geology of the Walla Walla Valley *appellation* is unusual, with vast Miocene plateau basalts, slackwater flood deposits, widespread loess, and a variety of derived soil types, it is the *integration* of these geological attributes with climatic factors and viticultural practices that demonstrates the importance of *terroir* with all of its components. This third Geology and Wine series paper is a further example of the kinds of papers we are hoping to publish in *Geoscience Canada* as the series develops. R.W. Macqueen, editor.

activity such as the well-known 1980 eruption of Mt. St. Helens continues to shape the oenological and geological landscape. Most Washington vineyards lie in the rain shadow of the Cascade volcanic arc, and many vineyard soils have a component of ash from Mt. St. Helens and other Cascade volcanic eruptions such as the much larger Mt. Mazama eruption (6850 years B.P.) which formed present-day Crater Lake in Oregon (Busacca *et al.*, 2000).

Washington State is second only to California in terms of wine produced in the United States (Table 1). This is somewhat surprising in that Washington has a relatively short history of wine production by international standards. Although the first *Vitis vinifera* grapes were planted in 1825 by the Hudson's Bay Company along the banks of the Columbia River in southwest Washington (Peterson-Nedry, 2000), commercial production extends back only about 100 years and most of the state's 155+ wineries were started in the past 15 years (there were only 19 wineries when the first author moved to Washington in 1981).

Most Washington State wineries are located between latitudes 45°N and 48°N, well to the north of the more widely known California vineyards but parallel to some of the great French wine regions such as Burgundy and Bordeaux. This northerly latitude provides about two hours more summer sunlight than occurs in California wine regions. Wine quality is also influenced by the arid conditions created by the large rain shadow of the Cascade Mountains; almost all commercial vineyards lie east of the Cascade Mountain Range (Fig. 1).

Although there is considerable local variability, most Washington vineyards are located on Quaternary

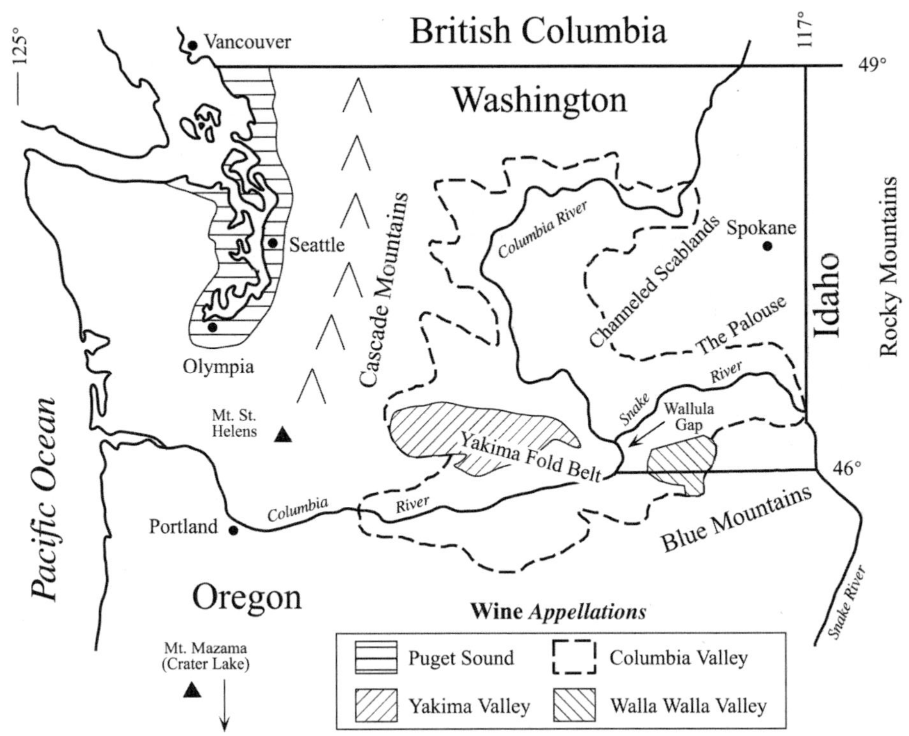

Figure 1 Location map of the Pacific Northwest showing wine *appellations* and major geologic features described in the text.

Table 1 US *vinifera* wine grape production 1997-1999. Data from http://www.nass.usda.gov/wa/graperpt.htm.

		Tons Utilized			Average Price (US$/ton)		
Rank	**Variety**	**1997**	**1998**	**1999**	**1997**	**1998**	**1999**
1	California	2,895,000	2,528,000	2,655,000	603	586	570
2	Washington	62,000	70,000	70,000	972	922	910
3	New York	44,000	36,000	50,000	328	392	348
4	Oregon *	18,500	14,700	17,900	1,120	1,180	1,310
5	Pennsylvania	6,000	11,500	13,000	350	303	325
6	Michigan	2,600	2,500	3,000	800	775	700
7	Missouri	1,800	2,050	2,600	480	535	583
8	Georgia	1,000	1,300	2,000	627	500	1,000
9	Ohio	1,000	800	2,000	492	679	725
	Other States	4,500	4,910	3,485	549	574	735
	United States	3,036,400	2,671,760	2,818,985	503	510	517

* Includes small quantities used for other processing (jam, jelly, *etc.*).

sediments and soils that overlie Miocene basaltic rocks of the Columbia River flood basalt province. Many of the Quaternary sediments are related to cataclysmic glacial outburst floods that formed the spectacular geomorphic features of the Channeled Scablands (cover photo). This in itself is one of the great geologic stories of all time, and the fact that it is intimately related to superlative wine makes a brief synopsis of the geologic history particularly worthwhile. The fact that some of the geological and soil features are unique to this part of the world suggests the possibility that Washington wines may develop flavour and quality characteristics that set them apart from other wine-producing areas.

OVERVIEW OF WASHINGTON STATE WINE

Washington State has four wine *appellations* called American Viticultural Areas (AVAs) by the Bureau of Alcohol, Tobacco, and Firearms (BATF), the chief regulatory agency of the wine industry in the United States (Fig. 1). Unlike *appellations* in many parts of the world, AVAs are not defined in terms of wine quality or even wine characteristics. An AVA is defined by the Bureau of Alcohol, Tobacco, and Firearms as "a delimited grape growing region distinguished by geographical features, the boundaries of which have been recognized and defined…" (http://www.atf.treas.gov/core/regulations/27cfr9.html). In Washington State the wine *appellations* (AVAs) are Columbia Valley, Yakima Valley, Walla Walla Valley, and Puget Sound (Fig. 1). Sub-*appellations*, that may someday become AVAs, include Alder Ridge, Canoe Ridge, Cold Creek, Columbia River Gorge, Horse Heaven Hills, Red Mountain, Wahluke Slope, and Zephyr Ridge (Peterson-Nedry, 2000). As with most other regions, AVAs can be nested such that the Columbia Valley *appellation*, which produces more than 90% of the state's wine grapes, includes the Yakima Valley and Walla Walla Valley *appellations*.

Total wine grape production in Washington State in 1999 was 70,000 tons from approximately 17,000 acres of bearing vineyards (Table 2). Wine grape production will continue to increase in that there are an additional 6000 acres of

Table 2 Washington State *vinifera* wine grape production 1995-1999. Data from http://www.nass.usda.gov/wa/graperpt.htm

	Tons Utilized					Acres Planted	Acres Bearing	Yield Tons/Acre	Average Price (US$/ton)				
Variety	**1995**	**1996**	**1997**	**1998**	**1999**	**1999**	**1999**	**1999**	**1995**	**1996**	**1997**	**1998**	**1999**
Chardonnay	15,000	8,000	17,700	20,500	22,900	6,100	5,030	4.6	732	1,240	1,055	971	882
White Riesling	11,000	8,800	10,100	11,500	9,800	1,900	1,780	5.5	385	478	489	533	547
Sauvignon Blanc	3,700	2,300	3,300	3,900	3,400	700	600	5.7	519	726	729	750	746
Semillon	3,600	1,800	2,800	3,100	2,600	600	590	4.4	502	642	661	625	598
Chenin Blanc	3,400	1,100	1,800	2,300	1,700	400	400	4.3	339	464	499	488	473
Gewürztraminer	1,700	700	1,700	1,600	1,300	400	310	4.2	413	541	552	629	706
Other (*)	600	300	600	600	800	400	290	2.8	546	755	789	952	777
Total White	39,000	23,000	38,000	43,500	42,500	10,500	9,000	4.7	542	786	794	772	753
Merlot	10,000	4,500	12,800	14,100	14,300	5,600	4,040	3.5	933	1,400	1,360	1,197	1,149
Cabernet Sauvignon	8,000	5,900	7,800	8,300	8,400	5,000	2,690	3.1	834	1,230	1,195	1,204	1,236
Cabernet Franc	500	600	1,100	1,700	2,000	700	510	3.9	834	1,170	1,120	1,102	1,066
Pinot Noir	1,200	400	1,000	1,100	1,000	200	200	5.0	520	698	653	661	647
Syrah	#	#	#	#	800	1500	290	2.8	#	#	#	#	1,398
Lemberger	400	300	500	500	500	150	110	4.5	565	844	757	760	785
Other (+)	900	300	800	800	500	350	160	3.1	917	1,110	1,400	1,386	1,152
Total Red	21,000	12,000	24,000	26,500	27,500	13,500	8,000	3.4	862	1,260	1,255	1,168	1,152
Washington total	60,000	35,000	62,000	70,000	70,000	24,000	17,000	4.1	654	948	972	922	910

* Includes muscat, pinot gris, and viognier
\+ Includes gamay beaujolais, grenache, nebbiolo, sangiovese, and zinfandel
\# Included in "Other" Red Varieties for 1995-1998

wine grapes planted that were not yet bearing fruit in 1999 (most grape varieties start producing in the third year). Projected production for 2000 is 88,500 tons (Roger Gamache, Washington Association of Wine Grape Growers, written communication, 2000). For comparison, to the north in British Columbia in 1999 there were 4200 acres of bearing vineyards, yielding 11,200 short tons of crops; just over 50% of the acreage under cultivation for wine grapes in British Columbia in 1999 was located in the Oliver-Osoyoos area of the Okanagan Valley region, about 320 km north of the Walla Walla Valley *appellation* (British Columbia Wine Institute, 2000).

In the early days of the Washington State wine industry most wine was made from Riesling and Gewürztraminer grapes because it was assumed that Washington was a cool-climate viticultural area due to its northern latitude, and therefore best suited to production of light fruity white wines. Although such grapes can produce excellent wine in Washington, it was quickly realized that varieties that benefit from warmer conditions also prosper in Washington (Clore and Nagel, 1969; Tukey and Clore, 1972; Nagel *et al.*, 1972). Of the wine produced in Washington State in 1999, 60% was white and 40% was red, a ratio which is almost identical in British Columbia (British Columbia Wine Institute, 2000). This ratio is likely to swing toward a predominance of red wine in Washington State in the future because of the increased plantings of red varietals. For example, in 1999 there were 9000 acres bearing white wine grapes and 8000 acres bearing red wine grapes, whereas in terms of planted but not-yet-bearing grapes, there are only 1500 acres of white wine grapes *versus* 5500 acres of red wine grapes. This trend appears to be accelerating with the newest plantings and those that are on the drawing boards (Norm McKibben, past president Washington Association of Wine Grape Growers, oral communication, 2000).

As shown in Table 2, white wine production is dominated by Chardonnay (54%) and Riesling (23%), whereas red wine production is led by Merlot (52%), Cabernet Sauvignon (31%), and Cabernet Franc (7%). Syrah currently is a fairly small part (3%) of Washington red wine production but is likely to grow in importance owing to spectacular early successes, as well as to the extensive new plantings of this grape. In 1999 only 290 acres of Syrah grapes were bearing commercial fruit but more than 1500 acres have been planted. This changing mix between red and white wine and among different grape varieties is to be expected in such a relatively young viticultural area. It is likely that further changes will occur as growers and winemakers learn more about which grapes are best suited to particular *appellations* and individual vineyards. Although at present about 85% of Washington State wines are blends of fruit from different vineyards, it is likely that single-vineyard wines will become more important in the future. This is one of the reasons why an understanding of *terroir* is so important as the Washington wine industry begins to mature. This is also the case in other Pacific Northwest viticultural areas such as the Okanagan Valley of British Columbia, where much experimentation with *vinifera* grape varieties is underway, following a major reorganization of the local wine industry, and the adoption of new high standards, in response to the Free Trade agreement of 1988 (*e.g.*, Aspler, 1999).

GEOLOGIC HISTORY

Columbia Plateau and Environs

The Columbia Plateau is bordered on the north and east by the Rocky Mountains, on the south by the Basin and Range Province, and on the west by the Cascade Mountains. The Walla Walla part of the Columbia Plateau is adjacent to the northwest flank of the Blue Mountains and is crossed by two of the largest rivers in North America: the Snake River that originates in Jackson Hole, Wyoming, and the Columbia River that starts in the Columbia Icefields of the Canadian Rockies.

The Blue Mountains (Fig. 1) are cored by an accreted terrain of late Paleozoic to early Mesozoic predominantly clastic and carbonate strata, intruded by late Mesozoic granitic plutons, and partly overlain by Tertiary volcanic rocks of the Columbia River Basalt Group (Hooper, 2000). The Columbia River Basalt Group covers an area of about 165,000 km^2. It was erupted mostly between 17-15 Ma (Early Miocene) from north-south fissures roughly paralleling the present-day Washington-Idaho border, and has individual flows with estimated eruptive volumes of at least 3000 km^3, making them the largest documented lava flows on earth (Baksi, 1989; Landon and Long, 1989; Tolan *et al.*, 1989). This dwarfs the erupted volumes of typical Cascade volcanoes: even the explosive eruption of Mt. St. Helens in 1980 yielded only about 1 km^3 of volcanic material (Pringle, 1993).

Stress Regime, South-central Washington State

Concurrent with Early Miocene volcanism were subsidence (*e.g.*, Pasco Basin), deformation (*e.g.*, Yakima fold belt), erosion (many of the valleys were later filled by intracanyon flows, Fig. 2a), and intraflow sedimentation (Fig. 2b). The north-south oriented compressional stress regime of south-central Washington State has existed from the Miocene to the present day. Some anticlines in the Yakima fold belt have developed as much as 100 m of structural relief in the past 10 m.y. (Reidel *et al.*, 1992), and the Walla Walla area experienced an intensity VII (approximate Richter magnitude 6) earthquake on 15 July 1936 (Brown, 1937), most likely caused by movement on the Wallula fault zone (Mann and Meyer, 1993), which is subparallel to the Yakima Fold Belt (Fig. 1).

Quaternary Processes and Products

The Quaternary Period brought a new group of processes. In the Pasco Basin were many sources for wind-blown detritus such as the Miocene Ellensburg Formation, the Pliocene Ringold Formation, sediment deposited by the Columbia and Snake rivers, and sediment from giant glacial outburst floods. The prevailing southwesterly winds transported silt from the Pasco Basin and deposited thick blankets of loess (Fig. 2c). Buried soils in the loess (Fig. 2d) indicate that deposition was intermittent during the Quaternary (Busacca, 1991).

Most of the basement geology has been masked and modified by a variety of features related to the Quaternary glaciation that affected much of the region. The vast continental ice sheet of the glacial maxima covered much of Canada, and the subsidiary Cordilleran Ice Sheet, which originated in the mountains of British Columbia, expanded southward into northern Washington, Idaho, and Montana. Of particular significance to the landscape of eastern Washington State was a lobe of the Cordilleran Ice Sheet which flowed along the Purcell Trench and blocked the northwest-flowing Clark Fork River near Cabinet Gorge on the Idaho-Montana border.

Lake Missoula and the Channeled Scabland

The Cordilleran Ice Sheet lobe dammed glacial Lake Missoula (Fig. 3), which covered 7800 km^2 of western Montana (Pardee, 1910). At the ice dam the water was approximately 600 m deep (Weis and Newman, 1989). The ice dam failed repeatedly, releasing the largest floods documented on earth (Baker and Nummedal, 1978). These floods overwhelmed the Columbia River drainage system and sent up to 2500 km^3 of water across the Columbia Plateau with each outburst (called jökulhlaups). The floods created a spectacular complex of anastomosing channels cut into southwest-dipping basalt surfaces. They also formed huge cataracts now seen as dry falls, "loess islands" that are erosional remnants of an early thick loess cover on the plateau, immense gravel bars, and ice-rafted erratic boulders at high elevations (Fig. 3, 12B).

In a series of early papers (Bretz, 1923, 1925, 1928a,b,c, 1932), J Harlen Bretz shocked the geological community and precipitated one of the most celebrated scientific debates in American geology with his studies of this enormous series of proglacial channels eroded into the loess and basalt of the Columbia Plateau. This region, which he named the "Channeled Scabland," contains erosional and depositional features that are unique among fluvial phenomena. With painstaking field work, before the advent of aerial photographs and modern topographic maps, Bretz documented the field relationships of the region. He argued that the landforms could only be explained as originating from a relatively brief, but enormous flood, which he called the "Spokane Flood." Considering

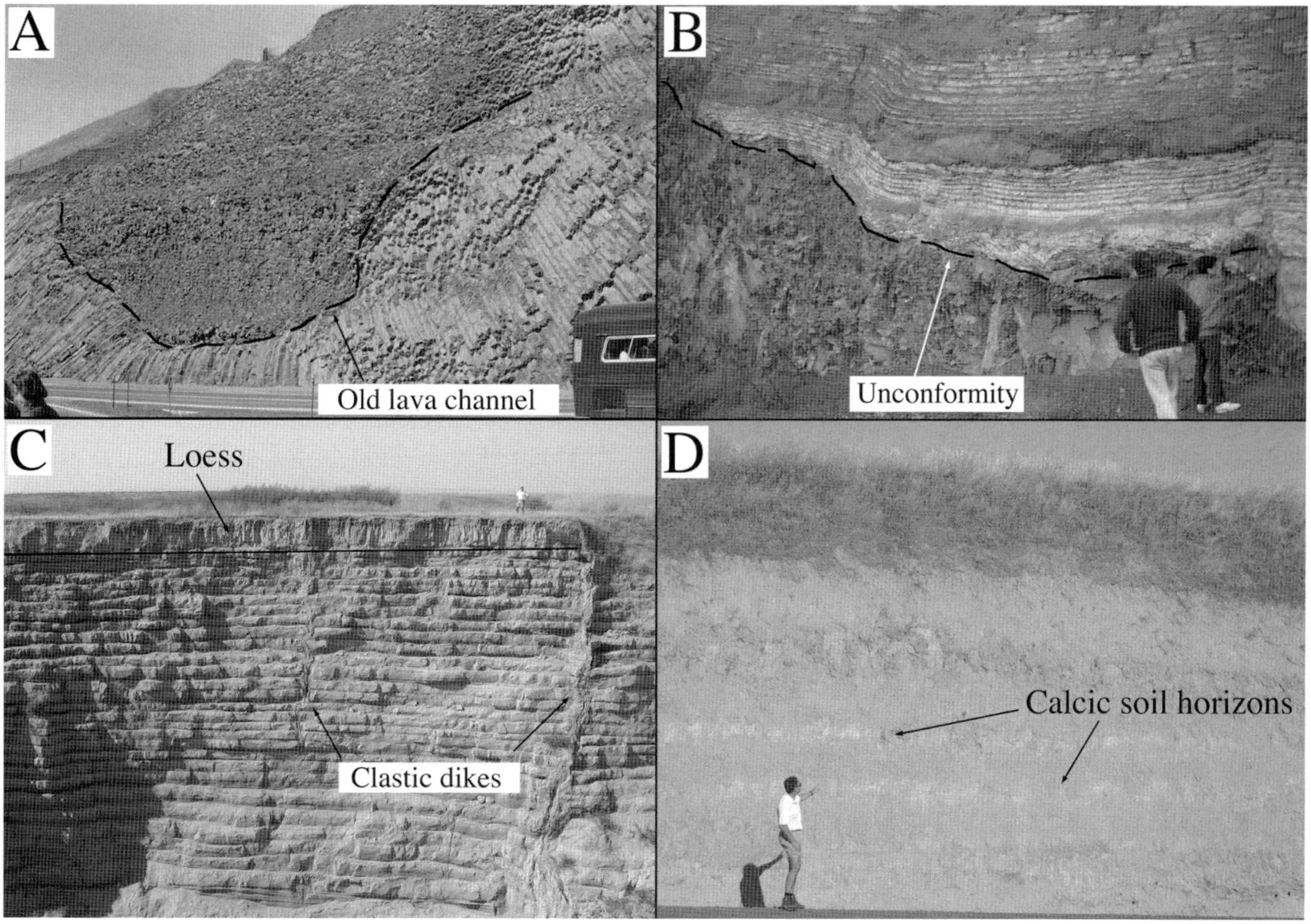

Figure 2 (**A**) Contact of intra-canyon lava flow with slightly older lava flow with well-developed columns. Both flows are part of the Miocene Columbia River Basalt Group. (**B**) Miocene sedimentary rocks deposited on unconformity developed within Columbia River Basalt Group. (**C**) Quaternary slackwater sediments (Touchet beds) deposited by backflooding of the Walla Walla Valley. Each horizontal layer represents a separate glacial outburst flood event. Touchet beds are overlain by 2 m thick loess bed with characteristic vertical prismatic structure. Entire sequence is cut by clastic dikes, thought to form by infill of surficial cracks (Alwin, 1970). Exposure is in Burlingame Canyon ~3km south of Touchet, Washington. (**D**) Multiple loess beds separated by calcic paleosol horizons. Exposure is on Hwy 12 ~3 km ENE of Walla Walla, Washington.

the nature and vehemence of the opposition to this outrageous hypothesis, the eventual triumph of his ideas constitutes one of the most fascinating episodes in the history of modern geomorphology (Bretz *et al.*, 1956; Bretz, 1969).

In south-central Washington State, the many paths of the onrushing floods converged on the Pasco Basin where floodwaters were slowed by the hydrologic constriction of Wallula Gap before draining out through the Columbia River Gorge to the Pacific Ocean (Figs. 1, 3). This caused *backflooding* up local river valleys and deposition of relatively fine-grained slackwater sediments characterized by rhythmically graded bedding (Bretz, 1928a,b,c); these graded rhythmites are locally called Touchet beds (Fig. 2c, cover), and multiple sets have been recognized (Flint, 1938). Waitt (1980, 1985) argued that each of about 40 Touchet beds resulted from separate catastrophic floods between 15,300 to 12,700 radiocarbon years B.P. Atwater (1986) determined that the interval between Missoula floods was 35-55 years based upon the number of flood deposits and varves in nearby glacial lakes. Mt. St. Helens "S" ash fell in eastern Washington about 13,000 years B.P., and is 10 rhythmites below the top of the Touchet beds, consistent with Waitt's (1985) age estimate. However, some researchers still prefer Bretz's original explanation of a single giant flood (Shaw *et al.*, 1999).

In recent decades, researchers have estimated the hydraulics of these giant flood flows (O'Connor and Baker, 1992; Smith, 1993) and explained the origin of many puzzling scabland features (Baker, 1973; Baker and Nummedal, 1978; Baker and Bunker, 1985). Other studies have shown that there were as many as 90 glacial floods (Waitt, 1980, 1984, 1985; Atwater, 1986) and that the floods were associated with not just one but at least six or seven major glacial advances in the Pleistocene (Patton and Baker, 1978). However, the detailed glacial stratigraphy is beyond the scope of the present paper.

Reworked Sediments/Loess

During the Holocene the prevailing southwesterly winds continued to rework Quaternary loess, slackwater, and other glacial sediments into the present thick blanket of loess (Fig. 2d) that covers much of the Columbia Plateau (45°30' to 48°N and 116°30' to 120°W). The core of this area, which is centered in eastern Washington, is called "The Palouse" (Fig. 1) and covers more than 10,000 km^2 with loess that is up to 75 m thick. Thinner and less continuous loess deposits cover an additional 30,000-40,000 km^2 in Washington, north central Oregon, and northern Idaho. As previously mentioned for the Quaternary, the presence of buried soils in the Holocene loess indicates that deposition was intermittent (Busacca, 1991).

Surface Soils, Columbia Plateau

Because of the Columbia Plateau's proximity to volcanoes of the Cascade Range, it is commonly assumed that the parent material of the loessial soils is predominantly volcanic tephra, or airborne volcanic material, and that the volcanic glass content of the soil is the cause of high susceptibility to wind erosion. However, the volcanic glass content is typically less than 8% in south central Washington (Fig. 4). In addition, the volcanic glass distribution in surface

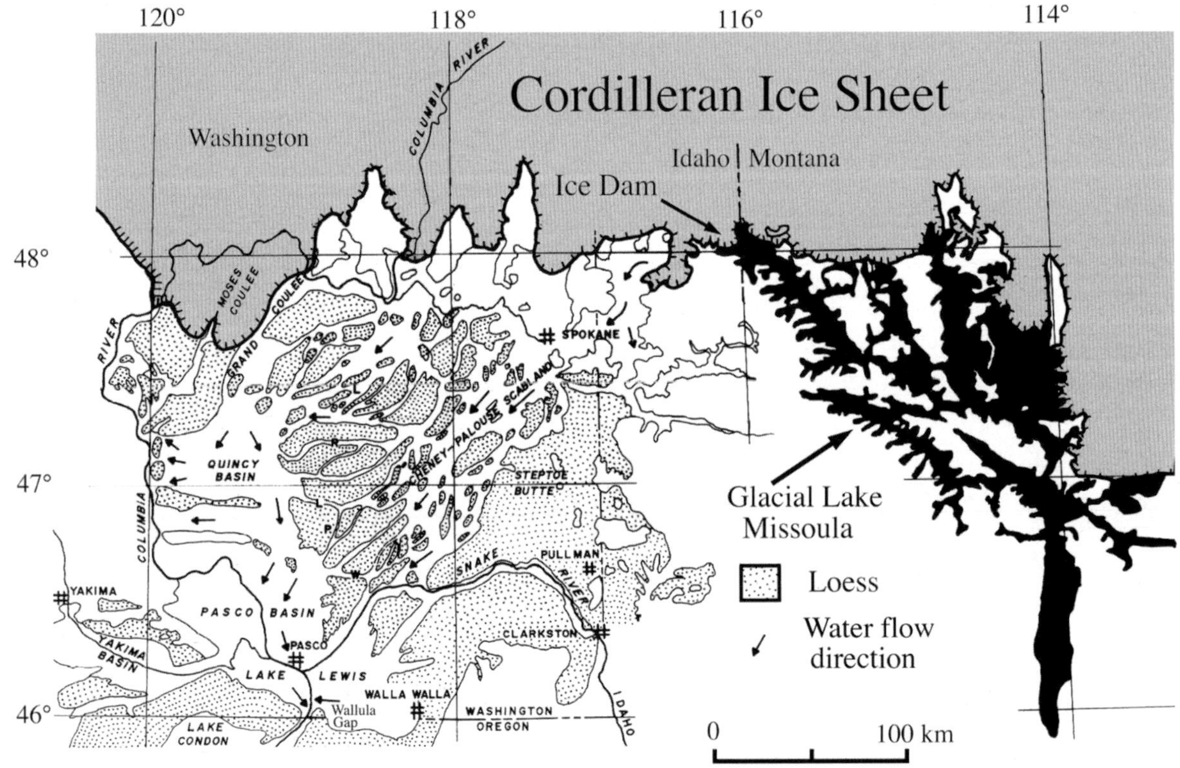

Figure 3 The Channeled Scabland of the Pacific Northwest showing Cordilleran ice sheet in its approximate position at 15,000 years before present, major outflow channels, and the distribution of loess. Modified from Baker and Nummedal (1978).

soils across the Plateau does not correspond to known patterns of tephra fallout from the 1980 Mt. St. Helens eruption. Electron microprobe analysis of glass composition confirms that the eruption of Mt. Mazama 6850 radiocarbon years B.P. contributed the majority of volcanic glass found in surface soils of the Columbia Plateau (Busacca *et al.*, 2000).

CLIMATE AND NATIVE VEGETATION

The climate of the Columbia Plateau is influenced to a great extent by prevailing westerly winds and by the Cascade and Rocky mountains. The Cascade Mountains create a rain shadow and as a result, the climate of the Columbia Plateau is arid to sub-humid (15-100 cm of mean annual precipitation; http://www.wrcc.dri.edu/climsum.html). The amount of precipitation closely correlates with elevation, generally increasing from west to east and southeast as elevation increases. The Rocky Mountains protect the Pacific Northwest from the coldest of the arctic storms that sweep down through Canada and this affects both vineyards and native vegetation.

During the summer, high pressure systems prevail in the Pacific Northwest, leading to dry, warm conditions and low relative humidity. Average afternoon temperatures in the summer range from 20°C to more than 35°C (http://www.wrcc.dri.edu/climsum.html). Most of the growing season is very dry and some vineyards experience no measurable precipitation during the summer months. The rainy season extends from October to late May or June, as frontal storms sweep across the area. In eastern Washington, most of the precipitation from mid-December to mid-February is in the form of snow.

Pre-agricultural Vegetation

As a result of the rainshadow and elevation effects, pre-agricultural vegetation in southeastern Washington ranged from sagebrush-steppe in the driest areas, to meadow steppe in areas of intermediate precipitation, to coniferous forest in areas of highest precipitation (Daubenmire, 1970). Xerophytic (drought tolerant) shrubs include several species of *Artemisia, Purshia,* and *Crysothamnus.* Perennial grasses include the major species bluebunch wheatgrass (*Agropyron spicatum*), Idaho fescue (*Festuca idahoensis*), Sandberg bluegrass (*Poa sandbergii*) and a host of less common annual and perennial grasses and forbs. Mesophytic (moisture loving) shrubs include *rosa* spp., Serviceberry (*Amelanchier alnifolia*), and Snowberry (*Symphoricarpos albus*). Several zones of conifer vegetation have been recognized with increasing effective moisture and decreasing temperature (Daubenmire and Daubenmire, 1984); no zone of hardwoods is interposed between steppe and conifer habitats. The semiarid climate and steppe vegetation which have persisted throughout the Quaternary Period, have produced the Aridisol and Mollisol soils that form the backbone of Washington State wine grape production.

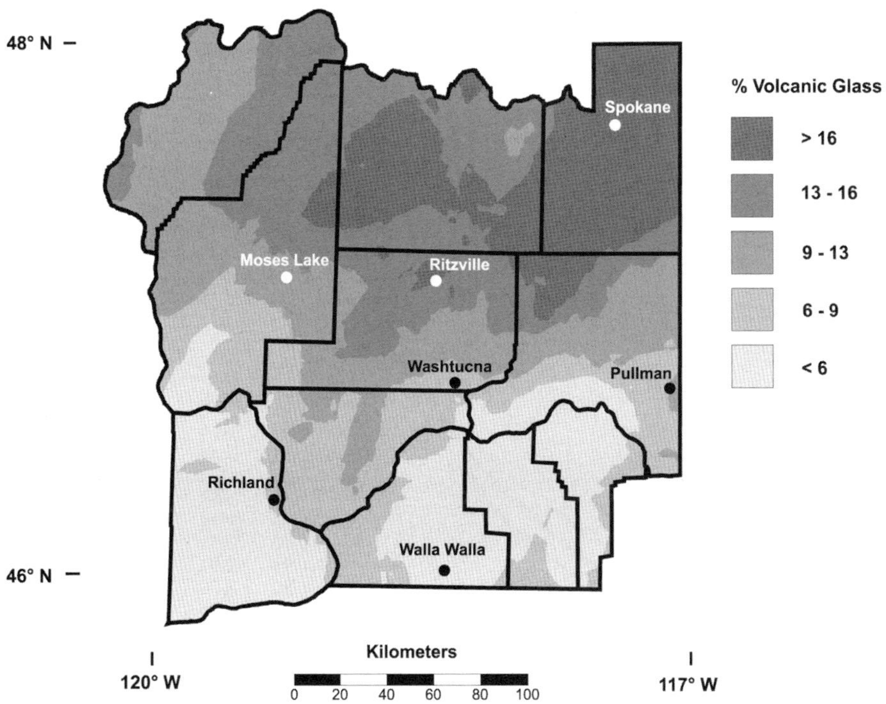

Figure 4 Map of volcanic glass content of Columbia Basin loess soils. From Busacca *et al.* (2000).

Climate and *Terroir*

Climate is one of the more important components of *terroir.* In some ways it is the most difficult to evaluate because it varies in both space and time. Adjacent vineyards may have quite different microclimates owing to factors such as wind direction, elevation, slope, and angle to the sun. Several of these factors can change within a single season and/or from year to year. For example, in years that are relatively warm and dry, vineyard "X" with a particular slope, elevation, sun angle, and soil type may produce better wine than vineyard "Y," whereas the reverse may be true in years that are cooler and wetter.

There are many weather variables that can be measured to describe the overall viticultural climate, such as average temperature during the growing season (warm temperatures promote plant growth and ripening), minimum temperature in winter (temperatures significantly below freezing can damage plant tissue and kill vines), and variation of temperature between day and night (cool nights preserve grape acidity).

Degree Days

Perhaps the most widely used temperature measurement is the concept of "degree days" (also called day-degrees, heat units, or heat summation) which is based on the observation, first made by de Candolle (1855), that relatively little plant growth and grape ripening occurs below 10°C (50°F).

Standard degree days are calculated based upon the amount of time above the 10°C (50°F) threshold. Conversion of degree days between °F and °C is difficult due to the variable number of days in a given summation. For example, if the mean temperature for a day, week, or month is 25°C, this would correspond to 15 (1x15), 105 (7x15), or 450 (30x15) degree days (°C), respectively (Winkler, 1936). The total degree days are summed for the entire growing season (sometimes approximated by April-October in the northern hemisphere) and the resulting number can be used to subdivide viticultural areas into five broad Regions: I (<2500°F day-degrees), II (2501-3000°F day-degrees), III (3001-3500°F day-degrees), IV (3501-4000°F day-degrees), and V (>4000°F day-degrees). Table 3 summarizes degree day information for representative viticultural regions of the world: Washington State vineyards are mostly in Winkler regions II and III. Annual variations for a particular Walla Walla Valley vineyard are shown in Figure 5. The concept of degree days can be further modified by excluding temperatures above 30°C when plant growth slows (Chas Nagel, written communication, 2000), or including only the time period from fruit ripening to harvest (Andrew Reynolds, written communication, 2000).

Table 3 Temperature summation during the growing season for representative viticultural regions of the world (data from Amerine *et al.*, 1980; Norm McKibben, written communication, 2000; Markus Keller, written communication, 2000; Dejan Tesic, written communication, 2000; and, for the Niagara Peninsula, the Niagara Agricultural Weather Network, St. Catharines, Ontario, Canada). Degree days °F and °C are not strictly convertible due to variable number of days of measurement, but approximately: degree days °C = degree days °F * 0.554.

Place	Degree Days (10°C)	Degree Days (50°F)	Winkler Region
Algiers, Algeria	2880	5200	V
Fresno, California	2590	4680	V
Hunter Valley, Australia	2340	4220	V
Naples, Italy	2220	4010	V
Davis, California	2000	3620	IV
Florence, Italy	1950	3530	IV
Walla Walla, Washington (high)	1830	3300	III
Napa Valley, California	1820	3280	III
Barossa Valley, Australia	1710	3090	III
Asti, Italy	1650	2980	II
Walla Walla, Washington (low)	1480	2670	II
Coonawarra, Australia	1450	2620	II
Niagara Peninsula, Canada	1440	2590	II
Bordeaux, France	1400	2520	II
Burgundy, France	1330	2400	I
Hawke's Bay, New Zealand	1300	2350	I
Otago, New Zealand	1000	1800	I
Geisenheim, Germany	950	1710	I

Related Climatic Factors

Minimum temperatures are also of interest in Washington State, such as during the depths of winter, during bud break and fruit set in the spring, and during grape ripening (*véraison*) in the summer. Temperatures that are too cold can kill new vine growth or even the entire plant. This occurs in some vineyards in Washington about once a decade, and was responsible for the drastic drop in grape production in 1996 (Table 2).

Another measure of climate is continentality, which is defined as the difference between mean July and January temperatures (reversed for the southern hemisphere). For most Washington vineyards the continentality index is about 23°C, which is on the high end of regions with a typical Mediterranean climate of winter rain and summer drought, that is generally favourable for wine grapes. Some wineries have used the latitudinal similarity of Washington vineyards to suggest similarities of wine characteristics to famous European viticultural areas such as Burgundy and

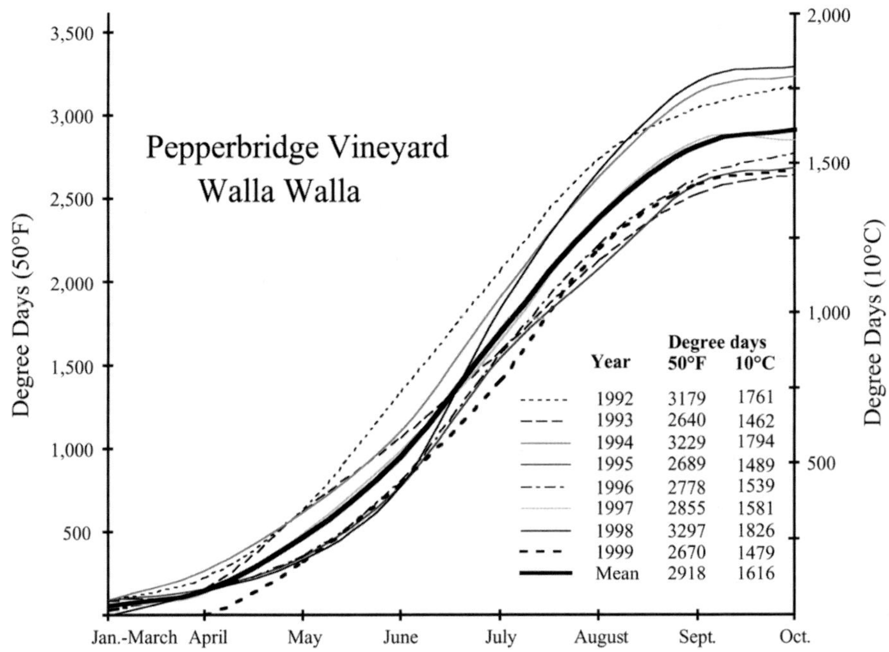

Figure 5 Annual variations in accumulated growing degree days for the Pepperbridge Vineyard in Walla Walla Valley. Data from Norm McKibben (co-owner, written communication, 2000).

Bordeaux or even Germany, but the degree day information in Table 3 makes it clear that even though the Walla Walla Valley of Washington State occurs at high northerly latitudes, it is in fact a warmer climate than comparable latitudes in Europe.

Most of the vineyard areas in Washington State are also blessed with abundant sunshine, due to the long summer days of northerly latitudes and to the generally cloud-free conditions during much of the growing season. During summer months most Washington vineyards experience more than 90% sunny days without measurable precipitation, in contrast to the generally cloudier conditions that prevail in much of northern Europe. Sunshine, precipitation and soil fertility are the natural controls on grapevine growth, but there are additional steps that can be taken in the vineyard to optimize the quality and/or quantity of the grapes produced. This is generally termed "canopy management" (control of the growing parts of the vine such as leaves and canes).

Importance of *Terroir*

There are a multiplicity of factors that contribute to the total quality of the grape that is then used as the raw ingredient in the process of making fine wine. To focus on a single factor of temperature, sunshine, rainfall, soil, bedrock, or viticultural practice, is to miss the essential synergy that marks the difference between an average wine and one that is exceptional. This is not to say that any one component is *insignificant*, but merely to assert that no one factor is *transcendent*. Thus, we return to the essential character of the term *terroir* in its influence on the quality of the grape and hence the quality of the wine. Lest this seem an impossible task, one has only to taste a single example of magnificent wine to be motivated to try to understand *terroir*. To better understand the concept of *terroir* in Washington State, the rest of this paper will focus on a single *appellation*, the Walla Walla Valley, and the features of geography, geology, pedology and climate that make it such a special place.

THE WALLA WALLA VALLEY *APPELLATION*

Walla Walla is a Native American name meaning "many waters" or "small, rapid streams." Many diverse tribes, including the Walla Walla, Nez Perce, Cayuse and Umatilla, inhabited the region prior to European settlement. The Walla Walla Valley was one of the first areas between the Rockies and Cascades to be permanently settled. Lewis and Clark traveled through the area in 1805 and 1806 as part of the initial exploration of the American West. Fur traders carved out early settlements, trading posts, and forts in the early 1800s. With the lure of nearby gold in the Blue Mountains in 1860, Walla Walla quickly became a commercial, banking and manufacturing center, and soon became the largest city in Washington Territory. The City of Walla Walla was incorporated in 1862. After the gold rush ended, farming anchored the community and remains vital to the current economy. Early settlers planted wine grapes for personal vinification and consumption but no commercial production was attempted. Some of these now-wild vines can still be found, such as near the junction of Prunedale Road and County Road 517 south of Walla Walla (Gary Figgins, oral communication, 2000). In addition to wine grapes, the Walla Walla Valley is a major production area of alfalfa seed in the United States. Ground-dwelling native alkali bees are required to pollinate the alfalfa flowers and may also be important for wine grapes. The Green Giant cannery (now owned by Seneca) in Dayton on the edge of the Walla Walla Valley is the world's largest asparagus cannery, and formerly was a world leader in peas as well (Donald Turner, Manager of the Port of Columbia, oral communication, 2000). Thus, the Walla Walla Valley is in many ways a hub of agricultural activity in the state of Washington.

Location and Geology of the Walla Walla Valley *Appellation*

The Walla Walla Valley *appellation* is defined by elevation contours on the south and east and by lines approximating drainage divides on the west and north (Fig. 6). The valley is underlain by Miocene basalt flows of the Columbia River Group, consisting of parts of the Saddle Mountains, Wanapum and Grande Ronde formations, from youngest (~8.5 Ma) to oldest (~16.9 Ma) (Schuster, 1994). These units are exposed only in deep stream-cut valleys, but do crop out to the south and east on the slopes rising to the Blue Mountains (Figs. 2a, 7).

Overlying the basalt in the Walla Walla Valley are Quaternary sediments consisting of loess, dune, and flood deposits related to glacial processes, and alluvium from the streams that drain the Blue Mountains and the valley itself. As described previously, the draining of glacial Lake Missoula in Montana created cataclysmic floods across eastern Washington that scoured much of the scabland topography and backflooded the Walla Walla Valley. Sediments deposited from these slackwaters range from gravels near Wallula Gap on the western side of the valley, to massive silts and sands with local bedding in the central and eastern parts of the valley (Figs. 2c, 7). On the higher slopes above the slackwater deposits are thick mantles of loess (Fig. 2d). Rivers and streams have downcut into the loess and slackwater deposits during the Holocene (last 10,000 years) and, in some cases, the underlying basalts, to yield an overall dendritic pattern of Quaternary sedimentation (Fig. 7). Terrace remnants of the rhythmically bedded slackwater deposits form "islands" within the modern drainage network (Fig. 11b). Present-day drainages and flood plains consist of a mixture of reworked loess, slackwater deposits, and basalt (Carson and Pogue, 1996). Most of the basaltic cobbles are restricted to the larger streams that can carry relatively fresh pieces of basalt from the surrounding hillsides.

Soils of the Walla Walla Valley

Each of the rock units in the Walla Walla Valley is associated with distinctive soils, some of which are excellent for the production of wine grapes and some of which are not suited at all to grape production. Because of this, geographically adjacent vineyards can exhibit radically different *terroir* even though their major climatic features may be very similar.

Figure 8 is a generalized map of the associations or general groupings of major soils in the Walla Walla Valley. Detailed soil surveys for Walla Walla County, Washington (Harrison *et al.*, 1964), and Umatilla County, Oregon (Johnson and Makinson, 1988) name more than 75 different soils in the area. Many of the soils share a number of properties such as silty textures and a mixed mineralogy. Silty textures are characteristic of some of the slackwater sediments, of the loessial soils that were derived by wind reworking of these sediments (Busacca and McDonald, 1994), and of some of the alluvial soils that were formed by water erosion and transport of the same materials into valley bottoms. Similarly, a mineralogy dominated by quartz, feldspars, micas, and illitic clays is characteristic of the slackwater sediments and thus of the loessial and alluvial soils as well (Busacca, 1991). Composition and mineralogy of typical loess and slackwater soils are contrasted with underlying Columbia River Group basalt and granite boulder erratics in Table 4. The lack of correlation of the composition and mineralogy of the basalt with the overlying loess and slackwater soils (Table 4) reinforces the concept that almost all of the eastern Washington soils are ultimately derived not from basaltic bedrock but from glacial material related to the outburst floods, including ice-rafted boulders, water-born sediments, and wind-reworked sands. Not each one of the 75 soils found in the region is different enough to create a distinctive *terroir* for grape production; nevertheless, when grouped into the eight associations shown in Figure 8, several contrasts in soil properties become apparent that affect the choice of optimal grape varietal and the qualities of the wine produced. Note that the general soil map units in Figure 8 contain multiple soil series and that in some cases, a soil series occurs in more than one general soil map unit.

Walla Walla Valley Appellation Soil Groups with Wine Production

Soils in **map unit 1** (Fig. 8) are cobbly, loamy, and silty soils formed in recent alluvium on flood plains and terraces of the Walla Walla River and its tributary streams such as Dry Creek, Mill Creek, and Cottonwood Creek. Soils such as the Freewater and Yakima are deep, well drained to excessively drained, and formed in cobbly alluvium along ancestral channels of the Walla Walla River. These have been planted intensively for many years to apples and other orchard crops. These soils are very low in organic matter, are very well drained, and have no water table, so grape vigour can be controlled successfully with water and nutrient management. Catherine and related soils are formed in silty alluvium,

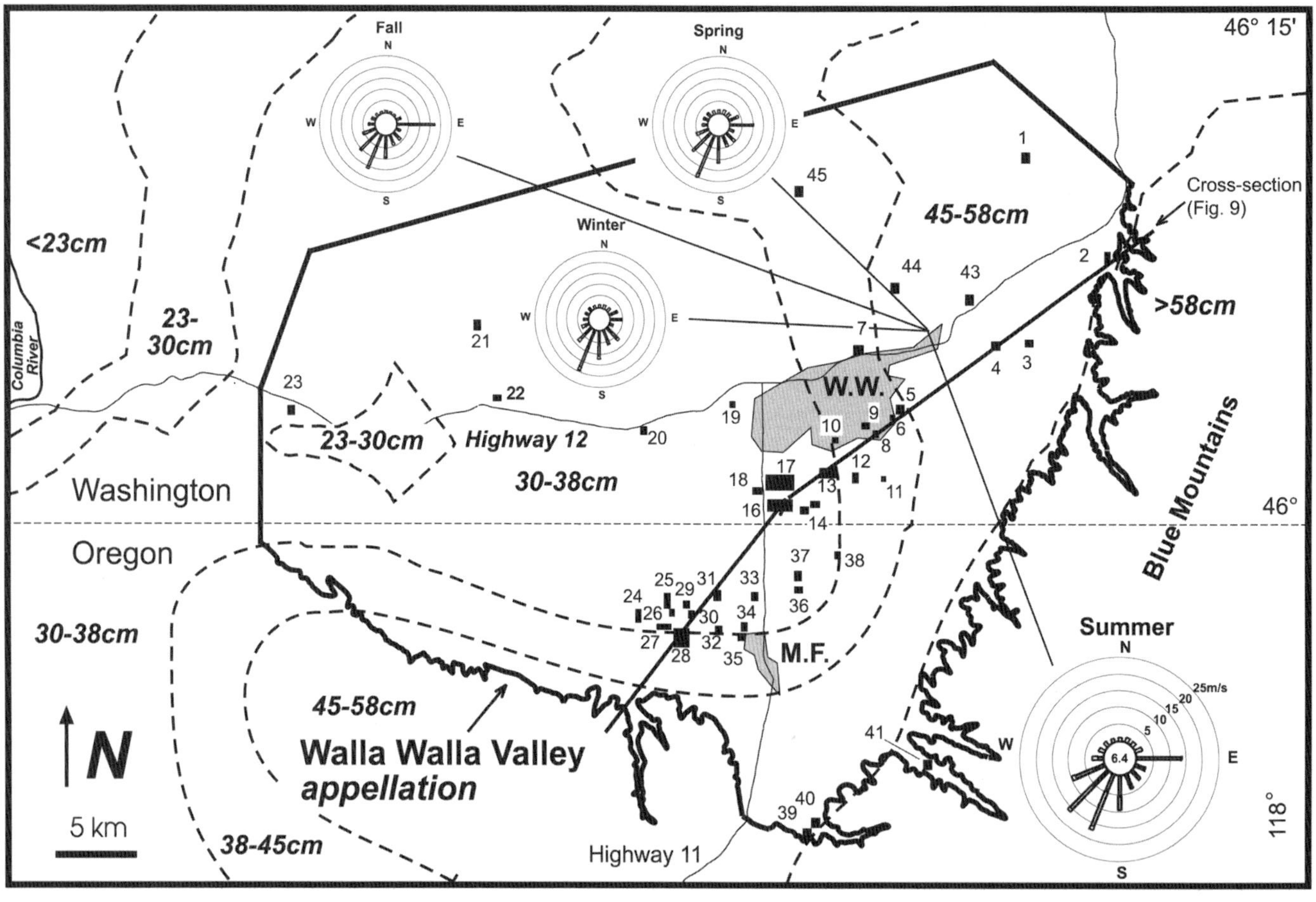

Figure 6 Precipitation, wind velocity in metres/second, and location of vineyards within the boundaries of the Walla Walla Valley *appellation*. Vineyard information in Table 4. Bars on wind roses represent percentage of the time that winds of a given velocity blew from the compass direction shown. Value in the centre of the wind rose is the percentage of time in the season that there was no sensible wind. Wind roses generated from PAWS (Public Agricultural Weather System) data at http://index.prosser.wsu.edu for the Walla Walla weather station at Latitude 46.06, Longitude 118.27, for the period July 1992 to December 1999. Precipitation contours modified from Oregon Climate Service (http://www.ocs.orst.edu/pub/maps/Precipitation/Total/Regional/West/westus_precip.GIF). WW = city of Walla Walla.

are somewhat poorly drained, have higher organic matter contents than do Freewater soils, and have a seasonal water table that has formed mottles in the lower part of the soil. Soils like the Ahtanum are similar to the Catherine soils but are not suited to grape production because of a saline-alkali condition and a hardpan at shallow depth.

Soils in **map unit 3** are silty and formed in loess that overlies stratified slackwater sediments from cataclysmic outburst floods. These soils are located mainly to the northwest of Walla Walla and generally receive between about 23 cm and 30 cm of mean annual precipitation (Fig. 6). Where the loess is greater than 150 cm thick, soils of the Ritzville series are recognized, which have deep, uniform silty profiles and low organic matter content. Where the loess is about 60 cm deep over stratified flood sediments, Ellisforde soils occur. The Woodward Canyon Vineyard (#21, Fig. 6, Table 5) is planted on Ellisforde and related soils. The characteristics of these soils are similar to those of the Sagemoor and Sagehill series in map unit 4. Cost of water development may limit the development of these soils for vineyards, as rainfall is insufficient for dryland production on these upland soils.

Soils in **map unit 4** have been and are being intensively developed for vineyards, particularly to the south and southeast of Walla Walla. The soils formed in thin to thick loess overlying slackwater sediments on flat-topped, dissected terraces. These soils are silty and uniform in texture down to the fluvial sediments, and consist of stratified sands and silts below that. They have low to moderate contents of organic matter. The soils receive from 30 cm to 45 cm of annual precipitation (Fig. 6). Soils of the Sagemoor, Ellisforde and Sagehill formed in only about 50-75 cm of loess over stratified sediments. Erosion on the sideslopes of terraces in these soils commonly exposes white subsoil carbonates in the slackwater materials. Soils of the Oliphant series formed on about 125 cm of loess overlying stratified sediments. Soils of the Walla Walla and Athena series formed on more than 150 cm of loess and may or may not be underlain by slackwater sediments.

The soils in **map unit 7** are also formed in deep silty loess on uplands. Soils in this map unit, such as the Athena, Palouse, and Calouse, are differentiated from those in unit 6 mainly in receiving more rainfall, about 38-58 cm annually (Fig. 6), which has promoted lush growth of the native bunchgrasses and allowed them to accumulate large amounts of organic matter. The Athena and Palouse soils are the classic Mollisols (prairie soils) of dryland grain production in eastern Washington State. Only one vineyard, the Mill Creek Upland Vineyard (#3, Fig. 6, Table 5), has been planted on these

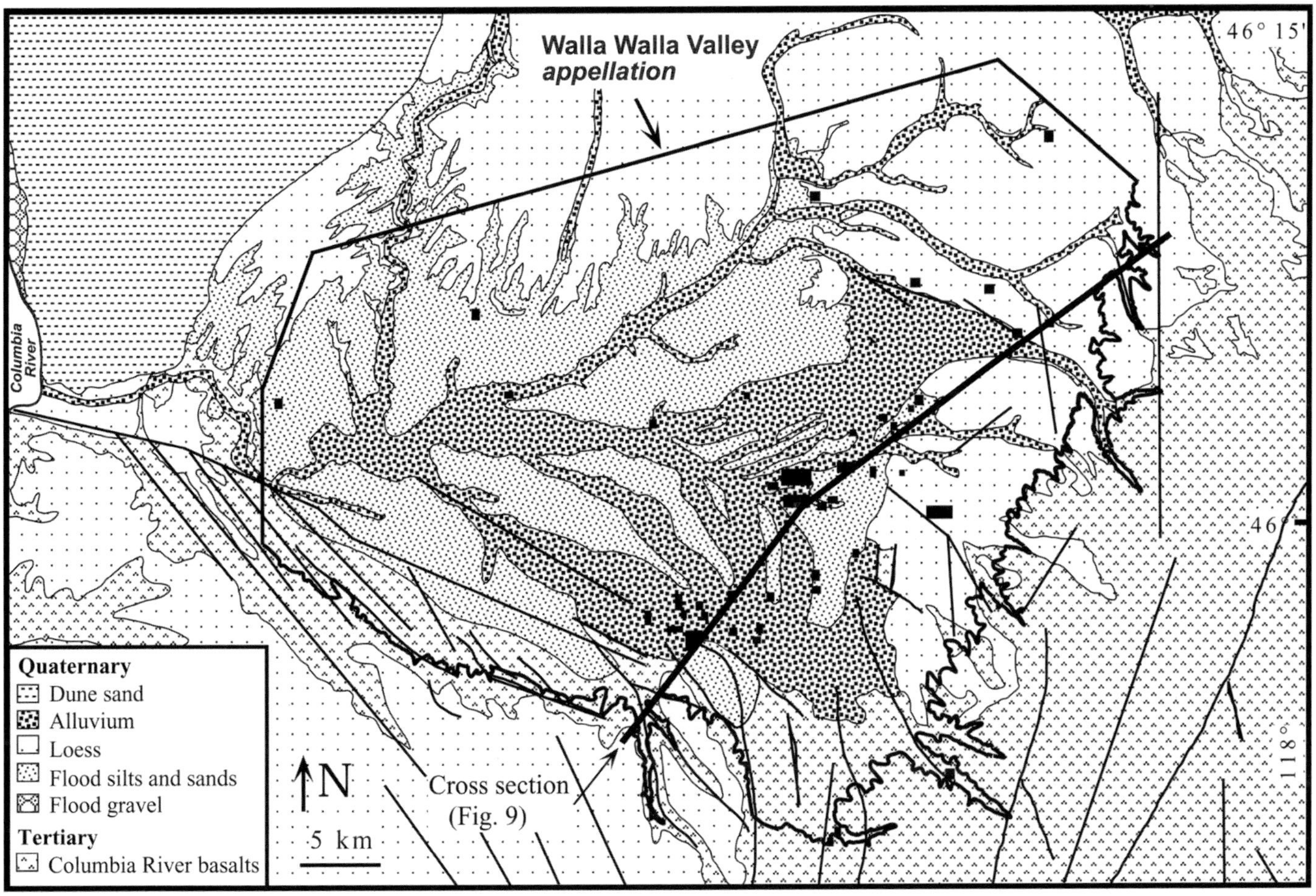

Figure 7 Surficial geology of the Walla Walla Valley region. Modified from Schuster (1994) and Schuster *et al.* (1997). Vineyards are identified in Figure 6 and Table 4.

soils at the present time; however, these soils may have potential for further vineyard development because of favourable soil moisture and other soil properties.

Walla Walla Valley Appellation Soil Groups Mostly Lacking Wine Production

Soils in **map unit 2** lie along floodplains to the west of Walla Walla and are poorly drained, saline-alkali soils unsuited to grape production. The saline-alkali conditions result from excessive evaporation and wicking of salts from a standing water table at shallow depth. The Three Rivers Vineyard (#20, Fig. 6, 12f, Table 5) is on well-drained terrace soils close to these marginal alkaline soils.

Adkins and Quincy soils dominate the area of **map unit 5** on the lower end of Eureka Flat to the northwest of Walla Walla (Fig. 8). These soils are well drained to excessively drained and formed in dune sands under less than 30 cm of rainfall (Fig. 6). No wine grapes are planted on these soils at the present time, although several thousand acres of table and juice grapes are planted on these soils along the Snake River outside the Walla Walla Valley *appellation* area.

Soil Groups Mainly or Completely Outside the Walla Walla Valley Appellation

Walla Walla and Ritzville series soils, formed in deep loess on uplands to the north and south of the Walla Walla Valley (Fig. 8), are the major soils in **map unit 6**. These soils receive from about 23 cm to 38 cm of annual rainfall, depending on location (Fig. 6). These soils have a uniform silty texture to more than 150

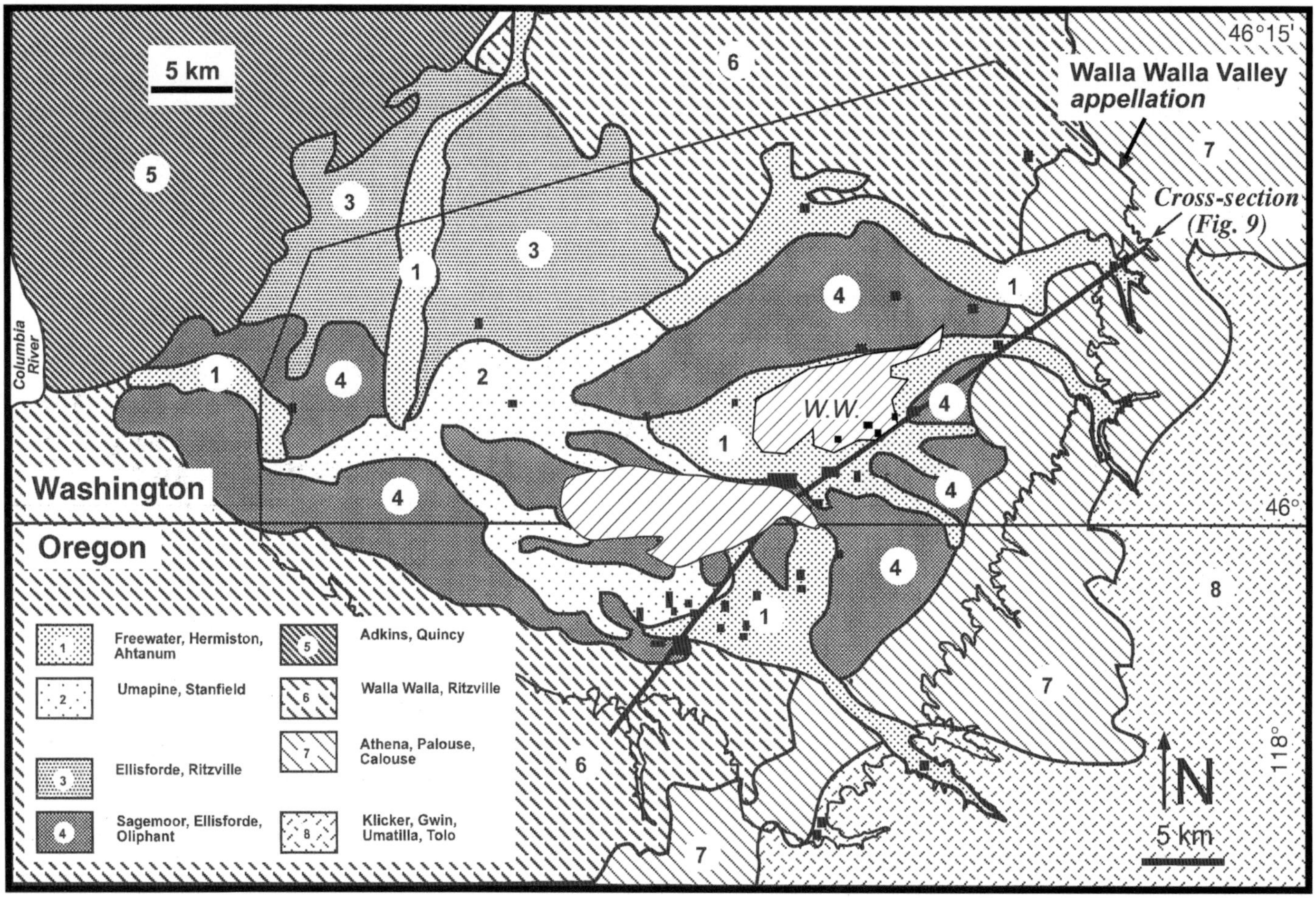

Figure 8 Generalized map of the soil associations of the Walla Walla Valley region. Modified from Harrison *et al.* (1964) and Johnson and Makinson (1988). Note that only one to three soil series are named for each association, although other soil series can and do occur in the associations. For example, a variant of the Athena soil series with slackwater sediments deeper than 150 cm occurs in unit 4, but it also is dominant in unit 7 where it is a named component. **Unit 1**: Freewater, Hermiston, Ahtanum: Cobbly to loamy, moderately well to well drained soils formed in alluvium, on floodplains and low terraces. **Unit 2**: Umapine, Stanfield: Loamy to silty, poorly drained saline and alkali soils formed in alluvium, on floodplains and low terraces. **Unit 3**: Ellisforde, Ritzville: Silty, well drained soils formed in thin to deep loess overlyng stratified silty to sandy outburst flood deposits, on dissected terraces and uplands. **Unit 4**: Sagemoor, Ellisforde, Oliphant: Silty, well drained soils formed in moderately deep to deep loess overlying stratified silty to sandy outburst flood deposits, on dissected terraces. **Unit 5**: Adkins, Quincy: Sandy, excessively drained deep soils formed mainly in dune deposits on high terraces. **Unit 6**: Walla Walla, Ritzville: Silty, well drained soils formed in deep loess on uplands in 20-38 cm rainfall zone. **Unit 7**: Athena, Palouse, Calouse: Silty, well drained soils formed in deep loess on uplands in 38-58 cm rainfall zone. **Unit 8**: Klicker, Gwin, Umatilla, Tolo: Loamy to cobbly, well drained soils formed mainly in loess, volcanic ash, and basaltic bedrock, on steep mountain slopes. WW = city of Walla Walla

Table 4 Comparison of chemical composition and mineralogy of representative samples of vineyard soil, loess, slackwater sediment, granitic glacial erratics, and Columbia River Group basalt. Data from Ross (1989), McCreery (1954), and the present study.

Unit	Basalt	Granite erratic	Touchet 2m below loess	Touchet 5m below loess	Loess regular	Loess calcic	Loess soil
Location	1	2	3	3	4	5	6
SiO_2	54.08	67.07	63.55	67.16	63.51	56.65	63.80
Al_2O_3	13.58	15.90	13.75	12.99	14.26	12.16	13.94
TiO_2	2.94	0.54	1.14	1.11	1.24	0.78	1.22
FeO(t)	13.20	3.90	5.64	2.44	6.73	4.32	6.16
MnO	0.21	0.11	0.11	0.04	0.12	0.07	0.12
CaO	6.45	4.20	4.50	2.49	2.88	10.00	3.14
MgO	3.13	1.54	2.43	1.15	1.96	1.87	2.02
K_2O	2.41	3.37	2.22	2.62	2.22	1.71	2.35
Na_2O	3.20	2.87	2.21	2.10	2.08	1.72	2.08
P_2O_5	0.80	0.26	0.25	0.47	0.24	0.15	0.26
L.O.I.	n.d.	n.d.	4.03	7.26	4.58	10.38	4.73
	100.00	99.77	99.81	99.83	99.82	99.83	99.82
Ni	n.d.	20	36	18	27	27	26
Cr	6	14	59	41	51	53	46
Sc	26	5	21	13	18	12	18
V	233	69	141	91	149	84	145
Ba	2848	848	704	636	668	626	673
Rb	43	119	82	82	84	63	81
Sr	268	754	280	246	263	355	265
Zr	415	186	277	282	268	262	264
Y	43	27	35	31	36	28	34
Nb	23	18	16	15	15	15	15
Ga	23	19	15	17	16	14	17
Cu	0	5	24	21	28	20	22
Zn	117	58	76	63	88	61	87
Pb	n.d.	12	17	10	16	9	17
La	n.d.	38	30	32	29	38	37
Ce	n.d.	58	60	74	70	72	69
Th	n.d.	10	13	11	12	8	10
quartz	0	25	30	30	40	40	40
orthoclase	0	20	0	0	12	10	12
plagioclase	48	43	1	1	2	1	2
olivine	2	0	0	0	0	0	0
pyroxene	21	3	<1	<1	<1	<1	<1
amphibole	0	3	<1	<1	2	2	2
muscovite	0	1	8	8	10	10	10
biotite	0	3	7	7	9	9	9
Fe-Ti oxides	11	1	0	0	1	1	1
apatite	1	1	0	0	n.d.	n.d.	n.d.
clay	2	0	10	10	10-15*	10-15*	10-15*
calcite	0	0	3	3	n.d.	10	n.d.
glass	14	0	n.d.	n.d.	1	0	1
rock fragments	0	0	35	35	n.d.	n.d.	n.d.
other	1	0	1	1	6	5	6

1) Bear Creek member of Saddle Mountains Basalt, Ross (1989), **2)** Seven Hills Vineyard (Fig. 15b), granitic glacial erratic, mineralogy estimated from CIPW norm, **3)** Seven Hills Vineyard, mineralogy from Alwin (1970), Touchet bed WW 3-13, **4)** Seven Hills Vineyard (Fig. 15a), mineralogy from McCreery (1954), **5)** Lower calcic paleosol (Fig. 2d) ~3 km ENE of Walla Walla, mineralogy estimated from CIPW norm, **6)** Seven Hills Vineyard, Block 16, * Montmorillonite>illite>kaolinite, **n.d.** = not determined cm depth and low to moderate organic matter contents. No vineyards are planted within map unit 6 at this time; however, with irrigation to supplement the scant rainfall, these soils should be capable of producing quality wine grapes because low soil moisture and a favourable rooting zone should allow the grower to control vine stress with judicious irrigation.

Map unit 8 consists of soils that are not in vineyard development at this time. This unit is formed on steep slopes of the foothills of the Blue Mountains, and a wide variety of soils developed in loess, volcanic ash, and basaltic bedrock have been recognized.

Soils Summary

In summary, the soils of the Walla Walla Valley *appellation* have formed from four different types of surficial sediments or bedrock. Various combinations of soil parent materials and a strong gradient of mean annual precipitation across the *appellation* are key to determining vineyard potential performance. Soils formed from young alluvium vary tremendously in their properties, such as texture (cobbly to clayey), salt effects, and presence or absence of a water table within the rooting zone of vines. Soils formed from loess more than 150 cm deep are found around the margins of the Walla Walla Valley *appellation* and have dominantly silty, uniform soil profiles. Mean annual rainfall varies widely depending on location in the *appellation*, and this, along with slope steepness and aspect, determine suitability or potential for development of dryland or irrigated vineyards. Soils formed from thin to moderately thick loess overlying slack-water sediments have been the main focus of vineyard development up to the present time in the *appellation*. Finally, soils located on steep slopes of the Blue Mountains that have bedrock at shallow depth have not been fully evaluated to determine their potential for vineyard development.

Walla Walla Valley Vineyards

The first commercial vineyard planting in the Walla Walla Valley was in the early 1950s by Bert Pesciallo who produced wine from Black Prince, Muscatel and Concord grapes under the Blue Mountain Vineyards label. This was a short-lived

Table 5 Vineyards in the Walla Walla Valley *appellation* (data from Myles Anderson, written communication, 2000; Norm McKibben, written communication, 2000)

Map #	Name	Grapes	First planted	Owner
1	Spring Valley	m, cf, sy, cs, pv	1994	Dean & Sherry Derby
2	Biscuit Ridge	7 g, 1 cf	1982	Duane & Mary Wollmuth
3	Mill Creek Upland	11cs, 3m, 1sg, 1pv	1997	Gary Figgins
4	Titus Creek	3cs, 3sy, 1v	1999	Ken Harrison
5	Leonetti estate	m, sy, sg, cf	1974	Gary Figgins
6	Patrick M. Paul	cs	*	Mike Paul
7	Waterbrook	chard	1998	Eric & Janet Rindal
8	Keiler	m	*	Fritz Keiler
9	Myson	cs, m	*	Steve Paul & Mark Colvin
10	Whitney	cf	*	Mike Paul
11	Morrison Lane	sy	1985	Dean & Verdi Morrison
12	Forgotten Hills	m, sy, cs	1999	Jeff and Kathyrn Hill
13	Cottonwood Creek	5 cf, 28 c	1989	Scott Byerly
14	Pheasant Run	cs, m	1998	Greg Basail
15	Via Piano	m, cs,	1999	Justin Wiley
16	Pepperbridge south	m, cs, sy	1991	Premier Partners IV
17	Pepperbridge north	m, cs, sy	1999	Premier Partners IV
18	Double River Ranch	*	2000	Kyle Mussman
19	Bunchgrass	m, cf, sy, cs, pv	*	Roger Cockerline
20	Three Rivers	c	2000	Steve Ahler, Bud Stocking, Duane Wollmuth
21	Woodward Canyon	cs, m, cf, c, b, d	1977	Rick & Darcie Small
22	Dunham	cs	1999	Eric Dunham & family
23	Ash Hollow	9 merlot, 1 cs	1999	Matt Tucker
24	Windrow	cs	*	Scott & Rebecca Hendricks
25	Veterinarian	*	*	California veterinarian
26	Hart	*	*	Ken Hart
27	Seven Hills west	cs, m, sy	1985	Walla Walla Valley *Appellation*
28	Seven Hills east	m, cs, sy, sg, gv, cf, s	1997	Blue Mountain Farm
29	Cayuse	syrah	1997	Christophe Barón
30	Seven Hills west	cs, m, sy	*	Blue Mountain Farm
31	Caillouxe	sy, v	1997	Christophe Barón
32	Cerise	sy, v	1997	Christophe Barón
33	Lefore	cs, sg	1987	Jack Lefore
34	Lefore	*	1999	John Lefore
35	Waliser	cs, cf, m	1998	Tom Waliser
36	Vanessa's	m, cs, s	1999	Terry & Diane Farley
37	Farley	m, cs, s	1999	Terry & Diane Farley
38	Mint Condition	cs, m	1999	Bob Buchannan
39	Couse Creek	pn	*	*
40	Alder Banks	pn, c	*	Mike & Maryann Banks
41	Wells	pn	*	Sam & Nancy Wells
42	Les Collines	cs, m, sy	2000	Norm McKibben
43	Homestead	g	*	Steve Ahler & Jeff Kolke
44	Minnick	cs, m, cf, sy	2000	Laura Minnick
45	Valley Grove	*	*	Henderson Orchard

Wine abbreviations: b = Barbera, ca = Carignane, cf = Cabernet Franc, cs = Cabernet Sauvignon, c = Chardonnay, d = Dolcetto, g = Gewürtztraminer, gv = Grand Vidure, m = Merlot, pn = Pinot Noir, pv = Petite Verdot, s = Semillon, sb = Sauvignon Blanc, sg = Sangiovese, sy = Syrah, v = Viognier, * not known effort as a deep freeze in 1955 killed all the vines, and the winery did not reopen (Bert Pesciallo, oral communication, 2000). Commercial production did not resume in the Walla Walla Valley until 1974 when half an acre of Cabernet Sauvignon grapes was planted at the Leonetti homestead. Early success was assured when the first vintage, a 1978 Leonetti Cellars Cabernet Sauvignon, was pronounced the "best Cabernet Sauvignon made in America" by the Wine and Spirits Buyers Guide in 1982. At present there are about 45 vineyards supplying grapes to 32 wineries in the Walla Walla Valley *appellation* (Table 5.) Another 15 vineyards have been planted, with more planned (Myles Anderson, Director of the Walla Walla Institute for Enology and Viticulture, oral communication, 2000).

The vineyards of the Walla Walla valley are located on four fundamentally different substrates including slackwater terrace (Fig. 2c), loess (Fig. 2d), river gravel, and flood plain silt (Fig. 8). Of the 45 vineyards listed in Table 5, the majority are planted in loess either overlying deep bedrock basalt or as a thin veneer on top of slackwater deposits (Fig. 2c), which in turn overlie basalt. Figure 9 is a cross section through six representative vineyards illustrating the considerable range of terroir and microclimate in the Walla Walla Valley. These six vineyards and their settings may be described as follows.

Pepperbridge Vineyard

Since no vineyards in the Walla Walla Valley are located directly on basalt, the Pepperbridge Vineyard (#16, #17, Fig. 6; Table 5), planted in a thin mantle of loess overlying slackwater terrace deposits, is the stratigraphically lowest vineyard (Figs. 7, 9). The slackwater terrace deposits in this area range from 10-50 m in thickness and are mantled by 0-2 m of loess (Fig. 10). Given the ability of grape roots to penetrate to depths of 10 m or more in such unconsolidated sediments (observation made in construction trenches, roots at 4 m depth shown in Fig. 11e), most Pepperbridge vines are rooted in the slackwater deposits. Soils of the Pepperbridge Vineyard are in the Oliphant, Walla Walla, and Sagemoor series (Fig. 8, map unit 4). The low organic content and good drainage of these soils make it relatively easy to control vine vigour and grape yield with modern viticultural techniques, particularly drip irrigation (Fig. 11a). This setting and these techniques contribute to the fruit intensity and concentration of grape flavours for which wines made from this vineyard are known. The main grape varieties planted at Pepperbridge are Merlot (83 acres), Cabernet Sauvignon (72 acres), and Syrah (12 acres) (Table 5). In contrast, the immediately adjacent but overly fertile flood plain soils (Fig. 11b) do not produce grapes of the same quality because of strong vegetal flavours (Norm McKibben, oral communication, 2000). This is one of several places in the Walla Walla Valley where a direct connection can be shown between soil characteristics and wine quality.

Like most Walla Walla Valley vineyards, Pepperbridge is relatively dry (<50 cm annual precipitation), sheltered from wind, with intense sun exposure (1,400 W/m^2 of annual solar radiation), and moderate wintertime temperatures (Table 6). The long growing season in most years exceeds 3000 (50°F) degree days and yields fully ripe fruit with good tannins (Fig. 5). In addition, the low humidity further concentrates fruit flavours through evaporation. A standard measure of humidity is the mean for July measured at 9 a.m. For the Pepperbridge vineyard in 1999 the mean humidity for July at 9 a.m. was 50.5% at an average temperature of 20.7°C (Table 6).

Another important feature of the Pepperbridge Vineyard is good air drainage. The slackwater terrace deposits have been dissected by strands of the Walla Walla River so that the vineyards are elevated above the valley floor on slackwater terrace "islands" (Fig. 11b). This allows cold air to drain away from the vines, thus avoiding excessive problems with frost during the growing season.

Cottonwood Creek Vineyard

Less than one kilometre east of the Pepperbridge Vineyard, the Cottonwood Creek Vineyard (#13, Fig. 6; Table 5) represents a completely different *terroir*. Cottonwood Creek is located on fine silty sediments of the floodplain of Cottonwood Creek (Figs. 8, 9, 11c). The floodplain soils contain more humified organic matter and have poorer drainage than the slackwater and loess soils (#1, Fig. 8; Fig. 10). In some places the water table and infiltration from Cottonwood Creek saturate the sediments at relatively shallow depths, such as in the Catherine series soils. This means that deep-rooted grapevines can tap ground water during the growing season, making supplemental irrigation unnecessary in most years, but also making it more difficult to control the natural plant vigour (Fig. 11c). Uncontrolled plant vigour makes it difficult to fully ripen the crop in some years and also increases vineyard management costs in terms of pruning, spraying, and general canopy management (Scott Byerly, vineyard owner, oral communication, 2000).

The wet soils also warm more slowly in the spring, due to the high heat capacity of water, delaying new plant growth and further exacerbating the problem of ripening the grapes in a normal growing season. The main variety planted at Cottonwood Creek vineyard is Chardonnay, which has produced award-wining wine in some years, proving that even in this difficult location exceptional wine can be produced. But in general, grapes from this vineyard produce wines with pronounced vegetal characteristics such that in some years the grapes are sold to the bulk juice market rather than being made into wine (Scott Byerly, vineyard owner, oral communication, 2000). This is another vineyard in which a direct connection can be shown between geological characteristics and wine quality.

Caillouxe Vineyard

The Caillouxe Vineyard (formerly called Cobblestone Vineyard) (#31, Fig. 6; Table 5) further to the West is perhaps the most unusual *terroir* in Walla Walla Valley (Figs. 7, 10, 11). At this site, very coarse but well rounded cobbles (1-10 cm) of basalt make up the gravel bed of the Walla Walla River. The surface of the vineyard is literally a pavement of stone (Fig. 11d). The stony nature of the soil forces the grapevine roots to go deep in search of water (Fig. 11e), and calcic soil horizons add further stress to the grape plant. These gravelly soils have extremely good drainage so that like Pepperbridge, supplemental drip irrigation can be used or withheld to control vine vigour and grape yield. Typical yield from this vineyard is kept to about 2 tons per acre to concentrate the fruit flavours. The basalt cobbles also perform another function in that they soak up the sun's heat and radiate it back to the vines. The Caillouxe Vineyard is planted predominately to a single grape variety, Syrah, because Syrah is a heat-loving grape. *Vigneron* Christophe Barón has accentuated these unique *terroir* features by trellising the plants very close to the ground and the hot basalt cobbles (Fig. 11f). The wine produced from this vineyard is an extremely concentrated, flavourful Syrah with excellent tannins. In only the vineyard's second year it has already won awards and promises great things for the future.

Seven Hills Vineyard

One of the oldest (first planted in 1981) and most prestigious vineyards in the Walla Walla Valley is Seven Hills (#27, #28, Fig. 6; Table 5). The vineyard is on a gentle north-facing slope of loess (Fig. 7, 12a) overlying slackwater terrace deposits (Figs. 9, 12b). This southernmost location within the Walla Walla Valley benefits from the strong southwesterly prevailing winds, which promotes excellent air drainage and avoids frost problems. The loess soil is well drained and, like Caillouxe and Pepperbridge, allows supplemental drip irrigation to be used or withheld to control vine vigour and grape yield. The main grape varieties planted at Seven Hills are Cabernet Sauvignon (91 acres), Merlot (61 acres), Syrah (19 acres), and Sangiovese (9 acres), with a further 2.5 acres of Viognier planted for blending with Syrah to make a classic Rhone-style wine. Grapes from this vineyard have produced many of the award-winning wines from L'Ecole No. 41, Leonetti Cellars, Seven Hills, and Woodward Canyon wineries. The Leonetti Cellars reserve Cabernet Sauvignon, which some feel is one of the best red wines made in the United States, draws most of its fruit from the oldest vines of the Seven Hills vineyard (Fig. 12c). This bodes well for future vintages from this vineyard as the younger vines mature.

Mill Creek Upland Vineyard

On the other side of the Walla Walla Valley and at significantly higher elevation lie two vineyards of completely dissimilar character, Mill Creek Upland Vineyard (500 m; #3, Fig. 6; Table 5) and Biscuit Ridge Vineyard (550 m; #2, Fig. 6; Table 5). Mill Creek Upland Vineyard is a new effort by Leonetti Cellars, one of the pioneer wineries of the Walla Walla Valley. It is located on a moderately steep south-facing slope of very deep (>50 m) loess (Figs. 7, 9). The soils are in the Athena series (Figs. 8, 10). The steep hillside provides excellent air drainage making it virtually a frost-free location (Fig. 12d). The occurrence on the eastern side of the Walla Walla valley means that there is more natural precipitation (56 cm per year, Fig. 6) but the well-drained loess soils still allow supplemental drip irrigation to be used to control vine vigour and grape yield. Natural grape yield at this site is about 3.5 tons per acre but this is usually reduced to 2 tons per acre with a combination of irrigation practice, canopy management, and bunch thinning (Chris Figgins, vineyard manager, oral communication, 2000). In most seasons, no additional water is needed. The water budget of the vineyard is monitored by neutron probe soil sensors that allow the vineyard manager to carefully control soil moisture for optimum grape maturity and flavour concentration (Fig. 13). The neutron probe measures soil water content with depth (usually in 0.3 m increments) and allows the infiltration of irrigation water to be tracked through the soil profile (Fig. 13).

Biscuit Ridge Vineyard

This vineyard is similar to the Cottonwood Creek Vineyard in that it is located on a silty flood plain (Figs. 7, 9, 10). The easterly location also means that it has the highest rainfall of any of the Walla Walla vineyards. Cold air collecting in the

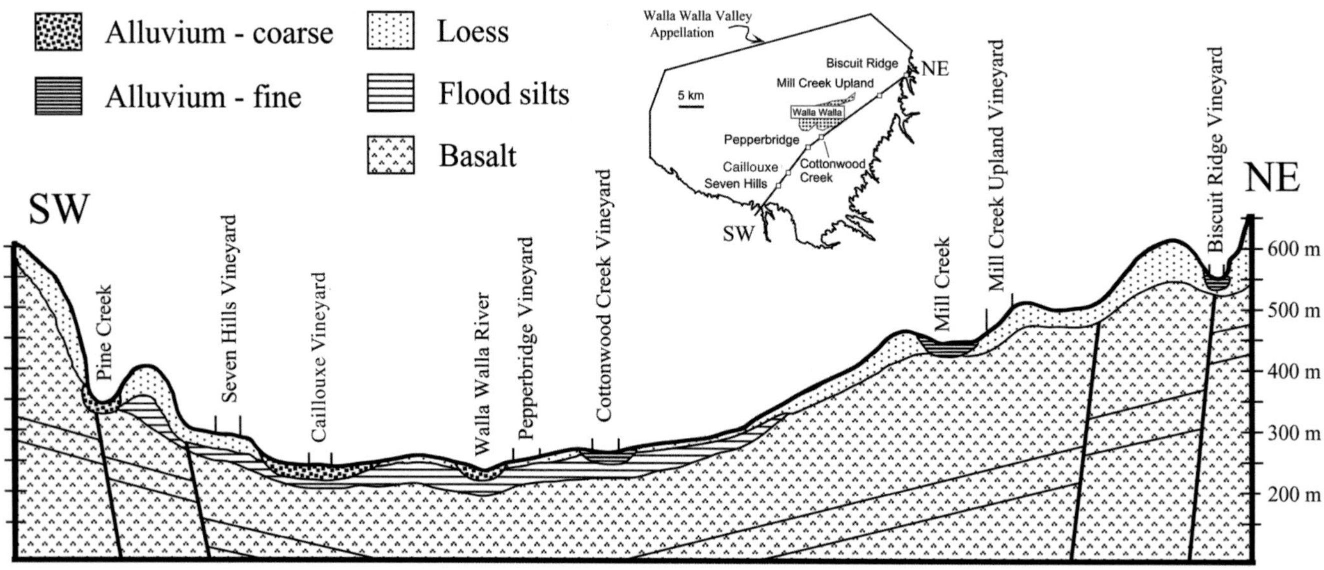

Figure 9 Cross section through Biscuit Ridge, Mill Creek Upland, Cottonwood Creek, Pepperbridge, Caillouxe and Seven Hills vineyards. Vertical exaggeration 25x. Thickness of surficial units is not to scale.

canyon bottom can be a problem for frost, and the high elevation (550 m) results in one of the coolest microclimates in the Walla Walla Valley (Fig. 12e). This vineyard is best known for Gewürztraminer, a grape that thrives in cool climates such as the Alsace region on the French-German border.

CONCLUSIONS

The range of grapes grown in the six vineyards summarized in the previous section illustrates the variability of microclimate and *terroir* of the Walla Walla Valley. In this relatively small *appellation* within the much larger Columbia Valley *appellation* of Washington State there is as much variation as is seen in entire regions, and in some cases, entire countries, in other parts of the world. It cannot be said that a single variable of bedrock, soil type, precipitation, or temperature is essential for fine wine in Walla Walla. But it is very clear that there is a direct correlation between the *terroir* and the grape varieties and wine styles that do best in a particular location. For example, the Biscuit Ridge Vineyard produces outstanding Gewürztraminer but probably could not fully ripen Cabernet Sauvignon or Syrah in even the warmest vintage of the century. Conversely, the Seven Hills vineyard, which is acclaimed for Cabernet Sauvignon, Merlot, and Syrah, is not as favorable for Chardonnay, much less for a cool climate grape such as Gewürztraminer.

Based upon the quality of wines produced from different *terroirs* in Walla Walla Valley there are a few generalizations that can be drawn to guide future grape plantings. One of the key considerations is controlling the natural vigour of the grape vine so that the resulting wine has concentrated fruit and a complex flavour profile. The generally low precipitation of the Walla Walla Valley, combined with soils formed from well-drained loess, slackwater sediments, and cobbly alluvium, provide the dominant control on plant vigour and yield. Judicious application or withholding of supplemental irrigation water allows the vineyard manager to restrict yields and optimize the quality of the grapes produced. The highest quality vineyards in Walla Walla Valley are managed for yields of only 2-3 tons per acre, in contrast to the state-wide average of 4-5 tons per acre (Table 2). The extended daylight hours due to the northerly location combined with generally sunny, arid conditions (1500-1800°C degree days) promote grape ripeness and the generation of intensely flavoured wines (Keller *et al.*, 1998). The cool nights, due both to elevation and latitude, preserve natural grape acidity to produce crisp wines with long complex finishes.

Within these overall characteristics of the Walla Walla Valley, the increase in precipitation from west to east and the differences in organic content, level of the water table, and drainage characteristics of the loess, slackwater, flood plain, and river gravel soils provide individual *terroirs* of distinction that vary in their suitability for particular grape varieties. As an example of this, the Pepperbridge Vineyard, developed on loess overlying slackwater deposits, is planted with classic Bordeaux varietals such as Cabernet Franc, Cabernet Sauvignon and Merlot, whereas immediately adjacent flood plain soils (Fig. 11b) do not produce grapes of the same quality. Instead, these flood plain soils are planted with vidalia onions, a crop known locally as Walla Walla Sweets, for which the region is also

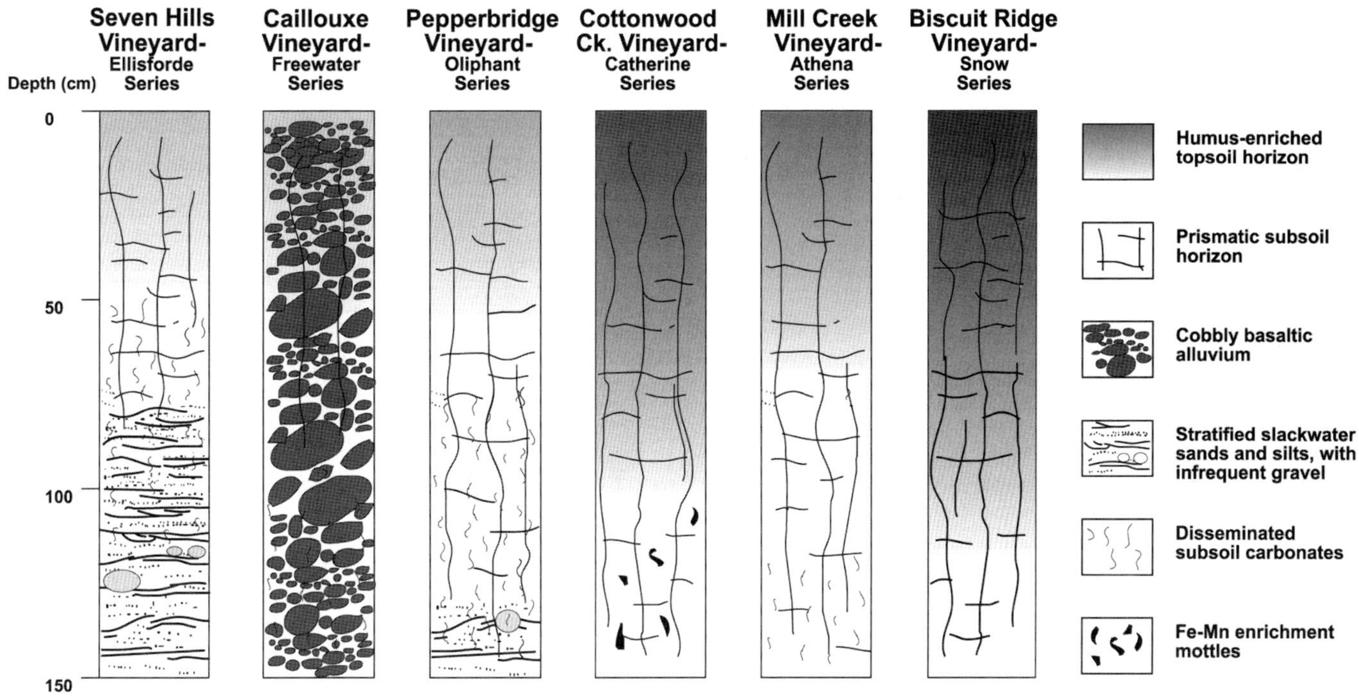

Figure 10 Selected representative soil profiles of the Biscuit Ridge, Mill Creek, Cottonwood Creek, Pepperbridge, Caillouxe, and Seven Hills vineyards. Degree of shading of the topsoil horizons is proportional to the content of humified organic matter. The irregular lines in the prismatic subsoil pattern represent the pattern of cracks that develops in this soil horizon.

Figure 11 Photographs of vineyard features in the Walla Walla Valley area. (**A**) Smart-Dyson trellis system at the Pepperbridge Vineyard with Cabernet Sauvignon grape vines planted on a thin loess mantle over bedded slackwater terrace deposits. (**B**) Vineyard-covered "islands" of slackwater sediments separated by alluvial flood plains of the Walla Walla River. (**C**) Verdant growth of Chardonnay grapes on water-saturated flood plain silts at the Cottonwood Creek Vineyard. Grape grower Scott Byerly is standing at the edge of Cottonwood Creek with the dry gravel creekbed in the foreground. (**D**) Basalt cobbles from Walla Walla River alluvium pave the surface of the Caillouxe (formerly Cobblestone) Vineyard. Newly planted Syrah grapes are not yet trellised. (**E**) Trench near Caillouxe Vineyard exposes roots several meters below surface in the Walla Walla River alluvium. Longest root is approximately 1 m in length. (**F**) Vigneron Christophe Baron shows how Syrah grapes are trellised close to the ground to take advantage of the radiant heat of the basalt cobbles in the Caillouxe Vineyard.

Figure 12 Photographs of vineyard features in the Walla Walla Valley area. (**A**) The Seven Hills vineyard is planted on ~2m of loess (in cliff exposure) overlying slackwater deposits (not visible here) similar to Figure 2c. (**B**) Further evidence of the glacial outburst flood activity that has shaped the *terroir* of the Walla Walla Valley are granitic erratics that were rafted in by floating glacial ice. Winemaker Gary Figgins is holding a 0.3 m granite boulder erratic that was found in the Seven Hills vineyard behind him. A chemical analysis of this rock is listed in Table 4. (**C**) The 20 year old vines, evidenced by the thick gnarled trunk, of this part of the Seven Hills vineyard have been the basis for some of the best wines produced by Casey McClellan (left), owner and winemaker at Seven Hills Winery, and Gary Figgins (right), owner and winemaker at Leonetti Cellars. (**D**) The Mill Creek Upland Vineyard is planted on thick loess soil on the eastern side of the Walla Walla Valley. Good air drainage off the moderate slopes prevents frost problems even though this vineyard at 500 m is higher than most in the valley. (**E**) Canyon-bottom flood plain soils are planted mainly with Gewürztraminer grapes at the Biscuit Ridge Vineyard. Cool air draining from the hillsides adds to the cool climate resulting from the 550 m elevation. (**F**) New grape vines (in grow tubes) were planted on flat flood plain soils surrounding the Three Rivers Winery in order to be close to the winery.

famous. Another negative example of *terroir* is the Cottonwood Creek Vineyard where poorly drained and overly fertile flood plain soils cause measurable sensory problems, including distinct vegetal characteristics.

Perhaps the purest illustration of *terroir* is the Syrah being produced from the Caillouxe Vineyard. Great Syrahs have been produced from several Walla Walla vineyards, but the Caillouxe Vineyard with its pavement of coarse basalt cobbles produces a wine distinct from neighbouring vineyards on loess or slackwater sediments. Although soils developed from loess, slackwater, flood plain, and river gravel sediments display considerable differences in mineralogy and chemistry (Table 4), the effect of each of these components on nutrient uptake and wine flavour is a subject of future research.

An appropriate place to end this discussion of Washington wine and *terroir* is with suggestions for future research. Much modern viticultural research has focussed on quantifiable aspects of plant physiology and the effects of *terroir* on wine composition and sensory attributes (*e.g.*, Reynolds *et al.*, 1995, 1996; Keller *et al.*, 1998). What remains to be done in a rigorous fashion is to examine the relationship between specific physical factors such as bedrock geology, hydrology, and pedology and the quality of wine produced from individual vineyards. One goal of this paper, and indeed of the *Geoscience Canada* series on Geology and Wine, is to explore the concept of *terroir* so that both producers and consumers of fine wine can continue the quest for quality. It is now time for the authors to get on with the latter part of this task. Truly indeed, it has been a pleasure conducting this research (Fig. 14).

ACKNOWLEDGMENTS

This study grew out of initial discussions with Rusty Figgins, owner and winemaker at Glen Fiona, about what factors were important in determining the quality of Walla Walla Valley wine. We are greatly indebted to all the vineyard owners and winemakers who so generously gave of their time and in many cases, their wine. In particular, Norm McKibben, Pepperbridge winery co-owner and past president of the Washington Association of Wine Grape Growers, was indispensable for his exhaustive knowledge of the Washington wine industry and viticultural practice. Chris Figgins, vineyard manager and assistant winemaker at Leonetti Cellars, deserves special mention for his energy and insight about connections between wine and the land. In particular, he inspired the authors to complete this study by coordinating several barrel tastings that showed not only how different vineyards contributed to a particular wine blend but also how a passion for quality at every stage of the winemaking process can result in wines of magnificence.

Earlier versions of this manuscript were improved by informal reviews by Myles Anderson, Patricia Bowen, Gary Figgins, Simon Haynes, Jeff Hedenquist, Roger Macqueen, Casey McClellan, Norm McKibben, Chas Nagel, and Kevin Pogue. Formal reviews for *Geoscience Canada* were conducted by Simon Haynes and Andrew Reynolds. At the Geological Survey of Canada in Calgary, David Sargent and John-Paul Zonneveld helped with final digital figure preparation. Any errors remaining are the responsibility of the authors.

Our motivation to prepare this paper was derived at least in part from a single vineyard, Seven Hills in the Walla Walla Valley *appellation* (Figs. 12a-c), and the wine made from that vineyard by a small group of vintners including L'Ecole No. 41, Leonetti Cellars, Seven Hills, and Woodward Canyon. A field examination of some of the old Seven Hills vines and

Table 6 Climate data for the Walla Walla Valley and Pepperbridge Vineyard. Data from Norm McKibben (written communication, 2000) and http://www.wrcc.dri.edu/climsum.html

Walla Walla Valley — Latitude 46° 02' Longitude 118° 20' Elevation 290 m Years of record: 40

	Jan	Feb	Mar	Apr	May	Jun	Jul	Aug	Sep	Oct	Nov	Dec	Annual
Mean max. temp.	4.1	8.4	12.6	17.1	21.8	26.4	31.4	30.3	25.0	17.7	9.5	5.4	17.5
Mean avg. temp.	0.9	4.6	7.9	11.5	15.7	19.9	24.0	23.3	18.4	12.3	5.9	2.3	12.2
Mean min. temp.	-2.3	0.9	3.2	5.8	9.7	13.3	16.7	16.3	11.8	6.8	2.3	-0.8	6.9
Precipitation (cm)	5.1	4.1	4.2	3.3	3.5	2.6	1.0	1.6	2.2	3.7	5.2	5.6	42.2
Snowfall (cm)	18.8	7.4	3.0	0.0	0.0	0.0	0.0	0.0	0.0	0.5	5.6	13.7	49.0
Degree days (50°F)	8	11	31	117	322	532	782	742	455	159	21	7	3187

Pepperbridge Vineyard — Latitude 46° 01' Longitude 118° 21' Elevation 250 m Year of record: 1999

	Jan	Feb	Mar	Apr	May	Jun	Jul	Aug	Sep	Oct	Nov	Dec	Annual
Max. temp. °C	12.8	11.7	17.8	25.6	30.0	34.4	40.0	37.2	32.8	26.1	27.2	14.4	40.0
Min. temp. °C	1.7	-1.7	-0.6	-0.6	1.7	3.3	7.2	11.1	-0.6	-1.7	-2.2	-3.3	-3.3
Mean 9:00 a.m. temp. °C	3.9	4.4	5.6	9.4	12.2	17.2	20.6	22.2	15.0	7.8	6.7	3.9	10.7
Mean 9:00 a.m. humidity (%)	77	71	71	62	61	57	50	58	55	73	83	83	67
Precipitation (cm)	3.2	6.9	11.4	2.6	3.6	7.3	0.8	0.0	1.0	3.9	4.1	4.3	49.1
Daylight hrs wind >24 kph	2	2	0	2	0	0	0	0	0	0	2	2	10
Solar radiation (W/m^2)	20	28	190	166	170	231	234	201	142	79	37	20	1518
Degree days (50°F)	0	0	0	196	147	440	1068	817	546	83	0	0	3297

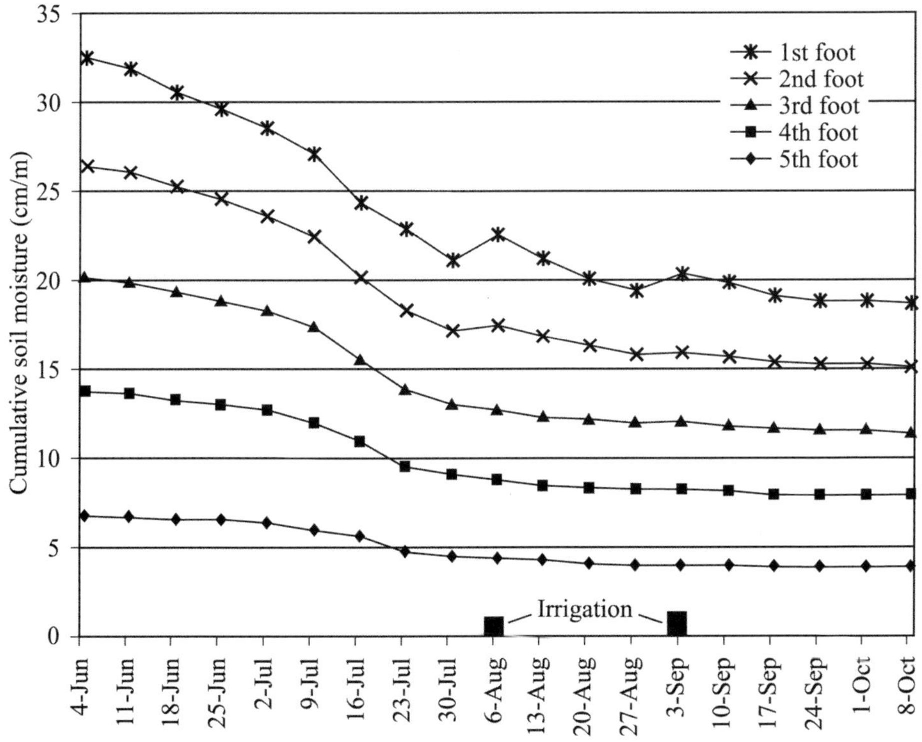

Figure 13 Neutron probe measurement of soil moisture for the Mill Creek Upland Vineyard (block 1, row 4) illustrating effects of irrigation and soil moisture retention properties. Drip irriation was applied on 6 August 1999 and 3 September 1999 (0.8 and 1.2 cm per acre, respectively) and affected the top metre of the soil profile. Data from Chris Figgins (written communication, 2000).

Figure 14 In order to fully understand the nuances of *terroir*, the authors, Larry Meinert (left) and Alan Busacca (right) tasted barrel samples of wine from many of the vineyards described in this paper. Some samples required repeated evaluation.

some of the wines made from them by Gary Figgins, winemaker at Leonetti, and Casey McClellan, winemaker at Seven Hills, provided further insight into the history of this vineyard and Walla Walla winemaking (Fig. 12c).

REFERENCES

Alwin, J.A., 1970, Clastic dikes of the Touchet beds, southwest Washington: unpublished M.Sc. thesis, Washington State University, Pullman, WA, 87 p.

Amerine, M.A., Berg, H.W., Kunkee, R.E., Ough, C.S., Singleton, V.L. and Webb, A.D., 1980, The Technology of Wine Making: AVI Publishing Company, Westport, CT, 794 p.

Aspler, T., 1999, Vintage Canada, The Complete Reference to Canadian Wines: McGraw-Hill Ryerson, Toronto, 225 p.

Atwater, B.F., 1986, Pleistocene glacial-lake deposits of the Sanpoil River valley, northeastern Washington: United States Geological Survey, Bulletin 1661, 39 p.

Baker, V.R. 1973, Paleohydrology and sedimentology of Lake Missoula flooding in eastern Washington: Geological Society of America, Special Paper 144, 79 p.

Baker, V.R. and R.C. Bunker, 1985, Cataclysmic late Pleistocene flooding from glacial Lake Missoula–a review: Quaternary Science Reviews, v. 4, p. 1-41.

Baker, V.R. and Nummedal, D., eds., 1978, The Channeled Scabland (a guide to the geomorphology of the Columbia Basin, Washington): National Aeronautics and Space Administration, 186 p.

Baksi, A.K., 1989, Reevaluation of the timing and duration of extrusion of the Imnaha, Picture Gorge, and Grande Ronde Basalts, Columbia River Basalt Group, *in* Reidel, S.P. and Hooper, P.R., eds., Volcanism and tectonism in the Columbia River flood-basalt province: Geological Society of America, Special Paper 239, p. 105-112.

Bretz, J H., 1923, The channeled scablands of the Columbia Plateau: Journal of Geology, v. 31, p. 617-649.

Bretz, J H., 1925, The Spokane flood beyond the channeled scablands: Journal of Geology, v. 33, p. 97-115, 312-341.

Bretz, J H., 1928a, Bars of channeled scabland: Geological Society of America, Bulletin, v. 39, p. 643-702.

Bretz, J H., 1928b, The channeled scabland of eastern Washington: Geographical Review, v. 18, p. 446-477.

Bretz, J H., 1928c, Alternate hypothesis for channeled scabland: Journal of Geology. v. 36, p.193-223, 312-341.

Bretz, J H., 1932, The Grand Coulee: American Geographical Society, Special Publication15, 89 p.

Bretz, J H., 1969, The Lake Missoula floods and the channeled scabland: Journal of Geology, v. 77, p. 505-543.

Bretz, J H., Smith, H.T.U. and Neff, G.E.. 1956, Channeled Scabland of Washington: New data and interpretations: Geological Society of America, Bulletin, v. 67, p. 957-1049.

British Columbia Wine Institute, 2000, Annual Report 1999-2000: British Columbia Wine Institute, Summerland, BC, 24 p.

Brown, B.H., 1937, The State-Line earthquake at Milton and Walla Walla: Seismological Society of America, Bulletin, v. 237, p. 205-209.

Busacca, A.J., 1991, Loess deposits and soils of the Palouse and vicinity, *in* Baker, V.R., Bjornstad, B.N., and others, Quaternary Geology of the Columbia Plateau: *in* Morrison, R.B., ed., Quaternary Nonglacial Geology–Conterminous United States: Geological Society of America, DNAG, Geology of North America, v. K-2, p. 216-228.

Busacca, A.J. and McDonald, E.V., 1994, Regional sedimentation of late Quaternary loess on the Columbia Plateau: sediment source areas and loess distribution patterns: Regional Geology of Washington State, Washington Division of Geology and Earth Resources, Bulletin, v. 80, p. 181-190.

Busacca, A.J., Marks, H.M. and Rossi, R., 2001, Volcanic glass in soils of the Columbia Plateau, Pacific Northwest US: Soil Science Society of America, Journal (in press).

Carson, R.J. and Pogue, K.R., 1996, Flood basalts and glacier floods: roadside geology of parts of Walla Walla, Franklin, and Columbia Counties, Washington: Washington Division of Geology and Earth Resources, Information Circular, 47 p.

Clore, W.J. and Nagel, C.W., 1969, Is there potential in Washington for growing *vinifera* grapes to make fine table and varietal wines?: Washington State Horticultural Association, Annual Meeting, Proceedings, v. 65, p. 194-197.

Daubenmire, R., 1970, Steppe vegetation of Washington: Washington Agricultural Experiment Station, Technical Bulletin 62, Washington State University, Pullman, WA, 130 p.

Daubenmire, R. and Daubenmire, J.B., 1984, Forest vegetation of eastern Washington and northern Idaho: Cooperative extension, Washington State University, Pullman, WA, 104 p.

de Candolle, Alphonse, 1855, Geographie Botanique Raisonnee; Ou, expositon des faits principaux et des lois concernant la distribution geographique des plantes de l'epoque actuelle: V. Masson, Paris, 2 volumes, 1366 p.

Flint, R.F., 1938, Origin of the Cheney-Palouse Scabland Tract, Washington: Geological Society of America, Bulletin, v. 49, p. 461-524.

Halliday, J., 1993, Wine Atlas of California: Penguin Books, New York, NY, 400 p.

Halliday, J., 1999, Wine Atlas of Australia and New Zealand: Harpercollins Australia, 496 p.

Harrison, E.T., Donaldson, N.C., McCreary, F.R. and Ness, A.O., 1964, Soil Survey of Walla Walla County, Washington: United States Government Printing Office, Washington, DC, 138 p.

Haynes, S.J., 1999, Geology and Wine 1. Concept of *terroir* and the role of geology: Geoscience Canada, v. 26, p. 190-194.

Haynes, S.J., 2000, Geology and Wine 2. A geological foundation for *terroirs* and potential sub-*appellations* of Niagara Peninsula wines, Ontario, Canada: Geoscience Canada, v. 27, p. 67-87.

Hooper, P.R., 2000, Flood basalt provinces, *in* Encyclopedia of Volcanoes, Sigurdsson-Haraldur, B.A., Houghton, B.F., McNutt, S.R., Rymer, H. and Stix, J., eds, Academic Press, San Diego, CA, p. 345-359.

Johnson, D.R. and Makinson, A.R., 1988, Soil Survey of Umatilla County Area, Oregon: United States Government Printing Office, Washington, DC, 388 p.

Keller M.K., Arnink, J. and Hrazdina, G., 1998, Interaction of nitrogen availability during bloom and light intensity during veraison: I. Effects on grapevine growth, fruit development, and ripening: American Journal of Enology and Viticulture, v. 49, p. 333-340.

Landon, R.D. and Long, P.E., 1989, Detailed stratigraphy of the N_2 Grande Ronde Basalt, Columbia River Basalt Group, in the central Columbia Plateau, *in* Reidel, S.P. and Hooper, P.R., eds., Volcanism and tectonism in the Columbia River flood-basalt province: Geological Society of America, Special Paper 239, p. 55-66.

Mann, G.M. and Meyer, C.E., 1993, Late Cenozoic structure and correlations to seismicity along the Olympic-Wallowa Lineament, northwest United States: Geological Society of America Bulletin, v. 105, p. 853-871.

McCreery, R.A., 1954, Mineralogy of Palouse and related series: unpublished Ph.D. thesis, Washington State University, Pullman, WA, 120 p.

Nagel, C.W., Atallah, N., Carter, G.H. and Clore, W.J., 1972, Evaluation of wine grapes grown in Washington: American Journal of Enology and Viticulture, v. 23, p. 14-17.

O'Connor, J.E. and Baker, V.R., 1992, Magnitudes and implications of peak discharges from glacial Lake Missoula: Geological Society of America Bulletin, v. 104, p. 267-291.

Pardee, J.T., 1910, The Glacial Lake Missoula: Journal of Geology, v. 18, p. 376-386.

Patton, P.C. and Baker, V.R., 1978, New evidence for pre-Wisconsin flooding in the Channeled Scabland of eastern Washington: Geology, v. 6, p. 567-571.

Peterson-Nedry, J., 2000, Washington Wine Country: Graphics Art Center Publishing, Portland, Oregon, 111 p.

Pringle, P.T., 1993, Roadside Geology of Mt. St. Helens National Volcanic Monument and Vicinity: Washington Department of Natural Resources, Information Circular 88, 120 p.

Reidel, S.P., Fecht, K.R. and Lindsey, K.A., 1992, Post-Columbia River basalt structure and stratigraphy of south-central Washington: Geological Society of America, Abstracts with Programs, v. 24, n. 5, p. 78.

Reynolds, A.G., Wardle, D.A., Hall, J.W. and Dever, M., 1995, Fruit maturation in four *Vitis vinifera* cultivars in response to vineyard location and basal leaf removal: American Journal of Enology and Viticulture, v. 46, p. 542-58.

Reynolds, A.G., Wardle, D.A. and Dever, M., 1996, Vine performance, fruit composition, and wine sensory attributes of Gewurztraminer in response to vineyard location and canopy manipulation: American Journal of Enology and Viticulture, v. 47, p. 77-92.

Ross, M.E., 1989, Stratigraphic relationships of subaerial, invasive, and intracanyon flows of Saddle Mountains Basalt in the Troy basin, Oregon and Washington, *in* Reidel, S.P., and Hooper, P.R., eds., Volcanism and tectonism in the Columbia River flood-basalt province: Geological Society of America, Special Paper 239, p. 131-142.

Schuster, J.E., 1994, Geologic map of the Walla Walla 1:100,000 quadrangle, Washington: Washington Department of Natural Resources, Open File Report 94-3, 18 p.

Schuster, J.E., Gulick, C.W., Reidel, S.P., Fecht, K.R. and Zurenko, S., 1997, Geologic map of Washington - Southeast quadrant: Washington Department of Natural Resources, GM-45, 20 p.

Shaw, J., Munro-Stasiuk, M., Sawyer, B., Beaney, C., Lesemann, J.E., Musacchio, A., Rains, B. and Young, R.R., 1999, The Channeled Scabland; back to Bretz?: Geology, v. 27, p. 605-608.

Smith, G.A., 1993, Missoula flood dynamics and magnitudes inferred from sedimentology of slack-water deposits on the Columbia Plateau, Washington: Geological Society of America, Bulletin, v. 105, p. 77-100.

Tesic, D., 2000, Ecophysiological Study of Cabernet Sauvignon in the Conditions of Hawke's Bay (New Zealand), http://members.nbci.com/_XMCM/dtesic/home.html.

Tolan, T.L. and Reidel, S.P., 1989, Structure map of a portion of the Columbia River flood-basalt province, *in* Reidel, S.P. and Hooper, P.R., eds., Volcanism and tectonism in the Columbia River flood-basalt province: Geological Society of America, Special Paper 239, map.
Tolan, T.L., Reidel, S.P., Beeson, M.H., Anderson, J.L., Fecht, K.R. and Swanson, D.A., 1989, Revisions to the estimates of the areal extent and volume of the Columbia River Basalt Group, *in* Reidel, S.P. and Hooper, P.R., eds., Volcanism and tectonism in the Columbia River flood-basalt province: Geological Society of America, Special Paper 239, p. 1-20.
Tukey, R. and Clore, W.J., 1972, Grapes - Their characteristics and suitability for production in Washington: Washington State University, publication EB 0635.
Waitt, R.B., 1980, About forty last-glacial Lake Missoula jökulhlaups through southern Washington: Journal of Geology, v. 88, p. 653-679.
Waitt, R.B., 1984, Periodic jökulhlaups from Pleistocene glacial Lake Missoula–new evidence from varved sediment in northern Idaho and Washington: Quaternary Research, v. 22, p. 46-58.
Waitt, R.B., 1985, Case for periodic, colossal jökulhlaups from Pleistocene glacial Lake Missoula: Geological Society of America Bulletin, v. 96, p. 1271-1286.
Weis, P.L. and Newman, W.L., 1989, The Channeled Scablands of eastern Washington –The geologic story of the Spokane Flood: Eastern Washington University Press, Cheney, WA, 25 p.
Wilson, J.E., 1998, *Terroir*: The Role of Geology, Climate, and Culture in the Making of French Wines: Mitchell Beazley, London, UK, 336 p.
Winkler, A.J., 1936, Temperature and varietal interrelations in Central Western Europe and Algeria: Wines and Vines, v. 17, p. 4-5.

Accepted as revised 24 November 2000

SERIES

Geology and Wine 4. The Origin and Odyssey of *Terroir*

James E. Wilson[1]
4248 South Hudson Parkway
Englewood, Colorado 80110 USA

SUMMARY

As a wine term *terroir* dates from the glorious years of 14th century wine-making in Burgundy, France. The term was first applied to certain top-quality wine-producing properties along the Côte d'Or, Burgundy. Outside France *terroir* is being applied more loosely, to properties producing less than top-quality wines, not the original purpose of the term. *Terroir* recognizes many of the physical elements of wine-making, including good drainage, soil structure, nature of clays, and the existence and interaction of chemical compounds such as calcium carbonate, iron, magnesium, and others in an almost mysterious alchemy in the wine-growing process. Although the physical and chemical interactions that affect the growing of wine are not well understood, the spread of the term *terroir* is encouraging better use of geology, soil, climate, and culture in producing better wines.

RÉSUMÉ

Dans son acception vinicole, le mot *terroir* date des années glorieuses de la culture du vin au 14e siècle, le long de la côte d'Or de Bourgogne en France. Ce terme a d'abord été utilisé pour désigner certaines propriétés vinicoles essentielles à production des meilleurs vins de la côte d'Or de Bourgogne. Hors France, le mot *terroir* est utilisé plus librement et pour des vins de qualité moindre que supérieure, contrairement à l'intention originale. Le mot *terroir* implique l'existence de nombreux facteurs physiques dans la production du vin, dont un bon drainage, une structure de sol, la nature des argiles, et l'existence de composés chimiques tels le carbonate de calcium, le fer, le magnésium, *etc.*, qui interagissent selon un procédé extrêmement complexe. Bien que les interactions physiques et chimiques de la production de vin ne soient pas bien comprises, l'usage du mot *terroir* pousse à un meilleur usage des connaissances géologiques, pédologiques, climatiques et des pratiques culturales dans la production vinicole.

TERROIR: ITS 14TH CENTURY ORIGIN

The 14th century encompassed glorious years of wine development in Burgundy, eastern France, under the Valois dukes (Fig. 1). For several centuries the Benedictine and Cistercian monks developed viticulture to a high standard. *Terroir* is a Burgundian wine term from that period. It was applied originally to certain wine-producing properties along the Côte d'Or that had established themselves as top quality; the guardians of that quality were the *gourmets*. The *gourmets* were responsible for the efficacy of the *terroir*, and were independent of the wine growers, mayors, and magistrates. The term "gourmet" is familiar today applying to judges of good wine and good food, but is no longer a professional group, having been replaced by interprofessional societies and the INAO in France (*Institut National des Appellations d'Origine de Vins et Eaux-de-Vie*). The *Appellation d'Origine*, usually shortened to *Appellation Contrôlée* (AOC or AC), on a French wine label is the consumer's guarantee that the wine in that bottle originated where the label says it does, was made from authorized grapes from that property, and vinified under very strict, legal controls. The category AOC is INAO's highest rank, and today would be equivalent to the *terroir*. The concept of *Appellation Contrôlée* has never been adopted in the United States and Canada; however, there are both state and federal bureaus who vie to regulate the wine industry in the United States. In Canada the Vintners Quality Alliance (VQA), established at the provincial level, serves as a quality control function similar to the *Appellation Contrôlée* of France.

The term *terroir* is being found more and more in the English wine vocabulary and appears in danger of being applied loosely to properties producing less than top-quality wines. That concern is the main thrust of this article. The principal source of the early history of these terms is an article by Monsieur Jacky Rigaux titled *Le Terroir et le Gourmet* in a recent issue of *Tastevin en Main*, the semi-

[1] James E. Wilson is the author of *Terroir: the Role of Geology, Climate, and Culture in the Making of French Wines,* published in 1998. This award-winning book, reviewed by Geology and Wine Series Editor Simon Haynes in *Geoscience Canada*, v. 27, p. 47, 48, introduced and publicized the French concept of *terroir,* with all its nuances of bedrock geology, soil, climate, and viticulture, to the English-speaking world. We are very pleased to publish Dr. Wilson's account of the origin and use of this important term. R.W. Macqueen, editor.

annual journal of the *Confrérie des Chevaliers du Tastevin* of France (Rigaux, 1999).

The development of such a term as *terroir* was almost a necessity in Burgundy because of the many small properties producing excellent wine along the sun-trapping, central curvature or "belly" of the Côte d'Or slopes. This belt of *terroirs* is now the avenue of the *Grands Crus* flanked by the *Premiers Crus*, with the communal wines on the higher slopes and the lower toe slopes.

The components making up a *terroir* were not defined in those days, but the use of *terroirs* was acknowledgment that *terroirs* did in fact produce quality wines. In *terroir*, we now recognize many of the physical elements such as good drainage, moisture-retaining structure of the soils, the nature of its clay, the existence of active calcium carbonate, iron, magnesium, and other chemical compounds in what has to remain a mysterious alchemy (Fig. 2).

The term *terroir* was rarely used outside of Burgundy until recent times. Although there were wines of outstanding quality produced in the other wine districts of France, they were usually relatively few in number and known by name and reputation. To protect that reputation, the producers of these wines were essentially their own "*gourmets.*"

CHÂTEAUX, AN ALTERNATE CONCEPT TO *TERROIR*

The vintners of the famous Bordeaux area of southwestern France felt that their "*châteaux*" defined their good vineyards sufficiently, even though many of its tracts might be widely separated: the *château* included all properties of the ownership, as well as buildings and wine-making equipment. Some of the scattered tracts themselves might qualify as a *terroir*; for example, in Château Margaux, the Cap de Haut and possibly the Puch Sem Peyre. In 1855, the Bordeaux Châteaux were given hierarchy ranking. This was brought about when the organizers of the grand exhibition in Paris invited Bordeaux to provide samples of its famous English claret, as Queen Victoria was to attend the exhibition. It was only at the insistence of the organizers that Bordeaux gave a ranking to its wines, finally submitting five or six categories that they called "growths." The validity of this ranking was based on a substantial history of market value. This ranking became the famous (or sometimes infamous) Bordeaux Classification of 1855. There have been numerous suggestions for changes or modifications, but by and large it has withstood the test of time.

MODERN USAGE OF *TERROIR*

The first widely publicized use of the term *terroir* was in 1984 with the appearance of *Terroirs et Vins de France, Itinéraires Oenologiques et Géologiques*

Figure 1 Château de Rully, dating from the 13th century, Côte Chalonnais region of central Burgundy, France. Côte Chalonnais is one of the five wine regions of Burgundy; the others are Chablis, Côte d'Or, Mâconnais and Beaujolais, all globally renowed in wine culture and production. Vineyard in the foreground extending to the castle wall is the Premier Cru vineyard of Bressande (Premier Cru = approximately First Growths, Médoc; see Wilson, 1998, p. 58 for discussion). Photographed in 1994 by Catherine Ponsot-Jacquin, a geologist and former student of Professor Noël Leneuf of the University of Dijon, France (see Fig. 2 caption, this paper).

Figure 2 Noël Leneuf, Emeritus Professor of Geology and Pedology, University of Dijon, France, standing in the Grand Cru vineyard of Chambertin, Côte d'Or region, Burgundy, France (Grand Cru = approximately Great Growths, Médoc; see Wilson, 1998, p. 58 for discussion). Professor Leneuf has long been a passionate advocate of *terroir*, and is convinced that "geology and its soils are the critical factors in the *terroirs* of Burgundy" (Wilson, 1998, p. 5). Professor Leneuf was a guide and correspondent in the development of my book *Terroir: the Role of Geology, Climate, and Culture in the Making of French Wines* (Wilson, 1998). Photographed 1994 by Catherine Ponsot-Jacquin.

(Pomerol *et al.*, 1984). This was essentially a field guide book for geologists and, although later translated into English, distribution was relatively limited. My book, *Terroir, The Role of Geology, Climate, and Culture in the Making of French Wines* was published in the United Kingdom in 1998 (Wilson, 1998), giving additional impetus to the use of the term *terroir*. The title of my original manuscript as submitted to Mitchell Beazley, the London publisher, was *Wine on the Rocks, With a Splash of History*. The publisher, however, insisted on using the title *Terroir*, feeling that my title was too American colloquial. The University of California Press in association with the Wine Appreciation Guild, as co-publishers of the book in the United States, in 1999, further spread familiarity of the term.

Terroir Defined

I defined the honoured word *terroir* in my book (Wilson, 1998, p. 55) as "... a concept ... not easily grasped but includes physical elements of the vineyard habitat — the vine, subsoil, siting, drainage, and microclimate. Beyond the measurable ecosystem, there is an additional dimension — the spiritual aspect that recognizes the joys, the heartbreaks, the pride, the sweat, and the frustrations of its history." I am pleased that the book (Wilson, 1998) has been recognized as a definitive reference for the proper meaning of *terroir.*

WINE QUALITY: CALIFORNIA CONCEPTS

Although I understand *terroir* is becoming something of a "buzzword" in wine parlance, the elements of the true meaning of the term have had a somewhat slow acceptance, especially in California. This is perhaps due, in part, to the long-held view that it was the, "heat summation," that was critical to the growing of good wine. I remember well a visit I had in 1985 with the late Professor A.J. Winkler at his retirement home near Davis, California. Professor Winkler knew that I am a geologist and therefore a likely disciple of the soil. I was hardly through the door of his small library to greet him when he waved his book at me saying, "It's all right here — it's not the soil but the heat summation." He related to me how in the late 1930s he and Professor Maynard Amerine, both professors of oenology and viticulture, had collected grapes from the various wine areas of California, making their own wine from these grapes. After studying the factors that might contribute to quality, Professor Winkler said they concluded that the critical feature was heat summation, which they defined as "degree-days," the sum of the hours when the temperature is above 50°F (10°C) for the months of the growing season. Several wine growers in the Napa Valley told me in 1985 that the hours of sunshine with the stimulation of the catalytic action in the plant were more critical than just the temperature. During the tenure of Professors Winkler and Amerine at the University of California at Davis, this hypothesis, that degree-days and hours of sunshine were definitive controls on wine quality, became essentially a doctrine to the exclusion of the importance of soil. Amerine and Winkler (1944; also see Amerine and Wagner, 1984) developed a five-part heat zonation of California with recommendations for the most adaptable grape varieties to those zones. Another feature about California wine growing is that irrigation is permitted, whereas it is forbidden in France. Production may be improved on a property by soil moisture monitoring and computer analysis of other physical variables. There is a service available for complete analyses and forecasting for properties here and abroad.

In recent years, there's been considerable exchange of students and a younger generation of wine growers between France and California. France has benefited greatly from the American wine-making technology; California has benefited from the French appreciation of the role of geology and soil. There's no question that California produces some "blockbuster" wines that have won first place in blind tastings with French and other foreign wines. But to really qualify for the honoured word *terroir*, a property must have a history of consistently producing fine wines for an appreciable period of time, perhaps a decade or more.

TERROIR: BENEFITS

With the more universal use of the term "*terroir,*" the debate also widens as to how much distinction can be made between the traits imparted by nature and those from the winemakers art. This argument may not be scientifically provable, but certainly the spread of the concept of *terroir* encourages the greater recognition of geology and geological processes, particularly as regards soil, drainage and siting, all to the benefit of wine quality.

ACKNOWLEDGMENTS

I thank Simon Haynes and Roger Macqueen who provided worthwhile suggestions as critical reviewers of this paper.

REFERENCES

Amerine, M.A. and Winkler, A.J., 1944, Composition and quality of must and wines of California grapes: Hilgardia, v. 15, p. 493-675.

Amerine, M.A. and Wagner, P.M., 1984, The vine and its environments, *in* Muscatine, D., Amerine, M.A. and Thompson, B., eds., Book of California Wine, Part lll-1: University of California Press/Sotheby Publications, Berkeley, CA, p. 99-105.

Pomerol, C., Combaz A. and Lautel R., eds., 1984, Terroirs et Vins de France, Itinéraires Oenologiques et Géologiques, Orléans and Paris: Editions du B.R.G.M.-Total Edition-Presse.

Rigaux, J., 1999, Le Terroir et le Gourmet: Tastevin en Main, n. 107, Confrérie des Chevaliers du Tastevin, Nuits-Saint-Georges Cedex–France, p. 31-34.

Wilson, J.E., 1998, Terroir: the Role of Geology, Climate and Culture in the Making of French Wines: Mitchell Beazley, London, UK, 336 p.

Accepted as revised 8 May 2001

SERIES

Geology and Wine 5. Provenance of Okanagan Valley Wines, British Columbia, Using Trace Elements: Promise and Limitations

Vivien F. Taylor[1, 2], Henry P. Longerich[1] and John D. Greenough[3]

[1]*Department of Earth Sciences and Centre for Earth Resources Research Memorial University of Newfoundland St. John's, Newfoundland A1B 3X5*

[2] *Present address: Watershed Ecosystems Trent University 1600 West Bank Drive Peterborough, Ontario K9J 7B8 vitaylor@trentu.ca*

[3]*Department of Earth and Environmental Sciences Okanagan University College 3333 College Way Kelowna, British Columbia V1V 1V7*

SUMMARY

The trace element compositions of 52 Okanagan wines from 26 vineyards were examined to test for correlations between wine and soil/bedrock composition, and to estimate total variability in Okanagan wine composition. Trace element concentrations in wines analysed are shown to provide verification of vineyard of origin, indicating that trace element compositions are influenced by vineyard geology, although geologic and environmental/climatic influences could not be clearly differentiated from possible wine processing influences in this reconnaissance study. Cluster analysis of Okanagan white wines using Al, Mn, V, Sr, Rb, Mo, Sb, Co, Zn, and U groups wines from the same vineyard, as well as showing a general north-south trend in wine chemistry related to vineyard location within the valley.

RÉSUMÉ

La composition en éléments-traces de 52 vins de l'Okanagan provenant de 26 vignobles a fait l'objet d'une étude visant à mesurer les corrélations entre la composition des vins et celle de la roche-mère ainsi que la variabilité totale de la composition des vins de l'Okanagan. Les concentrations en éléments-traces des vins analysés démontrent qu'elles peuvent servir à vérifier le vignoble d'origine, ce qui indique que les concentrations en éléments-traces sont influencées par la géologie du vignoble, quoiqu'il n'a pas été possible dans le cadre de cette étude préliminaire, de différencier clairement entre les influences de nature géologique, environnementale ou climatique, et des influences possibles reliées aux procédés de vinification utilisés. L'analyse typologique de vins blancs de l'Okanagan quant à leur teneur en Al, Mn, V, Sr, Rb, Mo, Sb, Co, Zn, et U permet de regrouper les vins par vignoble, et de démontrer l'existence d'une grande tendance nord-sud dans la composition chimique des vins liée à la localisation des vins dans la vallée.

INTRODUCTION

The Okanagan Valley has been a wine grape growing region since the 1860s, with extensive plantings near Kelowna in the 1920s (Schreiner, 1994). Until the mid-1970s, most vines in the Valley were *Vitis labrusca*, a North American grape species resistant to cold winters, but which yields wines with a strong musky flavour and aroma. Three great pullout programs (1979, 1984, and 1988) saw most *labrusca* replaced by *Vitis vinifera* by 1990 (Schreiner, 1994). The major grape varieties now grown in the valley are Chardonnay, Merlot, Pinot Gris, Cabernet Sauvignon, and Pinot Noir to the south; and Pinot Blanc, Pinot Noir, Pinot Gris, Riesling, and Gewürztraminer in the middle and northern Okanagan Valley (Schreiner, 1998).

Various studies indicate that trace element patterns can be used to "fingerprint" wines and that these patterns, at least in part, reflect the provenance, or site of origin, of wines (Baxter *et al.*, 1997; Danzer *et al.*, 1999). Preliminary studies of Okanagan Valley wines indicated that wines could be fingerprinted by vineyard (Greenough *et al.*, 1996, 1997). Given that many countries, including Canada, have labelling requirements that certify the geographic origin of grapes used in wine, fingerprinting may have both legal and regulatory uses.

Fingerprinting involves determining the concentrations of multiple elements in wines followed by statistical analysis. Okanagan Valley vineyards and wineries from which wines were sampled are listed in Table 1, along with grape variety, geographic subregion, and geologic classification (bedrock, glacial sediment, and soil). The strong influence of bedrock composition and soil chemistry on wine composition is supported by the discrimination of wines by subregion of origin, although viticultural practices and processing methods also have an impact. Okanagan vineyards are underlain by a wide variety of bedrock lithologies (Fig. 1a) and surficial geology (Fig. 1b), and there is a significant change in soil types (Fig. 1c) and climate, from relatively wet and cool in the north to more arid and hot in the south. Wines and vineyard soil samples from the Okanagan Valley (Fig. 1a) were collected in April 1999 and analysed later that year, to try to decipher geochemical influences on trace element composition of wine.

SUMMARY OF OKANAGAN VALLEY GEOLOGY

Bedrock

The Okanagan Valley is underlain by rocks of the Quesnellia terrane, which consists dominantly of upper Paleozoic and lower Mesozoic volcanic and sedimentary rocks accreted to North America and intruded by "granites" during the Mesozoic (Monger *et al.*, 1991). Six of the studied vineyards near Oliver (southern Okanagan, vineyards 21 to 26, Fig. 1a) are underlain by some of this terrane's oldest rocks, the Carboniferous or older, mafic metavolcanic rocks of the Anarchist and Kobau Groups. Three other vineyards (labelled 1, 10, and 11, north half of Fig.1a) have Jurassic granites as bedrock. Of these nine vineyards, only one (26) lies on the east side of what is perhaps the most fundamental structural feature of Okanagan Valley geology: a major, low-angle, normal fault that runs from north to south the length of the valley, with the west side the down-thrown block (Fig. 1a; Tempelman-Kluit, 1989; Roed *et al.*, 1995). The oldest rocks east of the fault tend to be Early Paleozoic orthogneisses that were last deformed during an Eocene mountain-building event. These gneisses underlie 12 of the 26 Okanagan vineyards studied (Fig. 1a). Eocene gneiss deformation was accompanied, on the surface, by intermediate to felsic volcanism represented by basins bearing volcanic rocks and volcanic-rich caldera-fill sediments west of the fault at Kelowna, Summerland, and Okanagan Falls (Church, 1973, 1980, 1981, 1982; Bardoux and Irving, 1988). At least three of the vineyards studied (2, 3, and 9, Fig. 1a) overlie these compositionally variable rocks. The last rock-forming event in the Okanagan Valley created the Pleistocene basaltic lava flow near Westbank that underlies one vineyard (7, Fig. 1a).

Pleistocene Sediments

Complicating the determination of the bedrock geology underlying many of the vineyards, and possibly mitigating the effect of bedrock on wine geochemistry, are glacial sediment deposits (Fig. 1b), which locally approach 100 m in thickness. During the Pleistocene, the advance and retreat of thick glaciers (3 km) scoured out the base and sides of the Okanagan Valley, deepening it to as much as 640 m below sea level (Eyles *et al.*, 1990). When the last ice sheet melted ~10,000 years ago, thick fluvioglacial and colluvial (glacial lake silt and clay) deposits formed along the sides of the

Table 1 Grape variety, subregion and geology of Okanagan vineyards sampled

Vineyard	Number (Fig. 1)	Grape Variety	Subregion	Bedrock (Fig. 1a)	Sediment (Fig. 1b)	Soil (Fig. 1c)
Gray Monk	1	Rotberger	Vernon	mJg	Morainal	Eutric Brunisolic
House of Rose	2	Verdelet	Kelowna	Em	Fluvioglacial	Gray Luvisolic
Pinot Reach	3	Riesling	Kelowna	Ewl	Alluvial	Gleysolic
Summerhill	4	Chardonnay	Kelowna	Egn	Fluvioglacial	Chernozemic
St. Hubertus	5	Riesling	Kelowna	Egn	Fluvioglacial	Chernozemic
Cedar Creek	6	Chardonnay	Kelowna	Egn	Fluvioglacial	Chernozemic
Slamka	7	Riesling	West Side	QPl	Morainal	Eutric Brunisolic
Quail's Gate	8	Riesling	West Side	Ewl	Morainal	Eutric Brunisolic
Hainle	9	Treminor	Peachland	Etr	Colluvial	Eutric Brunisolic
McKenzie	10	Pinot Blanc	Peachland	mJg	Fluvioglacial	Eutric Brunisolic
Scherzinger	11	Gewurztraminer	Summerland	mJg	Fluvioglacial	Chernozemic
Hillside	12	Muscat	Naramata	Egn	Colluvial	Eutric Brunisolic
Irvine	13	Chardonnay	Naramata	Egn	Fluvioglacial	Chernozemic
Lake Breeze	14	Pinot Blanc	Naramata	Egn	Fluvioglacial	Chernozemic
Kettle Valley	15	Chardonnay	Naramata	Egn	Fluvioglacial	Chernozemic
Red Rooster	16	Chardonnay	Naramata	Egn	Colluvial	Eutric Brunisolic
Lang	17	Riesling	Naramata	Egn	Colluvial	Eutric Brunisolic
Nichol's	18	Syrah	Naramata	Egn	Colluvial	Eutric Brunisolic
Wild Goose	19	Gewurztraminer	Okanagan Falls	Egn	Colluvial	Eutric Brunisolic
Stag's Hollow	20	Chardonnay	Okanagan Falls	Egn	Colluvial	Eutric Brunisolic
Tin Horn Creek	21	Gewurztraminer	Oliver	CPko	Morainal	Eutric Brunisolic
Gehringer Bros	22	Verdelet	Oliver	CPko	Morainal	Eutric Brunisolic
Hester Creek	23	Pinot Blanc	Oliver	CPko	Morainal	Eutric Brunisolic
Inniskillin	24	Pinot Blanc	Oliver	CPko	Morainal	Eutric Brunisolic
Gersighel	25	Pinot Blanc	Oliver	CPko	Morainal	Eutric Brunisolic
Inkameep	26	Riesling	Oliver	CPa	Fluvioglacial	Chernozemic

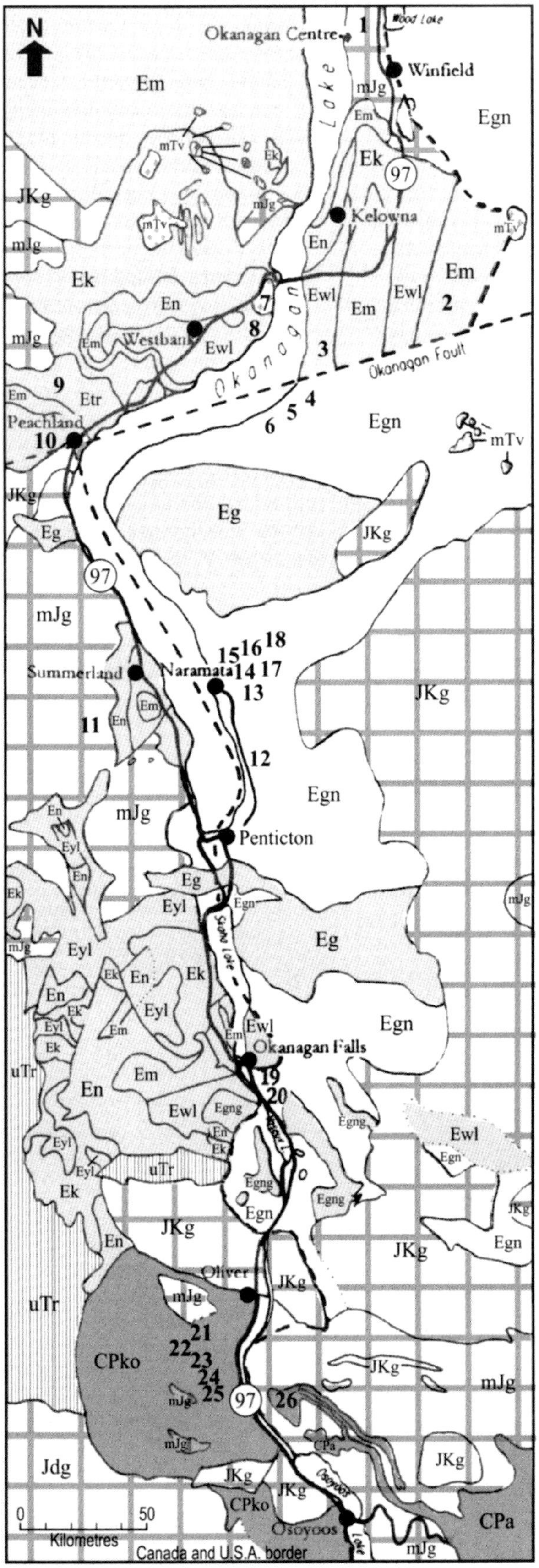

Geology Legend

Miocene to Pleistocene
mTv Plateau and valley basalts Basalt

Eocene

Egng	Skaha formation	Brecciated greenstones and cherts in fault sheets overlying the White Lake Formation
Ewl	White Lake Formation	Volcanic breccia, pyroclastic rocks, immature, Volcanic-rich clastic sedimentary rocks
Em	Marama Formation	Dalcite
En	Nimpit Lake Member	Amygdaloidal trachyandesite
Ek	Kitley Lake Formation	Trachyte to trachyandesite
Eyl	Yellow Lake Formation	Mafic phonolite
Etr	Trepanier Rhyolite	Rhyolite
Eg	Eocene Granodiorite	Mostly granodiorite in this map area

Jurassic to Cretaceous

JKg	Okanagan Batholith and Intrusives	Granite and granodiorite
mJg	Nelson Plutonic Rocks	Granodiorite
Jgd	Kruger Syenite	Syenite

Ordovician to Upper Iriassic

uTr	Undivided	Greenstones and cherts

Carboniferous or older

CPa	Anarchist group	Amphibolite, greenstone, chlorite schist, minor sepentinite
CPko	Kobau Group	Amphibolite, greenstone, mica schists, minor marble

Precambrian to Cenozoic

Egn	Okanagan Gneiss	Gneiss, last deformation Eocene, diverse protoliths but Okanagan Batholith dominant

Vineyard Legend

1. Gray Monk
2. House of Rose
3. Pinot Reach
4. Summerhill Estate
5. St. Hubertus Estate
6. Cedar Creek
7. Slamka Cellars
8. Quail's Gate
9. Hainle Vineyard
10. McKenzie
11. Scherzinger Vineyard
12. Hillside Estate
13. Irvine
14. Lake Breeze
15. Kettle Valley
16. Red Rooster
17. Lang Vineyard
18. Nichol Vineyard
19. Wild Goose
20. Stag's Hollow
21. Tin Horn Creek
22. Gehringer Bros.
23. Hester Creek
24. Inniskillin Vineyards
25. Gersighel Wineberg
26. Inkameep

Figure 1a Bedrock geology and vineyards of the Okanagan Valley (adapted from Tempelman-Kluit, 1989).

valley and adjacent to short-lived Glacial Lake Penticton, the larger and deeper precursor to present-day Okanagan Lake (Nasmith, 1981; Vanderburgh and Roberts, 1996).

Soils

Okanagan soils are glacially derived, sandy in the south Okanagan, and tend to have a higher clay content in the north of the valley. Soils in a number of the Okanagan vineyards are Eutric Brunisols (Fig. 1c), meaning that they have undergone moderate physical, chemical, and biological weathering, which has changed the morphology of the parent material, but no major translocations or transformations have occurred (Valentine and Lavkulich, 1999). In several of the central Okanagan vineyards (*e.g.*, 11, 13, 14, 15, Fig. 1a) soils are Dark Brown Chernozems, which are characterized by a dark organic matter in the topsoil, and an accumulation of carbonates in the lower horizons owing to high soil evaporation rates. Soils in two of the Kelowna vineyards (2, 3, Fig.1c), are Gray Luvisolic and Gleysolic, respectively, which form in areas having more precipitation than do the Chernozemic or Brunisolic soils, causing a greater amount of weathering in the upper horizons (Valentine and Lavkulich, 1999).

POSSIBLE EFFECTS OF WINE PROCESSING ON TRACE ELEMENT COMPOSITION OF WINE

The Okanagan Valley wine and vineyard soil samples obtained were analysed to test for correlations between wine and soil composition, and to estimate total variability in Okanagan wine composition. There are, however, other factors that may influence the trace element composition of wines as follows.

Methods of wine processing affect the inorganic composition of wines, and concentrations of certain elements are expected to be controlled by processing more than by geological and plant uptake factors. The addition of sulphite is a common practice, in which K or Na metabisulphite is added to the grapes just before crushing. This process is meant to inhibit the growth of microbes by the release of SO_2 gas (Vine, 1982). The precipitation of "wine diamonds" (calcium tartrate crystals) during wine storage is expected to control Ca concentration in wine, and is likely to co-precipitate other trace elements. Storage in stainless steel has been shown to increase Ni and Fe concentrations in wine (Muryani and Papp, 1997). Aging in oak is a source of tannins and phenols and also a possible source of trace elements, which would be expected to vary with oak species, barrel conditioning, and cooperage size (Jackson, 1994). Clarification by fining agents and/or filtration also may affect element concentrations in wine. The use of bentonite as a fining agent has been shown to increase the concentration of rare earth elements in wine (Jakubowski *et al.*, 1999), and is likely to affect other element concentrations. Casein (a milk protein) and egg albumin are also used to adsorb tannins in wines, and may affect trace element composition in wines. In red wine production, the macerate (crushed grapes, including seeds and skin) is fermented, but in white wine produc-

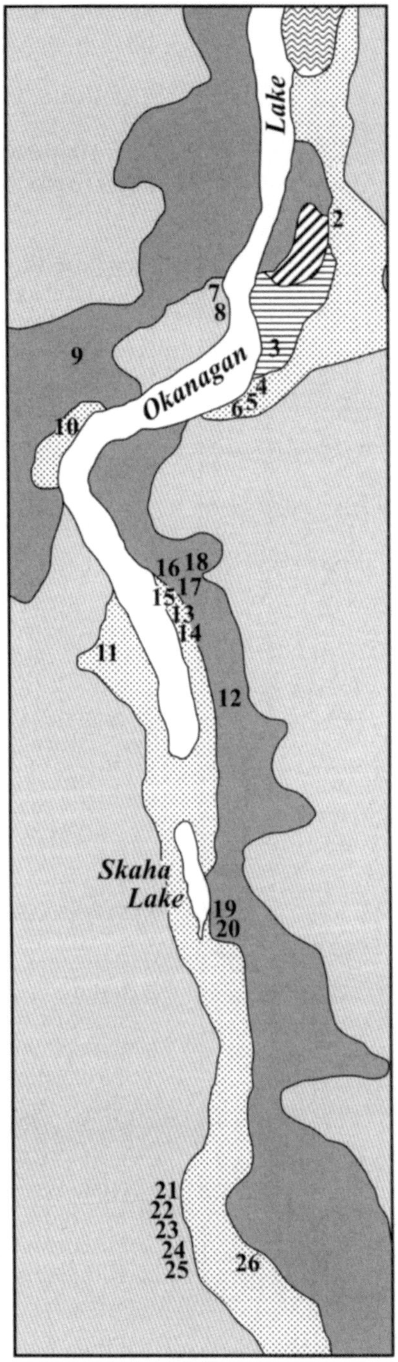

Figure 1b Surficial deposits of the Okanagan Valley (adapted from Canadian Soil Information Systems (CanSIS) Soil Landscapes of Canada, 2001). Units identified — morainal, fluvioglacial, alluvial, lacustrine, and colluvial — refer to the parent material mode of deposition (PMDEP) used in the National Soil Database of Agriculture and Agri-Food Canada. Unit descriptions given here are abbreviated from those given in the National Soil Database (see CanSIS, 2001).

tion, the skins and seeds are usually removed immediately by pressing, and the juice is fermented (Jackson, 1994). Red and white wines have been shown to be distinguishable by their trace element concentrations (Baxter *et al.*, 1997; Greenough *et al.*, 1997).

For this study, there are too many variables in wine processing, and insufficient data at the 26 vineyards/wineries from which wines were analysed to permit assessment of the significance of processing techniques in affecting the trace element compositions of the wines studied. Accordingly, this work should be regarded as a reconnaissance study of trace element compositions in the wines analysed.

SAMPLING AND ANALYTICAL METHODS

Sampling and Sample Preparation

Soils

Between five and eight soil samples were taken from a depth of 0.3 m to 0.6 m for each specific vineyard plot that produced a particular bottle of wine. Normally, this depth sampled the subsoil, which is less affected by microbial alteration and the addition of fertilizers than the topsoil. All Okanagan vineyards are irrigated (British Columbia Wine Institute, personal communication, 2000), and some have been tilled, which alters the soil horizons and chemical properties from natural soils of the area. Soils were sieved to <2 mm grain size and combined by weight to form a composite sample of each vineyard plot. A sub-sample was then sieved (unground) to a grain size of less than 0.074 mm (200 mesh) for each vineyard. This fine-grained fraction (<0.074 mm) was analysed for all vineyard soils, as trace elements are expected to be most concentrated in this silt and clay fraction, making differences in concentrations between soils more noticeable. To ensure that a bias was not given to vineyard soils with a high clay content, the <2 mm fraction was also analysed for the soils of 10 vineyards. Samples of both grain sizes were prepared as pressed pellets, and analysed by X-ray fluorescence for 28 elements/oxides as noted below.

Wines

All wines included in this study were produced from grapes grown on a single vineyard plot. White wines from the major *Vitis vinifera* varieties (Riesling, Chardonnay, Pinot Blanc, Gewürztraminer) were chosen preferentially, but were not always available. Previous studies of Okanagan wines indicated that wines from different vineyards can be distinguished chemically (Greenough *et al.*, 1997). As a further assessment of this finding, multiple (at least four) wines were obtained and analysed from five different vineyards. Sampling and analysis was also undertaken to permit evaluation of the effect of vintage on wine composition: five of the six wines from the House of Rose vineyard (Verdelet, 1992 to 1996 vintages), as well as five Wild Goose vineyard wines (Gewürztraminer, 1993 to 1997), and four Quail's Gate wines (Rieslings, 1995 to 1998) are from different vintage years. Also, five Rieslings sampled from Lang Vineyard are from the 1995 to 1997 vintage years, and of these two are late harvest wines (1995 and 1997) and one is an ice wine (1997). In order to determine variability in a single type of wine from one winery in one year,

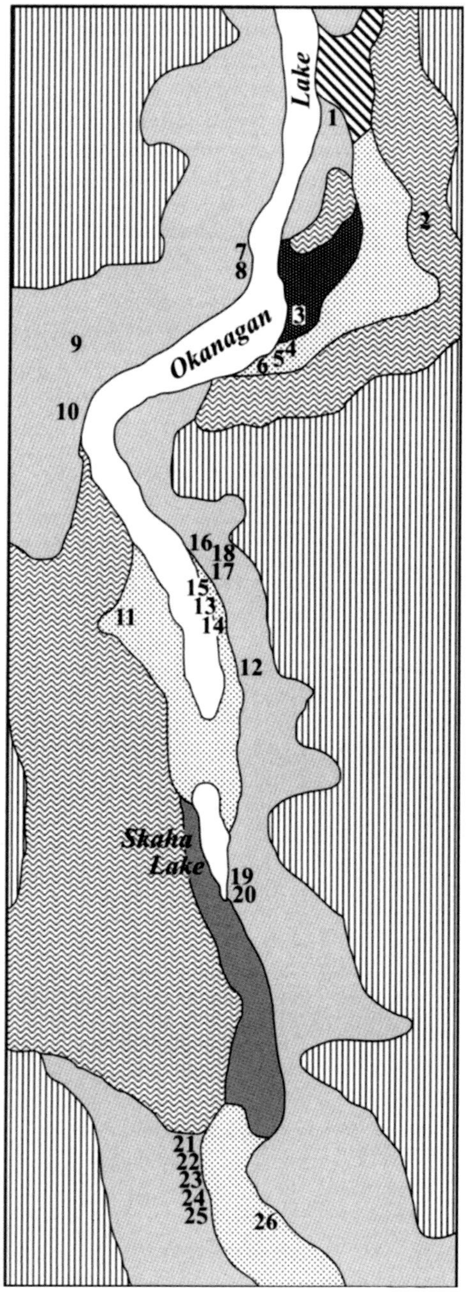

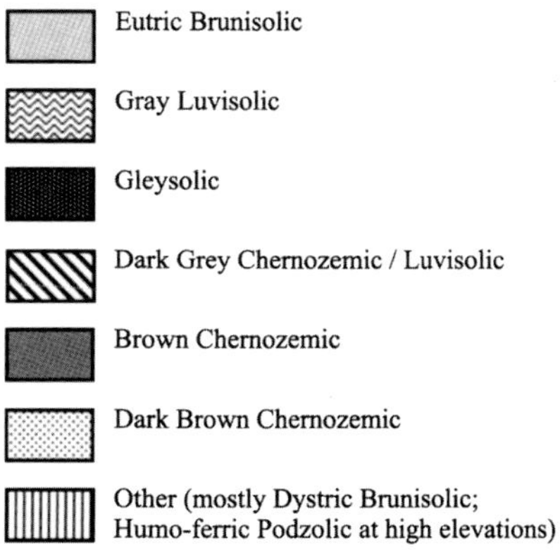

Figure 1c Soil classification of the Okanagan Valley (adapted from National Soil DataBase, Soil Landscapes of Canada, CanSIS, 2001).

three Lake Breeze samples were taken from different vats of the same batch of wine, and three were taken from the top, middle, and bottom of one of these vats. The Summerhill and McKenzie wines were also sampled from vats, whereas all other samples were taken from bottled wines.

Analysis

Soils

The pressed pellets of both grain sizes were analysed by X-ray fluorescence for the following 28 elements/oxides: Na_2O, MgO, Al_2O_3, SiO_2, P_2O_5, S, Cl, K_2O, CaO, Sc, TiO_2, V, Cr, MnO, Fe_2O_3, Ni, Cu, Zn, Ga, As, Rb, Sr, Y, Zr, Nb, Ba, Ce, Pb, Th, and U, according to the procedure in Longerich (1995). Precision and accuracy were monitored with geological reference materials [AGV-1(andesite, USGS), DNC-1(diabase, USGS), JG-2 (granite, JGS), and BCR-1(basalt, USGS))] and are better than ±2 % for most elements except Cl, As, and Pb, which are better than ±5%. This method of soil analysis determines total element concentrations, not plant-available element concentrations, and was chosen to examine differences in the total soil composition, derived from underlying sediments and parent rock. At present there is no single (*i.e.,* applicable to all elements), simple and reliable method of determining plant-available trace element concentrations in soils (Dr. Gerry Nielson, Summerland Research Station, British Columbia, personal communciation, 1999).

Wines

These were diluted 1:1 with 0.2 M HNO_3 and analysed for 34 elements using ICP-MS following a procedure similar to Baxter *et al.* (1997). Replicate analyses of two in-house reference wines indicated precision over a 5-month period, and five water reference materials [t-123, t-127, t-129 and t-135 (USGS) and AMW-3(DFC)] were matrix matched to the wine samples (0.1 M HNO_3 and 6% ethanol) to indicate precision and accuracy as follows: <5% for Cd, Sb, Ba, Tl, Pb, and U; <10% for As, Rb, Sr, Mo, Cs, La, Ce, and Th; <15% for V, Mn, Fe, Cu, Zn, Ag, and Bi; <20% for Mg, Al, Ca, Co, Ni, and Br; and <25% for Li, Be, Ti, Se, and I; and 27% for Cl and P.

RESULTS

Soils

Element concentrations were determined for each vineyard soil, and as found previously (Greenough *et al.*, 1996), little chemical variation exists between Okanagan vineyard soils. The mean concentration and standard deviation of the two measured particle-size fractions in all vineyard soils sampled are shown in Table 2. The fine-grained soil fraction (<0.074 mm) tends to contain similar or slightly higher concentrations of analytes than the <2 mm fraction, presumably reflecting the stronger sorption properties and higher surface area of electrostatically charged clay particles (Kabata-Pendias and Pendias, 1984). The use of soils as a direct indicator of bedrock composition is problematic because of glacial mixing, and because soil characteristics are highly dependent on climate (Krauskopf and Bird, 1995). Soils in the Oliver area (vineyards 21-25, Figs. 1a,c) were found to have the highest concentrations of Ni, Cu and Ba (see Table 3).

Wines

Mean trace element concentrations in wines from various Okanagan subregions, and in all Okanagan wines sampled, are given in Table 4. The large variation in element concentration within each group is partly due to the difficulty in defining subregions, as groups of wines become extremely small when subregions are defined by bedrock and overburden, making data from each group difficult to interpret. The data in Table 4 show that no one element can be used to discriminate wine regions, and as with other studies, the data were examined using multivariate analyses. Because multivariate statistical methods assume a

Table 2 Mean concentrations (in wt% for oxides, ppm for elements) and standard deviations of soils from 10 Okanagan vineyards (House of Rose, Pinot Reach, Slamka Cellars, McKenzie, Nichol, Irvine, Wild Goose, Inniskillin, Gersighel Brothers, and Inkameep).

Element	Mean 2 mm	SD	Mean 0.074 mm	SD
Al_2O_3	13.60	1.10	14.40	1.20
CaO	3.40	0.90	4.00	1.80
TiO_2	0.66	0.10	0.63	0.23
MnO	0.11	0.03	0.10	0.03
Fe_2O_3	5.10	1.00	6.60	4.70
Cl	158.00	72.00	200.00	113.00
V	97.00	23.00	96.00	18.00
Ni	17.90	18.70	18.40	18.10
Cu	21.80	10.10	27.20	10.20
Zn	33.30	8.70	50.20	8.50
As	8.90	5.10	16.40	9.20
Rb	70.00	9.00	70.00	9.00
Sr	456.00	130.00	459.00	83.00
Ba	1155.00	219.00	1026.00	156.00
Ce	47.00	24.00	63.00	32.00
Pb	18.30	18.00	27.60	21.30
Th	5.00	2.10	10.20	3.20
U	1.18	2.08	3.09	4.12

Table 3 Concentrations (ppm) of Ni, Cu and Ba in Oliver vineyard soils (Fig. 1, vineyard nos. 21-25) compared to all other Okanagan vineyard soils sampled Fig.1, vineyards nos. 1-20 and 26).

Element	Okanagan (except Oliver)	SD	Oliver	SD
Ni	10.5	7.7	47.0	8.1
Cu	22.0	7.0	52.0	12.0
Ba	998.0	313.0	1459.0	536.0

normal data distribution, and are easily skewed by outliers (Statsoft, 1999), element concentrations in wines were log transformed to better fit a normal distribution, and reduce the extremity of outliers. Log transformation also reduces the relative distribution of data, so that elements of higher abundance are brought into the same range as elements of lower abundance, which is important when elements of different abundance are included in the same analysis.

Elements with low analytical uncertainty were emphasized in comparing and "fingerprinting" wines, and those likely to be most affected by processing were eliminated, as follows. Elements which are important for discriminating wine provenance were determined by graphical analysis, and by F-statistics (between-group over within-group variability, where groups of wines were defined first by subregion, then by vineyard). The elements determined to be important to fingerprinting were Al, V, Mn, Co, Zn, Rb, Sr, Mo, Sb, and U. Exploratory cluster analysis was used as a method of visualizing data, by grouping objects (wine samples) into clusters so that objects in the same cluster are more similar to each other than to objects in other clusters (Hair, *et al.*, 1987). Cluster analysis of Okanagan white wines, using the 10 elements most important in fingerprinting wines as noted above, illustrates overall relationships among white wines from throughout the valley (Fig. 2). The analysis was found to group wines by vineyard of origin to a high degree, and there is an overall trend grouping wines from the north and south of the valley. Red and ice wines were not included in the cluster analysis due to the differences in fermenting and processing.

Variance within a batch of wine and wines from the same vineyard plot are compared in Fig. 3. The relative standard deviation (RSD) of element

Table 4 Mean trace element concentrations in wines (in ppb, except for the elements Mg, P, Cl, and Ca which are in ppm) and standard deviations (SD) for trace elements in North Okanagan (labelled N,1-8 in Fig. 1), Middle Okanagan: Peachland/Summerland (labelled MP, 9-11 in Fig. 1), Middle Okanagan: Naramata (labelled MN, 12-18 in Fig. 1), Middle Okanagan: Okanagan Falls (labelled MO, 19-20 in Fig. 1), South Okanagan (labelled S, 21-26 in Fig. 1) and for all wines analyzed in the Okanagan.

	Mean N	SD	Mean MP	SD	Mean MN	SD	Mean MO	SD	Mean S	SD	Mean Okanagan	SD
Li	9.20	7.90	3.30	2.30	6.20	3.20	8.00	2.40	4.90	2.50	7.20	5.70
Be	0.09	0.10	0.10	0.10	0.52	0.26	0.38	0.27	0.37	0.32	0.27	0.27
Mg	61.00	27.00	62.00	19.00	59.00	30.00	60.00	64.00	55.00	18.30	60.00	23.90
Al	146.00	114.00	205.00	37.00	593.00	365.00	254.00	158.00	391.00	279.00	309.00	286.00
P	131.00	122.00	140.00	55.00	96.00	38.00	93.00	28.00	114.00	55.00	116.00	84.00
Cl	12.90	6.80	18.90	12.40	17.20	16.30	21.90	23.40	11.80	6.30	15.40	12.90
Ca	72.00	30.00	80.00	64.00	86.00	61.00	65.00	13.00	75.00	33.00	76.00	41.00
Ti	5.50	3.20	6.80	4.60	6.40	4.20	10.90	8.70	8.40	4.90	6.90	4.90
V	6.80	8.90	3.10	2.30	8.60	13.60	3.60	2.90	26.00	46.60	9.20	19.90
Mn	678.00	216.00	896.00	629.00	631.00	272.00	712.00	214.00	697.00	367.00	692.00	300.00
Fe	1008.00	953.00	1611.00	390.00	1106.00	825.00	1178.00	850.00	890.00	374.00	1089.00	809.00
Co	1.68	0.59	1.64	1.03	3.34	1.24	1.78	0.61	2.60	1.73	2.23	1.22
Ni	17.90	8.70	69.30	85.90	18.20	9.10	10.00	4.30	18.50	7.20	21.60	28.50
Cu	45.00	61.00	95.00	85.00	67.00	125.00	39.00	41.00	56.00	52.00	56.00	80.00
Zn	456.00	291.00	743.00	479.00	322.00	164.00	415.00	156.00	555.00	446.00	457.00	311.00
As	1.90	1.40	2.00	1.40	5.20	3.90	1.80	0.90	3.50	2.90	2.90	2.70
Se	1.87	2.42	0.63	0.24	1.82	0.93	0.77	0.35	1.27	0.85	1.53	1.68
Br	166.00	81.00	203.00	77.00	198.00	79.00	168.00	30.00	197.00	59.00	182.00	72.00
Rb	367.00	161.00	376.00	102.00	504.00	145.00	765.00	141.00	372.00	133.00	451.00	194.00
Sr	764.00	424.00	721.00	395.00	796.00	271.00	791.00	171.00	687.00	198.00	761.00	329.00
Mo	4.79	2.91	27.01	21.18	9.75	7.20	3.79	1.09	4.54	2.19	7.80	9.45
Ag	0.01	0.01	0.02	0.02	0.02	0.01	0.01	0.01	0.01	0.01	0.01	0.01
Cd	0.38	0.41	1.57	2.82	0.28	0.11	0.44	0.45	0.42	0.21	0.47	0.89
Sb	0.32	0.18	0.24	0.13	0.36	0.29	0.30	0.16	0.80	0.58	0.39	0.33
I	1.60	0.70	2.90	2.10	3.70	1.60	1.00	0.20	2.40	1.30	2.30	1.50
Cs	0.81	0.46	1.08	0.57	1.77	1.09	1.83	0.70	1.31	0.62	1.27	0.82
Ba	132.00	128.00	217.00	79.00	131.00	56.00	113.00	12.00	156.00	94.00	140.00	97.00
La	0.24	0.41	0.69	1.17	0.35	0.32	0.47	0.41	0.26	0.19	0.34	0.48
Ce	0.46	0.75	1.32	2.17	0.91	0.84	0.96	0.89	0.54	0.40	0.72	0.95
Tl	0.07	0.03	0.06	0.02	0.21	0.13	0.11	0.01	0.10	0.05	0.11	0.09
Pb	7.80	6.40	12.40	7.50	8.90	4.10	21.10	13.20	19.40	10.00	11.70	9.10
Bi	0.14	0.18	0.15	0.16	0.15	0.11	0.40	0.31	0.26	0.21	0.19	0.20
Th	0.05	0.07	0.06	0.08	0.16	0.16	0.23	0.15	0.09	0.11	0.11	0.13
U	0.20	0.19	0.50	0.80	0.24	0.29	0.87	0.54	1.97	2.85	0.57	1.22

concentrations for six wines from the same batch of 1998 Lake Breeze Pinot Blancs; and for white wines from the same vineyard plot but from different vintage years (Quail's Gate Riesling, 1995-1998; Wild Goose Gewürztraminer, 1993-1997; House of Rose Verdelet, 1992-1996), are compared to the RSDs of element concentrations of all wines sampled in the valley (Fig. 3).

DISCUSSION

Soil Composition

Soils in the Oliver region have higher concentrations of Ba, Ni, and Cu than other Okanagan vineyard soils (Table 3), which may reflect the influence of climate on soil composition. In arid areas, metals can become enriched in the soil due to adsorption onto illuviated clay materials (Aubert and Pinta, 1977). The high Ni and Cu content may reflect derivation from bedrock of the Kobau Group amphibolites and greenstones that underlie other Oliver region vineyards. With the possible exception of basalts beneath the Slamka Cellars vineyard, these are the most mafic rocks known in the study area, and mafic rocks commonly have high Ni and Cu concentrations.

Despite some clustering of wines according to geography (Fig. 2), the correlation between wine and soil element concentrations is weak. The relationship between As (arsenic) in soil and wine reported by Greenough *et al.*, (1997), appears to be fortuitous when re-examined with the present, larger, data set. Similarly, element ratios (examined to account for variable element dilution in wine) yielded no strong correlations between wine and soil. Most of the Okanagan vineyards are underlain by deep glacial deposits, and soils formed from these parent materials were found to be fairly uniform in composition. Wine compositions possibly reflect bedrock rather than soil compositions, as the field of biogeochemical prospecting uses vegetation rather than soil samples in exploring for bedrock ore deposits (*e.g.*, Dunn, 1995). Grape roots, especially in arid areas (*e.g.*, southern Okanagan), can grow deep enough that they intersect the watertable, which tends to interact chemically with bedrock. Another factor mitigating a relationship between wine and soil composition is that plants do not

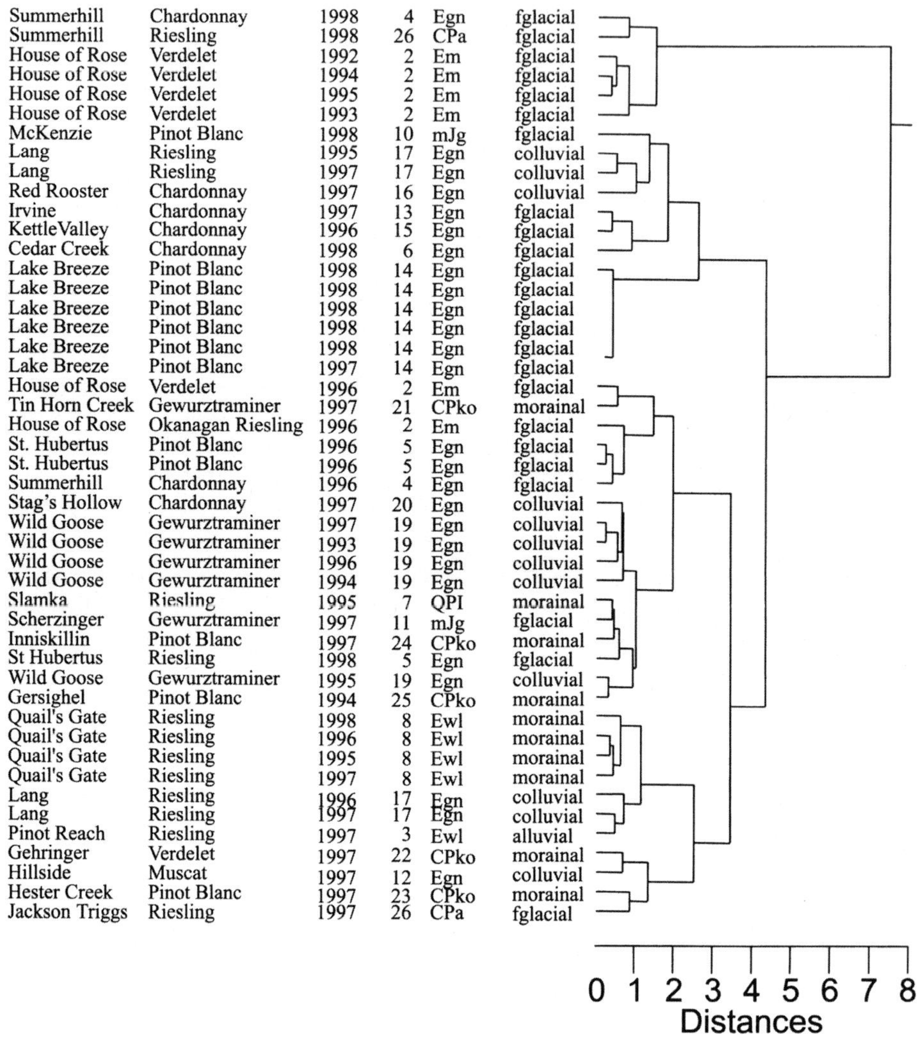

Figure 2 Cluster analysis of log transformed concentrations of Al, V, Mn, Co, Zn, Rb, Sr, Mo, Sb, and U in white Okanagan wines, using Euclidean distances and linkage by Ward's method. Wines are labelled by winery, grape variety, vintage year, vineyard number (see Fig. 1), bedrock type, and surficial geology.

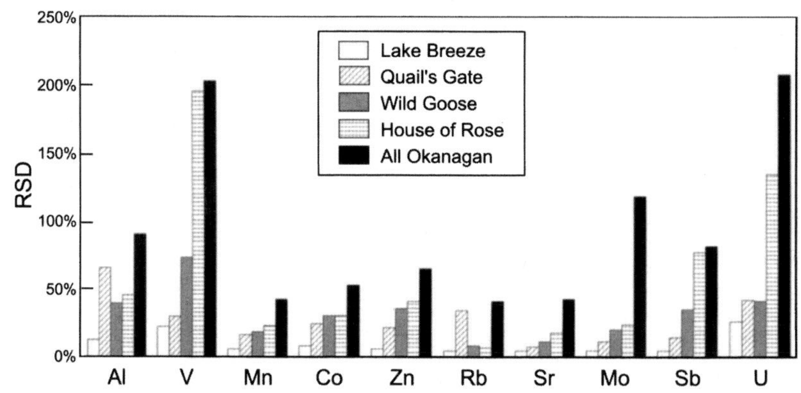

Figure 3 Percent relative standard deviation calculated for elements used to discriminate Okanagan wines by vineyard. Bars represent variance (RSD) in element concentrations for six wines from the same batch of 1998 Lake Breeze Pinot Blancs; Quail's Gate Riesling (1995-1998); Wild Goose Gewürztraminer (1993-1997); House of Rose Verdelet (1992-1996); and all Okanagan wines.

uptake essential nutrients at a constant rate (Kovalevsky, 1995).

Controls on Wine Composition

Although correlations between wine and soil chemistry are weak, it seems likely that geochemical factors influence the composition of Okanagan valley wines, based on trends seen in the cluster analysis results (Fig. 2). Okanagan wines are grouped according to vineyard of origin using the discriminating elements Al, V, Mn, Co, Zn, Rb, Sr, Mo, Sb, and U. For example, wines from the House of Rose, Lake Breeze, Wild Goose, and Quail's Gate vineyards form distinct groups. The two St. Hubertus 1996 Pinot Blancs were made from grapes from the same vineyard, but one was aged in oak for eight months and the other processed in a steel tank. That they plot together suggests that the different processing methods did not substantially affect trace element concentrations. The two Inkameep wines (vineyard 26) from two producers (Summerhill and Jackson-Triggs) do not plot together, but the Jackson-Triggs wine associates with other wines whose grapes originated in southern Okanagan vineyards. There is a tendency for north and northwestern Okanagan wines (Kelowna, Westbank, Peachland, and Summerland; vineyards 2-11 in Fig. 1a) to plot at the top of Figure 2, and southern Okanagan wines (vineyards 19-26) to fall in the lower half of the plot. This result suggests that wine composition might also reflect subtle changes in climate and its impact on soil chemistry.

The composition of irrigation water or variations in precipitation from year to year do not appear to affect wine composition, as wines of several vintage years from the same vineyard group together (Fig. 2) and have a low RSD of concentrations of the 10 fingerprint elements compared to wines from throughout the valley (Fig. 3). The wines from Quail's Gate, House of Rose, and Wild Goose vineyards/wineries represent several vintage years of the same wine, whereas the wines from Lake Breeze are from different processing tanks of the same batch, and show that within-batch variability is low for a single vintage year. Variability (RSD) in these individual vineyard wines is consistently lower than that of all wines sampled from the Okanagan. (Fig. 3). In spite of the fact that wine makers and their methods, supplies, and equipment might change from year to year and lead to variable compositions between vintages, the similarity in the trace element compositions of these wines suggests that vineyard of origin — including soil, bedrock composition, and possibly the long-term effects of climate on soil — affects the composition of wines more than the processing or viticultural methods used.

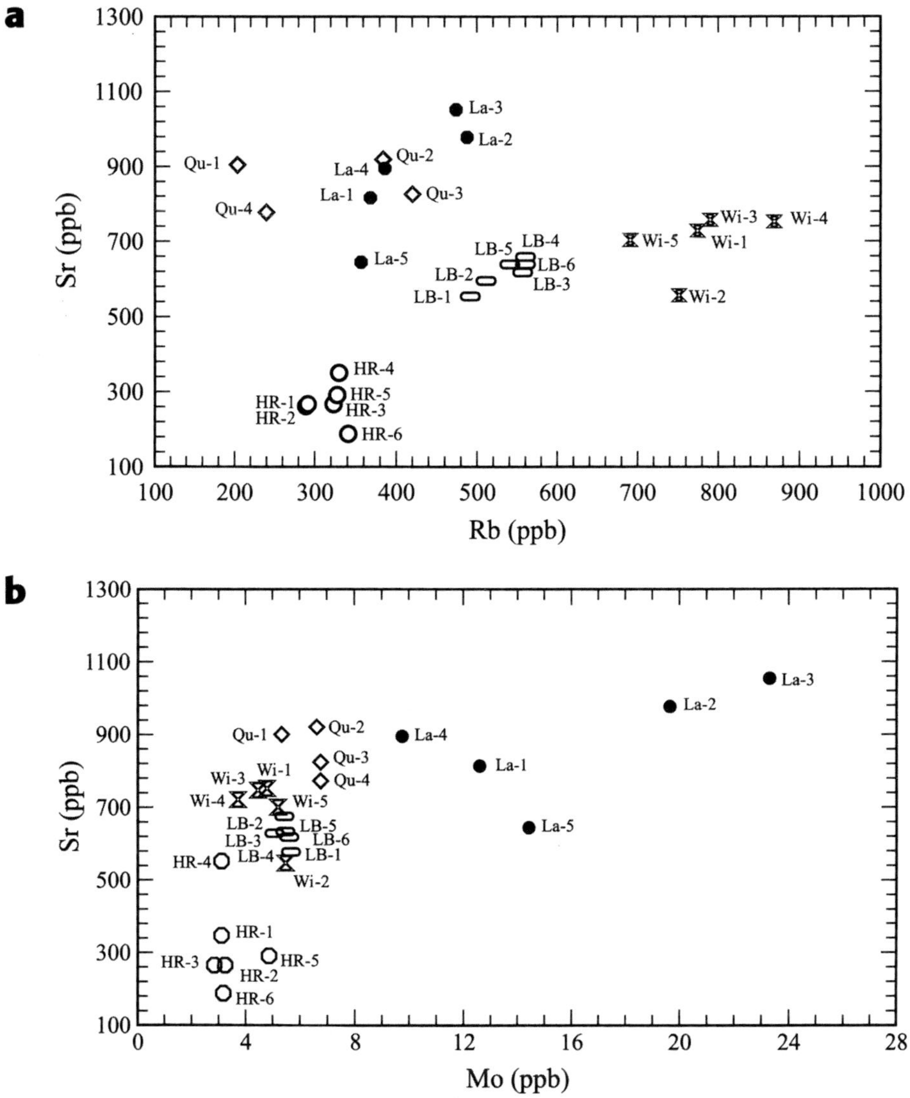

Figure 4(a)(b) Concentrations, in ppb, of Sr *versus* Rb and Sr *versus* Mo, used to discriminate wines from the same vineyard, but different vintage years (House of Rose Verdelet 1992 to 1996 (HR-1 to HR-5) and Okanagan Riesling 1996 (HR-6), Lake Breeze Pinot Blancs 1998: 6 different samples from the same batch (LB-1 to LB-6), Lang Riesling 1995 to 1997 (La-1 to La-5), including a late harvest (La-4) and an icewine, (La-3), Wild Goose Gewürztraminer 1993 to 1997 (Wi-1 to Wi-5) and Quail's Gate Riesling 1995 to 1998 (Qu-1 to Qu-4).

Fingerprints of Individual Okanagan Vineyards

Trace element fingerprinting studies of wines in other regions have used Cd, Cr, Cs, Er, Ga, Mn, and Sr to discriminate English and Spanish wines (Baxter *et al.*, 1997), and Al, B, Ba, Ca, Cr, Cu, Fe, Mg, Mn, P, Pb, Si, Sr, V, and Zn to group wines from six regions of Germany (Danzer *et al.*, 1999). In addition, wines from Galicia, Spain were classified using Li and Rb (Latorre *et al.*, 1994). These studies of other regions demonstrate both the success of wine fingerprinting analysis, and the necessity of using multiple elements.

In this study, cluster analysis using 10 elements (Fig. 2) indicates that there are natural patterns in the data that enable correlation of trace element composition with vineyard of origin. Wines from Quail's Gate, House of Rose, and Wild Goose wineries show variance of select elements in wines from several vintage years to be consistently lower than the relative standard deviation (RSD) for all Okanagan wines (Fig. 3). The RSD of element concentrations in the Lake Breeze samples is low, and indicates that element concentrations are fairly uniform within different processing tanks of the same batch of wine. This low RSD has implications for fingerprinting individual wines from particular vintages.

Vineyards with four or more wine samples (House of Rose, Quail's Gate, Wild Goose, Lake Breeze and Lang) were used in discriminant analysis experiments, which relied on both multivariate statistics and graphical analysis. Elemental Sr/Rb plots differentiated vineyards (Fig. 4a), except for overlap between the Lang and Quail's Gate vineyards. Wines from the Lang vineyard were completely discriminated from the other four vineyards using an Sr/Mo plot (Fig. 4b), however. The high discriminating power of Sr and Rb may reflect element mobility in the environment, defined by ionic charge and ionic radius (Greenough *et al.*, 1996), which determines Sr to be a moderately mobile element, and Rb a more mobile element. The mobility of Mo depends on its oxidation state, but tends to be high in soil.

The ability to distinguish among these wines even when grown over several vintage years implies that geology and vineyard environment have a strong influence on wine composition, and that Sr and Rb are important discriminating elements. Isotopic ratios of Sr have been used as a natural tracer between soil and wine, where wines grown on basaltic, mixed, and granitic soils could be differentiated (Barbaste *et al.*, 2002). As the Sr isotopic ratios between wine and soil are similar, we suggest that there is little addition of Sr during wine processing. This experiment provides further evidence that individual vineyards can be fingerprinted, but many more vineyards (wines) will have to be multiply sampled in order to determine the precision/certainty of the fingerprinting process and the possible degree of overlap in trace element composition among wines from different vineyards.

Multi-element analysis was unsuccessful at grouping wines by grape variety, colour (red or white), or vintage year. Greenough *et al.*, (1997) and Baxter *et al.*, (1997) were able to accurately discriminate wine chemically by colour, but in this study there were only six red wines, an insufficient number for statistical significance.

Discrimination of wines by vineyard geology was examined using the discriminating elements Sr and Rb. Plots of Sr/Rb ratios *versus* vineyard latitude (Fig. 5a,b) for one wine from each of the 26 vineyards sampled, show no strong trends relating wines to latitude, bedrock or surficial deposit. The variability of Sr and Rb concentrations may reflect the heterogeneity of the bedrock, particularly the Okanagan Gneiss (Egn), and/or the complexity of the wine-geology relationship.

SUMMARY

Trace element composition of wines was found to group wines by vineyard, to show some trends with geographic location, and to discriminate wines by vineyard. This preliminary study suggests that multi-element analysis may be a useful addition to current Canadian certification parameters. Although the relationship between geology and wine composition is complex, the success in verifying vineyard of origin, and the low variance in element concentrations in several vintage years of the same wine, indicates a correlation between vineyard environment and wine composition, although there are limitations to deciphering provenance. In order to further examine the relationship between wine and geology, a study of the effects of

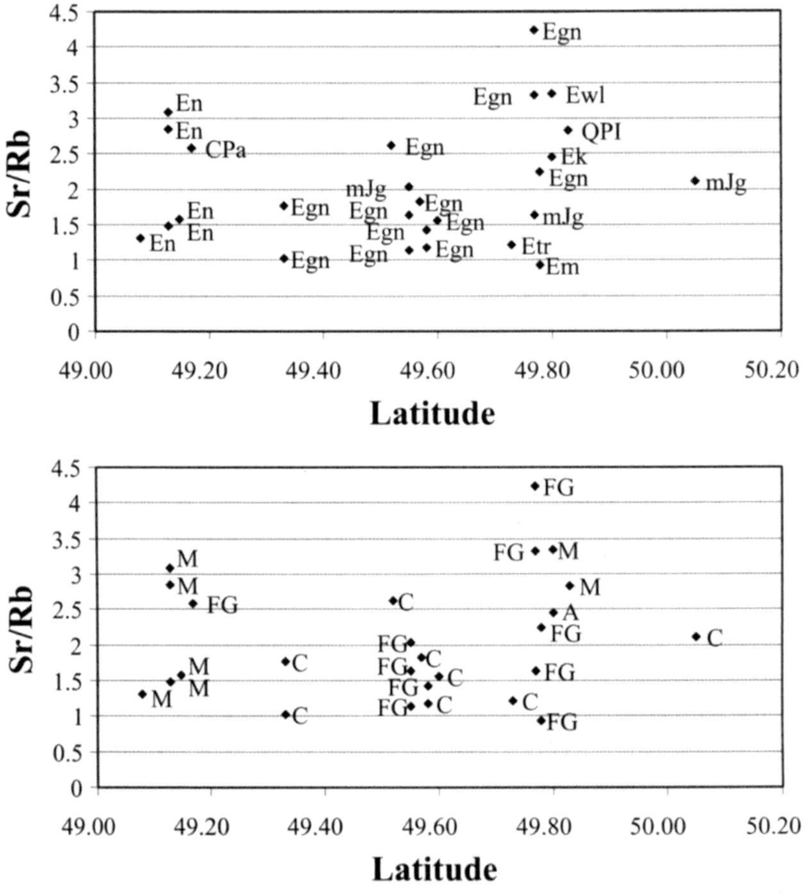

Figure 5 Plots of Sr/Rb concentrations in wine *versus* latitude, showing a bedrock geology (where bedrock abbreviations correspond to those of Figures 1a, b surficial geology (where M = morainal, C = colluvial, FG = fluvioglacial, and A = alluvial).

different wine production methods on wine composition is suggested.

ACKNOWLEDGMENTS

The authors thank the winemakers and viticulturists of the Okanagan for their time and assistance. The Natural Sciences and Engineering Research Council of Canada (NSERC) provided research grants to Professors John Greenough and Henry Longerich. Graduate Studies (Memorial University of Newfoundland) provided some of Vivien Taylor's stipend. Hanni Muggeridge prepared the map, and Lakmali Hewa, Mike Tubrett, and Pam King provided advice and assistance in the laboratory. We are grateful for thorough critical reading of this paper by the late Professor Simon Haynes, former editor of the *Geoscience Canada* Geology and Wine Series, to an anonymous reviewer for useful comments, and also to current Geology and Wine series editor Roger Macqueen for helpful organizational suggestions. David Sargent of GSC Calgary modified some of the digital figures for publication. Any errors remaining are the responsibility of the authors.

REFERENCES

Aubert, H. and Pinta, M., 1977, Trace Elements in Soils: Elsevier Scientific Publishing Company, New York, 395 p.

Barbaste, M., Robinson K, Guilfoyle S, Medina B. and Lobinski R, 2002, Precise determination of the strontium isotope ratios in wine by inductively coupled plasma sector field multicollector mass spectrometry (ICP-SF-MC-MS): Journal of Analytical Atomic Spectrometry v. 17, p. 135-137.

Bardoux, M. and Irving, E., 1988, Paleomagnetism of Eocene rocks of the Kelowna and Castlegar areas, British Columbia: studies in determining paleohorizontal: Canadian Journal of Earth Science, v. 26. p. 829-844.

Baxter, M.J., Crews, H.M., Dennis, J.M., Goodall, I. and Anderson, D. 1997, The determination of the authenticity of wine from its trace element composition: Food Chemistry, v. 60, p. 443-450.

CanSIS, 2001, National Soil Database (NSDB). Soil Landscapes of Canada (SLCs) Version 2.0. Agriculture and Agri-Food Canada. (http://sis.agr.gc.ca/cansis/nsdb/slc/v2.0/component/pmdep.html)

Church, B.N., 1973, Geology of the White Lake Basin: British Columbia Department of Mines and Petroleum Resources, Bulletin 61, 120 p.

Church, B.N., 1980, Geology of the Kelowna Tertiary Outlier (West Half): Province of British Columbia, Ministry of Energy, Mines and Petroleum Resources, Preliminary Map 39.

Church, B.N., 1981, Geology of the Kelowna Tertiary Outlier (East Half): Province of British Columbia, Ministry of Energy, Mines and Petroleum Resources, Preliminary Map 45.

Church, B.N., 1982, Volcanology and structure of Tertiary Outliers in South-Central British Columbia. Trip 5: Province of British Columbia, Ministry of Energy, Mines and Petroleum Resources, p. 5-1 to 5-46.

Danzer, K., De La Calle Garcia, D., Thiel, G. and Reichenbacher, M., 1999, Classification of wine samples according to origin and grape varieties on the basis of inorganic and organic trace analysis: American Laboratory, v. 31, p. 26-33.

Dunn, C.E., 1995, Introduction to biogeochemical prospecting *in* Brooks, R.R., Dunn, C.E. and Hall, G.E.M., eds., Biological Systems in Mineral Exploration and Processing: Ellis Horwood, Toronto, p. 233-266.

Eyles, N., Mullins, H.T. and Hine, A.C., 1990, Thick and fast: sedimentation in a pleistocene fiord lake of British Columbia, Canada: Geology, v. 18, p. 1153-1157.

Greenough, J.D., Longerich, H.P. and Jackson, S.E., 1996, Trace element concentrations in wines by ICP-MS: evidence for the role of solubility in determining uptake by plants: Canadian Journal of Applied Spectroscopy, v. 41, p. 76-80.

Greenough, J.D., Longerich, H.P. and Jackson, S.E., 1997, Element fingerprinting of Okanagan Valley wines using ICP-MS: relationships between wine composition, vineyard and wine colour: Australian Journal of Grape and Wine Research, v. 3, p. 75-83.

Hair, J.F., Anderson, R.E. and Tatham, R.L., 1987, Multivariate Data Analysis, 2nd Edition: Macmillan Publishing Company, New York, 449 p.

Jackson, R.S., 1994, Wine Science: Principles and Applications: Academic Press, Toronto, p. 227-332.

Jakubowski, N., Brandt, R., Stuewer, D., Eschnauer, H.R. and Gortges, S., 1999, Analysis of wines by ICP-MS: Is the pattern of the rare earth elements a reliable fingerprint of provenance?: Fresenius Journal of Analytical Chemistry, v. 364. p. 424-428.

Kabata-Pendias, A. and Pendias H., 1984, Trace Elements in Soils and Plants: CRC Press, Inc., Boca Raton, 315 p.

Kovalevsky, A.L., 1995, The biogeochemical parameters of mineral prospecting *in* Brooks, R.R., Dunn, C.E. and Hall, G.E.M., eds., Biological Systems in Mineral Exploration and Processing: Ellis Horwood, Toronto, p. 267-282.

Krauskopf, K.B. and Bird, D.K., 1995, Introduction to Geochemistry, 3rd Edition, McGraw-Hill, Inc., Toronto, p. 534-557.

Latorre, M.J., Garcia-Jares, C., Medina, B. and Herrero, C., 1994, Pattern Recognition Analysis Applied to Classification of Wines from Galicia (Northwest Spain) with Certified Brand of Origin: Journal of Agriculture and Food Chemistry, v. 42, p. 1451-1455.

Longerich, H.P., 1995, Analysis of Pressed Pellets of Geological Samples Using Wavelength- Dispersive X-Ray Fluorescence Spectrometry: X-Ray Spectrometry, v. 24, p. 123-136.

Monger, J.W.H., Wheeler, J.O., TIpper, H.W., Gabrielse, H., Harms, T. Struik, L.C., Campbell, R.B., Dodds, C.J., Gehrels, G.E. and O'Brien, J. 1991, Part B, Cordilleran Terranes *in* Gabrielse, H. and Yorath, C.J., eds., Upper Devonian to Middle Jurassic assemblages, Chapter 8 of Geology of the Cordilleran Orogen in Canada, Geology of Canada: Geological Survey of Canada, n. 4, p. 281-327.

Muranyi, Z. and Papp, L., 1997, ICP-AES metal content analysis of wines made with different technologies: ACH Models in Chemistry, v. 134, p. 529-537.

Nasmith, H., 1981, Late Glacial History and Surficial Deposits of the Okanagan Valley, British Columbia: British Columbia Ministry of Energy, Mines and Petroleum Research, Bulletin 46, 46 p.

Roed, M.A. and the Kelowna Geology Committee, 1995, Geology of the Kelowna Area and Origin of the Okanagan Valley British Columbia: Kelowna Geology Committee, Kelowna, BC, 183 p.

Schreiner, J., 1994, The wineries of British Columbia: Orca Book Publishers, Victoria, BC, 218 p.

Schreiner, J., 1998, Chardonnay and Friends: Orca Book Publishers, Victoria, BC, 218 p.

StatSoft, Inc., 1999, Electronic Statistics Textbook: Tulsa, OK.

Tempelman-Kluit, D.J., 1989, Geology, Penticton, British Columbia: Geological Survey of Canada, Map 1736A.

Valentine, K.G.W. and Lavkulich, L.M., 1999, The Soil Orders of British Columbia *in*, The Soil Landscapes of British Columbia: Ministry of Environment, Lands, and Parks, Government of British Columbia, 22 p.

Vanderburgh, S. and Roberts, M.C., 1996, Depositional systems and seismic stratigraphy of a Quaternary basin: north Okanagan Valley, British Columbia: Canadian Journal of Earth Science, v. 33, p. 917-927.

Vine, R.P., 1981, Commercial winemaking, processing and controls: Avi Publishing Company Inc., Westport, Connecticut, p. 102.

Accepted as revised 31 July 2002

SERIES

Geology and Wine Retrospect and Prospect

R.W. Macqueen
Geological Survey of Canada, Calgary
3303 33 Street N.W.
Calgary, AB T2L 2A7
rmacqueen@nrcan.gc.ca

Many readers will know of the untimely death of *Geology and Wine* series founder and editor Simon Haynes of cancer on April 6, 2002, and thus the need for a new series editor. At Simon's request, I am continuing the series, unfortunately without anything like the knowledge that Simon had of this subject, but with a willingness and enthusiasm to learn. A change of editor is a good time to look at where we've come from in the series and where we hope to go in future.

Retrospect: Why this series - what's geology *really* got to do with wine, other than the fact that many earth scientists enjoy wine? Simon pointed out in the first paper of the series (Haynes, 1999) that there are five main groups of factors that affect wine grapes during growth: meteorological, physiographic, pedological, geological, and viticultural. Simon also reviewed the French concept of *terroir*. In terroir, the French recognize the importance of *all* of these factors, based in part on 2000 years of experience of wine grape growing. Partly because "terroir" is commonly translated as "soil" in English, there has been controversy over what's important in wine grape growing beyond climate, soil, and the skill of viticulturalists. This brings us back to the question: what has geology – including physiography and bedrock geology – got to do with wine?

The origin of "terroir" also was reviewed in paper 4 of our *Geology and Wine* Series by James Wilson (2001), author of the classic English language book *Terroir: the Role of Geology, Climate and Culture in the Making of French Wines* (Wilson, 1998). Wilson noted that, following publication in the 1940s of the concept of "heat summation" – the sum of the hours when the temperature is above 50 degrees Fahrenheit during the growing season – this was regarded as *the* critical factor in wine quality. One result of the exaggerated importance of this one factor was that most new-world viticulturalists, particularly in California, paid little to no attention to the importance of physiographic, pedological (soil) or geological (bedrock) factors in wine grape growing. Wilson also noted (*ibid.*, 2001) that this view is changing with a younger generation of wine growers, and the exchange of student viticulturists between France and California. He also noted that the whole concept of terroir encourages greater recognition of the importance of geology and geological processes, particularly as regards soil, and bedrock, drainage and physiography including siting, in the growth of wine grapes.

Excellent examples of the concept and importance of terroir appeared in *Geology and Wine* Series papers 2 (Niagara Peninsula, Ontario; Haynes, 2000), 3 (Walla Valley Appellation, Washington State, U.S.A.; Meinert and Busacca, 2000), and 6 (Red Mountain Appellation, Washington State, U.S.A.; Meinert and Busacca, 2002). What these three papers do is outline the nature and significance of bedrock geology, overlying soil, and siting (physiographic/ geologic) factors in the growing of wine grapes in each of these settings. What these papers *don't* do is to claim that geological factors - soil, bedrock, physiography/siting - outweigh the other factors that lead to the production of quality wines, a claim that is sometimes raised as a critique of the concept of terroir. Viticulture, a key part of wine grape growing, involves factors from siting to choice of grape varietal, and management practices including trellising, irrigation, pruning, pest control, etc. Viticultural management practices can strongly affect plant vigour and fruit quality, key factors in successful wine grape production. Indeed the science (and literature) of viticulture is a huge field, not unlike the many fields of geology.

In terms of the relationship between terroir and viticultural practices, Larry Meinert says: "......(W)e have never, ever attributed good quality wine to site alone, although there are cases of bad wine quality for which the site is largely to blame. That being said.... (it) is absolutely right that different management practices can result in better or worse quality wine from a particular site. But the really important question is, what happens when the best possible management practices are applied as appropriate to different sites and the resulting wines differ in quality? This is what we are trying to document in our studies and it is the essence, I think, of terroir. Some sites really are better or worse than others. Or phrasing it another way, the best wine that can be produced from some sites really is better or worse than the best wine that can be coaxed from other sites. In both the Walla Walla and

Red Mountain papers we have cited examples of specific sites that have produced great wines and specific sites that "struggle". For the Red Mountain paper we even identified two different end members of "struggle", a vineyard on the floodplain whose rich soils and high water table lead to uncontrolled vigour and a vineyard on basalt outcrop that doesn't retain water. In the first case, there is too much water even with minimal or zero irrigation and in the second case irrigation water drains right through the root zone and, short of hydroponic growing conditions, the plants are stressed (have low vigour). Although best management practice can try to minimize the problems of a site, I don't think any management practice can result in great wine from a terrible site. Similarly, no matter how good the site, poor management or winemaking practices can result in bad wine.

"But our terroir studies are not aimed at understanding terrible sites or bad wines. Rather we have focussed on areas known to have produced great wines and then tried to document the terroir of those sites. In the future we will attempt to make it more clear in our manuscripts that viticultural and enological practice are very important, without losing sight of our main topic which is documenting the effect of terroir." (e-mail to R.W. Macqueen, October 9, 2002).

Paper 5 in the series dealt with another aspect of terroir, the trace element composition of wines and their geological environs in the Okanagan Valley, B.C. (Taylor *et al.*, 2002). There is scope for more studies of this kind, in the Okanagan and elsewhere.

Prospect: Where do we go from here in the *Geology and Wine* series? A forthcoming paper (currently in review) in the series will be on the regional geology and wine production of the Western Cape Province of South Africa, by Christopher Bargmann, one of the 48 current Cape Wine Masters of South Africa. This is an area with 300 years of wine production, and one with spectacular geology and scenery as well. We hope to return to the Okanagan Valley by publishing a paper on a Geographic Information System (GIS) approach to the characterization of Okanagan vineyards by Patricia Bowen (viticulturalist) and Scott Smith (geologist) of the Pacific Agricultural Research Centre (PARC) at Summerland, B.C. We are also hoping to attract a paper on the terroir of a California setting in the near future.

Meanwhile Larry Meinert and Alan Busacca of Washington State University are continuing their terroir studies, in part by really focussing down: they are, under controlled conditions, examining changes in the body and flavour of wines which may be specifically related to such geological factors as differing soil profiles. This should be an excellent paper in the series! Larry is also organizing a terroir symposium for the Geological Society of America meeting to be held in Seattle, Washington, November 2–5, 2003: we hope to obtain some of these papers for the series. It is planned to dedicate this symposium to the memory of Simon Haynes, whose vision, enthusiasm, considerable knowledge and dedication got this series underway.

We continue to be on the lookout for good papers on geology and wine. This is an international series, so we welcome papers from any region. Now that Canada is becoming established on the international wine scene, particularly in icewines, it would be great to see more papers from Canadian settings.

Raise your next glass of wine to Simon Haynes, our series founder: well done, Simon!

REFERENCES

Haynes, S.J., 1999, Geology and Wine 1. Concept of terroir and the role of geology: Geoscience Canada, v. 26, p. 190-194.

Haynes, S.J., 2000, Geology and Wine 2. A geological foundation for terroirs and potential sub-appellations of Niagara Peninsula wines, Ontario, Canada: Geoscience Canada, v. 27, p. 67-87.

Meinert, L.D., and Busacca, A.J., 2000, Geology and Wine 3. Terroirs of the Walla Walla Valley appellation, southeastern Washington State, U.S.A: Geoscience Canada, v. 27, p. 149-171.

Meinert, L.D., and Busacca, A.J., 2002, Geology and Wine 6. Terroir of the Red Mountain Appellation, Central Washington State, U.S.A.: Geoscience Canada, v. 29, p. 149-168.

Taylor, V.F., Longerich, H.P., and Greenough, J.D., 2002, Geology and Wine 5. Provenance of Okanagan Valley Wines, British Columbia, Using Trace Elements: Promise and Limitations: Geoscience Canada, v. 29, p. 110-120.

Wilson, J.E., 1998, Terroir, the Role of Geology, Climate and Culture in the Making of French Wines: Mitchell Beazley, London, U.K., 336 p.

Wilson, J.E., 2001, Geology and Wine 4. The Origin and Odyssey of Terroir: Geoscience Canada, v. 28, p. 139-141.

Accepted as revised 6 November 2002

SERIES

Geology and Wine 6. Terroir of the Red Mountain Appellation, Central Washington State, U.S.A.

Lawrence D. Meinert
Department of Geology
Washington State University
Pullman, WA, 99164-2812
meinert@wsu.edu

Alan J. Busacca
Department of Crop and Soil Sciences
Washington State University
Pullman, WA, 99164-6420
busacca@wsu.edu

SUMMARY

Red Mountain is the newest of five appellations in Washington State and like the majority of Washington vineyards its terroir is influenced by 1) the rain shadow effect and volcanic tephra of the Cascade Mountain Range, 2) soils derived from Quaternary glacial sediments and wind-blown loess overlying Miocene basalt, and 3) a warm, dry climate with abundant sunshine and cool nights due to high latitude (N 46°) and topography. Variations of rooting depth, textures, calcium carbonate content, and other properties of vineyard soils are directly influenced by Quaternary glacial flood deposits in the back-eddy of Red Mountain and by the variable nature of the loess and dune cover on the flood deposits. This is the first demonstration of the effect of paleohydrology on vineyard site characteristics. In the past two years, wines made from Red Mountain grapes have received nine scores of $\geq$ 94 out of 100 in independent blind tastings and merlot wine made from Red Mountain grapes has been ranked as the best in the United States for each of the past two years by two different national wine magazines.

Red Mountain is one of the warmest and driest viticultural sites in Washington State, having heat summation (growing degree days) and total sunshine similar to Napa Valley, California which is 1000 km farther south. Other wine regions that have been influenced by the worldwide glacial processes that were so important in the development of Washington vineyards include the gravel mounds that underlie most of the first growth vineyards of Graves-Médoc, Bordeaux, France and some of the outwash gravel plains of New Zealand.

RÉSUMÉ

L'appellation Red Mountain est la plus récente des cinq appellations de l'État de Washington et, comme c'est le cas pour la plupart des vignobles de l'État de Washington, son terroir dépend 1) de l'effet parapluie de la chaîne des monts Cascade et de ses cendres volcaniques, 2) des sols dérivés des sédiments glaciaires quaternaires et des loess éoliens recouvrant le basalte miocène et, 3) d'un climat chaud et sec avec un fort ensoleillement, avec des nuits fraîches dues à la latitude (N 46°) et à la topographie. Les variations de la profondeur d'enracinement, de texture, de contenu en carbonate de calcium et d'autres propriétés pédologiques des vignobles varient selon le patron de dépôt des alluvions glaciaires quaternaires par écoulement tourbillonnaire de Red Mountain et de la variation de composition du loess et de la couverture des dunes des dépôts alluvionnaires. Cela constitue la première démonstration des effets paléohydrologiques sur les propriétés des sites vinicoles. Au cours des deux dernières années, les vins des raisins de Red Mountain ont reçu neuf cotes de plus de 94% lors de dégustation anonyme, et les vins merlots provenant de raisins de Red Mountain ont été jugés les meilleurs aux USA pour chacune des deux dernières années, par deux revues spécialisées d'envergure nationale.

La région de Red Mountain constitue l'un des sites viticoles les plus chauds et les plus secs de l'État de Washington, possédant un bilan thermique (degré-jour de croissance) et un ensoleillement comparable à ceux de la vallée de Napa en Californie, laquelle est située à 1 000 km plus au sud. Parmi d'autres régions ayant subi l'influence de processus glaciairemondiaux et qui se sont avérés si déterminant dans le développement des vignobles de l'État de Washington, on retrouve celle des collines graveleuses constituant le lieux de croissance de la plupart des vignobles de première génération de Graves-Médoc de la région de Bordeaux en France, et d'autres sur des plaines d'alluvions en Nouvelle-Zélande.

INTRODUCTION

As described in other articles in this Geoscience Canada series (*e.g.*, Wilson, 2001), *terroir* involves the complex interplay of climate, soil, geology, and culture that influences the character and quality of wine. These factors are in addition to, or perhaps underlie, the substantial contribution of good viticultural practice and expert winemaking. One common illustration of the importance of terroir is the occurrence of adjacent or nearby vineyards that produce strikingly different wines, even where many of the measurable aspects of climate, viticulture, and winemaking technique are very similar. It is also common, although usually incorrect, to point to a single factor as the explanation: "It's the soil." "It's the water." "It's the limestone." Terroir is the integration of individual factors that contribute to wine quality, and to make matters even more complicated there is the variable of time. What may be good terroir in one year may be less so in another. For example, in years that are relatively warm and dry, vineyard "X" with a particular slope, elevation, sun angle, and soil type may produce better wine than vineyard "Y", whereas the reverse may be true in years that are cooler and wetter. Similarly, no matter how good the site, poor vineyard management or winemaking practices can result in bad wine.

Although the term originated in France, terroir increasingly is being used in other parts of the world to explore differences at the scale of appellations to individual vineyards to within-vineyard domains (Halliday, 1993, 1999; Wilson, 1998; Haynes, 1999, 2000). The goal of the present paper is to document the terroir of Washington's newest appellation, Red Mountain (Fig. 1) and to contrast it with other Washington appellations such as Walla Walla (Meinert and Busacca, 2000). At least one aspect of the terroir of the Red Mountain appellation is similar to that previously documented for Walla Walla, 75 km to the east: the strong control on soil materials and landforms exerted by cataclysmic glacial outburst floods. However, there are other factors that vary to different degrees, such as topographic patterns that control air drainage, soil texture, depth to bedrock, growing degree days, solar radiation, rainfall, and frost hazards.

Washington State has some of the world's largest and most spectacular flood basalts and glacial outburst flood deposits as well as widespread dune and loess fields. All of these contribute to the terroirs of Washington State wines. In addition, recent volcanic activity such as the well-known 1980 eruption of Mt. St. Helens continues to shape the oenological and geological landscape. Many Washington vineyard soils have a component of ash erupted from Mt. St. Helens and other Cascade volcanoes such as the much larger Mt. Mazama eruption (6,850 ^{14}C yr. BP), which formed present-day Crater Lake in Oregon (Busacca *et al.*, 2001).

Washington State is second only to California in terms of the volume of wine produced in the United States (www.nass.usda.gov/wa/grape02.pdf). Most Washington State vineyards are located between latitudes 45° and 48° N, well to the north of the more widely known California vineyards but parallel to the great French wine regions of Burgundy and Bordeaux. This northerly latitude provides up to two hours more summer sunlight than

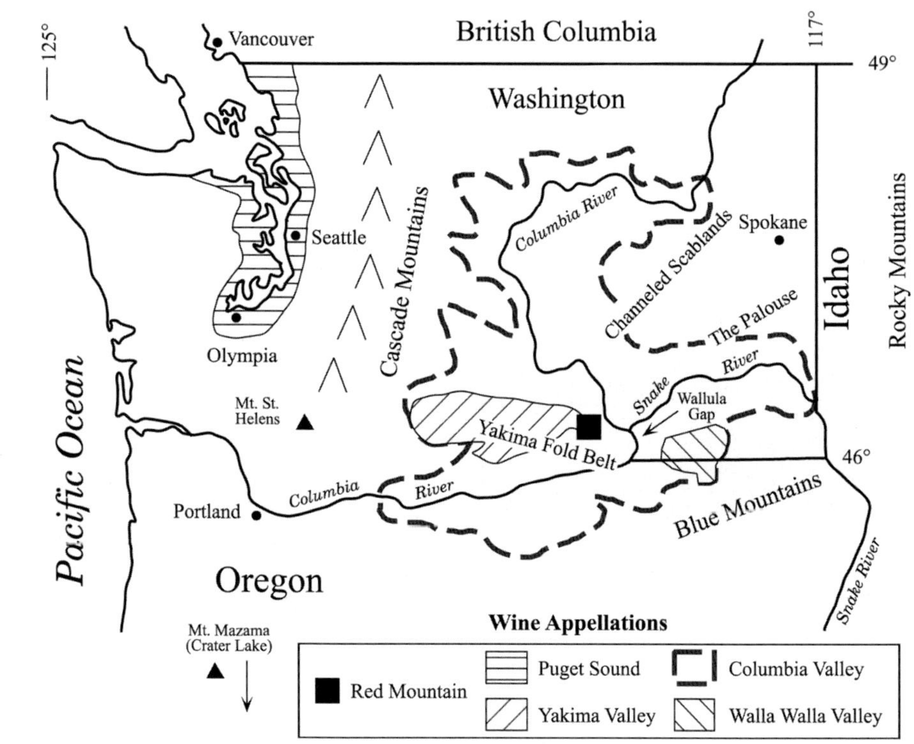

Figure 1 Location map of the Pacific Northwest showing wine appellations and major geological features described in the text.

occurs in California wine regions, somewhat offsetting the generally shorter growing season (Gladstones, 1992). Arid conditions created by the large rain shadow of the Cascade Mountains allow vineyardists a high degree of control of soil moisture. Almost all commercial vineyards lie east of the Cascade Mountain Range (Fig. 1).

Although there is considerable local variability, most Washington vineyards are sited on soils formed from Quaternary sediments that overlie Miocene basaltic rocks of the Columbia River flood basalt province. Many of the Quaternary sediments were deposited during cataclysmic glacial outburst floods that formed the spectacular geomorphic features of the Channeled Scabland (Fig. 2). In fact, comparison of Figures 1 and 2 demonstrates that more than 90% of Washington vineyards are located in areas affected by the glacial outburst floods. This element of terroir, combined with the rain shadow effect of the Cascade Mountains, distinguishes Washington vineyards from other grape growing areas of the world and suggests that Washington wines may develop flavour and quality characteristics that are equally distinct.

OVERVIEW OF WASHINGTON STATE WINE GRAPE PRODUCTION

Washington State has five wine appellations called American Viticultural Areas (AVAs) by the Bureau of Alcohol, Tobacco, and Firearms (BATF), the chief regulatory agency of the wine industry in the United States. The current Washington State appellations (AVAs) are Columbia Valley, Puget Sound, Red Mountain, Walla Walla, and Yakima Valley (Fig. 1). Sub-appellations, that may someday become AVAs, include Alder Ridge, Canoe Ridge, Cold Creek, Columbia River Gorge, Horse Heaven Hills, Wahluke Slope, and Zephyr Ridge (Peterson-Nedry, 2000). Other areas, such as the north bank of the Columbia River near the small town of Paterson, also are becoming known for growth of premium grapes owing to their south-facing slopes and the mitigation of temperature extremes by proximity to Lake Wallula on the Columbia River.

As with most other wine growing regions, Washington AVAs can be nested such that the Columbia Valley appellation, which produces more than 90% of the state's wine grapes, includes the Yakima Valley, Walla Walla Valley, and Red

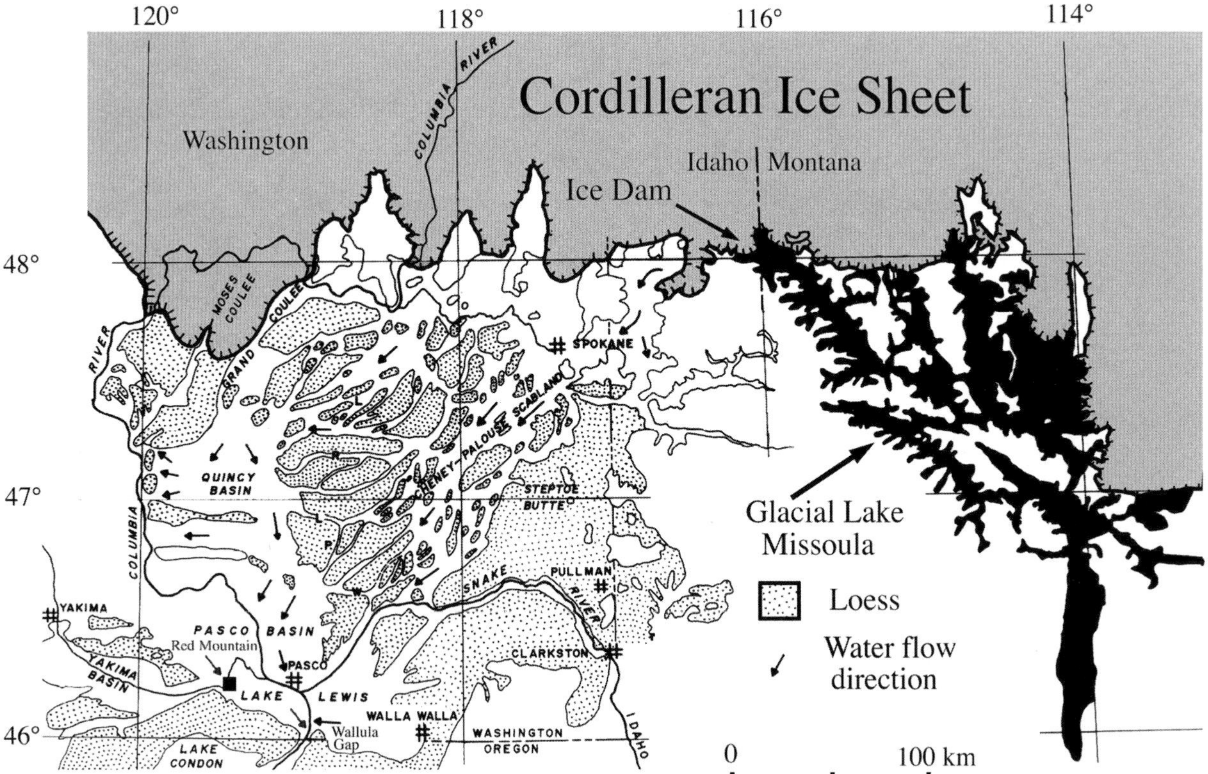

Figure 2 The Channeled Scabland of the Pacific Northwest showing the Cordilleran ice sheet in its approximate position at 15,000 years before present, glacial outburst flood channels, and the distribution of loess. Modified from Baker and Nummedal (1978).

Mountain appellations. The area available for future plantings is very large. In the 10.7 million-acre Columbia Valley appellation only about 16,000 acres are planted to wine grapes. Even the smallest appellation, Red Mountain, has room for expansion with approximately 710 acres planted to vines out of the 4040 acres of the AVA. In many cases the availability of water for irrigation is a larger limitation than the suitability of land for growing high-quality grapes.

Only about 18% of Washington's wine grapes are from vineyards more than 20 years old and from these older vineyards, white grapes (73%) predominate over red grapes (27%) (www.nass.usda.gov/wa/wine02.pdf). For example, prior to 1982 Riesling was the most widely planted white wine grape. In 1982, 54% of the current (2002) acreage was planted with Riesling grapes. In contrast, Cabernet Sauvignon, Merlot, and Syrah were the three most widely planted red grapes in 2002 and had only 12%, 5%, and 0%, respectively, of their current acreage planted prior to 1982.

In the summer of 2002, there were 212 wineries in Washington State. Total wine grape production in 2001 was 100,000 tons from 24,000 acres of bearing vineyards (Table 1). Wine grape production will continue to increase since there are an additional 6,000 acres of wine grapes planted that were not yet bearing fruit in 2001. Most grape vines start producing commercial yields in the third year. Of the wine produced in Washington State in 2001, there was an equal split between white and red wine, down from a 62% white wine majority in 1998. For example, the production of Semillon and Chenin Blanc in this three-year time period decreased 35% whereas the production of Cabernet Sauvignon, Merlot, and Syrah increased 200%. This trend towards a predominance of red wine production in Washington State likely will continue in the future because of the increased plantings of red varietals and the higher prices realized from red grapes in general (Table 1).

REGIONAL GEOLOGIC HISTORY

Red Mountain is located in the geographic center of the Columbia Plateau, which is bordered on the north and east by the Rocky Mountains, on the south by the Basin and Range Province, and on the west by the Cascade Mountains (Fig. 1). Red Mountain is part of a northwest-southeast-trending series of anticlinal ridges called the Olympic-Wallowa Lineament (OWL) that can be traced on topographic and geological maps for hundreds of kilometres (Fig. 3). The ridges result from the north-south-oriented compressional stress regime of south-central Washington State that has existed from the Miocene to the present day. Some anticlines in the Yakima fold belt have developed as much as 100 m of structural relief in the past 10 m.y. (Reidel *et al.*, 1992), and the Red Mountain area experienced an intensity VII (approximate Richter magnitude 6) earthquake on July 15, 1936 (Brown, 1937), most likely caused by movement on the Wallula fault zone (Mann and Meyer, 1993), which is subparallel to the Yakima Fold Belt (Fig. 1).

Erosion of some of the anticlinal ridgetops exposes bedrock of the Columbia River Basalt Group, which covers an area of about 165,000 km^2. The Columbia River Basalt Group was erupted mostly between

Table 1 Washington vinifera wine grape production 1997–2001 (data from http://www.nass.usda.gov/wa/wine02.pdf)

Variety	Tons Utilized					Acres planted	Acres bearing	Yield tons/acres	Average Price (US$/ton)				
	1997	1998	1999	2000	2001	2001	2001	2001	1997	1998	1999	2000	2001
Chardonnay	17,700	20,500	22,900	27,800	29,200	6,640	6,290	4.6	$1,055	$971	$882	$818	$788
White Riesling	10,100	11,500	9,800	10,100	10,600	2,200	1,940	5.5	$489	$533	$547	$590	$603
Sauvignon Blanc	3,300	3,900	3,400	3,400	3,300	710	680	4.9	$729	$750	$746	$728	$719
Gewurztraminer	1,700	1,600	1,300	1,600	2,200	670	540	4.1	$552	$629	$706	$684	$662
Semillon	2,800	3,100	2,600	2,700	2,100	550	540	3.9	$661	$625	$598	$571	$576
Chenin Blanc	1,800	2,300	1,700	1,500	1,400	450	450	3.1	$499	$488	$473	$494	$439
Other (*)	600	600	800	900	1,200	780	460	2.6	$789	$952	$777	$866	$834
Total White	38,000	43,500	42,500	48,000	50,000	12,000	10,900	4.6	$794	$772	$753	$736	$721
Merlot	12,800	14,100	14,300	21,400	23,400	5,980	5,630	4.2	$1,360	$1,197	$1,149	$1,060	$1,034
Cabernet Sauvign	7,800	8,300	8,400	13,000	16,700	6,050	5,130	3.3	$1,195	$1,204	$1,236	$1,144	$1,122
Syrah	#	#	800	2,200	4,400	2,100	1,610	2.7	#	#	$1,398	$1,343	$1,221
Cabernet Franc	1,100	1,700	2,000	3,300	3,300	750	720	4.6	$1,120	$1,102	$1,066	$994	$1,012
Pinot Noir	1,000	1,100	1,000	1,000	900	290	260	3.5	$653	$661	$647	$642	$689
Lemberger	500	500	500	500	500	230	180	2.8	$757	$760	$785	$790	$748
Other (+)	800	800	500	600	800	600	370	2.2	$1,400	$1,386	$1,152	$1,232	$1,286
Total Red	24,000	26,500	27,500	42,000	50,000	16,000	13,900	3.6	$1,255	$1,168	$1,152	$1,085	$1,073
Washington total	62,000	70,000	70,000	90,000	100,000	28,000	24,800	4.0	$972	$922	$910	$899	$897

* Includes muscat canelli, pinot gris, and viognier

+ Includes gamay beaujolais, grenache, nebbiolo, sangiovese, and zinfandel

Included in "Other" Red Varieties for 1997-1998

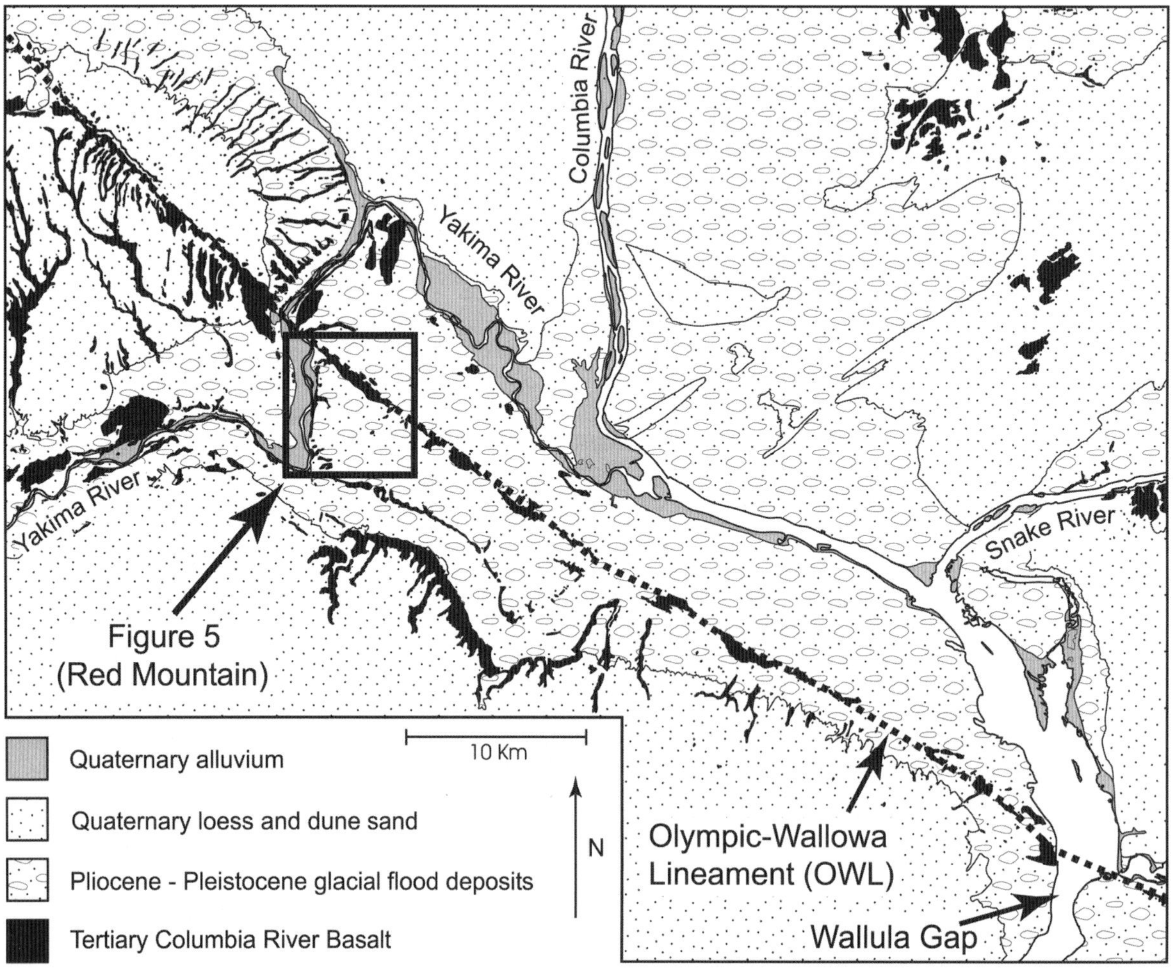

Figure 3 Regional geology of the Pasco Basin showing alignment of ridges along the Olympic-Wallowa Lineament (modified from Reidel and Fecht, 1994).

17 and 11 Ma (early Miocene) from north-south fissures roughly paralleling the present-day Washington-Idaho border. The Columbia River Basalt Group has individual flows with estimated eruptive volumes of at least 3,000 km^3, making them the largest documented lava flows on earth (Baksi, 1989; Landon and Long, 1989; Tolan *et al.*, 1989). This dwarfs the erupted volumes of typical Cascade volcanoes: even the explosive eruption of Mt. St. Helens in 1980 yielded only about 1 km^3 of volcanic material (Pringle, 1993). The basalt bedrock bedrock is overlain by unconsolidated sediments deposited by glacial outburst floods and eolian processes described in some detail in Meinert and Busacca (2000).

To briefly summarize: A lobe of the Cordilleran Ice Sheet blocked the Clark Fork River near the Canadian border in northern Idaho most recently about 18,000 Ka and created glacial Lake Missoula (Fig. 2), which covered 7800 km^2 of western Montana (Pardee, 1910). At the ice dam the water was approximately 600 m deep (Weis and Newman, 1989). The ice dam failed repeatedly, releasing the largest floods documented on earth (Baker and Nummedal, 1978). These floods overwhelmed the Columbia River drainage system and sent up to 2500 km^3 of water across the Columbia Plateau with each outburst (called jökulhlaups in Iceland where similar events occur). The floods eroded a spectacular complex of anastomosing channels, locally called 'coulees', into southwest-dipping basalt surfaces. They also eroded huge cataracts in the basalt now seen as dry falls and "loess islands" that are erosional remnants of an early thick loess cover on the plateau. The floods deposited immense gravel bars and ice-rafted erratic boulders at high elevations. Collectively these features make up the Channeled Scabland as detailed in the early work by Bretz (1923, 1925, 1928a,b, c, 1932).

In south-central Washington State, the many paths of the onrushing floods converged on the Pasco Basin where floodwaters were slowed by the hydrological constriction of Wallula Gap before draining out through the Columbia River Gorge to the Pacific Ocean (Figs. 1, 2). This constriction caused backflooding of local river valleys and basins that resulted in deposition of relatively fine-grained slackwater sediments characterized by rhythmically graded bedding; these graded rhythmites locally are called Touchet beds, and multiple sets, indicative of multiple floods, have been recognized (Flint, 1938; Waitt, 1980, 1985).

In the last stages of the Pleistocene (roughly from 15,000 to 10,000 yr. ago) and continuing through the Holocene (the last 10,000 yr., a time when climates worldwide were essentially similar to those of today), prevailing southwesterly winds eroded slackwater and other glacial sediments and redeposited them into the present thick blanket of loess that covers much of the Columbia Plateau (45°30' to 48°N and 116°30' to 120°W). These windblown soils form the backbone of agriculture in all of eastern Washington (Boling *et al.*, 1998). The loess is thickest in a 10,000 km^2 area northeast of Red Mountain called the Palouse (Busacca, 1991) (Fig. 1).

RED MOUNTAIN AVA

The Red Mountain American Viticultural Area (AVA) is defined by a combination of topographic contour lines and geographic markers (Fig. 4) and encompasses approximately 4,040 acres. Of this 4,040 acres there is potential for up to 2,700 bearing acres, of which 710 are currently planted. Red Mountain is home to Hedges Cellars, Kiona and several smaller wineries, including Blackwood Canyon, Sand Hill, Seth Ryan, and Terra Blanca. There are 16 vineyards in the AVA (Table 2) and 37 different Washington wineries currently use Red Mountain wine grapes, including Arc-en-Ciel, Andrake, Andrew Will Cellars, Apex Cellars, Barnard Griffin, Betz Family Winery, Bookwalter, Cadence, Camaraderie Cellars, Canoe Ridge, DeLille Cellars, Gibbons Lane, Foris Vineyards, Hedges Cellars, Hightower Cellars, JM Cellars, Januik, Kiona, Kestrel vintners, L'Ecole No 41, Matthew Cellars, McCrea Cellars, Mount Baker Vineyards, Owen-Sullivan, Quilceda Creek, Randall Harris, Ryan Cray, Sandhill, Seth Ryan, Seven Hills, Soos Creek, Terra Blanca, Three Rivers, Waterbrook, Wilridge, Woodward Canyon, and Yakima River Winery.

Compared to the rest of Washington state, Red Mountain has a much higher proportion of red (80%) versus white wine grapes (20%). Currently, there are 570 acres of red wine grapes (480 bearing) and 140 acres of white wine grapes (120 bearing). Most of these are relatively young plantings, with only 4% of the Red Mountain vineyards planted prior to 1982, comprising 10 acres each of Gewürztraminer, Riesling, and Merlot (www.nass.usda.gov/wa/wine02.pdf). Currently, the three most abundant grape varieties planted within the Red Mountain AVA are Cabernet Sauvignon (35%), Merlot (21%), and Syrah (10%).

RED MOUNTAIN GEOLOGY

Columbia River Basalt Group

In general, the vineyards of Red Mountain all share the same basic geology in that they are rooted in a stratigraphic package consisting of a thin but variable mantle of Pleistocene to Holocene loess and dune sands on top of Quaternary glacial flood sediments. These in turn overlie Tertiary Columbia River Basalt basement. In the Red Mountain area the Columbia River Basalt Group consists of the Saddle Mountains Formation, about 8.5 to 12 million years old, which can be subdivided into three members: the Ice Harbor (8.5 Ma), Elephant Mountain (10.5 Ma), and Pomona (12 Ma) members (Reidel and Fecht, 1994). Locally, these members can be subdivided further into individual flow units that are exposed along the ridgetop of Red Mountain and by incisions of the Yakima River on the west side of the Red Mountain AVA (Fig. 5). Compositions of representative rock units are illustrated in Table 3. As shown in Figure 6, subdivisions among individual flow units can be made on a compositional basis using a variety of major and trace elements (Hooper, 2000).

In drill intercepts, it is also possible to distinguish between individual basalt flows by the presence of locally persistent but regionally discontinuous shaly interbeds (Fig. 7) that developed during the 1 to 2 million year intervals between eruption of major basalt units. These interbeds, along with relatively porous flow tops and basal flow breccias, form the main aquifers in the Columbia River Basalt Group. Static water level is about 100 m below the surface in most of these wells, coincident with the upper shaly interbed. Because of this hydrological regime, the rooting zone of all Red Mountain vineyards is well above the water table.

Structure

The basalt bedrock in Central Washington has been folded into a series of NW-SE-trending anticlines and synclines. On a regional scale (Fig. 3) the anticlines are exposed in ridges that form elongate

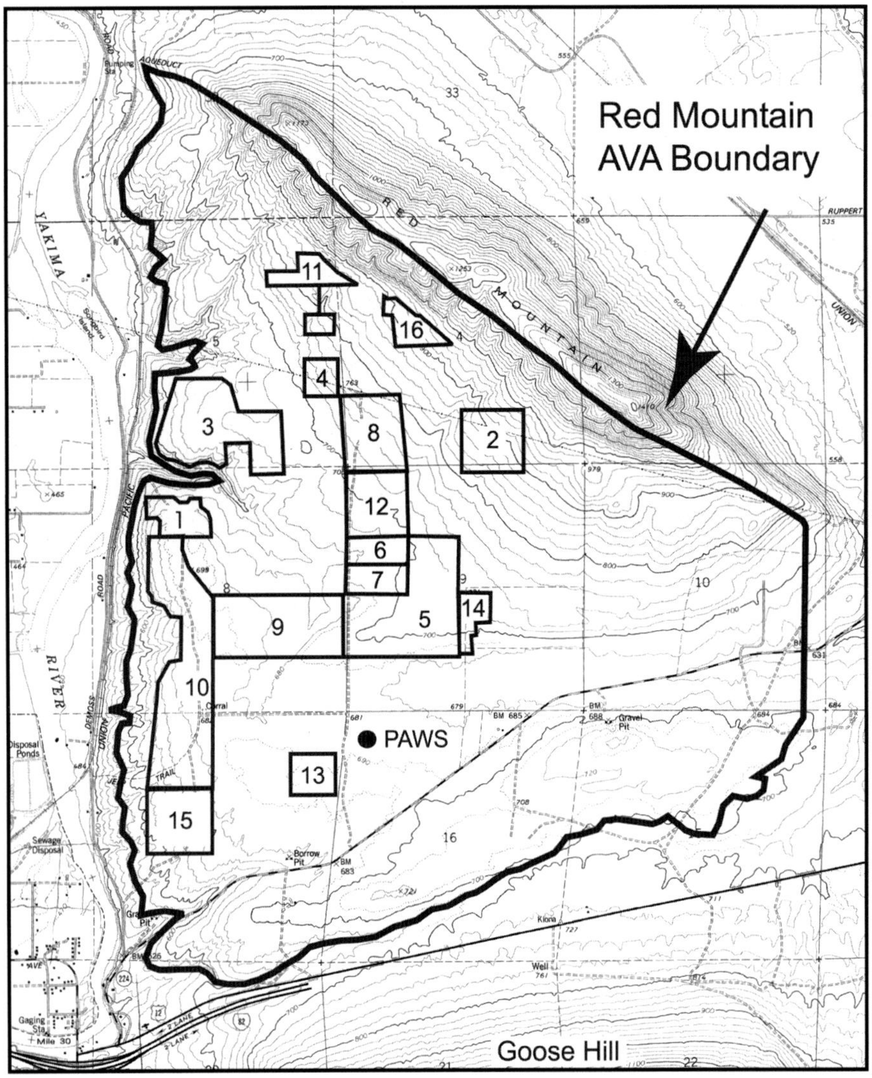

Figure 4 Location of vineyards and topography of the Red Mountain AVA. Base map is the 1:24,000 scale U.S.G.S. Benton City topographic map. Also shown is the location of the Washington State University Public Agricultural Weather System (PAWS) station that collected the data in Table 5. Information about numbered vineyards is presented in Table 2.

topographic features such as the Olympic-Wallowa lineament whereas the synclines generally form basins that are covered by younger sediments (Fig. 7). The Red Mountain and Goose Hill anticlines form the northern and southern geological boundaries of the Red Mountain AVA. Most of the vineyards are underlain by the intervening syncline, which has been called both the Goose Gap syncline and the Benton City syncline (Reidel and Fecht, 1994). According to Reidel and Fecht (1994) the southern limb of the Red Mountain anticline is steeper than the northern limb and like many such structures in the Yakima Fold Belt, may have local reverse faults (West *et al.*, 1996). Such faulting, as illustrated in the upper part of Figure 7, may explain the great differences in yield between otherwise similar water wells in the Red Mountain area.

Glacial Sediments

Except on ridgetops above about 400 m in elevation and where incised by modern drainage, the basalt in the Red Mountain area is everywhere overlain by unconsolidated sediments (labeled as Qht in Fig. 5) that were deposited by the series of glacial outburst floods related to the drainage of glacial Lake Missoula, described above. As the peak of Red Mountain is 430 m and the mean slackwater elevation of Lake Lewis, which temporarily filled the Pasco Basin (Fig. 2), is estimated at about 390 m, it is likely that the floodwater torrent surrounded Red Mountain, forming a very strong back-eddy behind (to the south of) Red Mountain as illustrated in Figure 8. The paleohydrology of this back-eddy resulted in extreme heterogeneity of the sediments that were deposited to the south of Red Mountain and that make up the deeper soil levels of the vineyards within the Red Mountain AVA.

Perhaps the most noticeable sedimentary features are the glacial erratics, large boulders that were carried by icebergs from the broken ice dam. These icebergs must have ranged up to 10s and perhaps 100s of metres in size, as the boulders that were dropped when these icebergs grounded, melted and deposited their entrained load, range up to several metres in size (Fig. 9A, B). Erratics are particularly prevalent along the high water strand of the slackwaters (Fig. 9A), but also are distributed throughout the Red Mountain AVA as illustrated in Figure 8D and evidenced by the rows, walls, and boxes of boulders (Fig. 9C) removed from the vineyards during initial planting and subsequent ploughing. Most of the boulders are granitic and metamorphic rock types that are not found within several hundred kilometres of Red Mountain, but which are abundant in the Rocky Mountain source region of the flood waters and the impounding glaciers. Figure 9B shows one such boulder that is about 3 m in diameter and consists of gleaming white marble with layers of brown garnet skarn, rock types that do not occur in outcrop anywhere in Central Washington.

Although the largest boulders could not have been carried directly by the flood waters, the peak velocity and discharge of the floods, estimated at 200 m/sec and 2 x 10^7 m^3/sec, respectively, (O'Connor and Baker, 1992) were sufficient to carry large amounts of relatively coarse sediment that was subsequently deposited when the water slowed near obstructions, as would be the case in the eddies behind Red Mountain (illustrated in Fig. 8D). This resulted in lenses of coarse gravels (Fig. 9D, F) in the generally finer grained silt and sand deposited by the ponded slackwater of Lake Lewis (Fig. 2).

In addition to the multiple pulses of floodwater that resulted in complex interbeds of gravel, sand, and silt, periodic eruption of Cascade volcanoes to the West

Table 2 Vineyards in the Red Mountain appellation[1]

Map no.	Name	Acres	Grapes	First Planted	Owner
1	Artz Vinneyard	20	8 cs, 5 m, 4 sb, 1 cf, 1 s, 1 c	1995	Fred Artz
2	Belle Ville	31	15 cs, 5 sy, 3 cf	1997	Pete Hedges
3	Blackwood Canyon	80	25 cs, 20 c, 15 m, 10 s, 5 sy, 2 n, 1 ca, 1 t, 1 cf	1984	Mike Moore
4	Cadence Vineyards	11	4 cf, 3 cs, 1 pv	2003	Ben Smith & Gaye L. McNutt
5	Ciel du Cheval	95	25 cs, 15 m, 14 sy, 9 cf, 7 sg, 5 ri, 5 ro, 4 v, 4 pv, 3 cb, 2 g, 2n, 1 co	1976	Jim Holmes
6	Grand Ciel Vineyard	9	7 cs, 2 sy	2001	Jim Holmes, Delille LLC
7	Golitzen Vineyard	17	17 cs	2001	Jim Holmes, Alex & Jeanette Golitzen
8	Hedges Cellars	36	18 m, 16 cs, 1 cf, 1 pv, 1 so	1990	Pete Hedges
9	Kiona Vineyards	70	28 cs, 15 l, 8 m, 5 cb, 4 c, 3 cf, 3 sg, 3 ri	1975	John and Ann Williams
10	Klipsun Vineyard	120	60 cs, 30 m, 13 sb, 9 s, 6 sy, 1 n	1984	Patricia & David Gelles
11	Tapteil	25	16 cs, 8 m, 2 cf	1985	Larry Pearson
12	Sand Hill (RMV)	40	24 cs, 13 m, 2 cf, 1 pg	1989	John Dingethal
13	Seth Ryan	7	2 cs, 2 m, 2 cf, 1 c	1996	Ron and Jo Brodzinski
14	Shaw Vineyard	12	7 cs, 4 m, 1 cf	1998	Ed and Eve Shaw
15	Terra Blanca	80	20 cs, 20 sy, 15 m, 5 c, 1 ma, 1 pv	1993	Keith Pilgrim
16	Williams	21	12 sy, 5 g, 3 sg	1978	Scott Williams

Grape abbreviations: ca = Carmenere, cf = Cabernet Franc, cs = Cabernet Sauvignon, c = Chardonnay, cb = Chenin Blanc, co = Counoise, g = Gewurtztraminer, l = lemberger, ma = malbec, m = Merlot, n = Nebbiolo, pg = Pinot Gris, pn = Pinot Noir, pv = Petite Verdot, ri = Riesling, ro = Roussane s = Semillon, sb = Sauvignon Blanc, sg = Sangiovese, so = Souzao, sy = Syrah, t = Tannat, v = Viognier

[1]Data from Jim Holmes and individual vineyard owners, written communication, 2002. Vineyard locations are shown in Figure 4.

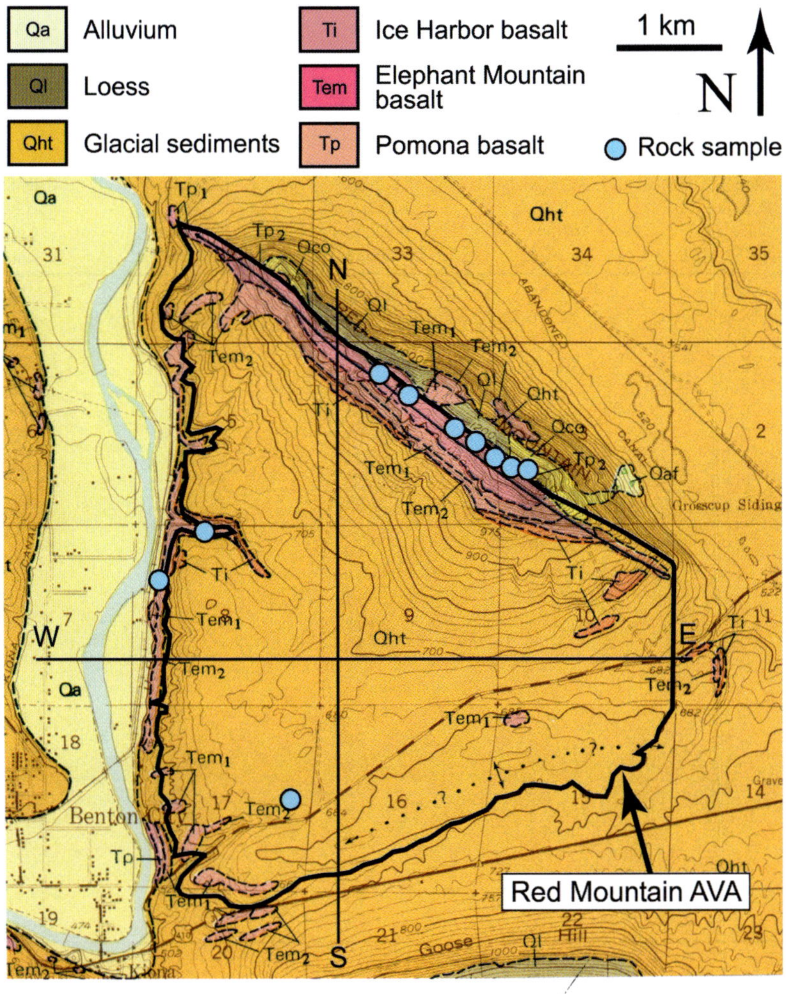

Figure 5 Geology of the Red Mountain AVA showing location of analyzed rocks (Table 3) and location of N–S and E–W cross-sections of Figure 7 (modified from Reidel and Fecht, 1994).

added a component of volcanic ash to the sediment mixture. Ash layers, such as Mt. St. Helens ash "S" (13,000 ^{14}C yr.; Mullineaux, 1996) are preserved in some roadcut and vineyard exposures (Fig. 9E). Although volumetrically minor, such ash layers provide evidence of periods of quiescence between flood pulses, necessary for the preservation of the delicate ash horizon.

WIND-BLOWN SEDIMENTS

Even though it is not shown generally on the geological map of the Red Mountain AVA (Fig. 5), loess or dune sand is everywhere present as a thin mantle ranging in thickness from decimetres to several metres on top of the slackwater sediments. As described above, the loess and the dunes were created by southwesterly winds that eroded slackwater and other glacial sediments and redeposited them into the present blanket of loess that covers much of the Columbia Plateau (Busacca, 1991). Buried soils in older loess at nearby sites not inundated by the floods indicate that deposition was intermittent during the Quaternary. The generally arid climatic conditions that continue to the present day resulted in limited chemical weathering and leaching, causing formation of calcic and petrocalcic soil horizons (Busacca, 1991; Soil Survey Staff, 1999).

CALCIUM CARBONATE CONTENT

Although as much as 90% of incoming precipitation is returned to the atmosphere

Table 3 Chemical composition of representative Red Mountain rocks[1]

Sample no.	RM-1	RM-9	RM-11	RM-2	RM-3	RM-4	RM-5
Unit	Elephant Mtn	Elephant Mtn	Elephant Mtn	Pomona	Pomona	Pomona	Pomona
Type	Basalt	Basalt	Basalt	Basalt	Basalt	Basalt	Basalt
Location	RM top	Artz-BC gully	Demoss Rd.	50m W RM	100m W RM	150m W RM	200m W RM
SiO_2	51.47	51.17	51.50	52.48	52.29	52.53	52.78
Al_2O_3	13.23	13.00	12.89	15.34	14.98	15.22	15.27
TiO_2	3.69	3.62	3.55	1.65	1.68	1.72	1.72
FeO(t)	14.13	14.53	14.79	9.32	10.37	9.90	9.94
MnO	0.20	0.21	0.20	0.17	0.18	0.17	0.17
CaO	8.71	8.96	8.64	11.31	11.01	11.28	11.12
MgO	4.06	4.36	4.26	6.55	6.22	5.91	5.68
K_2O	1.42	1.30	1.16	0.59	0.66	0.59	0.66
Na_2O	2.51	2.32	2.47	2.36	2.38	2.44	2.43
P_2O_5	0.58	0.53	0.54	0.24	0.24	0.24	0.24
Ni	11	13	14	37	47	60	47
Cr	33	36	33	118	102	109	107
Sc	46	33	37	35	37	39	37
V	426	426	432	264	279	293	273
Ba	531	480	476	941	568	828	339
Rb	29	26	28	10	11	9	9
Sr	238	242	230	262	254	253	246
Zr	256	244	253	127	135	140	140
Y	50	49	51	29	29	30	31
Nb	26	24	24	12	13	14	13
Ga	26	25	22	20	18	20	17
Cu	7	7	5	28	36	43	40
Zn	147	142	142	81	82	88	93
Pb	12	3	11	6	5	4	0
La	39	31	31	41	3	9	13
Ce	73	61	82	28	23	40	47
Th	8	0	4	3	2	2	0

Sample no.	RM-6	RM-7	RM-8	Average	Average	Average	RM-10
Unit	Pomona	Pomona	Pomona	Ice	Elephant		Scooteney
Type	Basalt	Basalt	Basalt	Harbor	Mtn	Pomona	Caliche
Location	300m W RM	400m W RM	500m W RM	Basalt[2]	Basalt[2]	Basalt[2]	Quarry
SiO_2	52.29	52.29	52.81	47.98	51.68	52.19	38.48
Al_2O_3	14.81	14.89	15.05	11.70	13.16	14.76	4.82
TiO_2	1.70	1.60	1.67	3.78	3.54	1.64	0.41
FeO(t)	10.50	10.42	9.80	17.24	14.17	10.38	2.27
MnO	0.17	0.17	0.17	0.28	0.21	0.18	0.03
CaO	10.99	10.97	10.97	8.91	8.79	10.78	49.19
MgO	6.34	6.46	6.31	4.23	4.08	6.88	3.25
K_2O	0.68	0.69	0.58	1.31	1.24	0.57	0.56
Na_2O	2.29	2.30	2.40	2.71	2.58	2.39	0.73
P_2O_5	0.24	0.22	0.24	1.85	0.55	0.23	0.24
							28.29*
Ni	45	44	36	7	10	43	0
Cr	107	107	105	33	30	106	23
Sc	39	33	33	41	35	37	0
V	284	265	276	214	408	272	56
Ba	368	264	430	838	552	191	297
Rb	11	11	13	25	31	13	19
Sr	239	234	237	240	243	231	427
Zr	138	131	143	467	242	130	166
Y	31	30	30	108	50	31	15
Nb	13	12	12	55	28	13	7
Ga	23	20	20	26	23	19	9
Cu	38	32	30	6	18	50	10
Zn	87	87	88	223	154	88	20
Pb	0	0	1	9	6	7	4
La	37	17	26	72	26	9	21
Ce	7	47	28	160	72	43	27
Th	0	0	0	5	6	2	0

* Loss on Ignition

[1]Analyses by XRF and ICPMS in the Geoanalytical Lab at Washington State University.

[2]Averages for Ice Harbor, Elephant Mountain, and Pomona members from Hooper (2000).

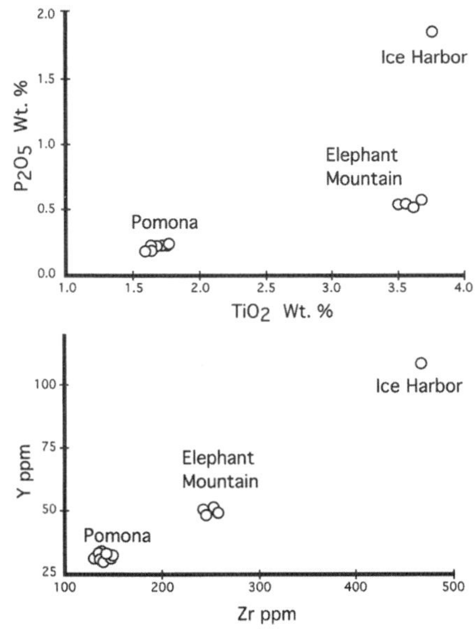

Figure 6 Compositional discrimination diagrams for Columbia River basalts using a) P_2O_5 versus TiO_2 and b) Y versus Zr. Data from Table 3 and Hooper (2000).

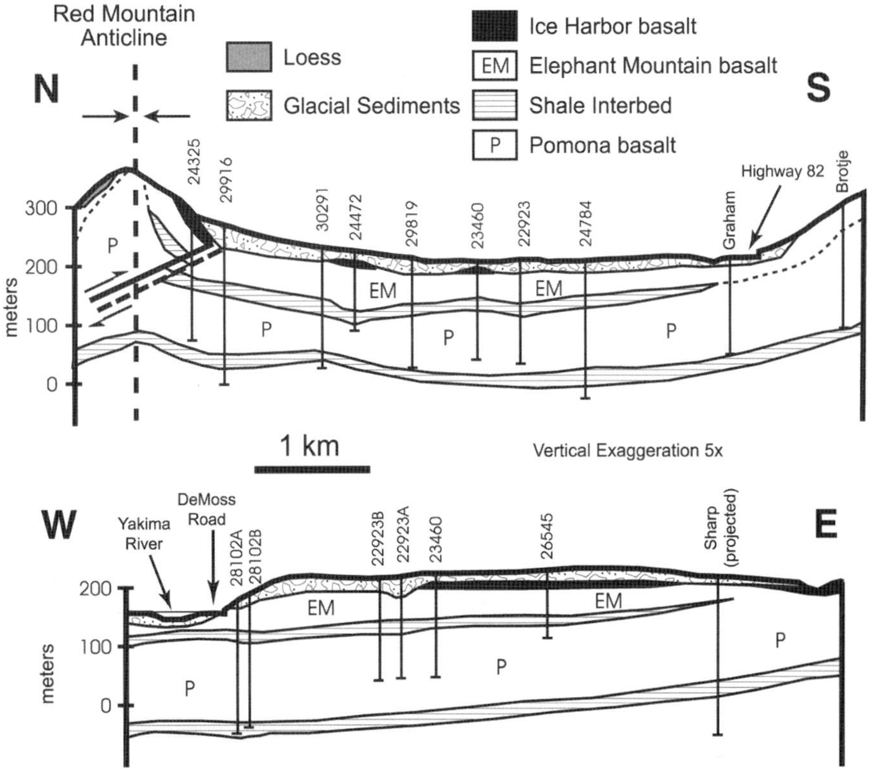

Figure 7 Geological cross-sections through the Red Mountain area (vertical exaggeration 5:1). Well log data courtesy of Lorne Jacobsen, written communication, 2001. Location of cross-sections is illustrated in Figure 5.

as evapotranspiration, the coarse gravel lenses intercept and channel the infiltration of precipitation descending through overlying loess and dune materials. Thus, shallow gravel lenses can contain significant calcium carbonate. An analysis of one such petrocalcic horizon (RM-10, Table 3) shows that it contains up to 50% CaO, an order of magnitude more calcium than average soil and rock values (Tables 3, 4). Although limestone is absent in the Red Mountain AVA, other than as erratic boulders, the calcite-cemented gravel lenses form significant reservoirs of calcium carbonate that can affect vineyard performance. For example, calcium is known to inhibit vine uptake of such essential nutrients as nitrogen and potassium (Winkler *et al.*, 1974; Ribéreau-Gayon *et al.*, 2000). Also, Fe is particularly affected if water pools on calcic layers in the root zone (Sara Spayd, written communication, 2002). Some Red Mountain wineries such as Terra Blanca, Spanish for white earth, point to these calcic layers as an important and sometimes negative part of the local terroir.

RED MOUNTAIN SOILS

The layered stratigraphy of bedrock basalt and overlying glacial floodwater and eolian sediments forms the substrate for soil development in the Red Mountain AVA. Given the heterogeneity of these sediments, it is not surprising that there are a number of different soils in the Red Mountain AVA (Fig. 10), and that these soils perform very differently under grape production.

A common theme to the soils of the benched area of Red Mountain is that they formed in eolian materials (loess or dune) over slackwater sediments from giant glacial outburst floods (Meinert and Busacca, 2000). Yet within this landscape, no fewer than eight different soil series have been mapped, and these can have very different textures (Table 4) and profile morphology (Fig. 11): Hezel (Xeric Torriorthents), Quincy (Xeric Torripsamments), and Finley, Scooteney, Prosser, Starbuck, Kiona, and Warden (all Xeric Haplocambids; Soil Survey Staff, 1999). Vineyards are planted on most of these, with development planned for the remainder.

All but two of the principal soils are classified as Aridisols in soil taxonomy (formative word Arid; Soil Survey Staff, 1999) based primarily on an aridic soil moisture regime. The two soils that formed in dune materials (Hezel and Quincy) are Entisols (formative word Recent) because the shifting sands lack most soil profile features. In sharp contrast to the Aridisols and Entisols of the appellation, soils on the floodplain of the Yakima River less than one kilometre outside the appellation and planted to *Vinifera* are Mollisols of the Pasco series (Fig. 11; Cumulic Endoaquolls), which are wet soils with very dark, thick, humus-enriched topsoils. These soils have a permanent water table whose height in the soil profile fluctuates seasonally with stages of the Yakima River. The generally high water table results in carbonates to the surface and uncontrolled access to water during the growing season.

Large areas of the benchlands of the Red Mountain appellation are underlain by the Warden series soils, formed in about 50 cm of loess or mixed loess and eolian sand over stratified flood sediments (Fig. 11), whereas adjacent areas, even within the same vineyard, are underlain by the Hezel series soils (Fig. 14b), which formed in a cover of about 50 cm of dune sand over sandy stratified flood sediments. In contrast, areas of Scooteney soils grade downward from an eolian sandy loam or loam at the surface to a fluvial unit of very cobbly sandy loam at

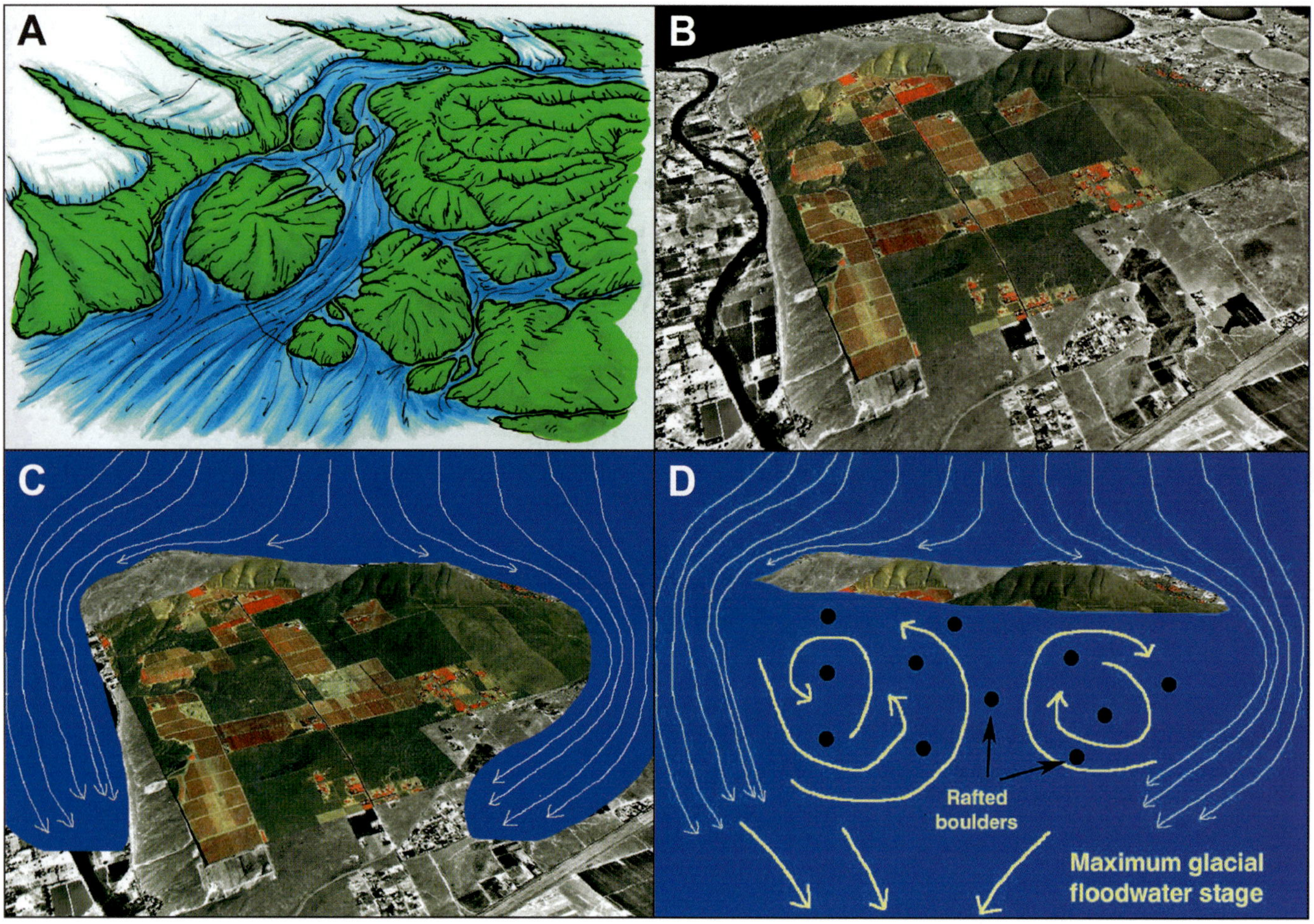

Figure 8 Cartoon illustrating the sequence of events during flooding caused by the catastrophic draining of glacial Lake Missoula. A) Artist's rendition of outburst flooding caused by failure of a glacial ice dam. Note torrent of water rushing around isolated hills analogous to Red Mountain. Modified from Molenaar (1988). B) 3-D perspective view of Red Mountain area with color infrared imagery draped over black and white ortho aerial photograph. Note that in infrared images, green vegetation such as leafy vineyards are red whereas arid range lands are dark green. Images courtesy of Francis Pierce, Center for Precision Agricultural Systems, WSU-IAREC. C) Approach of floodwaters with flowlines around Red Mountain. Even though standing water, as evidenced by the strandline of ice-deposited erratics, did not cover Red Mountain, it is possible that the initial flood surge may have overtopped the peak creating a temporary but rather large standing wave. D) Maximum flood stage as evidenced by the strandline of ice-deposited erratics, and schematic location of boulders deposited by melting of rafts of ice. See Figure 9A, B, C for examples of such ice-rafted boulders.

150 cm. All three of these soils can be tens of metres thick above hard basalt bedrock. A variant of Scooteney series soils, which was not recognized in the previous soil survey (Rasmussen, 1971), was found in our field reconnaissance to have a strongly lime-silica cemented duripan in flood gravel at depths of 50 cm to 100 cm. Thus vineyards in the Red Mountain AVA have soils that range from loess to dune sand to gravel to slackwater sediments in the lower part of the vine rooting zone.

Still other areas in this same landscape are underlain by Prosser and Starbuck series soils (Fig. 11), with bedrock at less than 40 cm to 80 cm depth. Small areas have Quincy soils that formed in dune sand more than 150 cm deep. Soils of the Kiona series (Fig. 11) that occupy the steep south face of Red Mountain (with slopes up to 60 percent) have formed in slope colluvium of fractured basalt mixed with loess and are cobbly loams to more than 150 cm. It appears that no vineyards have yet been planted on areas of Kiona soils.

The majority of the soils in the Red Mountain appellation are thicker than several metres (Fig. 7). The most important vineyard soils formed as the result of two end-member eolian processes, dune saltation and loess suspension. Dominantly sandy dune materials accumulated over either flood gravels or stratified flood slackwater sediments (*e.g.,* Hezel series, Fig. 11) in some places on Red Mountain, whereas dominantly silty loess materials have accumulated over flood materials in other places (*e.g.,* Warden series, Fig. 11). This is evident in the high total sand content of a Hezel sample (75.9%, CC-3, Table 4) versus the lower total sand content of a Warden sample (19.6%, BV-3, Table 4). However, mixing of variable amounts of saltated sand and suspended silt occurred in most of the sampled soils, leading to highly variable soil textures. In general, the soils are more sandy in the surface layer and more silty at 1 m depth (Table 4), suggesting that loess deposition dominated early in the post-flood history of Red Mountain and that dune sands have more recently covered or

Figure 9 Photographs of vineyard features in the Red Mountain area. A) South side of Red Mountain about 100 m below the peak showing location of ice-rafted boulders. For scale, sagebrush is about 1 m high. B) Example of an ice-rafted boulder. The polished and rounded stone is gleaming white marble with layers of brown garnet skarn. C) Glacial erratic boulders gleaned from the Kiona vineyard are collected in wooden boxes (grape picking bins). D) Cut bank in Tapteil vineyard exposing 3–4 m cross-section of Quaternary loess, slackwater deposits, gravel lenses, and loess-colluvium. Gravel in lenses is similar in size to that in photo F. E) 2–4 cm white band of Mt. St. Helens "S" ash in Quaternary slackwater and channel deposits exposed in roadcut on south side of Red Mountain AVA. F) Lime-cemented gravels of the Scooteney soil association.

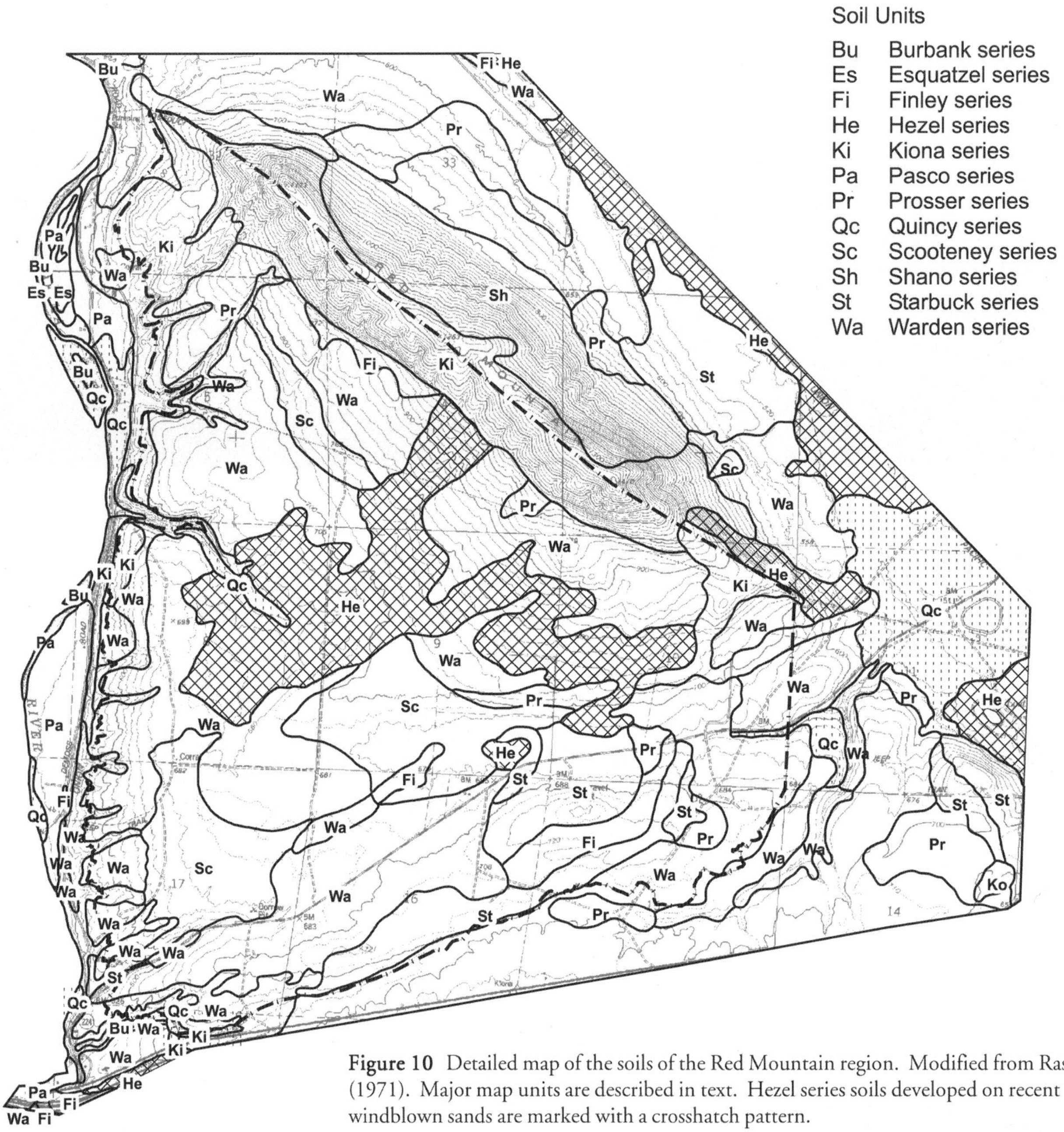

Figure 10 Detailed map of the soils of the Red Mountain region. Modified from Rasmussen (1971). Major map units are described in text. Hezel series soils developed on recent windblown sands are marked with a crosshatch pattern.

influenced most of the soils. With such marked variation in texture within and among soil profiles at Red Mountain, the volumetric water content at field capacity ranges from 7% for soils like the Hezel at the Ciel du Cheval vineyard to as high as 24% for soils like the Warden at the Belle Ville vineyard.

X-ray fluorescence and inductively coupled plasma mass spectrometry (ICPMS) total elemental analyses of selected soil samples from Red Mountain (Table 4) show that these soils have a remarkably uniform composition, ranging from 61 to 67% SiO_2, 12 to 14% Al_2O_3, and 6 to 8% FeO. The most variable element is calcium, ranging from about 4 to 8%. This is explained by the presence of variable amounts of pedogenic lime coatings in the samples. The major-element composition of the soils appears to be partly related to clay content; that is, soils with higher clay contents tend to have higher SiO_2 contents.

Soil profiles like the Scooteney hardpan variant strongly restrict rooting depth, whereas Scooteney, Warden, and Hezel pose only moderate restrictions on rooting depth. We have observed a vineyard just outside the appellation that is partly planted on the Starbuck soil (Fig. 11) that is less than 50 cm to bedrock. Vines in this soil show stunted development and severe late-season water stress compared to vines on deeper soils in the same vineyard.

Virtually all of the soils have free lime both at the surface and at one-metre depth (Table 4), which is a reflection both of the very low rainfall for leaching and the continual addition of carbonate-bearing dust to these soils. All of the soils have pH values of about 8 at the surface, reflecting a pH control by free calcium carbonate. Interestingly, all of the samples from a depth

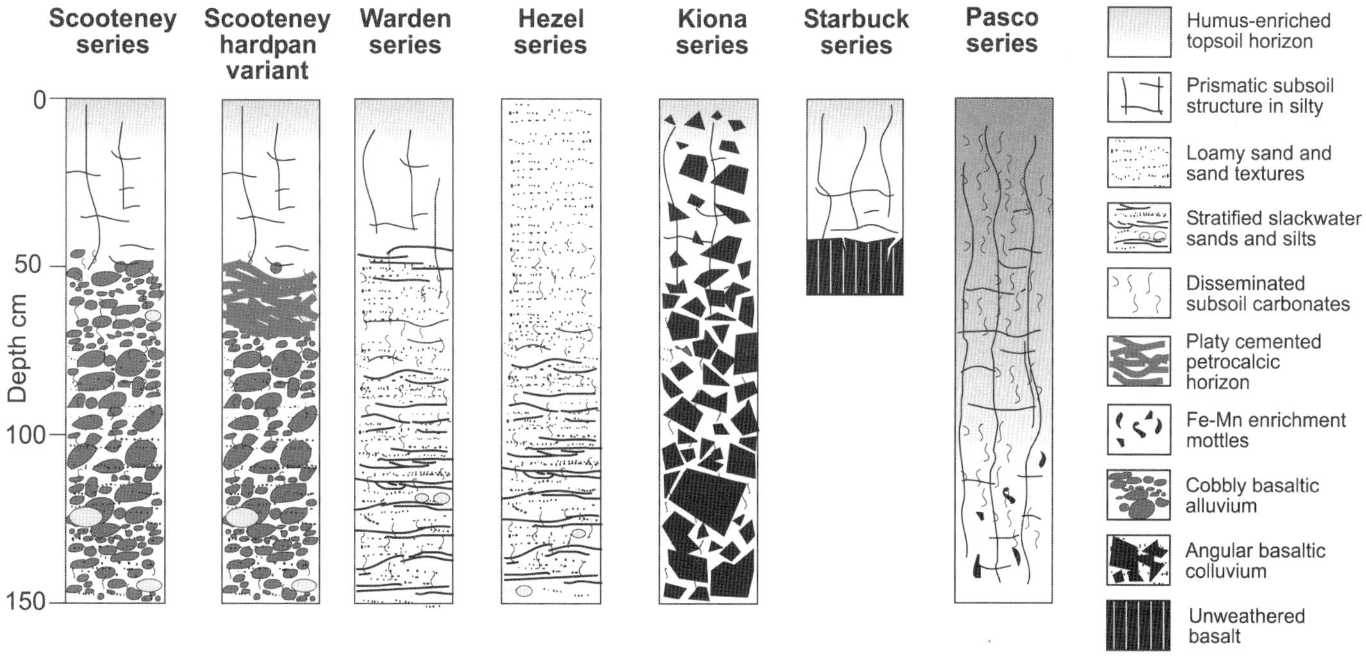

Figure 11 Representative soil profiles of the main soil series in the Red Mountain area. Degree of shading of the topsoil horizons is proportional to the content of humified organic matter. The irregular lines in the prismatic subsoil pattern represent the pattern of soil structure that develops in these silty soil horizons.

of one metre have pH values from 8.5 to 8.8, suggesting that these soils contain moderate amounts of exchangeable sodium. No evidence for a true sodic soil condition has been observed in the vineyards, however.

Vineyards planted on the wide range of soils that occur within the bounds of the AVA have large differences in water-holding capacity, infiltration, and potentially runoff. All of the soils share a strong soil-moisture deficit during the growing season, which allows growers to control water uptake by the grape vines and thus, one determinant of vine vigour. Some of the Scooteney hardpan variant soils may have a risk of shallow saturation and waterlogging under irrigation whereas others nearby are susceptible to drought.

RED MOUNTAIN CLIMATE

Red Mountain is one of the warmer vineyard sites in Washington State with 3409 degree days (50°F) recorded in 1998 and an average of 3016 degree days for the years of record (Table 5). For comparison, the Napa Valley in California and the Barossa Valley in Australia average 3280 and 3090 degree days, respectively [see the broader discussion by Meinert and Busacca (2000) of climatic measures in Washington and by Gladstones (1992, 2001) of general climatic measures relative to viticulture]. Red Mountain also may be the driest viticultural area in Washington State, with an average annual precipitation of 17.8 cm and a low in 1999 of 8.4 cm (Table 5). Comparison of monthly temperatures and precipitation (Fig. 12) shows that the lowest precipitation coincides with the highest temperatures, and because of the low water-holding capacity of the soil and low water table, this creates a moisture deficit that requires irrigation in all vineyards. With the high evapotranspiration rates in such conditions, drip irrigation is the dominant method of supplying supplemental water.

Red Mountain, at the eastern end of the Yakima Valley, is exposed to fairly steady westerly winds, averaging 6.8 km/hr, with relatively little monthly variation (5.5–7.9 km/hr). Strong winds (>24 km/hr) mostly occur in the spring, with only three months (March–May) having on average 10 or more daylight hours of winds >24 km/hr. During the prime ripening season of July–October there are only 4 daylight hours total with winds >24 km/hr (Table 5).

Another variable of critical importance in ripening of grapes is sunlight exposure (see Fig. 2 in Gladstones, 1992) which at Red Mountain averages 1229 kW/m^2 from April to October (Table 5). Both the heat summation (degree days) and the total sunlight exposure are nearly identical for vineyards in Red Mountain and Napa Valley, California even though Red Mountain is about 1000 km farther north (Gladstones, 1992). However, there are many other variables that distinguish between these regions, such as number of frost-free days and diurnal temperature range.

When standing in Red Mountain vineyards on a typical summer day, another aspect of the mesoclimate can be observed. Generally, because of regional air circulation patterns, there is a high pressure zone over Red Mountain that results in clear skies directly overhead even when partial cloud cover may obstruct areas only 10 km away (Fred Artz, personal communication, 2002). These clear skies over Red Mountain can have a negative aspect during the cooler months, as both frost damage and winter kill can be a problem in some years (Jim Holmes, personal communication, 2001). Winter temperatures drop below –10°C in most years and –27.6°C was recorded in 1996 (Table 5), a year of devastating winter kill in many Pacific Northwest vineyards.

VINEYARDS AND WINES OF THE RED MOUNTAIN AVA

The Red Mountain AVA is not only Washington's newest AVA (approved in 2001), but many of the plantings are

Table 4 Composition and grain size of representative Red Mountain soils[1]

Sample #	CC-3	CC-4	CC-5	CC-6	CC-8	CC-9	CC-10	CC-1	CC-2	CC-7
Vineyard	Ciel Cheval	Ciel Cheval	Ciel Cheval	Ciel Cheval	Ciel Cheval	Ciel Cheval	Ciel Cheval	Ciel Cheval	Ciel Cheval	Ciel Cheval
Soil type	Hezel	Hezel	Hezel	Hezel	Warden	Warden	Warden	Scooteney	Scooteney	Scooteney
Depth	Surface	1 m	Surface	1 m	Surface	1 m	Surface	Surface	1 m	Surface
Grape type	Cab Sauv	Cab Sauv	Cab Sauv	Cab Sauv	Cab Sauv	Cab Sauv	Syrah	Merlot	Merlot	Cab Sauv
SiO_2	66.40	64.40	66.65	65.53	66.68	64.98	66.23	66.10	64.98	66.28
Al_2O_3	13.22	13.44	13.49	13.22	13.83	13.52	13.89	13.17	13.19	13.87
TiO_2	1.37	1.49	1.36	1.35	1.25	1.21	1.33	1.36	1.42	1.35
FeO(t)	6.98	7.09	6.77	6.69	6.84	6.81	7.16	6.93	7.34	6.95
MnO	0.12	0.13	0.12	0.12	0.12	0.12	0.12	0.12	0.13	0.12
CaO	4.62	6.03	4.31	5.72	4.00	6.16	4.00	5.00	5.61	4.12
MgO	2.35	2.71	2.29	2.51	2.23	2.54	2.30	2.54	2.55	2.28
K_2O	1.99	1.83	1.95	1.80	2.09	1.99	2.06	2.00	1.80	2.07
Na_2O	2.51	2.42	2.61	2.63	2.55	2.24	2.47	2.32	2.52	2.51
P_2O_5	0.26	0.28	0.25	0.25	0.24	0.24	0.25	0.27	0.27	0.25
LOI	3.58	3.88	3.17	3.91	3.89	5.44	4.82	4.01	3.19	4.22
Ni	18	24	22	22	20	23	21	19	19	22
Cr	47	55	51	51	41	40	48	47	45	53
Sc	17	19	23	19	22	15	18	16	16	20
V	183	210	184	190	156	161	155	182	199	172
Ba	647	651	653	677	651	659	659	627	655	657
Rb	63	62	61	58	69	74	70	68	60	67
Sr	355	354	359	393	330	317	319	337	390	336
Zr	245	262	245	250	252	223	271	262	250	263
Y	31	33	32	32	35	34	35	32	31	35
Nb	15	17	16	16	16	16	16	13	14	17
Ga	14	16	20	14	16	17	17	17	15	15
Cu	6	9	4	4	6	7	7	10	8	9
Zn	84	84	73	73	86	80	83	82	84	80
Pb	14	13	17	17	21	19	25	19	14	19
La	29	22	55	32	32	27	31	26	45	28
Ce	55	70	62	81	63	52	56	66	51	86
Th	10	13	12	11	11	9	7	8	12	11
very coarse sand	1.0	2.4	1.2	1.3	0.6	0.2	1.7	n.d.	0.9	n.d.
coarse sand	2.5	1.8	2.4	2.2	1.3	0.5	2.1	n.d.	1.5	n.d.
medium sand	3.9	2.6	3.7	3.3	2.0	0.5	2.8	n.d.	2.0	n.d.
fine sand	1.3	0.1	0.1	1.4	7.0	0.1	0.4	n.d.	0.1	n.d.
very fine sand	67.2	43.9	65.0	69.2	46.2	30.3	56.2	n.d.	68.5	n.d.
% total sand	75.9	50.9	72.4	77.4	57.1	31.7	63.2	n.d.	73.0	n.d.
coarse silt	8.7	36.0	8.5	15.6	17.9	47.2	16.5	n.d.	20.8	n.d.
fine silt	4.1	11.4	5.1	4.5	10.1	18.4	7.7	n.d.	5.1	n.d.
% total silt	12.8	47.4	13.7	20.0	28.0	65.6	24.2	n.d.	25.9	n.d.
% total clay	11.3	1.8	13.9	2.6	14.9	2.7	12.6	n.d.	1.1	n.d.
total	100.0	100.0	100.0	100.0	100.0	100.0	100.0	n.d.	100.0	n.d.
pH 1:1	8.2	8.5	8.0	8.7	8.2	8.5	8.2	8.4	8.8	8.1
$CaCO_3$	0.7	3.0	2.7	3.3	0.4	4.8	0.4	1.1	2.2	0.4
$MgCO_3$	0.38	0.21	0.12	0.22	0.15	0.21	0.59	0.19	0.20	0.13

[1]All samples sieved to exclude >2mm fraction. Major and trace elements analyses by XRF and ICPMS in the Geoanalytical Lab at Washington State University. Grain size, pH, carbonate, and nirtogen analyses by Department of Crop and Soil Sciences at Washinton State University

relatively young, making up only about 30% of the AVA's potential acreage in cultivation. Irrigated vineyards form a striking contrast to the natural xerophytic (drought tolerant) vegetation, including species of *Artemisia, Purshia,* and *Crysothamnus* (Fig. 13A). While most of the older vineyards are planted in N–S rows, some of the younger plantings have a N10°E orientation (Fig. 13B) to maximize sun exposure (60% morning, 40% afternoon) at this latitude (46°20') and provide greater shading of fruit on the west side of the canopy in the later afternoon (Jim Holmes, personal communication, 2001).

Although the Red Mountain AVA is bounded on the west by the Yakima River (Fig. 13C), none of this water is used directly for irrigation within the Red Mountain AVA. Rather, all irrigation water in the Red Mountain AVA comes from wells that typically are about 200 m deep in order to tap intraflow aquifers (Fig. 7). Static water level is about 100 m below the surface in most of these wells, except for those closest to the Yakima River on the western margin of the Red Mountain AVA. Because of this hydrological regime, the rooting zone of all Red Mountain vineyards is well above the water table and most available soil moisture comes from drip irrigation augmented by stored soil moisture from winter rainfall and by relatively rare rainshowers during the growing season (about 5 cm total from April to October; Table 5).

As previously described, the soils and bedrock geometry of the Red Mountain AVA are well suited for vineyard production and differ in several respects from those in areas immediately surrounding the AVA. For example, the soils in vineyards such as Klipsun (Fig. 13D) within the Red Mountain AVA are quite different from those in vineyards that are only a few hundred metres outside the AVA, such as on

Table 4 Continued

Sample #	CC-11	CC-12	CC-13	K-1	BV-1	BV-2	BV-3	O-1	O-2
Vineyard	Ciel Cheval	Ciel Cheval	Ciel Cheval	Klipsun	Belle Ville	Belle Ville	Belle Ville	Oakwood	Oakwood
Soil type	Scooteney	Scooteney	Scooteney	Scooteney	Warden	Warden	Warden	Pasco	Pasco
Depth	1 m	Surface	1 m	Surface	Surface	Surface	1 m	Surface	1 m
Grape type	Syrah	Syrah	Syrah	Cab Sauv	Sagebrush	Syrah	Syrah	Riesling	Riesling
SiO_2	65.20	66.91	63.67	67.31	66.94	64.05	65.72	66.39	67.02
Al_2O_3	13.40	13.60	12.88	14.12	13.71	13.18	13.52	13.91	14.39
TiO_2	1.43	1.26	1.31	1.15	1.29	1.36	1.23	1.37	1.13
FeO(t)	7.73	6.74	7.14	6.38	6.69	7.21	6.87	6.65	5.93
MnO	0.13	0.12	0.12	0.11	0.12	0.12	0.12	0.12	0.11
CaO	4.99	4.10	7.80	3.84	4.07	7.06	5.39	4.16	4.05
MgO	2.59	2.31	2.60	2.04	2.15	2.46	2.57	2.25	2.17
K_2O	1.82	2.06	1.84	2.00	1.95	1.90	1.93	1.88	1.94
Na_2O	2.29	2.48	2.21	2.64	2.67	2.20	2.23	2.81	2.85
P_2O_5	0.25	0.24	0.24	0.21	0.22	0.26	0.22	0.26	0.23
LOI	4.18	3.79	6.12	3.72	2.78	5.79	4.79	4.10	2.93
Ni	25	22	20	29	21	20	23	29	32
Cr	45	50	46	61	44	43	49	97	81
Sc	24	19	17	18	21	16	19	16	18
V	173	152	170	134	170	171	168	161	136
Ba	635	623	626	643	639	645	682	619	643
Rb	68	68	65	67	63	69	72	59	62
Sr	314	335	333	331	354	321	308	341	330
Zr	288	272	286	285	256	283	321	262	216
Y	37	34	36	31	32	36	37	32	29
Nb	18	16	16	16	16	18	18	15	13
Ga	17	17	18	17	13	15	18	18	19
Cu	11	8	11	9	3	11	9	7	10
Zn	79	78	76	82	74	81	77	84	75
Pb	23	21	26	23	20	22	23	22	20
La	29	29	48	35	55	24	32	55	19
Ce	63	61	72	99	64	69	69	82	47
Th	9	10	14	11	11	11	13	10	8
very coarse sand	2.4	0.5	1.2	2.5	3.8	1.8	0.2	1.8	2.2
coarse sand	2.2	1.2	1.4	3.1	4.6	2.2	0.3	2.8	2.7
medium sand	0.9	2.4	1.9	3.2	4.6	1.9	0.2	4.0	3.5
fine sand	0.1	1.0	0.1	6.8	0.1	0.1	0.0	12.4	0.3
very fine sand	29.9	51.0	34.6	34.8	61.2	34.9	18.8	41.3	53.5
% total sand	35.5	56.1	39.1	50.4	69.9	40.9	19.6	66.6	62.2
coarse silt	48.2	20.9	39.6	23.2	18.7	40.8	64.9	18.1	22.2
fine silt	15.2	9.2	17.0	10.5	5.9	15.2	13.8	8.1	9.3
% total silt	63.4	30.1	56.6	33.7	24.6	56.0	78.7	26.2	31.5
% total clay	1.1	13.8	4.2	15.9	5.5	3.1	1.7	7.2	6.3
total	100.0	100.0	100.0	100.0	100.0	100.0	100.0	100.0	100.0
pH 1:1	8.7	8.4	8.6	8.7	7.9	8.7	8.8	7.8	8.6
$CaCO_3$	2.0	0.5	7.8	0.6	0.3	6.5	3.9	0.3	0.4
$MgCO_3$	0.14	0.10	0.18	0.11	0.09	0.23	0.27	0.08	0.12

a high basalt knob to the Southwest (Fig. 13E) or on the Yakima River floodplain to the West (Fig. 13F).

The pictured vineyard (Fig. 13F) on the Yakima River flood plain is about 100 m west of the Red Mountain AVA boundary and about 200 m from the Klipsun vineyard (Fig. 13D). The rich Mollisol soils (Table 4, Fig. 11) and abundant water (even without supplemental irrigation) in the flood plain of the Yakima River result in more vigourous vegetative growth of both grape vines and ground cover relative to the nearby Red Mountain vineyards, as illustrated in photos taken on the same day of the growing season (compare Fig. 13D and 13F).

About 700 m southwest of the Red Mountain AVA there is a vineyard planted on a basalt hill that has less than a metre of loess and Quaternary glacial sediments on top of the basalt to form soils. The vine roots have penetrated into the basalt and the free drainage of the fractured basalt and thin sediment cover results in a rapid loss of soil moisture. In contrast to the vineyards on the Yakima River floodplain that have perhaps an overabundance of moisture and plant vigour, the shallow soils on this basalt knob hold little water, even after irrigation, and the resulting moisture deficit can result in reduced plant vigour (Fig. 13E).

Although the above examples illustrate differences between Red Mountain and surrounding areas, there also are important differences within the Red Mountain AVA. As previously discussed, there are variations in soil types, thickness and homogeneity of Quaternary glacial sediments, and depth to bedrock among Red Mountain vineyards (Figs. 5, 7, 10, 11, 14).

For example, Ciel du Cheval (Fig. 14B) lies in the geographic center of the Red Mountain AVA (Fig. 4) and is one

Table 5 Red Mountain 1995–2001 climate data (Red Mountain PAWS*, Lat. 46°20', Long. 119°40', Elev. 194 m)

	Jan	Feb	Mar	Apr	May	Jun	Jul	Aug	Sep	Oct	Nov	Dec	Annual
Mean max. temp. °C	5.3	8.8	14.2	18.8	23.3	27.5	32.9	32.5	27.1	18.0	10.2	4.4	18.6
Mean avg. temp. °C	1.5	3.7	7.8	11.6	15.9	19.8	24.2	23.4	18.3	10.7	5.5	1.0	11.9
Mean min. temp. °C	-2.1	-1.0	1.3	3.6	7.4	10.7	14.3	13.8	9.8	3.6	1.0	-2.3	5.0
Precipitation (cm)	2.4	2.1	1.8	0.8	1.2	0.7	0.8	0.1	0.7	1.8	3.4	2.0	17.8
Snowfall (cm)	10.9	4.3	0.8	0.0	0.0	0.0	0.0	0.0	0.0	0.0	0.8	6.4	23.1
Mean wind speed (kph)	7.6	7.9	7.5	7.2	7.4	7.2	6.1	5.7	5.5	5.7	6.9	7.2	6.8
Daylight hrs wind >24 kph	2	8	11	12	10	5	1	3	0	0	0	2	54
Solar radiation (kW/m²)	38	63	112	161	195	222	216	201	142	93	44	28	1514
Mean 9 am temp. °C	0.4	2.5	6.9	11.6	16.3	20.1	24.0	23.2	18.1	10.4	4.6	0.3	11.5
Mean 9 am relative humidity	86	81	69	55	49	45	41	44	53	67	85	86	63
Mean 2 pm temp. °C	4.2	7.6	12.8	17.3	21.6	25.8	30.8	30.7	25.7	16.9	9.0	3.2	17.1
Mean 2 pm relative humidity	71	60	45	36	33	28	24	24	31	42	68	75	45
Degree days (50°F)	2	8	26	161	320	529	772	700	389	91	16	2	3016

	1995	1996	1997	1998	1999	2000	2001	Mean
Maximum air temperature °C	40.1	41.5	39.4	42.7	39.5	39.8	40.2	40.4
Minimum air temperature °C	-17.0	-27.6	-13.3	-16.8	-8.2	-10.5	-6.6	-14.3
Precipitation (cm)		24.8	19.4	22.9	8.4	19.3	11.8	17.8
Degree days (50°F)		2836	3039	3409	2786	2920	3106	3016

* Data from Washington State University Public Agricultural Weather System (PAWS) http://frost.prosser.wsu.edu/ Courtesy of Francis Pierce, Center for Precision Agricultural Systems, WSU-IAREC

of the older vineyards, first planted in 1976. The Ciel du Cheval vineyard has a relatively low slope (Fig. 4), with homogeneous air drainage and mesoclimate. Thus, most climatic variables are approximately constant throughout the vineyard. In contrast, there are three different soil units present in the vineyard (Fig. 14B) that cut across the N–S rows of vines. The three soil units are Hezel, Scooteney, and Warden. As previously discussed, these differ in texture and composition. The Ciel du Cheval vineyard has been owned and managed by a single person, Jim Holmes, for nine years and thus has a consistency of management style that should allow for examination of viticultural and enological variations as a reflection of soil type. Initial results of grape analyses suggest that there are differences that may correlate with soil types and an ongoing study of wine sensory analysis over a three-year period has been designed to test whether different soil types can be correlated with statistically significant wine flavor profiles (Sara Spayd, written communication, 2002). This will be reported in a future publication. Such sensory differences have been correlated previously with a number of terroir variables (*e.g.*, Douglas *et al.*, 2001).

In contrast to Ciel du Cheval, the Belle Ville vineyard is relatively young, first planted in 1997. It is the northeasternmost vineyard in the Red Mountain AVA (Fig. 4), is underlain entirely by Warden series soils, and lies on a moderate slope that was extensively graded to smooth the original topography (Steve Lessard, personal communication, 2001). Expressions of topography and drainage are readily visible in Figure 14C, an orthogonal infrared image taken in August, 2001. The variations visible in the image correspond to zones of vine stress that cut across rows of a single cultivar and management regime and therefore are unlikely to result from differences in canopy management, irrigation, or other viticultural practice (Steve Lessard, personal communication, 2001). It is hypothesized that grading of the Warden series soils may have disrupted the natural drainage patterns, grain size distribution, or lime content that vary systematically with depth in the other non-graded vineyards (Table 4). All of these factors may be interrelated, making it difficult to pinpoint a single variable that may be most responsible for differences in vineyard vigour. Previous studies have correlated the onset of black leaf and other problems in Washington State to within vineyard variations of soil moisture, drainage, and grading (Ahmedullah and Dow, 1982; Silbernagel *et al.*, 1998). Whatever the cause of the observed vine stress in the Belle Ville vineyard, the apparent correlation with topography and drainage patterns suggests that environmental variation can have at least as significant an effect as can viticultural practices in vineyard performance.

DISCUSSION

The vineyards in the Red Mountain AVA share a common history of soils developed on eolian materials (loess or dune) over variable sediments that derive from giant glacial outburst floods, which in turn overlie basalts of the Columbia River Group. Variations in soil type within the appellation or even within individual vineyards, such as Ciel du Cheval, are significant but pale in comparison to that of vineyards just outside of the appellation such as on adjacent basalt knobs (Fig. 14E) or on the Yakima River floodplain (Fig. 14F). This illustrates the importance of terroir in defining a sense of place, that with appropriate vinification can lead to a distinct wine style. The Red

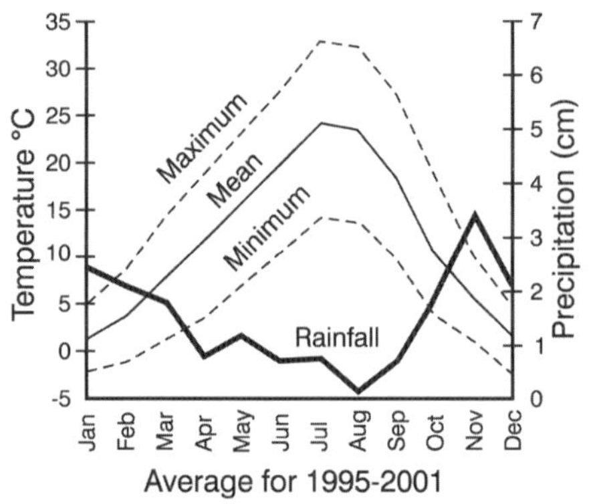

Figure 12 Graph of monthly maximum, mean, and minimum temperatures in contrast to precipitation for the Red Mountain area. Data from Table 5.

Figure 13 Photographs of vineyard features in the Red Mountain area. A) View looking northeast from the Klipsun vineyard (#10) across the Hedges Cellars (#8) and Bell Ville (#2) vineyards towards Red Mountain. B) Newly planted vines of the Golitzen vineyard (#7) have a northeasterly trend relative to the older N–S plantings of the Sand Hill (#12) and Hedges Cellars (#8) vineyards. C) View from Artz Vineyard (#1) northwest across the Yakima River towards Rattlesnake Mountain. D) View from Klipsun vineyard (#10) south towards Goose Hill. E) Vineyard planted on very shallow (<1 m, locally <10 cm) soils on top of basalt knob approximately 700 m southwest of Red Mountain AVA illustrating difference in plant vigour relative to Klipsun vineyard (#10). F) Vineyard planted on Yakima River floodplain approximately 100 m west of Red Mountain AVA illustrating difference in vine and ground cover vigour relative to Klipsun vineyard (#10). Photos D and F taken on the same day.

Mountain wine style has been described by wine critics as powerful and intense (Antrim, 2002). Some terroir elements that contribute to this wine style include high heat summation, intense growing-season solar energy from nearly cloud-free skies, very low annual precipitation, and large growing-season soil-moisture deficits.

The numerous awards won by wines made from the Red Mountain appellation suggest that this grape-growing region is indeed becoming recognized. For example, in the past two years wines made from Red Mountain grapes have received nine scores of 94 or above out of 100 in independent blind tastings by national wine publications (Jim Holmes, written communication, 2002). In addition, the Andrew Will Merlot made from grapes from the Ciel Du Cheval Vineyard was named the best merlot in the United States by Food and Wine Magazine in 2000 (Anon., 2000) and by Wine & Spirits Magazine in 2002 (Teague, 2002). Finally, the Klipsun vineyard of Red Mountain was recognized in 2002 as one of the top 25 vineyards in the world by Wine & Spirits Magazine (Antrim, 2002).

Comparisons with other appellations in Washington State (*e.g.,* Meinert and Busacca, 2000), North America, and other wine regions in the world show some common threads and also some important differences. As noted at the onset of this paper, more than 90% of Washington vineyards are located in areas affected by glacial outburst floods. In the Red Mountain area, these flood sediments were mostly deposited from the swirling back-eddies behind Red Mountain and include numerous lenses of relatively coarse gravel. In the Walla Walla area, the flood sediments are generally finer grained as a result of their deposition from ponded floodwaters behind Wallula Gap, although there are some zones of coarse gravels in modern river channels (Fig. 9C, D, E; Meinert and Busacca, 2000). Even with these differences in grain size, however, both Red Mountain and Walla Walla have in common a mantle of loess and more recent wind-blown sands over the glacial flood sediments.

The common glacial history of most Washington vineyards is offset by differences in a variety of climate measures. Red Mountain is near the extremes for Washington State in terms of high heat summation (growing-degree days) and low precipitation. For example, most of the Yakima Valley has 200 to 400 fewer growing-degree days than Red Mountain, and Walla Walla has more than twice the precipitation on average (Meinert and Busacca, 2000). Like many Washington appellations, Red Mountain is susceptible to both spring frosts and winter kill. According to the optimum climatic criteria for winegrape growing identified by Gladstones (1992, 2001), Red Mountain is near the top in growing-degree days, growing-season solar radiation, mean temperature, and rainfall. However, Red Mountain is outside of the optimum ranges identified by Gladstones for relative humidity, high maximum temperature, and temperature variability index (roughly the daily variation between highest and lowest temperatures), although the very high-quality wines that have been produced from Red Mountain grapes suggest that the subtle interactions of climatic factors are not completely understood.

With the clear evidence of glacial influence on Washington state vineyards, it seems appropriate to examine how other viticultural regions have been affected by glaciation. Although the complex sedimentary patterns caused by the glacial outburst flooding in the back-eddy of Red Mountain may be unique among world vineyards, many other wine-producing areas of the world also have links to Pleistocene glaciation. This is primarily due to two factors: worldwide lowering of sea level during maximum glaciation, and the locally abundant clastic sediments produced by alpine and continental glaciers.

At the glacial maximum, sea level was lowered by about 130 m relative to today, exposing many coastal areas to increased erosion and changing sedimentation patterns, such as wind-blown loess and sand dunes from the newly exposed coastal shelves. Many viticultural areas within 100 km of the ocean have been so affected. Concurrent with this sea level drawdown was an increased sediment load in both rivers and valley glaciers fed by mountain ranges that intercepted moisture-laden maritime air and consequently built up extensive icefields.

A prime example of this is the Graves-Médoc region of Bordeaux, France (Wilson, 1998). Outwash gravels from alpine glaciation in the Pyrenees along the French-Spanish border and the Massif Central in central France overloaded the Garonne and Dordogne rivers leading to the Gironde Estuary, which itself had been enlarged and deepened in response to the lowering of sea level. Each period of

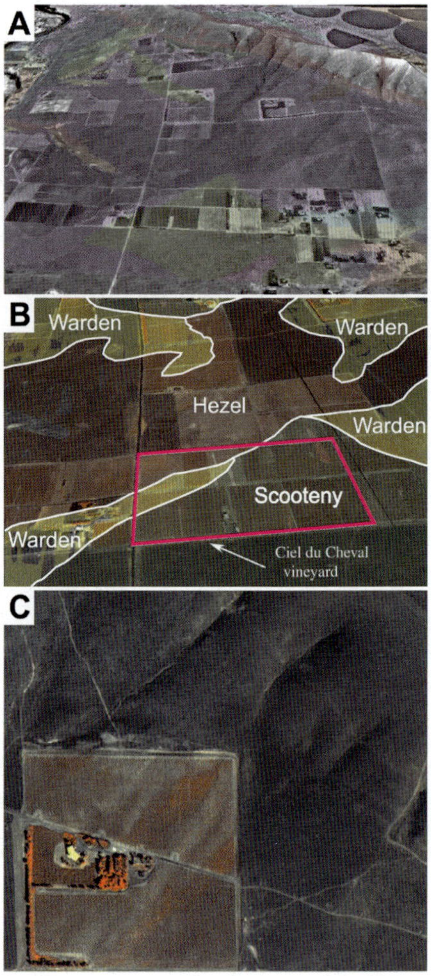

Figure 14 Soil and vegetation patterns for vineyards in the Red Mountain AVA. A) 3-D perspective view of Red Mountain area with soil series draped over black and white ortho aerial photograph. Data from Rasmussen (1971). B) Close-up of 14A showing soil series patterns within individual vineyards. In the Ciel du Cheval vineyard (#5) some vine rows cross three different soil types, providing an opportunity to study the effect of different soil characteristics on grape vine performance with little or no change in climate or viticultural practice. C) Aerial color infrared imagery showing differences in plant vigour relative to topography and soil distribution in the Belle Ville vineyard (#2). Note that in infrared images, green vegetation such as leafy vineyards is red whereas arid range lands are dark green. Images courtesy of Francis Pierce, Center for Precision Agricultural Systems, WSU-IAREC.

glaciation produced its own series of gravel outwash floodplains along the rivers. Of the four glacial stages identified by Wilson (1998), the Günz stage produced the gravel mounds that are synonymous with the best (First Growth) vineyards (Wilson, 1998). Wineries such as Chateaux Lafite-Rothschild, Haut-Brion, Latour, and Mouton-Rothschild are well known to wine-lovers throughout the world, and each of these wineries and their estate vineyards are located on gravel mounds (the geographical name 'Graves' means gravel in French) formed during a Pleistocene glacial maximum. Wilson (1998, his Table 5.1) provided an extended analysis of the various glacial and interglacial epochs of the Bordeaux region along with correlations to archeological timelines that predate settlement of North America by tens of thousands of years, such as the prehistoric cave art of Les Eyzies, Lascaux, and Pech Merle. Even without the common terroir connection to Washington State this makes for fascinating reading!

Less well known are the gravel outwash plains of the South Island of New Zealand. These were fed by the extensive alpine glaciation of the Southern Alps mountain range that transects the island. These gravels form the substrate for many of the vineyards in the Marlborough area of New Zealand, and some of the wineries of this region focus on the coarse gravels, such as the Rapaura Series, for their best vineyards (http://www.stoneleigh.co.nz/). Another area in New Zealand, Gimblett Gravels, is perhaps the first viticultural region in the world to specifically define itself on the basis of the gravel. Legally, wines from this appellation have to consist of at least 95% grapes grown on the Gimblett Gravels, specifically the Omahu Gravels (http://www.gimblettgravels.com/index.htm). These gravels are somewhat different from those in the Marlborough area in that they have been extensively reworked by present and paleoflows of the Ngaruroro River. Within the Gimblett Gravels are lenses of sand, silt, and clay at various depths. These lenses contain up to 20% silt and 9% clay. In general, the gravels are free draining with little water-holding capacity, mostly between 9 and 20% soil moisture (http://www.gimblettgravels.com/terroir.htm). These features are similar to those documented for the gravels of Bordeaux, France (Figure 6.3 of Wilson, 1998) and some of the soil and gravel lenses of Red Mountain, Washington (Table 4, Fig. 9D, F).

Even though some vineyards in Washington State, Bordeaux, and New Zealand share elements of a common Pleistocene glacial history, there still are large differences in climate, viticultural, and oenological practices that make these areas distinct. The present study does not aim to minimize these other factors or the importance of human ingenuity in making great wine from great vineyards (Moran, 2001), but seeks to illustrate the importance of understanding the physical environment as one essential element of terroir. From both an oenological and geological perspective, the terroir of Red Mountain is truly excellent.

ACKNOWLEDGEMENTS

This study grew out of initial discussions with Tom Hedges and Lorne Jacobsen of Hedges Cellars. We are greatly indebted to all the vineyard owners and winemakers who so generously gave of their time and in many cases, their wine. In particular, Fred Artz, Patricia Gelles, and Jim Holmes were indispensable for their exhaustive knowledge of Red Mountain history and viticultural practice. Doug McCrea, owner and winemaker at McCrea Cellars, deserves special mention for his understanding of terroir in guiding production of his magnificent Syrah wines. We also acknowledge funding from the Washington Wine Commission for part of this study. Thanks to Francis Pierce, Center for Precision Agricultural Systems, WSU-IAREC for arranging and funding the aerial overflight of Red Mountain that produced the infrared image of the vineyards as seen in Figures 8 and 14. Earlier versions of this manuscript were improved by informal reviews by John Dingenthal, David Gelles, Jeff Hedenquist, Larry Pearson, and Kathleen Willemsen. In addition, Patricia Bowen and Roger Macqueen provided formal reviews that further improved the manuscript. Any errors remaining are the responsibility of the authors.

REFERENCES

Ahmedullah, M. and Dow, A.I., 1982, Black leaf of grapes in Washington: Washington State University Cooperative Extension Bulletin EB 0745, 2 p.

Anon., 2002, Top 10 Merlot: Wine & Spirits Magazine, New York City, NY, v. 21, June, p. 98-99.

Antrim, T., 2002, 25 Great Vineyards of the World: Wine & Spirits Magazine, New York City, NY, v. 21, p. 86-115.

Baker, V.R. and Nummedal, D., eds., 1978, The Channeled Scabland (a guide to the geomorphology of the Columbia Basin, Washington): National Aeronautics and Space Administration, 186 p.

Baksi, A.K., 1989, Reevaluation of the timing and duration of extrusion of the Imnaha, Picture Gorge, and Grande Ronde Basalts, Columbia River Basalt Group: *in* Reidel, S.P., and Hooper, P.R., eds., Volcanism and tectonism in the Columbia River flood-basalt province: Geological Society of America Special Paper 239, p. 105-112.

Boling, M., Frazier, B. and Busacca, A., 1998, General Soil Map, Washington: Department of Crop and Soil Sciences, Washington State University, Pullman and USDA Natural Resources Conservation Service, scale 1:750,000.

Bretz, J.H., 1923, The channeled scablands of the Columbia Plateau: Journal of Geology, v. 31, p. 617-649.

Bretz, J.H., 1925, The Spokane flood beyond the channeled scablands: Journal of Geology, v. 33, p. 97-115, 312-341.

Bretz, J.H., 1928a, Bars of channeled scabland: Geological Society of America Bulletin, v. 39, p. 643-702.

Bretz, J.H., 1928b, The channeled scabland of eastern Washington: Geographical Review. v. 18, p. 446-477.

Bretz, J.H., 1928c, Alternate hypothesis for channeled scabland: Journal of Geology. v. 36, p. 193-223, 312-341.

Bretz, J.H., 1932, The Grand Coulee: American Geographical Society Special Publication, n.15, 89 p.

Brown, B.H., 1937, The State-Line earthquake at Milton and Walla Walla: Seismological Society of America Bulletin, v. 237, p. 205-209.

Busacca, A.J., 1991, Loess deposits and soils of the Palouse and vicinity, *in* Baker, V.R., Bjornstad, B.N., and others, Quaternary Geology of the Columbia Plateau: *in* Morrison, R.B., ed., Quaternary Nonglacial Geology - Conterminous United States: Geological Society of America, DNAG, Geology of North America, v. K-2, p. 216-228.

Busacca, A. J., Marks, H. M., and Rossi, R., 2001, Volcanic glass in soils of the Columbia Plateau, Pacific Northwest U.S.: Soil Science Society of America Journal, v. 65, p. 161-168.

Douglas, D., Cliff, M. A., and Reynolds, A.G., 2001, Canadian terroir: characterization of Riesling wines from the Niagara Peninsula: Food Research International, v. 34, p. 559-563.

Flint, R.F., 1938, Origin of the Cheney-Palouse Scabland Tract, Washington: Geological Society of America Bulletin, v. 49, p. 461-524.

Gladstones, J., 1992, Viticulture and environment: Winetitles, Underdale, Australia, 310 p.

Gladstones, J., 2001, Climatic indicators guide site selection: Practical Winery and Vineyard, v. 23, p. 9-18.
Halliday, J., 1993, Wine Atlas of California: Penguin Books, NY, NY, 400 p.
Halliday, J., 1999, Wine Atlas of Australia & New Zealand: Harpercollins Australia, 496 p.
Haynes, S.J., 1999, Geology and Wine 1. Concept of *terroir* and the role of geology: Geoscience Canada, v. 26, p. 190-194.
Haynes, S.J., 2000, Geology and Wine 2. A geological foundation for *terroirs* and potential sub-*appellations* of Niagara Peninsula wines, Ontario, Canada: Geoscience Canada, v. 27, p. 67-87.
Hooper, P.R., 2000, Chemical discrimination of Columbia River basalt flows: Geochemistry, Geophysics, Geosystems: v. 1, Paper number 2000GC000040, 16 p.
Landon, R.D. and Long, P.E., 1989, Detailed stratigraphy of the N_2 Grande Ronde Basalt, Columbia River Basalt Group, in the central Columbia Plateau: *in* Reidel, S.P. and Hooper, P.R., eds., Volcanism and tectonism in the Columbia River flood-basalt province: Geological Society of America Special Paper 239, p. 55-66.
Mann, G.M. and Meyer, C.E., 1993, Late Cenozoic structure and correlations to seismicity along the Olympic-Wallowa Lineament, northwest United States: Geological Society of America Bulletin, v. 105, p. 853-871.
Meinert, L.D. and Busacca, A.J., 2000, Geology and Wine 3. *Terroirs* of the Walla Walla Valley *appellation*, southeastern Washington state, USA: Geoscience Canada, v. 27, p. 149-171.
Molinaar, D., 1988, The Spokane Aquifer, Washington: Its geologic origin and water-bearing and water-quality characteristics: U.S. Geological Survey Water Supply Paper 2265, 74 p.
Moran, W., 2001, Terroir: the human factor: The Australian and New Zealand Wine Industry Journal, v. 16, p. 32-51.
Mullineaux, D.R., 1996, Pre-1980 tephra-fall deposits erupted from Mount St. Helens, Washington: U.S. Geological Survey Professional Paper 1563.
O'Connor, J.E. and Baker, V.R., 1992, Magnitudes and implications of peak discharges from glacial Lake Missoula: Geological Society of America Bulletin, v. 104, p. 267-291.
Pardee, J.T., 1910, The Glacial Lake Missoula: Journal of Geology, v. 18, p. 376-386.
Peterson-Nedry, J., 2000, Washington Wine Country: Graphics Art Center Publishing, Portland, Oregon, 111 p.
Pringle, P.T., 1993, Roadside Geology of Mt. St. Helens National Volcanic Monument and Vicinity: Washington Department of Natural Resources Information Circular 88, 120 p.
Rasmussen, J., 1971, Soil Survey of Benton County Area, Washington: USDA Soil Conservation Service, US Government Printing Office, Washington, DC. 72 p. + map sheets.
Reidel, S.P. and Fecht, K.R., 1994, Geologic Map of the Richland 1:100,000 Quadrangle, Washington: Washington Division of Geology and Earth Resources Open File Report 94-8, 21 p., 1 plate.
Reidel, S.P., Fecht, K.R. and Lindsey, K.A., 1992, Post-Columbia River basalt structure and stratigraphy of south-central Washington: Geological Society of America, Abstracts with Programs, v. 24, p. 78.
Ribéreau-Gayon, P., Dubourdieu, D., Donéche, B. and Lonvaud, A., 2000, Handbook of Enology – the microbiology of wine and vinification: John Wiley and Sons, New York, 454 p.
Silbernagel, J., Lang, N.S., Wample, R.L., Smithyman, R. and Hendrix, W.G., 1998, Geospatial analysis of blackleaf in Vitis: First International Conference on Geospatial Information in Agriculture and Forestry, Lake Buena Vista, Florida, p. 257-264.
Soil Survey Staff, 1999, Soil Taxonomy. 2nd Ed.: Agricultural Handbook Number 436, US Department of Agriculture, Natural Resources Conservation Service. US Government Printing Office. Washington, DC. 869 p.
Teague, Lettie, 2002, Drinks-Honors, American Wine Awards: Food and Wine Magazine, October, 2002, p. 151-153.
Tolan, T.L., Reidel, S.P., Beeson, M.H., Anderson, J.L., Fecht, K.R. and Swanson, D.A., 1989, Revisions to the estimates of the areal extent and volume of the Columbia River Basalt Group: *in* Reidel, S.P. and Hooper, P.R., eds., Volcanism and tectonism in the Columbia River flood-basalt province: Geological Society of America Special Paper 239, p. 1-20.
Waitt, R.B., 1980, About forty last-glacial Lake Missoula jökulhlaups through southern Washington: Journal of Geology, v. 88, p. 653-679.
Waitt, R.B., 1985, Case for periodic, colossal jökulhlaups from Pleistocene glacial Lake Missoula: Geological Society of America Bulletin, v. 96, p. 1271-1286.
Weis, P.L. and Newman, W.L., 1989, The Channeled Scablands of eastern Washington - The geologic story of the Spokane Flood: Eastern Washington University Press, Cheney, WA, 25 p.
West, M.W., Ashland, F.X., Busacca, A.J., Berger, G.W., and Shaffer, M.E., 1996, Late Quaternary deformation, Saddle Mountains anticline, south-central Washington: Geology, v. 24, p. 1123-1126.
Wilson, J.E., 1998, *Terroir*: The role of Geology, Climate, and Culture in the Making of French Wines: Mitchell Beazley, London, UK, 336 p.
Wilson, J.E., 2001, Geology and Wine 4. The origin and odyssey of Terroir: Geoscience Canada, v. 28, p. 139-142.
Winkler, A.J., Cook, J., Kliewer, W. and Lider, L., 1974, General viticulture: University of California Press, Berkeley, California, 710 p.

Accepted as revised 23 October 2002

SERIES

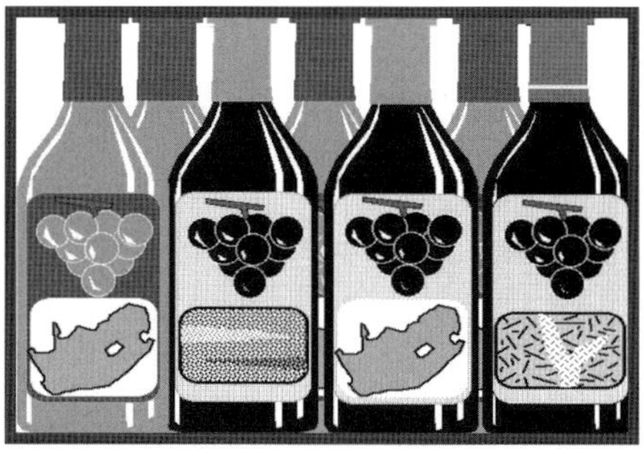

Geology and Wine 7. Geology and Wine Production in the Coastal Region, Western Cape Province, South Africa

Christopher J. Bargmann
Cape Wine Master
5, Allt-y-Wennol, Pontprennau
Cardiff CF23 8AS, United Kingdom
c.bargmann@virgin.net

SUMMARY
South Africa's wine industry, founded in 1659, ranks 16th in the world in terms of vineyard acreage and 7th in terms of wine production, making the country a significant player in the world wine industry. The industry saw early success with the wines of Constantia in the 18th century and is currently recovering from the influence of legislative controls that restricted the industry over much of the 20th century.

The Coastal Region, which includes the important Stellenbosch and Paarl Districts, forms the heartland of the industry and accounts for approximately 45% of the country's wine production. Geologically this important region is underlain by argillaceous sediments of the Malmesbury Group (late Proterozoic); granitic intrusives of the Cape Granite Suite (Late Proterozoic–Cambrian); conglomerates and mudstones of the Klipheuwel Group (Ordovician); and arenaceous sediments of the Table Mountain Group (Middle Ordovician–Early Devonian). In addition, Cenozoic deposits – soils, river gravels, aeolian sands and bedrock weathering products including saprolites and kaolin – are important factors. Soils consisting of loam or containing clay layers are important owing to their ability to retain moisture that can be utilized by vines during the dry summer months.

The two main subdivisions of the Coastal Region, the Stellenbosch and Paarl districts, are discussed below in more detail. Their geological, topographic and general climatic characteristics illustrate the wide range of terroirs present in districts and are features to be considered as the country ponders over changes to its Wine of Origin legislation.

SOMMAIRE
Fondée en 1659, l'industrie vinicole d'Afrique du Sud est aujourd'hui la 16e en importance au monde quant à la superficie cultivée et la 7e quant au volume de vin produit, ce qui en fait un producteur significatif de l'industrie vinicole planétaire. L'industrie a connu ses premiers succès avec les vins de Constance au 18e siècle, et de nos jours, elle se remet progressivement des effets législatifs contraignants qui ont été en vigueur au cours d'une grande partie du 20e siècle.

La région Côtière, en particulier les importants districts de Stellenbosch et de Paarl, constituent le coeur de l'industrie avec 45 % de la production de vin du pays. Le substratum géologique de cette importante région est constitué des sédiments argileux du Groupe de Malmesbury (fin du Protérozoïque); des masses intrusives de la suite granitique du Cap (fin du Protérozoïque-début du Cambrien); des conglomérats et des lutites du Groupe de Klipheuwel (Ordovicien), et; des sédiments arénacés du Groupe de Table Mountain (Ordovicien moyen-début Dévonien). Les couches cénozoïques de sols, graviers fluviatiles, sables éoliens ainsi que les produits d'altération comme les saprolites et les kaolins sont également des facteurs importants. Les sols constitués de loam ou qui renferment des couches d'argiles sont importants pour la survie de la vigne durant les mois secs de l'été, à cause de leur capacité à retenir l'eau.

Les deux principales zones de la région Côtière, soit les districts de Stellenbosch, et celui de Paarl, font l'objet d'une discussion détaillée ci-dessous. Leurs caractéristiques géologiques, topographiques et climatiques générales permettent de comprendre le large éventail de terroirs de ces districts et constituent autant de facteurs à considérer au moment oj le pays considère la possibilité d'apporter des changements à sa législation sur les vins d'origine.

INTRODUCTION

South Africa's wine industry, based mainly around the towns of Stellenbosch and Paarl, some 40–50 km east of Cape Town, has a long tradition dating back to 1659. The country's wine production is a relatively small but significant contributor to the world wine scene. Since the return to democracy in 1994 the country has experienced booming wine exports, new cellars coming into production, and impressive increases in the value of historic vineyard properties in the heart of the winelands. Today the wine industry is poised to compete alongside other new world producers such as the Californians, Chileans, Argentines and Australians. Three hundred years of winemaking history also provides an important component to Cape Town's tourist industry, particularly through the presence of historic estates such as Groot Constantia (Fig. 1), and stunning geological scenery (Fig. 2). As Canadian wine writer Tony Aspler (1998, p. 226) stated: "Without doubt, the Cape is the world's most spectacular wine region in terms of its scenery".

For a country rich in geological heritage, surprisingly little has been written specifically on the geology of South Africa's winelands, a fact that this paper seeks to address. The aim of this paper is to introduce the reader to the geology of the Coastal

Figure 1 Groot Constantia Estate, established in 1680 by Simon van der Stel. Photo C. Bargmann.

Figure 2 Mountains at the southern end of Franschhoek Ward, Paarl District of the Coastal Region. Photo C. Bargmann.

Region and highlight some of the geological components that make the wine growing conditions of the area unique.

Two noteworthy previous publications are field trip excursions organised by Lawrie Minter from the University of Cape Town as part of the 6th International Conference on Fluvial Sedimentology, 1997, and the International Volcanological Congress, 1998 (Joubert et al., 1997; Minter, 1998). Additionally, Wooldridge (2000) detailed some aspects of geology affecting vineyards in South Africa, and Saayman (1977) discussed the influence of soil and climate on wine quality. The following M.Sc. theses also detail aspects of geology in South African vineyards. Carey (2001) described the definition of natural terroir units in the Stellenbosch District, and a summary of the principles involved is given in Carey et al. (2002). Various aspects of vineyard soils were described by van Schoor (2001) while Wooldridge (1988) demonstrated how the geological origins of soils affect clay mineralogy and potassium content.

The Infrutec-Nietvoorbij Institute of the Agricultural Research Council (ARC) in Stellenbosch is currently conducting a research project on terroir in the wine regions of South Africa. According to John Wooldridge of ARC, "the South African wine industry is committed to the process of terroir identification and demarcation, and is making every effort to do so on the basis of sound scientific principles including geology, which was previously thought by many to be irrelevant" (e-mail of 24 March 2003). This relatively new project is ongoing and only limited results have been published to date. These include the effects of soil and climate on Sauvignon Blanc grapes (Conradie et al., 2002), and the possible influence of soil types on Cabernet Sauvignon grapes (Conradie, 2002). Bonnardot et al. (2002) detailed the influence and effects of sea breezes, while the effects of the mountains around Stellenbosch on topography are discussed by Wooldridge and Beukes (2003a, b).

Early History

The first European settlement in Southern Africa was established by the Dutch East India Company as a re-supply point for their fleet on the sea journey between Holland and the Dutch colonies in the East Indies (modern Indonesia). This settlement was established on 6th April, 1652, and grew to become the bustling modern city of Cape Town.

One of the earliest decisions made by settlement commander Jan van Riebeeck was to cultivate vines for fruit and wine production. Van Riebeeck sought to supply wine to the ships as a means of reducing the death rate of sailors owing to scurvy. Observing that the climate was similar to that of France and Spain, he reasoned correctly that vines would survive on the southern tip of Africa. The first vine cuttings arrived from Europe in 1655; the first South African wine was made in February, 1659.

The appointment of Simon van der Stel as governor in 1679 resulted in expansion of the fledgling wine industry. The town of Stellenbosch was founded some 40 km east of Cape Town and a model wine and fruit farm established on 770 hectares of land at Constantia to the south of Cape Town. By 1761 the Constantia estate was exporting its red and white wine to Europe and in due course these Muscat based wines were widely acclaimed, possibly the earliest success for wines from the "new world" (Hughes et al., 1992; Hands and Hughes, 1997).

Unfortunately this early success did not continue into the 19th century, as outlined by the wine writer Hugh Johnson (1989). By 1814 the British had taken possession of the Cape, and the industry initially benefited from secure exports to Britain. Unfortunately the wines produced had declined from the former high quality to a greatly inferior standard and the industry suffered when protective duties in Britain were removed in the mid 19th century. Further misfortune struck in the late 19th century when epidemics of powdery mildew as well as the aphid-like insect Phylloxera attacked the vineyards.

Modern Wine Industry

By the early 1900s the wine industry was suffering from overproduction and low prices and in response farmers began to establish co-operative wineries. The structure of the modern South African wine industry owes much to the largest of these, the De Ko-operative Wijnbouwers Vereniging van Zuid Afrika (KWV), which was formed in January, 1918. KWV was given powers to sell the products of members' grape production in hopes of stabilizing farm revenues. Subsequently these powers were strengthened by act of parliament, giving KWV a controlling position over South Africa's wine industry. This situation lasted until the mid 1980s when KWV's controls were progressively relaxed. Over the last twenty years the industry has developed considerably, resulting in new producers, grape cultivars and expansion into new areas. The controlling influence of KWV finally ended in 1997 when the organization converted to a public company (Hughes et al., 1992; Hands and Hughes, 1997; du Plessis, 1999).

SOUTH AFRICAN WINE PRODUCTION

Conditions suitable for the growth of vines and grapes for wine production occur in the hinterland of Cape Town, an area that extends north to Vredendal, and east to Bonnievale and Calitzdorp (Fig. 3). A separate region of vineyards is located along the Orange River in the north of the country, in the Northern Cape Province (Fig. 3).

According to official figures published by the South African Wine Industry Information and Systems organization (SAWIS), the surface area occupied by vineyards totaled 105,566 hectares in year 2000. The Paarl, Stellenbosch and Malmesbury districts, as defined by the KWV and used by SAWIS (Fig. 3), account for 45% of South Africa's vineyard area (Fig. 4). This area corresponds approximately to the Coastal Region Wine of Origin area (see below). These three districts also contain a significant proportion of the country's red grape varietals (Fig. 5). The remaining districts are generally hotter, lower rainfall areas that traditionally have supplied grapes for brandy and port production, dried fruit and table grapes.

On a world scale South Africa ranked 16th in terms of hectares planted and

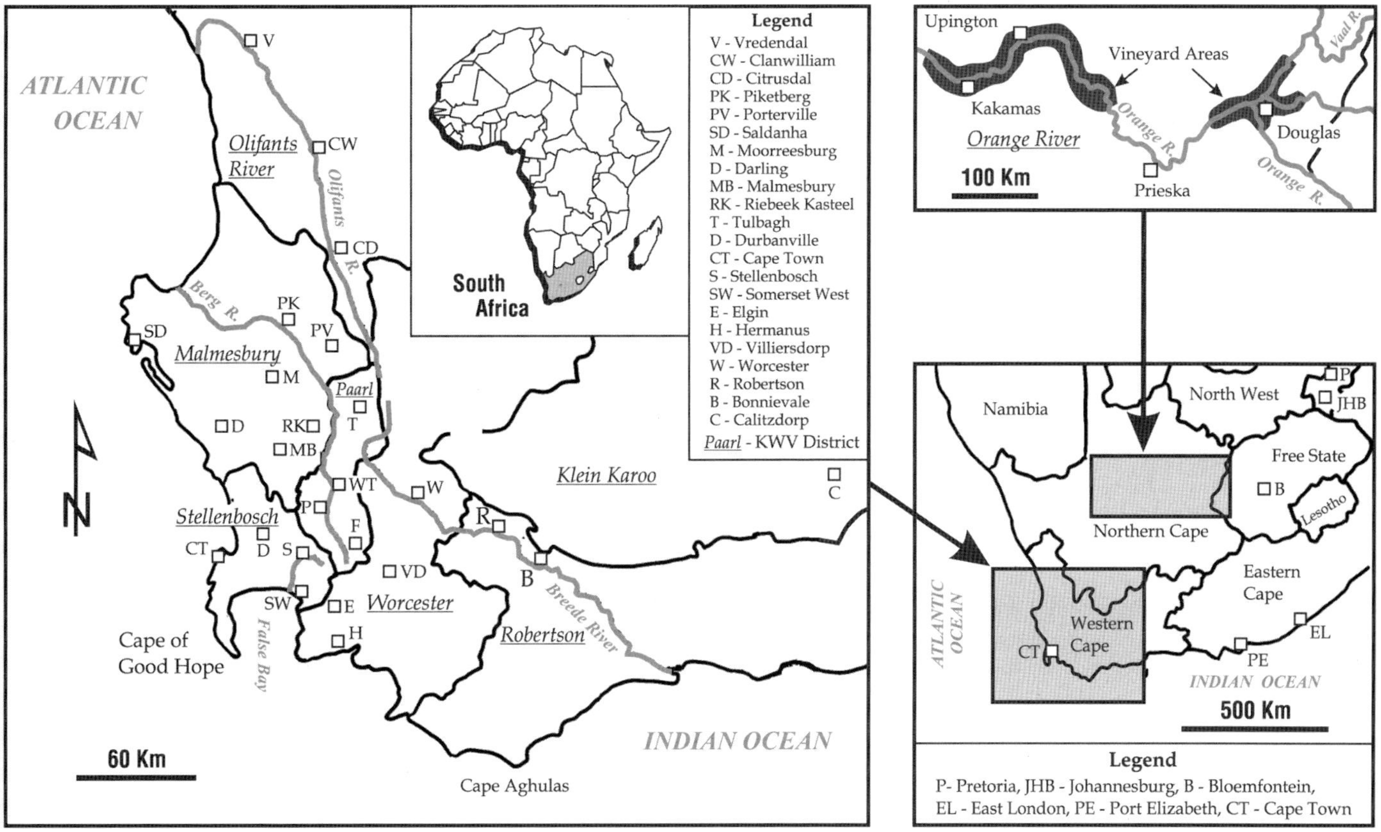

Figure 3 Wine-producing regions of the Western Cape and Northern Cape Provinces, South Africa. KWV regions labelled in underlined italics; cities and towns shown as solid squares with symbols. Modified from KWV (1997).

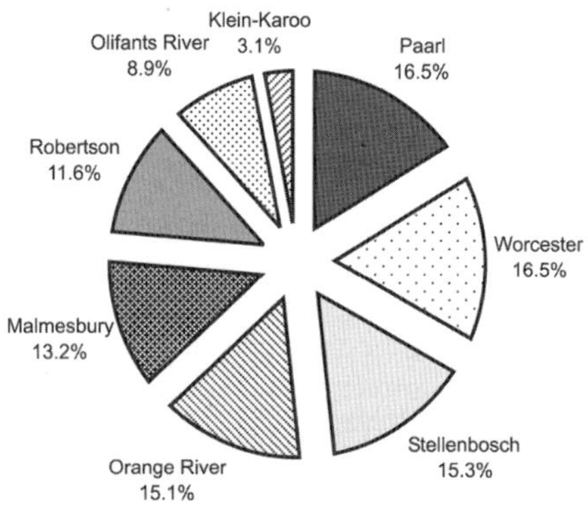

Figure 4 Distribution of South African Vineyards in year 2000; data from SAWIS (2001).

7th in terms of production in 1998, accounting for 1.4% of global vineyards and 3.2% of global wine production (Table 1). In comparison, France, Italy and Spain dominate the wine industry in terms of both surface area planted and production, with the USA and Argentina being the largest players amongst new world producers. For a relatively warm country, South Africa has a curious distribution of vine cultivars, 67.7% being white and 32.3% red in 2000 (Fig. 6). As discussed below, in terms of degree days South Africa's winelands compare with areas such as the Napa Valley, California, and Tuscany, Italy, both of which are noted for their red wine production. Present varietal composition in South Africa reflects the recent history of the wine industry, which encouraged quantity rather than quality, and the production of the country's favorite spirit, brandy. The presence of the table grape white varietals Sultana and Hanepoot shows the importance of the table grape and dried fruit industries.

These varietal percentages (Fig. 6) do not reflect the dramatic changes in South Africa's vineyards in recent years, however. Between 1985 and 2000 the surface area planted increased from 92,142 to 105,566 Ha, and there has also been a significant change in the cultivar composition. Chardonnay and Merlot plantings have increased from virtually zero in 1985 to 5.7% (6017 Ha) and 4.6% (4856 Ha) respectively in 2000 (Fig. 7). Over the same time period, Cabernet Sauvignon, Sauvignon Blanc, Shiraz and Pinotage, South Africa's own varietal, have all shown increases in plantings as the industry moves towards the production of quality cultivars. Colombar has also increased, and has proved

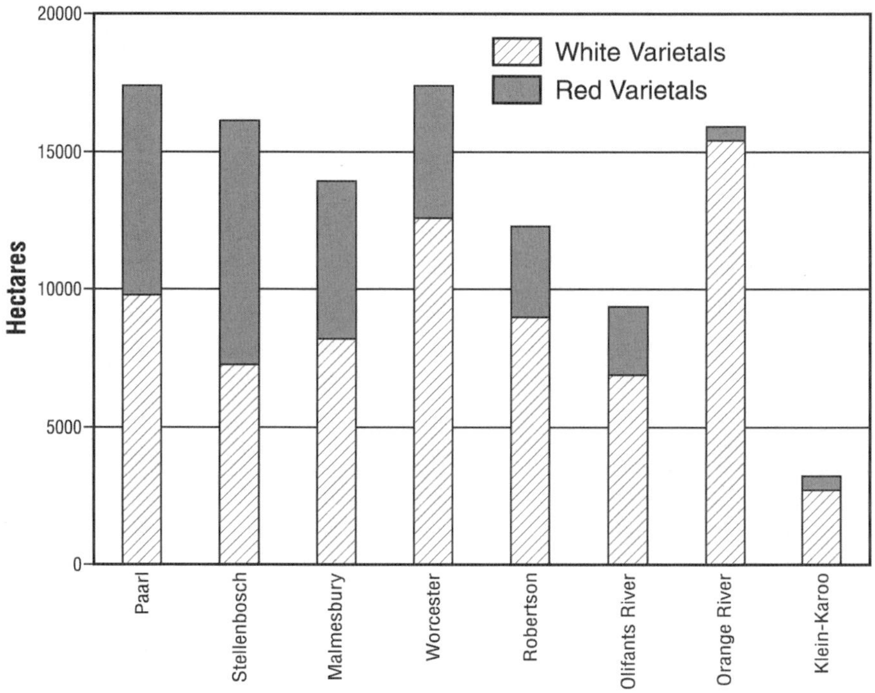

Figure 5 Distribution of red and white grape varietals in South Africa in year 2000. Data from SAWIS (2001).

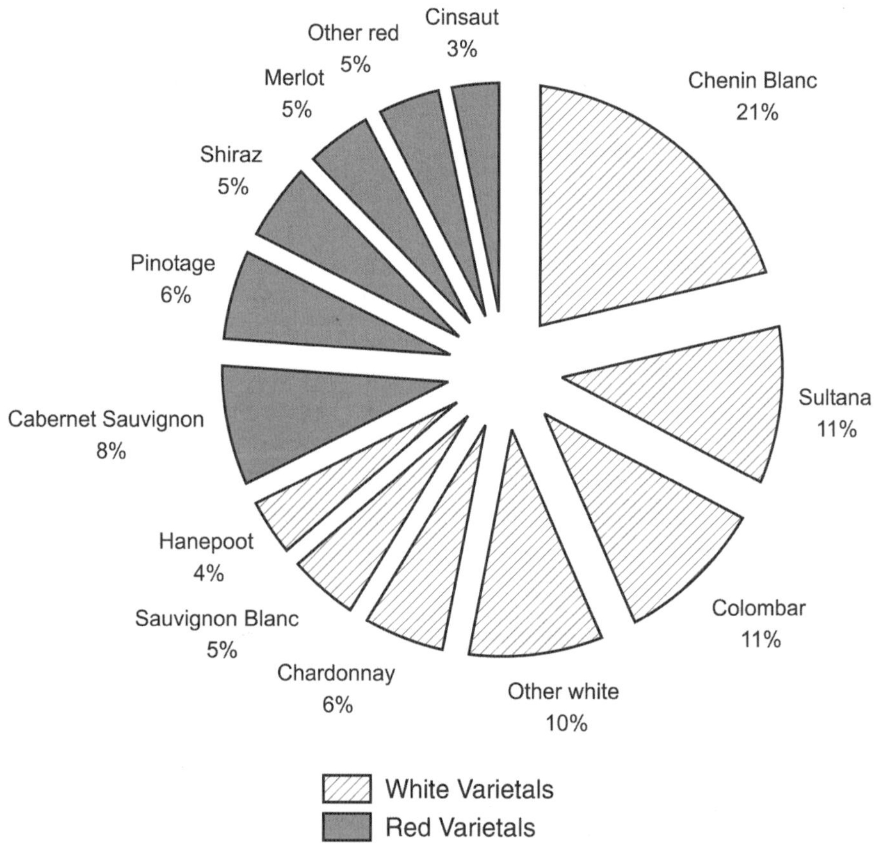

Figure 6 Grape varietal percentage composition of South African vineyards in year 2000. Data from SAWIS (2001).

Table 1 International comparison of areas under vines and wine production (data from KWV 1996, 1997; du Plessis 2000; SAWIS)

Country	Hectares planted with vines 1994	1995	1996	1997	1998	Percentage of world total vineyard area	Ranks (1998)
Spain	1280000	1235000	1224000	1155000	1180000	15.1	1
France	929000	926000	917000	914000	914000	11.7	2
Italy	956000	927000	922000	914000	899000	11.5	3
Turkey	567000	570000	567000	602000	602000	7.7	4
USA	310000	305000	311000	315000	364000	4.7	5
Argentina	-	-	211000	209000	210000	2.7	8
Chile	-	-	116000	132000	144000	1.8	11
South Africa*	103000	103000	106000	108000	111000	1.4	16
Australia	67000	73000	81000	90000	98000	1.3	19

* All vineyards including wine grapes, table grapes, currents and rootstock vineyards.

Country	Wine production (million litres) 1994	1995	1996	1997	1998	Percentage of world total vineyard area	Ranks (1998)
Italy	5927.6	5629.4	5965.0	5084.7	5416.8	20.9	1
France	5464.0	5561.0	5877.3	5361.2	5267.1	20.4	2
Spain	1895.4	1964.0	3267.5	3388.7	3032.0	11.7	3
USA	1617.5	1580.0	1864.3	2500.0	2045.0	7.9	4
Argentina	1817.3	1644.3	1268.1	1350.0	1267.3	4.9	5
South Africa	866.4	754.6	833.3	870.2	815.6	3.2	7
Australia	587.4	502.8	678.4	617.4	741.5	2.9	8
Chile	-	-	382.4	454.9	547.5	2.1	9

capable of making respectable white wines in addition to its role in brandy production. These increases have been largely at the expense of the country's traditional "workhorse" varieties of Chenin Blanc and Cinsault and the "other white" category which includes Palomino, Clairette Blanche and Cape Riesling.

WINE OF ORIGIN LEGISLATION IN SOUTH AFRICA

South Africa's wine regions are defined under the country's *Wine of Origin* legislation passed in 1972. This legislation stipulates that any indication of origin is prohibited unless the area has been officially demarcated (Saayman, 1999). The largest of the demarcated areas are *regions*, followed by smaller *districts* and finally *wards*. Region and district boundaries are based largely on administrative and magisterial divisions, whilst wards, in theory, take into account soil, geographical and climatic factors, although in practice wards commonly are based on the existing farm boundaries. Wine farms may also register as estates and label their wines as *Estate* wine. This requires that the estate wine producer uses grapes sourced solely from his/her property and the wine is made on site. Many wine farms choose not to register as estates, enabling them to use grapes from other vineyards – these wines may be labeled according to the applicable wine of origin rules but cannot claim estate status. The current (2001) demarcated Wine of Origin areas are listed in Table 2 along with the corresponding KWV regions used in the SAWIS information noted above.

In 1993 the Wine of Origin legislation was amended to allow for the creation of "super regions" forming geographical units. As a result, the Western Cape geographical unit has been defined to cover the entire wine producing area of the Western Cape Province (Table 2).

The focus of this paper is the Coastal Region, and in particular the Stellenbosch and Paarl districts shown in Figure 8. The Coastal Region covers a large area and includes the city of Cape Town and the towns of Durbanville, Stellenbosch, Paarl, Wellington and Malmesbury. Six districts are located in the Coastal Region: Paarl, Stellenbosch, Swartland, Tulbagh, Cape Point and Tygerberg. With the exception of the Tulbagh District, these are situated within a broad coastal plain between the Atlantic Ocean in the west and the mountains to the east of the towns of Stellenbosch, Paarl and Malmesbury. The Tulbagh district is an anomaly, as this area lies within an enclosed inland valley that is geographically an extension of the Breede River Region situated to the east of the Coastal Region (Fig. 8). As shown below, the Tulbagh District also is climatically warmer than the rest of the Coastal Region, and for these reasons will not be discussed in this paper.

CLIMATIC CONDITIONS

The Coastal Region benefits from a Mediterranean-style climate with warm to hot, relatively dry summers and cool, wet

Table 2 South Africa wine of origin classification for natural and sparkling wines in 2001 (data from SAWIS, KWV 1997)

Geographical Unit	Region	District	Ward	KWV Region
Western Cape	Coastal Region	Paarl	Franschhoek	Paarl
			Wellington	
		Tulbagh	-	
		Stellenbosch	Bottelary	Stellenbosch
			Devon Valley	
			Jonkershoek Valley	
			Papegaaiberg	
			Simonsberg-Stellenbosch	
		Tygerberg	Durbanville	
		Cape Point	-	
		No District	Constantia	
		Swartland	Groenekloof	Malmesbury
			Riebeekberg	
	No Region (Swartland area)	Piketberg	-	
	No Region (Overberg area)	Overberg	Elgin	Worcester
			Walker Bay	
	No Region (Ceres area)	*No District*	Ceres	
	Breede River	Worcester	Aan-de-Doorns	
			Goudini	
			Nuy	
			Scherpenheuwel	
			Slanghoek	
		Robertson	Agterkliphoogte	Robertson
			Boesmans River	
			Bonnievale	
			Eilandia	
			Hoops River	
			Klaasvoogds	
			Le Chasseur	
			McGregor	
			Vink River	
		Swellendam	Buffeljags	
			Stormsvlei	
	No Region (Bredasdorp District)	*No District*	Elim	
	Klein Karoo	Calitzdorp	-	Klein Karoo
		No District	Montagu	
			Tradouw	
	No Region (Klein Karoo area)	*No District*	Herbertsdale	
			Ruiterbosch	
			Swartberg	
	Olifants River	Lutzville Valley	Koekenaap	Olifants River
		No District	Spruitdrift	
			Vredendal	
			Piekenierskloof	
	No Region (Olifants River area)	*No District*	Cederberg	
	No Region (Northern Cape & Free State are	Douglas	-	Orange River
		No District	Hartswater	
			Lower Orange	
			Riet River	

Boberg Region is introduced for fortified wines produced in the Paarl and Tulbagh districts.

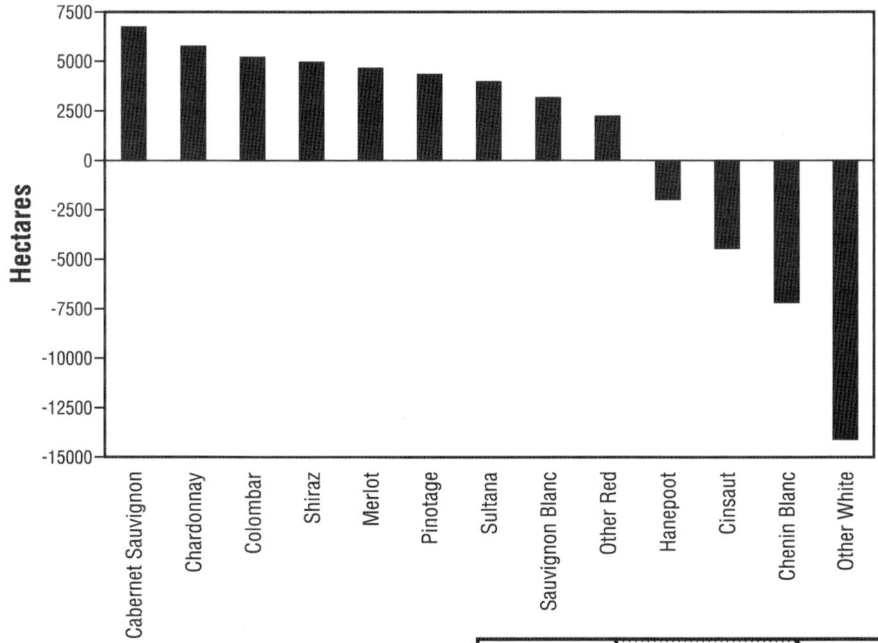

Figure 7 Change in cultivar composition of South African vineyards between 1985 and 2000. Based on data from SAWIS (2001).

winters, conditions that traditionally favour the production of wine grapes. General climatic characteristics will be described using data from the Cape Town, Stellenbosch, Franschhoek and Wellington areas, the locations of which are shown in Figure 8. Conditions become progressively warmer inland with average growing season temperatures (September–March) ranging from 18°C at Kirstenbosch (near Cape Town) to 20.9°C in Wellington, the most inland location. Winter temperatures (April–August) are fairly consistent across the Coastal Region at approximately 14°C (Fig. 9, 10).

Rainfall is concentrated in the southern hemisphere winter months of April–August (Fig. 11). The highest overall rainfall occurs in the Franschhoek area. Here the Groot Drakenstein weather station and Boschendal wine estate receive an average of 932 and 1355 mm/year, respectively (Fig. 12). The influence of the mountainous topography of the Franschhoek area is a significant factor in rainfall distribution. Cape Town, Stellenbosch and Wellington record annual average rainfalls that range between 500 and 700 mm/year (Fig. 12).

Degree Day System

One of the best known systems for international comparison of vine-growing regions used is that of degree days. This system operates by calculation of the heat summation over the growing season. Only days with average temperatures above 10°C are considered and the degree day figure is obtained by multiplying the average daily temperature by the number of days. The methodology used is described by Meinert and Busacca (2000) in an earlier paper in this series.

The system does have its limits. For example, it fails to account for total daylight hours, a factor in high-latitude wine regions.

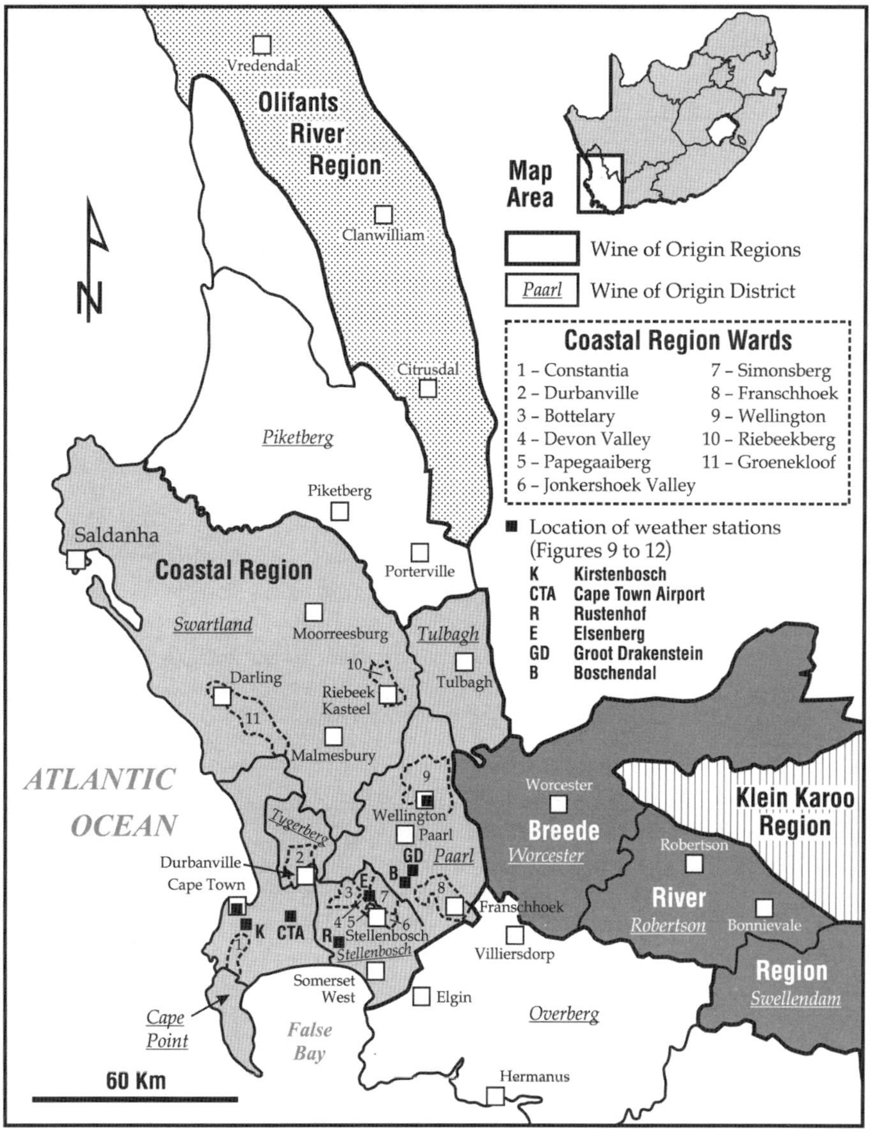

Figure 8 Wine of Origin subdivisions, Western Cape Province, South Africa. Coastal Region contains the Paarl, Tulbagh, Stellenbosch, Tygerberg, Cape Point and Swartland districts; Table 4 contains additional information. Based on data from SAWIS (2001).

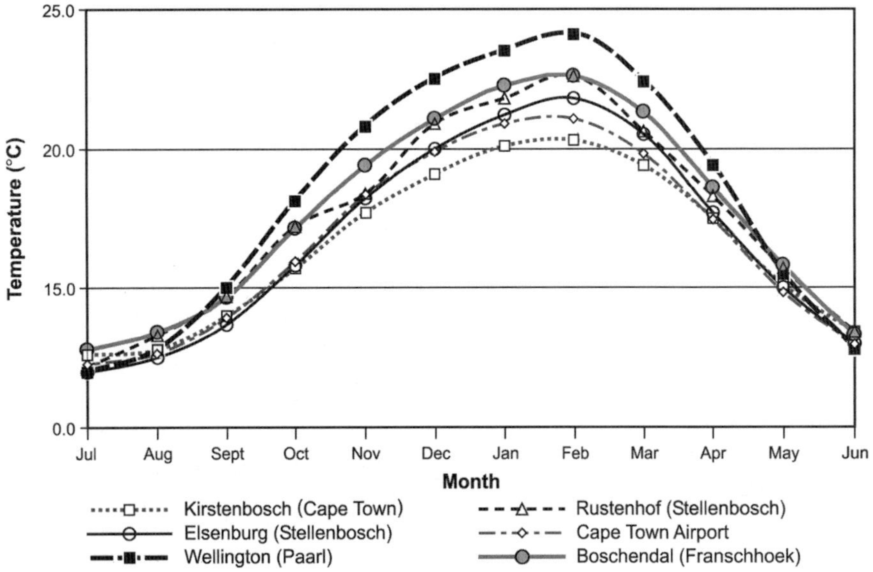

Figure 9 Average monthly temperatures in the Coastal Region of South Africa, July – June. Data from www.worldclimate.com; www.weathersa.co.za; also Agricultural Research Council - Infruitec (pers. comm., 2001; and Boschendal wine estate (pers. comm., 1995).

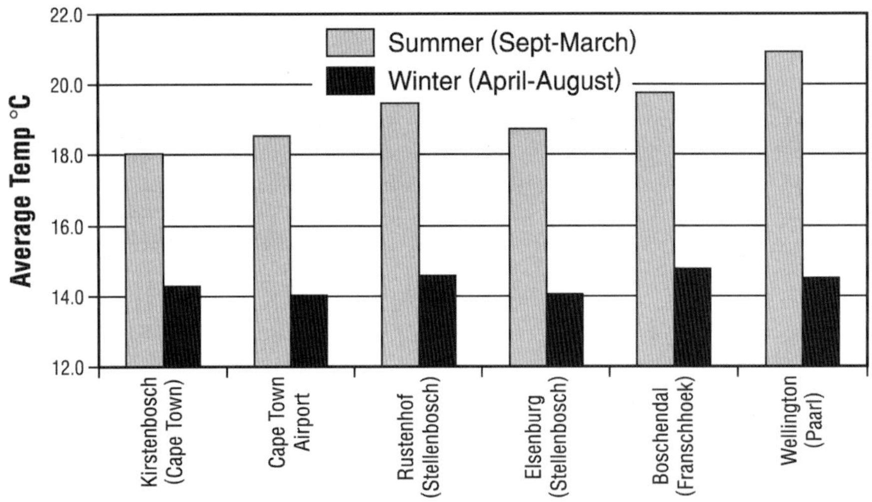

Figure 10 Average growing season and winter temperatures in the Coastal Region of South Africa. Data from www.worldclimate.com; www.weathersa.co.za; also Agricultural Research Council - Infruitec (pers. comm., 2001), and Boschendal wine estate (pers. comm., 1995).

Table 3 Regions defined by the Winkler degree-day classification (data from Gladstones 1994a, Knox 1982)

Region	Degree days (above 50°F)	Degree days (above 10°C)
1	<2500	<1389
2	2501-3000	1389-1667
3	3001-3500	1668-1944
4	3501-4000	1945-2222
5	>4000	>2222

However, degree days offer a useful guide for comparing wine-growing areas. Five categories are identified, with Region I being the coolest and Region V the warmest (Table 3). Gladstones (1994) observed that when this research was published by Amerine and Winkler (1944), the best dry wines were produced in Regions I and II, whilst Region III produced full-bodied dry and sweet wines. Region IV was identified as best for fortified wine production as table wine was usually of inferior quality; whilst Region V, usually irrigated, produced table grapes, dried fruit and poor quality table wines.

DEGREE DAY COMPARISONS

Table 4 presents a compilation of degree day information from various wine regions around the world. Eight locations in the South African Coastal Region are highlighted in bold face and show a degree day range of 1720–2230 using °C. Constantia, Durbanville (to the south and east of Cape Town, respectively), Firgrove and Koelenhof (both in the Stellenbosch District) are the coolest areas, falling within Region III, with degree days ranging from 1720–1906. Durbanville is widely regarded as a colder wine growing area than Koelenhof and the degree day figure of 1906 may not be representative of the Durbanville area as a whole. The two Franschhoek locations, Groot Drakenstein and Franschhoek itself, are warmer, 1967–2014 degree days, thus falling within Region IV. Wolseley and Tulbagh, within the Tulbagh District, in the northeast of the Coastal Region, are noticeably warmer areas with degree day figures of 2104 and 2230, thus falling within Region IV and V, respectively. As would be expected from the more inland location, the Tulbagh district records the warmest conditions. As shown in Table 4 the Stellenbosch and Paarl Districts are comparable in degree day terms to locations in the Napa and Sonoma valleys in California, and Tuscany in Italy.

GEOLOGICAL FRAMEWORK OF THE COASTAL REGION

Western Cape Province has been sculpted by almost continuous ice-free erosion since the breakup of Gondwanaland began ~200 million years ago; as a result the landscape closely reflects the underlying geology. Although weathering is deep and soils are widely developed, bedrock units are

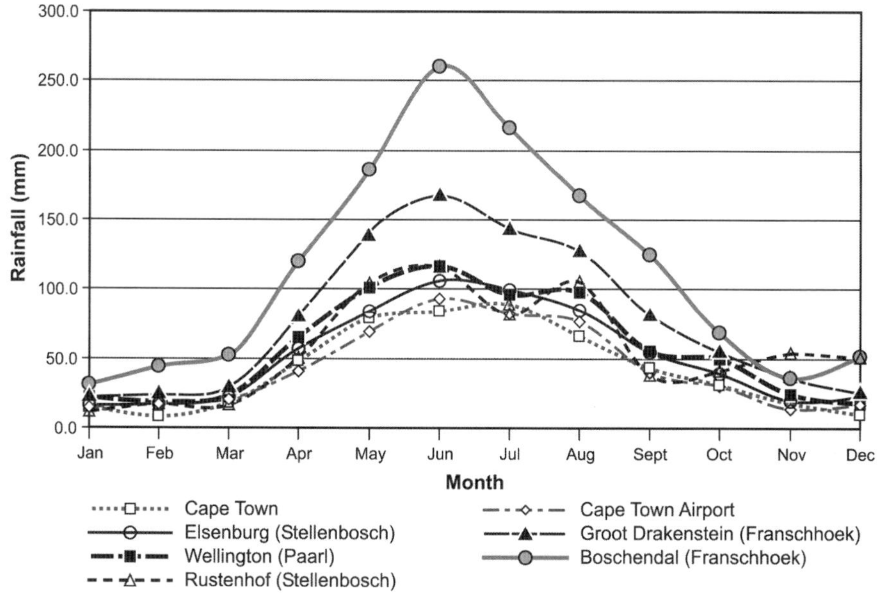

Figure 11 Monthly rainfall in the Coastal Region of South Africa. Data from www.atlasweather.com; www.weathersa.co.za; www.worldclimate.com; also Agricultural Research Council - Infruitec (pers. comm., 2001), and Boschendal wine estate (pers. comm., 1995).

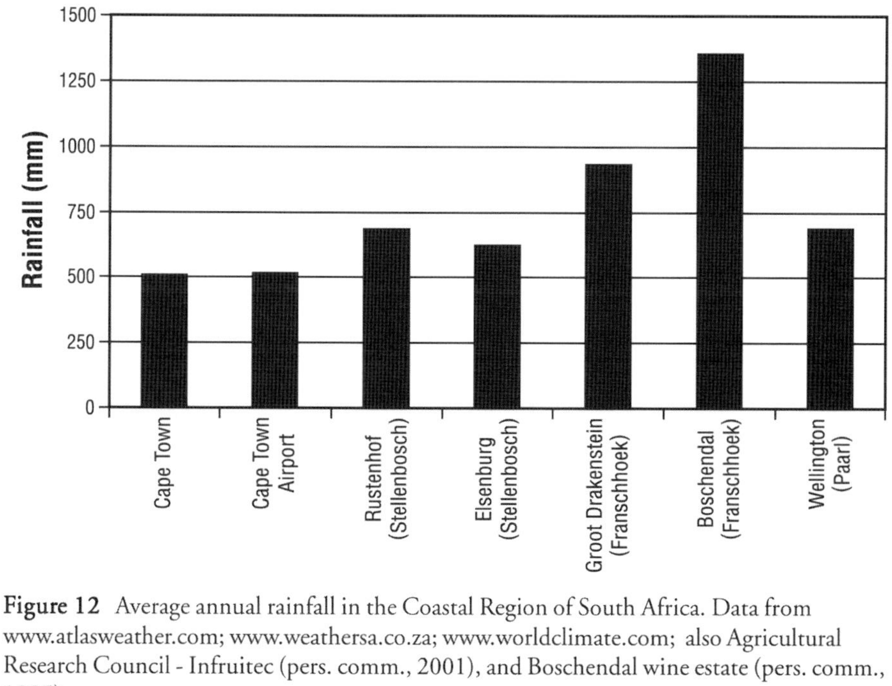

Figure 12 Average annual rainfall in the Coastal Region of South Africa. Data from www.atlasweather.com; www.weathersa.co.za; www.worldclimate.com; also Agricultural Research Council - Infruitec (pers. comm., 2001), and Boschendal wine estate (pers. comm., 1995).

reasonably well known through geological mapping. The Coastal Region includes geological units ranging from Late Proterozoic to Cretaceous in age, capped by more recent Cenozoic deposits. Geology, including the main rock types, is summarized in Tables 5, 6, 7, and a simplified geological map of the Coastal Region is provided in Figure 13. South Africa's vineyards have developed over time in areas where water supply is sufficient for vine growth rather than by specifically matching vines to geology, soils and climate. In terms of the distribution of vineyards the argillaceous sediments of the Malmesbury group and the granites of the Cape Granite Suite are the most important in the Coastal Region. These rocks have an extensive outcrop area and form clay rich soils, which are important for their water retention capabilities. In hotter areas, such as the Breede River and Olifants River regions (Fig. 8), where irrigation using water supplied by the rivers is more widely practiced, alluvial vineyard soils are significant for vineyards.

Malmesbury Group

The Malmesbury Group (Late Proterozoic) is the oldest geological unit of the Coastal Region. The Group formed as part of a pan-African mobile belt system that fringed a stable cratonic area, the Kalahari Craton, which occupied a large portion of Southern Africa in Late Proterozoic time. Sedimentation is considered to have begun at +/-950 Ma and terminated at approximately 570 Ma. The Malmesbury Group is believed to be have been deposited in a geosynclinal basin flanking the western and southern margin of the Kalahari Craton. As the top and base of the Malmesbury Group have not been identified, the stratigraphic thickness has only been estimated as being in excess of 6000 m (Theron et al., 1992; Gresse and Theron, 1992; Rozendaal et al., 1994; Gresse and Scheepers, 1993).

During deposition the Malmesbury Group was deformed by the Saldanian Orogeny, an event which started at +/-700 Ma and continued to +/-500 Ma. Based on the characteristics of the facies, lithology and tectonic character preserved, three distinct terranes – Tygerberg, Swartland and Boland – are identified. These are separated by two major northwest–southeast-trending fault structures known as the Saldanha/Franschhoek and the Wellington/Piketberg faults (Fig. 13).

The Tygerberg Terrane consists of medium- to fine-grained greywacke, phyllitic shales, siltstones and immature quartzites of the Tygerberg Formation. Hornfels, produced as a result of contact metamorphism owing to the later intrusion of the Cape Granite Suite, is also a feature (Theron et al., 1992; Gresse and Theron, 1992).

The Swartland Terrane lies between the Saldanha/Franschhoek and Wellington/Piketberg faults and contains the Berg River, Klipplaat, Moorreesburg and Franschhoek formations. The Berg River, Klipplaat and

Table 4 Degree-day totals for international wine regions (data from Knox, 1999; Meinert and Busacca, 2000; Stevenson, 1991; Paul Cluver Wines, Hamilton Russell Vineyards)

Location	Country/area	Degree days	Winkler Region
Algiers	Algeria	2880	V
Frenso	California	2590	V
Hunter Valley	Australia	2340	V
Calitzdorp	Klein Karoo, SA	2232	V
Tulbagh	**Coastal Region, SA**	**2230**	**V**
Naples	Italy	2220	V
Robertson	Breede River Region, SA	2164	IV
Wolseley	**Coastal Region, SA**	**2104**	**IV**
Lodi	California	2061	IV
Rawsonville	Breede River Region, SA	2032	IV
Franschhoek	**Coastal Region, SA**	**2014**	**IV**
Davis	California	2000	IV
Russian River	California	2000	IV
Ashton	Breede River Region, SA	1995	IV
Groot Drakenstein	**Coastal Region, SA**	**1967**	**IV**
Florence	Tuscany, Italy	1950	IV
Durbanville	**Coastal Region, SA**	**1906**	**III**
Koelenhof	**Coastal Region, SA**	**1898**	**III**
Healdsburg	Sonoma, USA	1767	III
Firgrove	**Coastal Region, SA**	**1756**	**III**
Calistoga	Napa Valley, USA	1745	III
Constantia	**Coastal Region, SA**	**1720**	**III**
Oakville	Napa Valley, USA	1717	III
Barossa Valley	Australia	1710	III
Asti	Piedmonte, Italy	1650	II
Santa Rosa	Sonoma, USA	1634	II
Hamel en Aarde Valley	Walker Bay, SA	1620	II
Walla Walla	Washington, USA	1616	II
Napa	Napa Valley, USA	1595	II
San Luis Obispo	California	1451	II
Coonawarra	Australia	1450	II
Niagara	Canada	1440	II
Bordeaux	France	1400	II
Elgin	Overberg, SA	1387	I
Mclaren Vale	Australia	1350	I
Burgundy	France	1330	I
Hawkes Bay	New Zealand	1300	I
Santa Cruz	California	1285	I
Reims	Champagne, France	1011	I
Otago	New Zealand	1000	I
Geisenheim	Germany	950	I
Trier	Germany	945	I

Moorreesburg formations consist of mica schist, greywacke, minor limestone, and quartz schist. The uppermost Franschhoek Formation consists of conglomerates, grits and quartzites with intermittent shales. The relative age of the Franschhoek Formation is subject to debate, but it is believed to be significantly younger and unconformable upon the rest of the Swartland Terrane (Theron et al., 1992; Gresse and Theron, 1992).

The final component of the Malmesbury Group is the Boland Terrane, which lies to the east of the Wellington/ Piketberg Fault. Gresse and Theron (1992) proposed that the Franschhoek Formation preserved in the Swartland Terrane represents only the basal remnant of a younger sequence now preserved as the Boland Terrane. The grits, conglomerates and greywackes of the Piketberg Formation occur to the north of the Coastal Region, in the vicinity of the town of Piketberg (Fig. 13), and are regarded as a facies variation of the Franschhoek Formation. Gresse and Theron (1992) further proposed that the Franschhoek Formation is overlain by the Norree and Porterville formations. The Norree Formation contains medium-grained to gritty greywacke and quartzites, and occurs in the Breede River valley to the east of the Coastal Region. The Porterville Formation contains predominantly phyllitic shale, schist and fine- to medium-grained greywacke, and occurs in the Wellington area.

Of the successions present in the Malmesbury Group it is the Tygerberg, Moorreesburg, Franschhoek, and Porterville formations that are most significant in vineyard terms in the Coastal Region.

Cape Granite Suite

Plutons of the Cape Granite Suite intrude the sedimentary rocks of the Malmesbury Group: the main plutons are detailed in Table 6 and shown in Figure 13. Various ages of intrusion have been recorded, ranging between 632 and 502 Ma (Kent, 1980; Theron et al., 1992; Rozendaal et al., 1994). These intrusive events are Late Proterozoic to Early Cambrian in age, and partly contemporaneous with the deposition of the Malmesbury Group. The Cape Granite Suite also shows folding and structures related to the Saldanian Orogeny.

Scheepers (1995) interpreted the Cape Granite Suite as being subduction related and intruded in three phases. Phase I, between 600–540 Ma, includes mafic and intermediate intrusives and S-type granites. The granites occur west of the Saldanha/Franschhoek Fault and include the Stellenbosch, Kuils River/Helderberg, Darling and Cape Peninsula plutons, whilst gabbro - diorite bodies occur to the east of the fault in the vicinity of the town of Malmesbury. Phase II occurred between 560–520 Ma and consists of I-type granites situated east of the Saldanha/Franschhoek Fault. These include the Wellington, Perdeberg and Paarl plutons. Phase III, 520–500 Ma, consists of small A-type granites, gabbros and granodiorites, the principal ones being the Klipberg and Boterberg plutons.

Minor intrusive events are also observed in some areas. These range in age from Cape Granite Suite (522 +/-15 Ma) for quartz porphyry dykes to Early Cretaceous (132 +/-6 Ma) for the

Table 5 Pre-Cenozoic stratigraphy of the Coastal Region (based on Reid et al., 1991; Theron et al., 1992; Gresse and Theron, 1992; Gresse and Scheepers, 1993; Scheepers, 1995; Rozendaal and Scheepers 1995)

Age	Group	Formation	Main rocktypes
Early Cretaceous (132 Ma)			Dolerite
Ordovician - Early Devonian (450 - 409 Ma)	Table Mountain Group	Rietvlei	Sandstone, siltstone, shale
		Skurweberg	Sandstone
		Goudini	Sandstone, siltstone, shale
		Cederberg	Shale
		Pakhuis	Diamictite, sandstone
		Peninsula	Sandstone
		Graafwater	Sandstone, siltstone, shale
		Piekenierskloof	Sandstone, conglomerate
Early -Mid Ordovician? (500-450Ma)	Kilpheuwel Group	Populierbos	Mudstone, shale
		Magrug	Conglomerate, sandstone
Late Proterozoic - Early Cambrian (600-500Ma)	Cape Granite Suite		Granite, mafic, granodiorite and alkali granite intrusives
Late Proterozoic (~950-570Ma)	Malmesbury Group	Porterville	Shale, greywacke
		Norree	Greywacke, quartzite
		Piketberg	Grit, conglomerate, greywacke
		Franschhoek	Conglomerate, quartzite
		Moorreesberg	Greywacke
		Klipplaat	Quartz schist
		Berg River	Mica schist, greywacke
		Tygerberg	Greywacke, shale, siltstone, quartzite

Table 6 Cape Granite Suite plutons in the Coastal Region (based on Theron et al., 1992; Gresse and Theron, 1992; Rozendaal et al., 1994; Scheepers 1995; Rozendaal and Scheepers 1995)

Pluton	Plutonic phase and association	Age range	Principle rock types
Boterberg	A Type	520 - 500Ma	Granodiorite
Klipberg			Quartz syenite - alkali granite
Paarl	I Type	560 - 520Ma	Porphyritic granite
Perdeberg			Coarse porphyritic granite
Wellington			Coarse porphyritic to coarse granite
Cape Peninsula	S Type	600 - 540Ma	Porphyritic biotite granite
Darling			Porphyritic and even-grained biotite granites
Kuils River/Helderberg			Coarse porphyritic granite
Stellenbosch			Porphyritic biotite granite, medium- and coarse-grained granite
Mafic Intrusives	Theolitic		Olivine gabbro, gabbro, diorite, granodiorite

northwest–southeast trending dykes known as the False Bay Dolerites (Theron et al., 1992; Reid et al., 1991).

Klipheuwel Group

Clastics of the Klipheuwel Group (Early-Mid Ordovician; Table 5) have limited aerial extent and are generally found in northwest-trending outcrops commonly flanked by similarly trending faults (Fig. 13). An unconformable relationship exists between the Klipheuwel Group and underlying Cape Granite Suite and Malmesbury Formation rocks. The Group is divided into two formations: the Magrug Formation, a lower conglomerate, grit and sandstone unit; and the Populierbos Formation, an upper unit of red shale and mudstone. The Klipheuwel Group postdates the Cape Granite Suite and is overlain by Cape Supergroup rocks, giving an age range of 500–450 Ma for Klipheuwel deposition (Gresse and Scheepers, 1993). The depositional environment is envisaged as an alluvial fan system for the basal conglomerates, overlain by lacustrine sediments. The Klipheuwel Group attains a maximum thickness of approximately 2000 m. The very limited extent of these sediments is the controlling factor in their use for vineyards.

Table Mountain Group

The Table Mountain Group (Mid-Ordovician – Early Devonian; Table 5) provides the most significant contribution to the scenery of the Coastal Region as it makes up most of the majestic mountains of the area (e.g., Fig. 2). The base of the group is marked by a classic unconformity that is exceptionally well-exposed on the Atlantic coast of the Cape Peninsula (Fig. 14). Here the sandstones, siltstones and mudstones of the Graafwater Formation overlie granites of the Cape Peninsula Pluton.

The Table Mountain Group is the basal group within the Cape Supergroup, a 6000–10,000 m thick sedimentary succession that accumulated from Middle Ordovician to Early Carboniferous as deltaic and shallow-marine sediments on a passive continental margin. The Table Mountain Group itself attains a maximum thickness of about 4000 m (Cotter, 2000; Theron et al., 1992; Kent, 1980). Overlying the Table Mountain Group are the Bokkeveld and Witteberg groups, which outcrop predominantly to the east of the Coastal Region.

The Table Mountain Group is subdivided into eight formations composed mainly of quartzitic sandstones as shown in Table 5. The following formations are the exceptions: the basal Piekenierskloof Formation is a conglomerate package that grades into quartzitic sandstones and is developed in the northeast of the Coastal Region; the Graafwater Formation is typically developed as interbedded reddish sandstone, siltstone and shale bands that are best developed on the Cape Peninsula to the south of Cape Town (Fig. 14). Glacial deposits consisting of diamictite and sandstone form the Pakhuis Formation, which is overlain by the Cederberg Formation, containing the only significant shales and siltstones within the Table Mountain Group. The Ordovician-Silurian boundary is placed within the shales of the Cederberg Formation based on the fossils preserved (Visser and Theron, 1973; Theron et al., 1992; Gresse and Theron, 1992).

The predominance of sandstone is a limiting factor in terms of vineyard development. Poor soils and steep topography result in a negligible number of vineyards being planted on Table Mountain Group sandstones. Few vineyards are located on the Piekenierskloof, Graafwater, Pakhuis and Cederberg formations owing to limited aerial extent and poor accessibility in the Coastal Region. These formations do, however, have potential for vineyards, as the Piekenierskloof and Cederberg formations have planted successfully in the Olifants River Region.

Cenozoic Deposits

Cenozoic-aged deposits in the form of soils, and fluvial and aeolian sediments all have an influence on Coastal Region winelands. These units are summarized in Table 7 using information from Söhnge (1991) and Theron et al. (1992).

Table 7 Cenozoic stratigraphy of the Coastal Region (based on Sohnge 1991; Theron et al., 1992)

Era	Period/Age	Stratigraphy	Lithology	Soils/river gravels/palaeosurface	Alteration	Marine, tectonic and river history
Quaternary	Holocene (0.01Ma)	Witzand Formation	Aeolian sand	Reworked alluvium	Weathering of granites and potential for saprolite and kaolin formation	Stone Age artefacts. Palaeosurfaces modified by headward watercourse incision
	Late Pleistocene (0.1-0.01 Ma)	Langebaan Formation Velddrif Formation	Aeolian calcareous sand Calcrete	Reworked alluvium		Stone Age artefacts. Palaeosurfaces modified by headward watercourse incision
	Early-Mid Pleistocene (1.65-0.1Ma)	Milnerton Formation Springfontein Formation	Fluvial gravel, sand and clay Aeolian sand	Reworked younger gravel and sand Younger gravels		Sea level restored. First Stone Age artefacts (Acheulian Man)
	Late Pliocene (3.8-1.65Ma)			Duplex soils (50-100masl), Older gravels (12-37marb)		Marine regression
Tertiary	Middle Pliocene (4.2-3.8Ma)			Duplex soils (50-100masl), Older gravels (55marb)		Marine regression
	Early Pliocene (5.3-4.2Ma)	Varswater Formation	Sand and shelly gravel	Residual soil (80-140 masl) Pools surface (150masl)		Marine transgression, deepening of Eerste River palaeovalley
	Middle - Late Miocene (14.5-5.3Ma)	Saldana Formation	Conglomeratic sand phosphorite			Marine regression and uplift, deepening of Eerste River palaeovalley
		Elandsfontein Formation	Sandy soils and peaty clay			
	Early - Middle Miocene (23.5-14.5Ma)			Residual soils (150-200masl), Chilton surface (180-280masl)		Marine transgression
	Oligocene (34-23.5Ma)			Red apedal soils (200-300masl)		Eerste River captures Jonkeshoek River. Major regression, continental shelf exposed
	Eocene - Palaeocene (65-34Ma)			Red apedal soils (200-300masl). Schoongezicht surface (415-500masl)		Jonkershoek River flows north. Major marine transgression, current coastal plane submerged

Soils

Soil types are commonly represented by duplex soils (sand rich or loamy topsoils over clay), structureless apedal soils, structured soils, shallow soils over bedrock, and wet, dark-coloured soils (Toombs and Archer, 1999). Theron et al., (1992) distinguished between loam and sandy loam soils formed as in-situ weathering products of Malmesbury Group rocks, and gravelly clay/loam soils derived from weathering of the Cape Granite Suite.

Soils derived from the Malmesbury Group are yellow, red or brown, clay rich, and commonly contain ferricrete nodules and vein quartz pebbles. The sand content is variable; these soils tend to become sandier near modern rivers. Cape Granite Suite soils are light brown to reddish in colour, commonly sandy to gritty, with pebble layers present locally. Proximity to outcrops of the Table Mountain Group is a significant factor in soil formation in the Coastal Region, resulting in increased sand content. Colluvial soils and scree are also developed. (Gresse and Theron, 1992; Theron et al., 1992; Hughes et al., 1992). The most sought-after soils for vineyards in the Coastal Region are clay-rich loams, as these have the capacity to retain water that can be utilized by vines during the dry summer months. Such clay layers are a feature that is regularly mentioned in wine literature.

Soil classification in South Africa is based on a series of soil forms and families. The type and colour of the soil layers present and the properties of these layers are specific to particular soil forms and families, and each is given a name based on the type location. The names of certain soil forms, for example reddish brown Hutton loams, have become entrenched in South African wine literature. It is beyond the scope of this study to provide a detailed review of South African soil classification: interested readers are referred to the Soil Classification Working Group (1991) for more detailed information.

Saprolites and Kaolin

An additional feature in areas of granite bedrock is the presence of saprolites and kaolin. Saprolites were described by Joubert

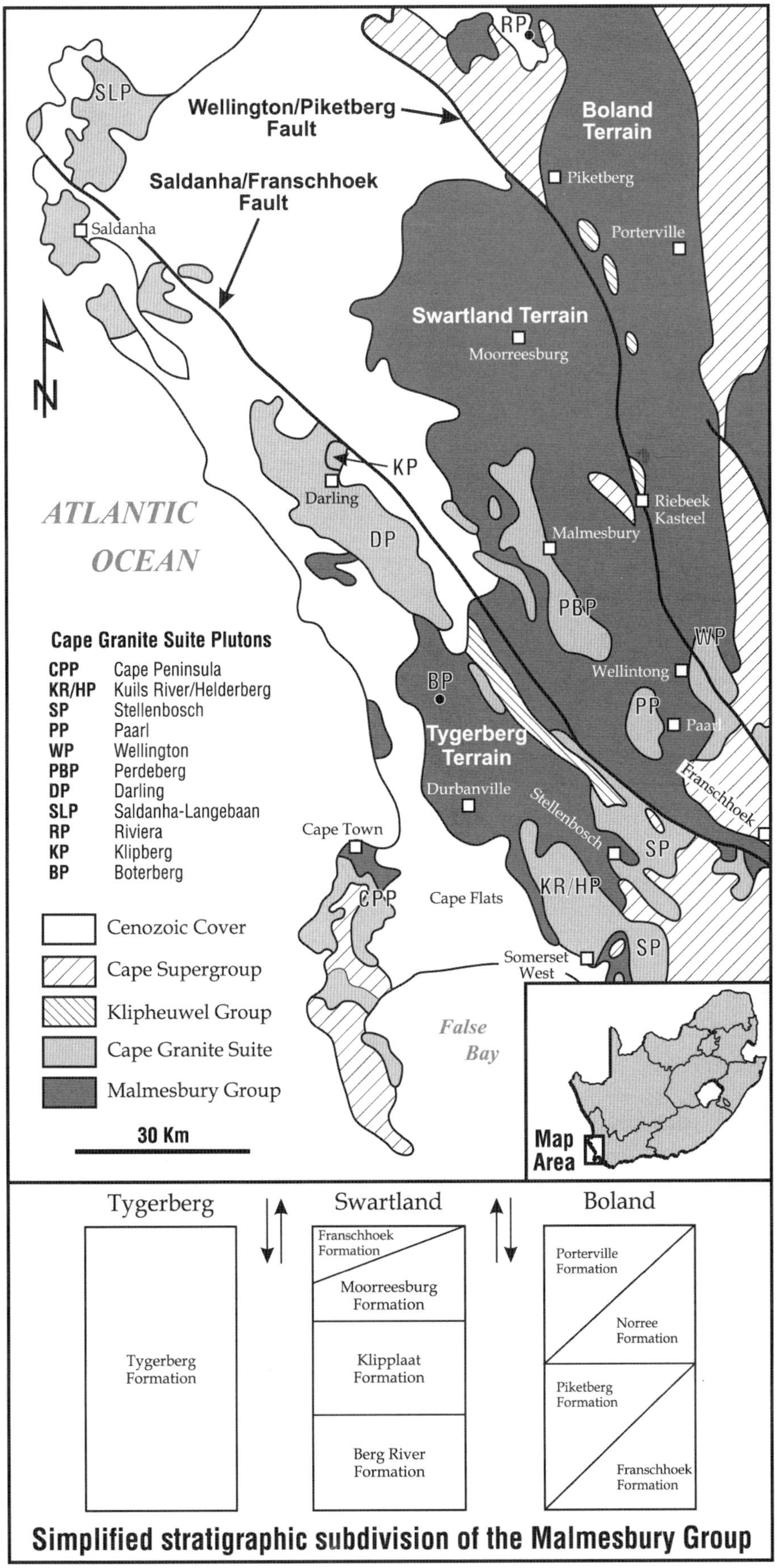

Figure 13 Generalized geology of the Coastal Region of South Africa. Modified from Rozendal and Scheepers (1995); Theron et al. (1992); and Gresse and Theron (1992).

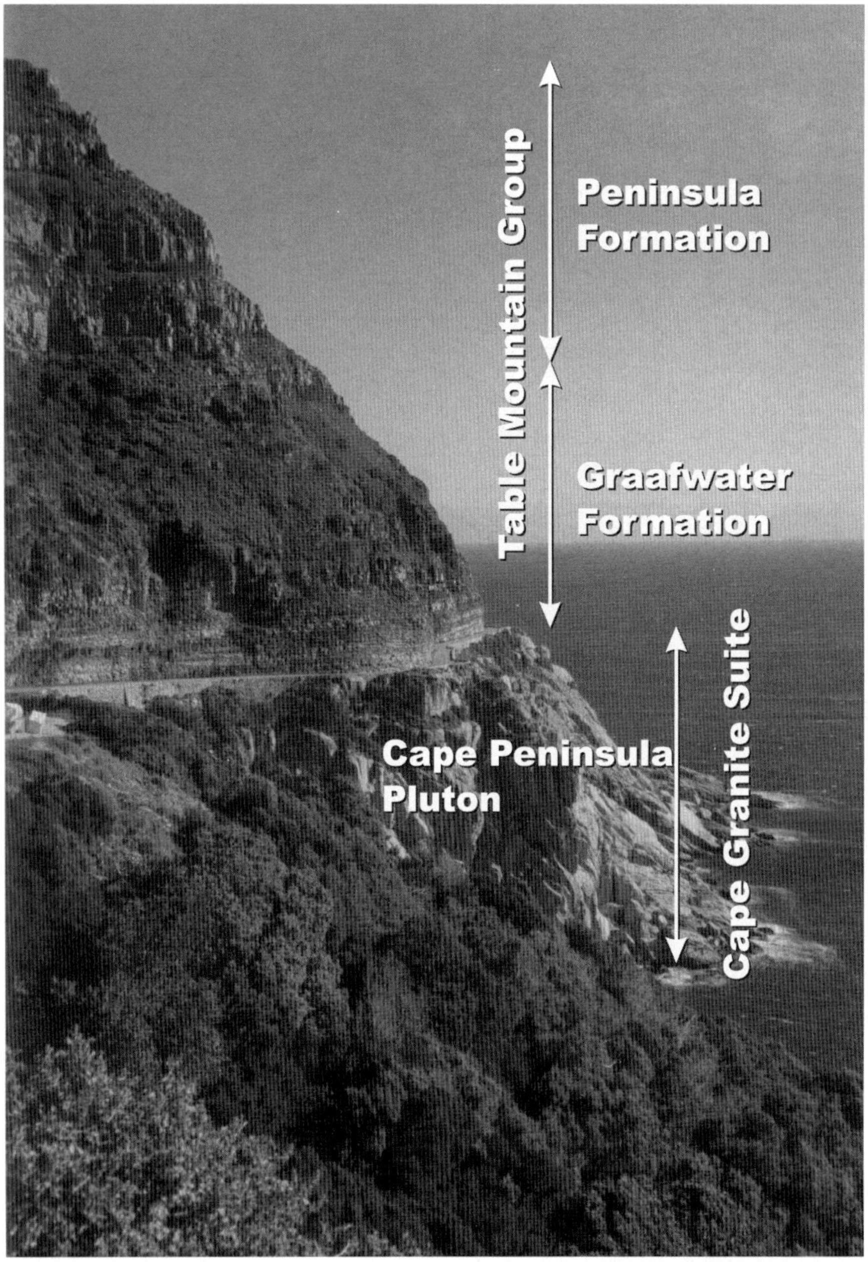

Figure 14 Basal contact of Graafwater Formation sandstones, siltstones and shales, Table Mountain Group (Mid-Ordovician to Early Devonian) overlying granite of the Cape Peninsula Pluton (Late Proterozoic). Photo C. Bargmann.

et al. (1997) in the soils of Groot Constantia and a Shiraz vineyard on the farm Langkloof near Wellington. According to Lambrechts (1982), saprolite horizons may reach depths in excess of 50 m. Kaolin has been identified at several important wine-producing locations in the Coastal Region, and is locally called "pot clay". Granite of the Kuils River/ Helderberg Pluton hosts the most extensive deposits of kaolin associated with vineyards. Commercially exploited kaolin deposits are located in the Cape Point District on the Cape Peninsula, where rehabilitation of the kaolin workings has included the planting of vineyards. Kaolin is also noted in the Stellenbosch pluton (Heckroodt, 1992; Theron et al., 1992; Söhnge, 1991; Knox, 1982).

Alluvium and Terrace Gravels

River alluvium and gravels are present in the Coastal Region's two main river drainage basins, the Eerste and Berg rivers. The composition of the river alluvium varies between fine silt, sand and gravel. Theron et al. (1992) observed that sandy alluvium, with organic material, is prominent in the Cape Peninsula and Stellenbosch areas.

Fluvial terraces are present in the Eerste River valley in the vicinity of Stellenbosch and were described in detail by Söhnge (1991). The best preserved terraces are divided into two groups: older gravels which range between 12–55 marb (metres above river bed), and younger gravels at 5–9 marb. Middle to Late Pliocene and Early Pleistocene ages were proposed by Söhnge (1991) for the older and younger gravels, respectively. The original river system is considerably older and may have started as a northerly flowing river during Late Cretaceous time, with river capture occurring during the Oligocene to produce the present westerly flowing drainage pattern (Table 7).

Marine and Aeolian Sediments

The Sandveld group is the proposed name for the Miocene to Holocene age marine and aeolian sediments deposited in the Cape Town hinterland (Theron et al., 1992). Deposition is concentrated in coastal areas and the Cape Flats, a broad low-lying area between Cape Town and Stellenbosch (Fig. 13). Aeolian sands regarded as Springfontein Formation occur as far inland as the Berg River in the vicinity of Paarl and are a feature of some vineyard areas.

TERROIR: STELLENBOSCH AND PAARL DISTRICTS

The Stellenbosch and Paarl districts are bounded on the east by mountains composed of Table Mountain Group sandstones; to the south is the coast of False Bay. To the west of these districts lie the Cenozoic sands of the Cape Flats and the towns of Brackenfell, Eerste River, Kuils River and Durbanville. The northern and western boundary of the Paarl District is based on the municipal boundary of the town of Paarl (Fig. 15).

Other features of the two districts include Bottelary hills; the Paarlberg, Stellenboschberg, Simonsberg, and Helderberg mountains; and two river drainages. Eerste River drainage basin covers much of the Stellenbosch District, flowing through the town of Stellenbosch in a southwesterly direction and eventually reaching the coast of False Bay. The Berg River valley is situated in the Paarl Distict; Berg River flows north from its source near Franschhoek, passing through the towns of Paarl and Wellington.

Role of Geology in Terroir

The French term "terroir" encompasses the whole natural environment of the vine: the climate, topography, soil, and bedrock geology, and includes viticultural practices such as cultivar selection and trellising systems, as described by Haynes (1999) and Wilson (2001) in earlier papers in this series. Traditionally in South Africa, research was limited to established wine areas because historical controls placed on the wine industry by the KWV prevented experimentation and expansion until the mid 1980s. Since then the country's winemakers have developed new wine growing areas, introduced additional grape varietals and become less dependent on the co-operative wineries. As a result of this process an understanding of the best combinations of grape varietal, climate, soils and bedrock geology, the fundamentals of terroir in South Africa and elsewhere, has begun to develop. Thus the concept of terroir is only beginning to be understood in Coastal Region settings.

Geology is a significant component of the local terroir. Perhaps most basically, geology plays an important part in the origin of the topography of the region as illustrated in Figure 16, which shows a section through the Eerste River basin to the south of the town of Stellenbosch. It is immediately clear from this diagram that river valleys are underlain by recessive-weathering granites of the Cape Granite Suite, whereas mountain peaks are composed of resistant-weathering sandstones of the Table Mountain Group, responsible, as noted above, for much of the spectacular scenery of the Cape. Faulted outcrops of Stellenbosch Pluton granites and Tygerberg Formation greywackes occur in the Jonkershoek Valley, which is flanked by the Stellenboschberg and Jonkershoek mountains composed of the resistant sandstones of the Table Mountain Group. Tygerberg Formation greywackes, intermediate in weathering character, form the slopes of the mountains. In the extreme west, the topographically more subdued Bottelary Hills mark the contact between Kuils River/Helderberg granites and Tygerberg Formation greywackes, which locally have been hornfelsed.

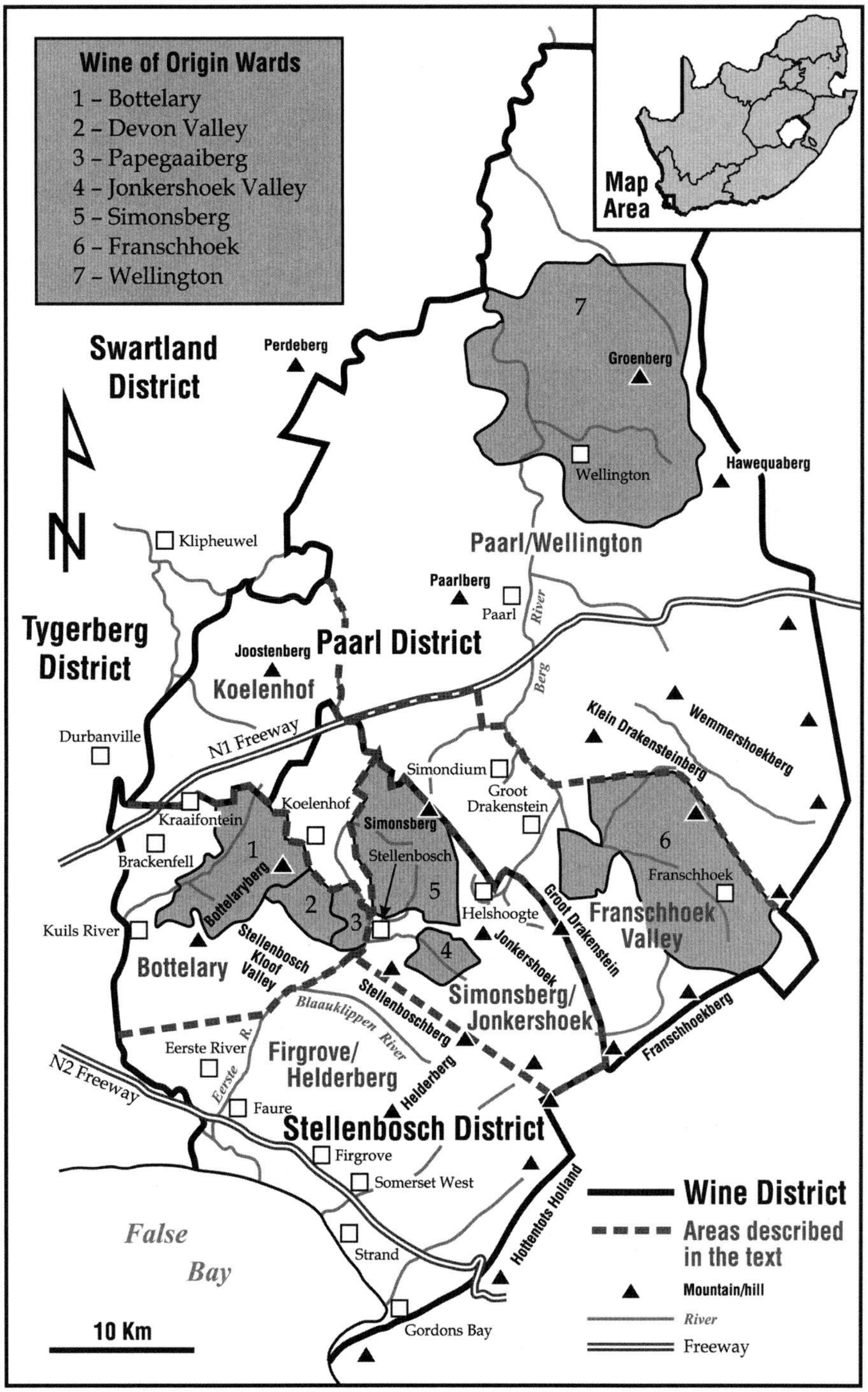

Figure 15 Stellenbosch and Paarl Wine of Origin Districts, Coastal Region of South Africa. Towns, e.g. Brackenfell, are shown as solid squares; mountains are shown as solid triangles with prominent mountains named in bold face, e.g. **Groenberg**. Modified from Domine (2001).

Clearly the differing slope orientations that occur are expressions of topography, and this is governed by bedrock geology. Topography is a key influence on grape vine growth and grape ripening through factors such as higher rainfall, exposure to morning or afternoon sunlight, and sea breezes. Additional geological influences on terroir include the site-specific influence of bedrock and associated colluvial and residual soils. Aspects such as weathering products, soil structure, texture, pH and retention of moisture, among others, are important considerations in the overall terroir. The recognition and importance of many of these site-specific variations are only now becoming apparent in the Coastal Region.

Informal Areas

The Wine of Origin boundaries of the Coastal Region and the Stellenbosch and Paarl districts are based on magisterial boundaries and as such do not reflect geographic and geological characteristics such as topography, river drainages, geology and soil type, or the varied local climates and weather patterns present. These "natural" characteristics are, in theory, defined by the boundaries of the wards. Unfortunately the wards within the Coastal Region cover a limited extent of the region as a whole (Fig. 8). In order to provide an overall picture of the Stellenbosch and Paarl districts the author has elected to use informal subdivisions convenient for the purpose of discussion at this early stage of understanding of terroir of the Coastal Region. These informal areas are described below and shown in Figure 15. Within these informal areas there are many vineyard-scale variations that reflect such components of terroir as altitude, slope orientation, bedrock geology and overlying soil type, which will become better understood with time.

Firgrove/Helderberg Area

This informal area covers the southern coastal portion of the Stellenbosch District (Fig. 15). The topography varies between the slopes and foothills of the Helderberg (1137 m) and Stellenboschberg (1167 m) mountains, and the lower-lying Eerste River valley (Fig. 17). To the east of the Helderberg Mountain ridge there are vineyards in the vicinity of the town of Somerset West (Fig. 15).

The village of Firgrove (Fig. 15) records a degree day total of 1756, which places this area in Winkler Region III (Tables 3, 4). Dew, totaling 100 mm, is noted as an important component of the local climate of the Meerlust Estate near the town of Faure (Fig. 15), and can remain on the vine leaves until late in the morning (Knox, 1982). The influence of the coast is strong, and sea breezes have a significant effect on the climate of the area. Vineyards are actively being planted on the foothills and slopes of Helderberg Mountain where cooler conditions result from both altitude and the effects of the sea breezes.

In terms of geology, Firgrove/ Helderberg area is underlain by the Kuils River/Helderberg Pluton granite, which is fringed by sediments of the Tygerberg Formation (Fig. 18). The Somerset West vineyards show distinctive geology with both Tygerberg Formation sedimentary

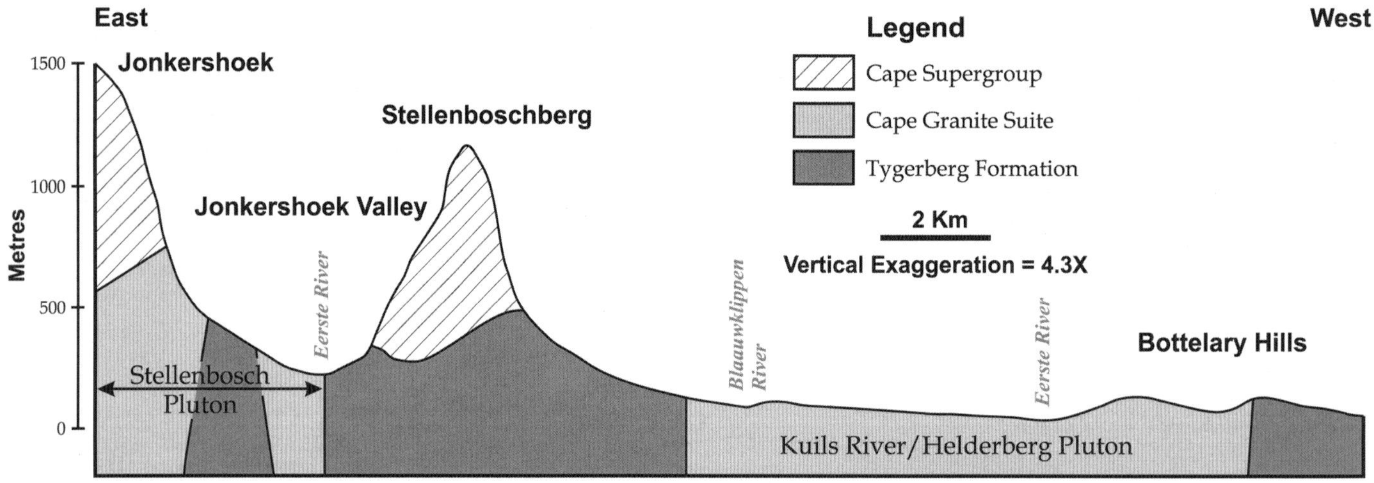

Figure 16 Schematic geological section through southern part of Stellenbosch District; Figure 18 shows location of east-west cross-section.

Figure 17 Firgrove/Helderberg informal area viewed from the Bottelary Hills, looking southeast from the Jordan Winery at the head of the Stellenboschkloof Valley. Eerste River basin is seen in the middle distance; Helderberg Mountain the far distance. Photo C. Bargmann.

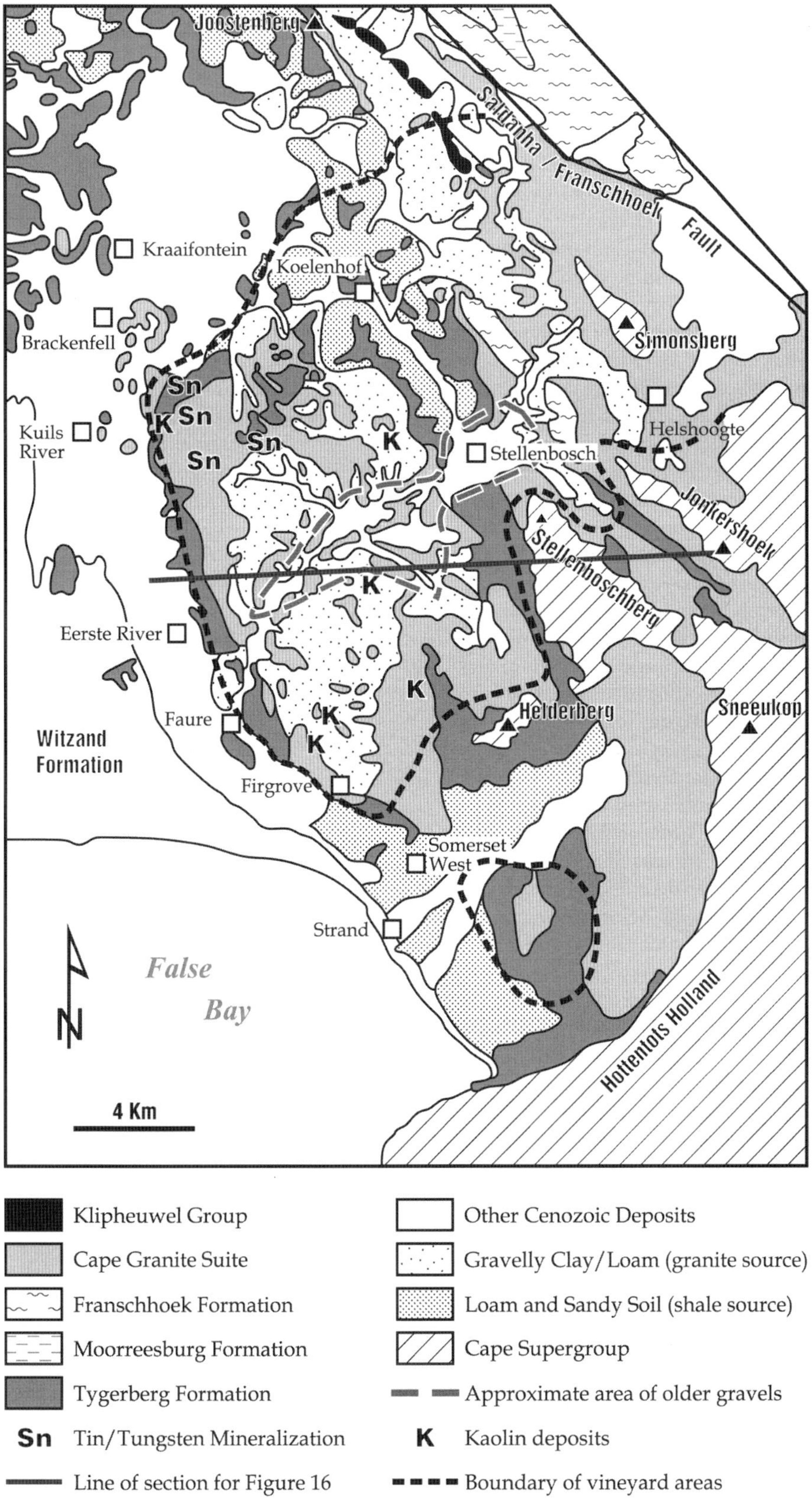

Figure 18 Generalized geology of the Stellenbosch District and the Joostenberg area, Coastal Region of South Africa. Line in centre of figure indicates location of schematic east-west cross-section seen in Figure 16. Modified from Geological Survey of South Africa (1990).

rocks and Stellenbosch Pluton granite present. Loamy soils are developed on the more mountainous areas near the Helderberg and Stellenboschberg mountains. In contrast, flatter areas in the Eerste River basin have a distinct sandy soil over a kaolin-rich sublayer. Heckroodt (1992) described the presence of kaolin over a wide area between the towns of Stellenbosch and Firgrove (Fig. 18). In areas with the Tygerberg Formation forming the bedrock, clay layers are present in overlying soils. Alluvial soils are present in areas close to the rivers, and deposits of the older gravels are present in the north of the area. In the vicinity of Somerset West there is significant alluvium, scree and sandy loam soils.

Surprisingly for an area with distinct local climates and geology, at present there are no wards defined in the Firgrove/Helderberg informal area. Based on topography, climate and geology, sub-areas can be identified covering Somerset West, Firgrove/Faure and the foothills of the Helderberg, which could all be considered for ward status.

Bottelary Area

This informal area covers the northwest part of the Stellenbosch District (Fig. 15). The hilly topography and associated valleys on the flanks of the Bottelary and Papegaaiberg Hills provide numerous slopes and aspects suitable for vineyards.

No degree day figures are available for the Bottelary area, however the village of Koelenhof, to the east, has a degree day total of 1898 and the Durbanville, to the west, has a degree day figure of 1906. This indicates that the Bottelary area would have a degree day figure of approximately 1900 and a warmer overall climate than Firgrove/Helderberg at 1756 (Table 4). The effects of sea breezes from both False Bay and the Atlantic are felt in the Bottelary area, although the coastal influence is lower than that felt by the Firgrove/Helderberg area discussed above.

The bulk of the Bottelary area is underlain by granites of the Kuils River/Helderberg Pluton (Fig. 18). Soils here commonly are loams, although sandy loams and some sandy soils occur. Alluvial soils and some local occurrences of older river gravel exist along the Eerste River. Clay layers were noted and kaolin deposits were described by Heckroodt (1992) in the Devon valley and the Kuils River area. In contrast, near the towns of Brackenfell and Kuils River in the north and west of the district (Fig. 15), the geology comprises Tygerberg Formation greywackes in contact with granites of the Kuils River/Helderberg Pluton. Tin and tungsten mineralization is present in this area, and vein deposits are located on the Hazendal and Zevenwacht wine farms (Theron et al., 1992). The soils here are variable, ranging from loams and sandy loams on higher areas to aeolian sands and alluvial soils in the lower-lying areas. Soils can be gravelly and clay sublayers, important in providing moisture during the dry summer months, are present.

Three wards are currently demarcated within the Bottelary area as shown in Figure 15. The Bottelary Ward covers the northern slopes of the Bottelary Hills; Devon Valley Ward lies to the southeast and covers a distinct valley of the same name; whilst Papegaaiberg Ward is centered on the Papegaaiberg Hills in the east of the Bottelary area, just west of Stellenbosch town. There is scope for further subdivision of the Bottelary area. The most significant undemarcated area is the Stellenboschkloof valley located on the southern flanks of the Bottelary Hills, to the south of the current Bottelary Ward (Fig. 15).

Simonsberg/Jonkershoek Area

This informal area is situated to the northeast and southeast of the town of Stellenbosch (Fig. 15). The area covers the western slopes and foothills of the Simonsberg and Jonkershoek mountains, including the Jonkershoek valley shown in Figure 16. The area also covers the high altitude vineyards of Helshoogte (Fig. 15) in the neck between the Simonsberg and Jonkershoek mountains.

In terms of climate, the effects of altitude are again noticed here, with rainfall, particularly, being higher. This increase in rainfall is illustrated in Figure 12, which shows high totals recorded from the eastern side of the mountains at the Boschendal estate. The effects of altitude, slope orientation and exposure to sea breezes all play an important role in providing cooler growing conditions: this is particularly true of the Helshoogte area where vineyards are at elevations above 300 m. The slopes of the Simonsberg mountains provide a good example of the effects of slope orientation on vineyards. West-facing slopes of the Simonsberg vineyards are exposed to full afternoon sunshine, in contrast to east-facing slopes such as those on the other side of the Simonsberg mountains in the Simondium/Groot Drakenstein area (Fig. 15) (see below), where cooler afternoon conditions result as the sun passes behind the mountains. The Simonsberg/Jonkershoek area lies between Koelenhof – degree days 1898 – and Groot Drakenstein – degree days 1967 (Table 4) – and similar degree day figures can be anticipated for the area.

The geology of the Simonsberg/Jonkershoek area is the most complex of all of the Stellenbosch District informal areas (Fig. 18). The area in the vicinity of the Simonsberg mountains is underlain by granites of the Stellenbosch Pluton, with Table Mountain Group clastic sediments forming the mountain slopes. A small outcrop of Klipheuwel Group clastics overlying Stellenbosch granites occurs to the northwest of the Simonsberg mountains whilst the Malmesbury Group, in the form of conglomerates and sandstones of the Franschhoek Formation, is developed in a northwest-trending zone beginning near Stellenbosch. Loamy soils are common and sandy soils occur in valleys.

The Jonkershoek valley is underlain by fault-bounded units of both the Stellenbosch Pluton granites and Tygerberg Formation sediments (Fig. 16). Soils range from loams and decomposed granite to sandy alluvial soils. Scree is present in the valley, which gives a pebbly texture to the soils.

Two wards are defined in this area. The Simonsberg-Stellenbosch Ward covers the western slopes and foothills of the Simonsberg mountains, and the Jonkershoek Valley Ward covers the vineyards of the valley of the same name to the southeast of Stellenbosch (Fig. 15).

Koelenhof Area

The Koelenhof informal area covers a low-lying area to the north of the town of Stellenbosch between the Bottelary hills in the west and the Simonsberg mountains to the east (Fig. 15). Geographically and geologically this area extends north into the Paarl District in the vicinity of the 291 m high hill called Joostenberg.

The village of Koelenhof has a degree day total of 1898, showing that the overall climate is warmer than Firgrove/Helderberg to the south. In climatic terms, breezes are important here: Van Zyl (2000) described the area as being "air conditioned" by breezes from both the Atlantic to the west and False Bay to the south (Fig. 13).

Koelenhof provides a geological contrast to the rest of the Stellenbosch District. The geology here is dominated by the shales and greywackes of the Tygerberg Formation (Fig. 18). Soils are sandy, may be aeolian or of transported granitic origin, and clay sublayers are developed. Tygerberg Formation greywackes and minor granite form the bedrock in the vicinity of Joostenberg, with much of this area covered by aeolian sands.

No wards are currently defined in this area. The distinctive nature, particularly of the geology, makes the area important enough to warrant an official classification. The Joostenberg portion in the northern part of the area currently falls within the Paarl District and highlights the problems of using municipal boundaries to define wine areas.

Franschhoek Valley area

Franschhoek Valley informal area forms the southern portion of Paarl District and is enclosed on three sides by mountains (Fig. 2). Between the Simonsberg and the Groot Drakenstein mountains is the northeast–southwest trending Groot Drakenstein valley with the Helshoogte vineyards (mentioned above) at its head (Fig. 15). Based on differing topography, the valley can be divided into two sub-areas: the Simondium/Groot Drakenstein sub-area and the Franschhoek sub-area. The Simondium/Groot Drakenstein sub-area covers the east-facing slopes of the Simonsberg mountains, and the Franschhoek sub-area occupies the more enclosed southern end of the Berg River Valley around the town of Franschhoek itself (Fig. 19). Topography in the Franschhoek valley creates a wide variety of slope orientations, ranging from the valley floor to west- and east-facing mountain slopes.

Climatic conditions are as follows. Average growing season temperatures at the Boschendal wine estate are slightly warmer than the Stellenbosch locations to the west, and cooler than the Wellington area to the north (Fig. 10). Degree day values for the Franschhoek Valley show a variation between 1967 and 2014 for Groot Drakenstein and Franschhoek, respectively. Both fall into Region IV, and again indicate that the Franschhoek Valley is a warmer wine region than Stellenbosch to the west

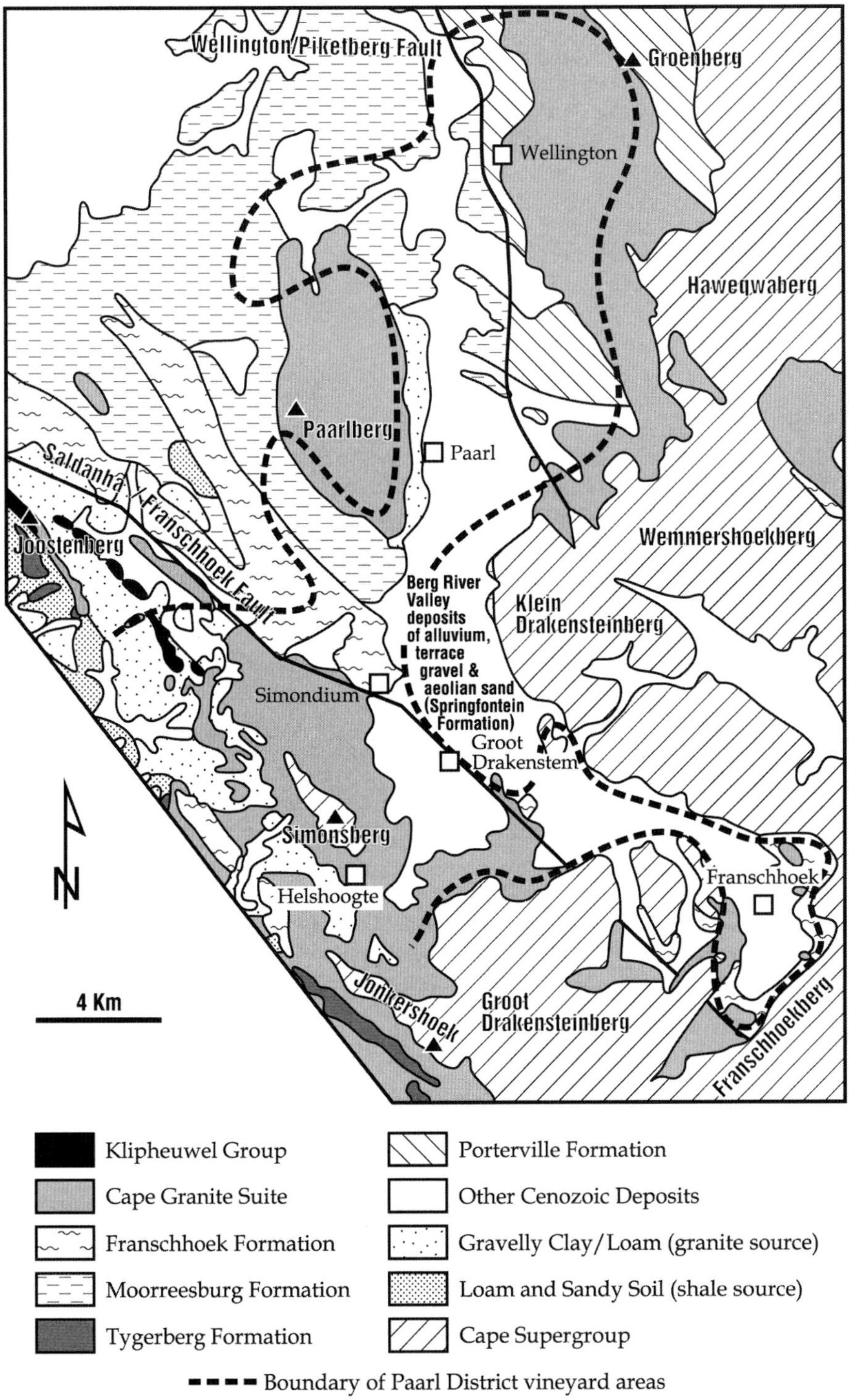

Figure 19 Generalized geology of the Paarl District, Coastal Region of South Africa. Modified from Geological Survey of South Africa (1990, 1997).

(Table 4). Average rainfall for the Franschhoek Valley is the highest of the areas described in this paper: the Boschendal wine estate and Groot Drakenstein weather station record 1355 and 931 mm of rainfall respectively (Fig. 12). These figures also indicate that there is considerable variation in rainfall dependent on location within the valley. Such variation in rainfall and temperature is to be expected in an area with varied topography that strongly influences weather and climate.

Bedrock geology consists of granites of the Stellenbosch Pluton and sediments of the Franschhoek Formation, both of which are faulted. The Simondium/Groot Drakenstein sub-area is mainly underlain by granites of the Stellenbosch Pluton and is bounded to the northeast by Franschhoek Formation conglomerates and quartzites. Loam and sandy loam soils occur on higher slopes with sandy and alluvial soils in the valleys (Fig. 19). The Franschhoek portion of the valley is dominated by alluvium along the valley floor and scree on mountain slopes. Sandy and alluvial soils sometimes gravelly are common, whilst loams and sandy loam occur in areas close to bedrock outcrop. The bedrock of the Franschhoek valley consists of a mixture of Stellenbosch Pluton granites and Franschhoek Formation clastics preserved below Cape Supergroup rocks.

Although two distinct sub-areas exist as described herein, only Franschhoek Ward is defined as a ward in terms of the Wine of Origin legislation in 2001 (Table 2). The eastern slopes of the Simonsberg mountains in the Simondium/Groot Drakenstein sub-area is a good example of an area with distinct topography and geology suitable for ward status. The variety of slope orientations in the Franschhoek Valley lends itself to the possibility of defining several discrete, small subdivisions with particular characteristics based on slope orientation, sunlight exposure, geology and rainfall.

Paarl/Wellington area

The final informal area covers the northern portion of the Paarl District (Fig. 15). Geographically this area is dominated by Paarlberg Mountain (729 m) to the east of Paarl, and the valley of the Berg River. Farther north near Wellington the slopes of the Groenberg (942 m) and Hawequaberg (1597 m) mountains are important vineyard areas.

As would be expected of the most inland location of the areas described, the maximum temperature over the growing season is the warmest, reaching 20.9°C at Wellington (Fig. 10). The presence of sea breezes from the Atlantic, approximately 60 km to the west, is noted as a factor in the local climate around Paarl (Hughes et al., 1992; Pienaar, undated). As a result relatively cooler conditions can be expected on the western slopes of Paarlberg Mountain, which are oriented favorably to benefit from these breezes. The Paarl/Wellington area probably falls within Region

IV in terms of degree day classification. This is based on Wellington being warmer than Franschhoek and Stellenbosch over the growing season, and the more inland locations of Wolseley and Tulbagh returning degree day figures falling between Regions IV and V (Table 4; Fig. 10).

Geologically this area is varied, and once again distinct sub-areas can be observed. Possibly the best known area of vineyards is located on the southwestern slopes of Paarlberg Mountain. This area also has distinctive geology, consisting of Paarl Pluton granites and Moorreesburg Formation sediments (Theron et al., 1992) (Fig. 19). The latter are locally cut by quartz porphyry dykes. The geology of this area is well represented at the Fairview wine farm. Here the highest altitude soils are derived from granitic and shaly bedrock. On midslopes the soils are colluvial, of granite origin, and cover a clay layer above the Moorreesburg Formation. At lower levels, sandy alluvial soils occur above a clay sublayer. In the extreme southwest of Fairview the land rises slightly and contains clay-based soils above the Moorreesburg Formation (Fairview, pers. comm.).

In contrast, the Berg River valley in the vicinity of the town of Paarl contains sandy alluvial soils that overlie Malmesbury Group sediments (Fig. 19). To the north near Wellington, vineyards are located on I-type granites of both the Paarl and Wellington plutons, and on sediments of the Moorreesburg and Porterville formations. Many of the vineyards on the Wellington Pluton are located on the western slopes of Groenberg Mountain and will benefit from the effects of altitude and slope orientation. Decomposed granite, Hutton-type loams and colluvial soils containing Malmesbury clasts were noted in this area by Joubert et al. (1997), and the Wellington Pluton was described as being deeply weathered and forming saprolites by Gresse and Theron (1992).

Wellington Ward is the only ward defined in this large area, and it only covers a discrete northern portion of the informal Paarl/Wellington area. Possibly the strongest terroir influence here is the warm climate, and further wards are likely to be defined where local conditions mitigate to provide a cooler climate. The southwest-facing slopes of Paarl Mountain and the vineyards of the Berg River valley east of Paarl are perhaps the most discrete areas that have not already been classified.

GEOLOGICAL INFLUENCES ON COASTAL REGION VINEYARDS

As can be seen from the foregoing discussion, the interrelationship between geology and climate is a key element in the terroir of the Coastal Region. Bedrock geology commonly governs topography, and topography in turn influences climate, including temperature, sunshine and rainfall. The bedrock geology also has a role in the type of soil present in areas of residual and colluvial soil. In terms of wine production the principal rock types and Cenozoic deposits in the Stellenbosch and Paarl districts are given in Table 8. For wine production the most important rock types are granites and greywacke dominated sediments. These rocks are particularly from the Stellenbosch and Kuils River/Helderberg plutons; and the Tygerberg and Moorreesburg formations. Cenozoic deposits are important in all areas, particularly loam soils and kaolin. The most significant alluvial soils on which vineyards are developed occur in the Berg River valley of the Paarl District.

Significant Geological Influences on Vineyards

Clays

The presence of clay in soils is one of the most important factors of the overall terroir of the Coastal Region. Clay-rich soils, and soils containing distinct clay layers, have high water retention capabilities, highly desirable during the generally dry summer months: this feature is regularly mentioned in South African wine literature. Successful vineyards with sandy soils commonly are duplex in character, with a clay sub-layer that assists in supporting the vine. The negative side of clay soils is their potential to become waterlogged, an undesirable feature for vines (Saayman, 1977).

The presence of significant quantities of kaolin has also been noted, particularly in the vineyards near Stellenbosch, although little has been written on its possible effects on vines and

Table 8 Principal rock types and Cenozoic deposits in wine producing areas of the Coastal Region

District	Informal area described in this paper	Wine of Origin Ward	Non classified subarea mentioned in this paper	Principal vineyard rocktype	Important Cenozoic deposits
Stellenbosch	Bottelary	Bottelary		Granite, greywacke	Loam soils, kaolin
		Devon Valley		Granite	Loam soils, kaolin
		Papegaaiberg		Granite	Loam soils
			Stellenboschkloof	Granite	
	Simonsberg/Jonkershoek	Simonsberg-Stellenbosch		Granite, conglomerate	Loam soils
		Jonkershoek Valley		Granite, greywacke	Alluvial gravel, scree
	Firgrove/Helderberg		Firgrove	Granite	Sandy soils, clay, kaolin
			Helderberg	Granite, greywacke	Loam soils
			Somerset West	Greywacke, granite	Alluvial gravels, scree
	Koelenhof			Greywacke	Sand, loam and clay soils
Paarl	Franshhoek Valley	Franschhoek		Conglomerate, granite	Alluvial gravels
			Simondium/Groot Drakenstein	Granite, conglomerate	
	Paarl/Wellington	Wellington		Granite	
			SW Paarlberg	Greywacke, granite	Colluvial soil
			Berg River Valley (Paarl)	Greywacke?	Alluvial gravels, sand

wine. Kaolin-rich soils may be less fertile and poor water sources – compared to soils containing clays such as montmorillonite and vermiculite – because kaolin has a low cation exchange capacity and a poor ability to absorb water molecules (Wilson, 1998; Dubbin, 2001).

Potassium-rich Settings

An important chemical characteristic of many shale- and granite-derived soils in the Coastal Region is the presence of potassium. Wooldridge (2000) noted that soils on the flanks of Paarlberg Mountain contain potassium released from granites, and that these soils have little capacity to buffer the uptake of potassium by vines. When potassium in the soil is transferred to the vine and grape, an increase in the pH of the grape results. In red wines produced from these settings the excess potassium can result in poor colour, and lower acidity wines with a "flabby" character (Saayman, 1992).

Soil Acidity

Acidity is another important chemical characteristic of many Coastal Region soils, and particularly the sub-soil. Acidic sub-soils can generate toxic aluminium levels which in turn restrict vine root growth. Deep soils, and potentially saprolite horizons, can be affected by this chemical restriction on vine growth. The addition of lime during soil preparation prior to planting a new vineyard is widely practiced in South Africa and elsewhere to counter the effects of low soil pH (Saayman, 1992).

CONCLUDING COMMENTS

My hope is that this paper introduces the reader to some of the history and geology of one of the most scenic wine regions of the world, the Coastal Region of South Africa. I particularly hope that my fellow Cape Wine Masters, and future students for the qualification, will benefit from the information provided herein.

At first glance Cape geology appears to be relatively simple, consisting as it does of four main geological units: the Malmesbury Group, Cape Granite Suite, Klipheuwel Group and Table Mountain Group. On further investigation a more complex scenario develops, with factors such as soil type, the presence of alluvium and sand, saprolite weathering, and kaolin formation all playing a role in the development of the geological and soil environment of Coastal Region vineyards.

In the Stellenbosch and Paarl districts the influence of climate and topography must not be understated, as both of these factors are critical in providing cooler conditions and rainfall in a wine-growing area that falls within the relatively warm degree day Regions III and IV. Geologically these districts are marked by strong influence from granites and, in the case of the Stellenbosch District, the unusual presence (in wine terms) of kaolin. Potassium, which has negative impacts on wine quality, is an unfortunate factor in the overall terroir resulting from the granite- and greywacke-based geology.

Owing to the legal restrictions and the co-operative dominated wine industry, experimentation with new areas and different grape cultivars was limited prior to the 1980s. It is only in the last two decades that South Africa's winemakers have been given the freedom to expand into new areas and experiment with matching cultivars and vineyard sites, which will ultimately define the unique elements of terroir in South Africa. The ongoing terroir study at the Agricultural Research Council in Stellenbosch will undoubtedly add considerably to this goal. As the industry develops, particularly into new viticultural areas, it is clear that further understanding of the role of geological factors in the local terroir will be established, to the continuing benefit of growers and consumers of wines from this historic area.

ACKNOWLEDGEMENTS

The late Simon Haynes is acknowledged for his encouragement and advice during the writing of the original draft of this paper. Simon and his family's hospitality during a visit I made to Canada in March 2001 is fondly remembered. I thank Kobus Conradie and John Wooldridge of the Agricultural Research Council in Stellenbosch, South Africa, for their comments as critical readers of this paper, as well as their summary of and references to the new work on terroir underway in South Africa as noted herein. Roger Macqueen is also thanked for his editorial comments and ideas that have greatly improved the final paper. The efforts of David Sargent, Geological Survey of Canada, Calgary, in redrafting or modifying a number the figures is gratefully acknowledged. I am also indebted to Debbie Wait at SAWIS for the information provided during this study.

EDITOR'S NOTE

Christopher Bargmann (B.Sc., Geology, Leicester University, England, 1983; M.Sc., Mineral Reserouces, Cardiff University, Wales, 1999) has spent much of his working career in gold and diamond exploration with JCI and De Beers in South Africa. Currently he is a geological consultant based in Wales. While living in Johannesburg, South Africa, he developed an interest in wine and began to study the subject through the Cape Wine Academy, the main wine education body in South Africa. In 2000 he became a Cape Wine Master (CWM). This qualification is similar to the U.K. Master of Wine (MW) qualification and requires the passing of written theory and tasting exams, and the completion of a thesis. Christopher Bargmann is one of only 48 CWMs at present.

REFERENCES

Amerine, M.A., and Winkler, A.J., 1944, Composition and quality of musts and wines of Californian grape: Hilgardia, v. 15, p. 493-575.

Aspler, T., 1998, Wine Lover's Companion: McGraw-Hill Ryerson, Toronto, New York (3rd edition), 256 p.

Bonnardot, V., Planchon, O., Carey, V., and Cautenet, S., 2002, Diurnal wind, relative humidity and temperature variation in the Stellenbosch-Groot Drakenstein wine growing area: South African Journal of Enology and Viticulture, V. 23, p. 62-71.

Carey, V. A., 2001, Spatial characterisation of natural terroir units for viticulture in the Bottelaryberg-Simonsberg-Helderberg winegrowing area: M.Sc. Agriculture Thesis, University of Stellenbosch, Stellenbosch, South Africa, 90 p.

Carey, V. A., Archer, E., and Saayman, D., 2002, Natural terroir units: What are they? How can they help the wine farmer: Wynboer, February 2002. (Text obtained from http://www.wynboer.co.za/recentarticles/0202terroir. php3).

Conradie, W. J., Carey, V. A., Bonnardot, V., Saayman, D., and van Schoor, L. H., 2002, Effects of different environmental factors on the performance of sauvignon Blanc grapevines in the Stellenbosch/Durbanville Districts of South Africa. 1. Geology, soil, climate, phenology and grape composition: South African Journal of Enology and Viticulture, V. 23, p. 78-91.

Conradie, W. J., 2002, Soil type may influence wine style: Cabernet sauvignon from Durbanville and Robertson: Wynboer, November 2002. (Text obtained from http://www.wynboer.co.za/recentarticles/1102soil. php3).

Cotter, E., 2000, Depositional setting and cyclic development of the lower part of the

Witteberg Group (Mid to Upper Devonian), Cape Supergroup, Western Cape, South Africa: South African Journal of Geology, v. 103, p. 1-14.

Cox, J., 1999, From Vines to Wines, the complete guide to growing grapes and making your own wine: Storey Books, Pownal, VT, 235p.

Domine, A., 2001, South Africa, *in* Domine, A., (ed.), Wine: English edition: Könemann, Cologne, p. 752-777.

Dubbin, W., 2001, Soils: The Natural History Museum, London, 110 p.

du Plessis, C. (ed.), 1999, South African (SA) wine industry directory 2000: Ampersand Press, Wynberg, South Africa, 408 p.

Geological Survey of South Africa, 1990, 1:250000 scale geological map sheet 3318 - Cape Town: Geological Survey of South Africa, Pretoria.

Geological Survey of South Africa, 1997, 1:250000 scale geological map sheet 3319 - Worcester: Geological Survey of South Africa, Pretoria.

Gladstones, J., 1994, Climate classification, *in* Robinson, J., (ed.), The Oxford Companion to Wine: Oxford University Press, Oxford, p. 256-257.

Gresse, P. G., and Theron, J. N., 1992, The geology of the Worcester area, explanation of sheet 3319: Geological Survey of South Africa, Pretoria, 89 p.

Gresse, P. G., and Scheepers, R., 1993, Neoproterozoic to Cambrian (Namibian) rocks of South Africa: a geochronological and geotectonic review: Journal of African Earth Sciences, v.16, p. 375-393.

Hands, P., and Hughes, D., 1997, Wines and brandies of the Cape of Good Hope: Stephan Phillips, Somerset West, South Africa, 192 p.

Haynes, S.J., 1999, Geology and Wine 1. Concept of terroir and the role of geology: Geoscience Canada, v. 26, p. 190-194.

Heckroodt, R. O., 1992, Kaolin Resources of the Republic of South Africa: Handbook of the Geological Survey of South Africa, 13, Geological Survey of South Africa, Pretoria, 102 p.

Hughes, D., Hands, P., and Kench, J., 1992, South African wine: Struik, Cape Town, 288p.

Johnson, H., 1989, Hugh Johnson's Story of Wine: Mitchell Beazley, London, 480 p.

Joubert, P., Toerien, W., Fey, M. V., and Winter, W. E. L. M., 1997, Geology of the Cape Winelands: Guidebook, 6th International Conference on Fluvial Sedimentology, University of Cape Town, South Africa, 35 p.

Kent, L. E. (ed.), 1980, Stratigraphy of South Africa, Part 1: Lithostratigraphy of the Republic of South Africa, South West Africa/Namibia and the Republics of Bophuthatwana, Transkei and Venda: The South African Committee for Stratigraphy, Geological Survey of South Africa, Handbook 8, 690 p.

Knox, G., 1982, Estate wines of South Africa: David Philip, Cape Town, 240 p.

KWV, 1991, SA wine industry statistics no. 15: KWV, Suider Paarl, 43 p.

KWV, 1996, SA wine industry statistics no. 20: KWV, Suider Paarl, 39 p.

KWV, 1997, SA wine industry statistics no. 21: KWV, Suider Paarl, 39 p.

Lambrechts, J. J. N., 1982, Soils, soil processes and soil distribution in the fynbos region: an introduction, *in* Deacon, H. J., Hendey, Q. B., and Lambrechts, J. J. N., (eds.), Fynbos palaeoecology: a preliminary synthesis: South African National Scientific Programmes Report no. 75, December 1982, p. 61-69.

Minter, L., (ed.), 1998, Geology of the Cape Winelands: International Volcanological Congress, Cape Town, July 1998, Field Excursion B3, Guide Book, 24 p.

Meinert, L., D., and Busacca, A., J., 2000, Geology and Wine 3, Terroirs of the Walla Walla Valley appellation, southeastern Washington State, USA: Geoscience Canada, v. 27, p. 149-171.

Pienaar, J., (undated), Introduction to the wine growing regions of South Africa: KWV, Stellenbosch, 92 p.

Reid, D. A., Erlank, A. J., and Rex, D. C., 1991, Age and correlation of the False Bay dolerite dyke swarm, south-western Cape, Cape Province: South African Journal of Geology, v. 94(2/3), p.155-158.

Rozendaal, A., Gresse, P. G., Scheepers, R., and Brown, C. A., 1994, Structural setting of the Riviera W-Mo deposit, Western Cape, South Africa: South African Journal of Geology, v. 97, p.184-195.

Rozendaal, A., and Scheepers, R., 1995, Magmatic and related deposits of the Pan-African Saldania belt in the Western Cape Province, South Africa: Journal of African Earth Sciences, v. 21, p. 107-126.

Saayman, D., 1977, The effect of soil and climate on wine quality: International symposium on the quality of the vintage, Cape Town, p. 197-208.

Saayman, D., 1992, Natural influences and wine quality, part 2: the role of soil: Wynboer, August 1992, p. 49 - 51.

Saayman, D., 1999, The development of vineyard zonation and demarcation in South Africa: Wynboer, January 1999. (Text obtained from http://www.wynboer.co.za/ recentarticles/ 0100zoneation.php3).

Scheepers, R., 1995, Geology, geochemistry and petrogenisis of Late Precambrian S-, I-, and A-type granitoids in the Saldania belt, Western Cape Province, South Africa: Journal of African Earth Sciences, v. 21, p. 35-58.

Söhnge, A. P. G., 1991, Alluvial history of the Eerste River, Stellenbosch: South African Journal of Geology, v. 94, p. 299-312.

Soil Classification Working Group, 1991, Soil classification: a taxonomic system for South Africa: Memoirs of the Agricultural Natural Resources of South Africa, 15, 257p.

Stevenson, T., 1991, Sotheby's world wine encyclopaedia: Dorling Kindersley, London, 480p.

Theron, J. N., Gresse P. G., Siegfried H. P., and Rodgers J., 1992, The geology of the Cape Town area, explanation of sheet 3318: Geological Survey of South Africa, Pretoria, 140 p.

Toombs, L., and Archer, E., 1999, A general comparison between soil and climate factors in South Australia and the Western Cape in South Africa: Wynboer, February 1999. (Text obtained from http:// www.wynboer.co.za/recentarticles/ 0299comparison. php3).

van Schoor, J. H., 2001, Geology, particle size distribution and clay fraction mineralogy of selected vineyard soils in South Africa and the possible relationship with grapevine performance: M.Sc. Agriculture Thesis, University of Stellenbosch, Stellenbosch, South Africa, 113 p.

van Zyl, P., 2000, Stellenbosch: appellations and flagships. Supplement to Wine, June 2000, Ramsay Son and Parker, Cape Town, 33 p.

Visser, H. N., and Theron, J. N., 1973, Explanation notes for sheet 3218: 1:250000 sheet 3218 – Clanwilliam: Geological Survey of South Africa, Pretoria.

Wilson, J. E., 1998, Terroir: the role of geology, climate, and culture in the making of French wines: Mitchell Beazley, London, 336p.

Wilson, J.E., 2001, Geology and Wine 4. The Origin and Odyssey of *Terroir*: Geoscience Canada, v. 28, p. 139-141.

Wooldridge, J., 1988, The potassium supplying power of certain virgin upland soils in the western Cape: M.Sc. Thesis, University of Stellenbosch, Stellenbosch, South Africa.

Wooldridge, J., 2000, Geology: A central aspect of terroir: Wineland, December 2000, p. 87-90.

Wooldridge, J., and Beukes, H., 2003a, Topography and solar energy interception in the Stellenbosch District. A geographic information system approach. Part 1, landscape, slope, and aspect: Wineland, February 2003, p. 74-76.

Wooldridge, J., and Beukes, H., 2003b, Topography and solar energy interception in the Stellenbosch District. A geographic information system approach. Part 2, effect of time of day: Wineland, March 2003, p. 73-74.

Accepted as revised July 2003

SERIES

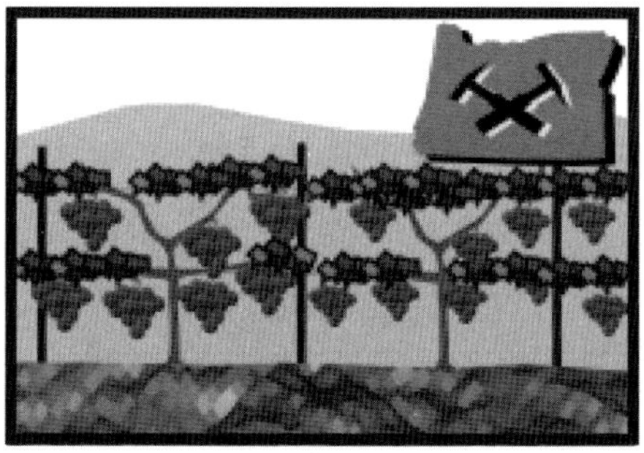

Geology and Wine 8. Modeling Viticultural Landscapes: A GIS Analysis of the Terroir Potential in the Umpqua Valley of Oregon

Gregory V. Jones[1], Nicholas Snead[1,2], and Peder Nelson[3]
[1]*Department of Geography, Southern Oregon University, 1250 Siskiyou Blvd Ashland, Oregon 97520 U.S.A.*
gjones@sou.edu
[2]*Department of Planning Public Policy & Management, School of Architecture and Allied Arts, 119 Hendricks Hall, 1209 University of Oregon, Eugene, Oregon 97403-1209 U.S.A.*
[3]*Environmental Education Program Southern Oregon University 1250 Siskiyou Blvd, Ashland, Oregon 97520 U.S.A.*

SUMMARY
Terroir is a holistic concept that relates to both environmental and cultural factors that together influence the grape growing to wine production continuum. The physical factors that influence the process include matching a given grape variety to its ideal climate along with optimum site characteristics of elevation, slope, aspect, and soil. While some regions have had hundreds and even thousands of years to define, develop, and understand their best terroir, newer regions typically face a trial and error stage of finding the best variety and terroir match. This research facilitates the process by modeling the climate and landscape in a relatively young grape growing region in Oregon, the Umpqua Valley appellation. The result is an inventory of land suitability that provides both existing and new growers greater insight into the best terroirs of the region.

SOMMAIRE
Le terroir est un concept holiste de facteurs environnementaux et culturels agissant sur un continuum s'étendant de la croissance de la vigne à la vinification. Dans le domaine des facteurs physiques, il faut trouver la combinaison idéale entre la variété du raisin d'une part, et le climat et les caractéristiques du site de culture comme l'élévation, la pente, l'aspect et le type de sol, d'autre part. Alors qu'en certaines régions, on a eu des centaines, voire des milliers d'année pour définir, développer et définir le terroir idéal, dans les régions nouvelles, on doit procéder par essais et erreur pour trouver le meilleur appariement raisin et terroir. La recherche décrite ci-contre entend faciliter ce processus de mariage idéal en modélisant les facteurs du climat et du paysage dans une région viticole relativement jeune de l'Orégon, celle de la vallée de Umpqua. Le résultat obtenu est un inventaire des terrains propices, ce qui fait aussi bien l'affaire des vignerons établis que des nouveaux vignerons dans leur quête des meilleurs terroirs de la région.

INTRODUCTION

Overall, the quality of wine produced in any viticultural region comes primarily from the high quality of the grapes, which are carefully vinified through often long-held cultural practices in the winery. The quality of the grape, however, is the result of the combination of five main factors: the climate, the site or local topography, the nature of the geology and soil, the choice of the grape variety, and how they are together managed to produce the best crop (Fig. 1). The French have named this interaction between cultural practices, the local environment, and the vines, the "terroir." While there will always be some disagreement over which aspects of the terroir are most influential, it is clear that the prudent grape grower must understand their interactions and controls on grape growth and quality (for a good review of the concept of terroir see Vaudour, 2002).

Numerous researchers have examined various aspects of terroir at different spatial scales providing insights into the complex inter-relationships between grape growing, quality wine production, and the site. In a comprehensive book on terroir, Wilson (1998) discusses the debate of which terroir factor is more important – nature or the winemaker's art – with the general conclusion that the debate may never be scientifically understood enough to differentiate the component contributions, but that its discourse has provided greater recognition of the influence that geology and geological processes have relative to wine quality traits. Haynes (1999) further discusses the importance of the interrelationships of all the terroir-related factors in the production of quality of wine, however the author makes a sound case of the soil and subsoil being the "determining factor controlling the taste and bouquet of a wine" by examining vineyard designation differences in the Côte d'Or region of France. Applied terroir-related research has resulted in examinations of individual elements such as soil (Becker, 1988), plant growth (Tesic et al., 2002a), viability of specific varieties (De Villars,

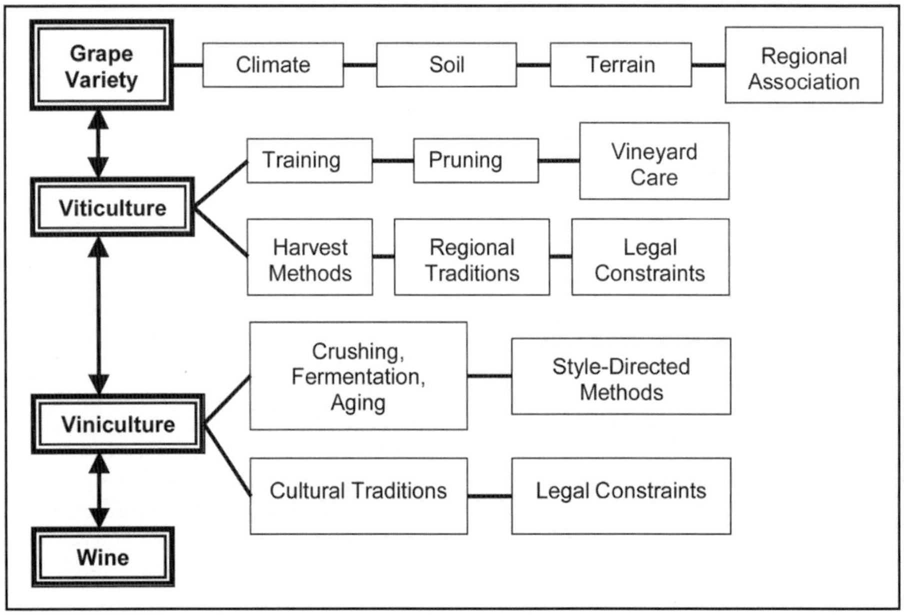

Figure 1 Factors encountered in the grape growing to wine production continuum. Note that viniculture is the study or science of making wines.

1997), cultural practices (Jordan et al., 1980), climate (Smart and Dry, 1980; Jackson and Cherry, 1988), and the zoning of viticultural environments (Carey, 2001; Carey et al., 2001). Others have addressed site suitability issues as a collection of factors that allow insights into a region's unknown potential (Tukey and Clore, 1972; Sayed, 1992; Boyer and Wolf, 2000; Jones, 2001) or as a measure of prediction for new areas to plant in existing regions (Davis et al., 1984; Margary et al., 1998; Tesic et al., 2002b). From a geology perspective, applied research has helped redefine sub-appellations in the Niagara Peninsula of Canada based on a combination of bedrock geology, soil, and climate, not simple climate zones as originally delineated (Haynes, 2000). In addition, two detailed studies by Meinert and Busacca (2000 and 2002) in the Walla Walla and Red Mountain appellations of Washington have yielded great insight into how, over a relatively small area, geology combines with climate to yield a broad range of wine styles and consistently high-rated wines.

Assessing the Physical Aspects of Terroir

Assessing a site's physical terroir aspects is arguably the single most important process that any potential grape grower will encounter when starting out (Jones and Hellman, 2003). Combined with matching the site to a grape variety, this decision will ultimately affect the vineyard's yield, the quality of the wine produced, and the vineyard's long-term profitability (Wolf, 1997). In addition, site selection normally involves compromises, in that few sites will possess ideal landscape and climate characteristics in every respect. Terroir assessment (site assessment) represents a complex suite of issues that must be factored into any plan to establish a successful vineyard operation. Numerous overviews exist that detail site selection in general (e.g., Dry and Smart, 1988; Gladstones, 1992) or for specific regions (e.g., Shaulis and Dethier, 1970; Wolf, 1997; Jones and Hellman, 2003) and focus mostly on climate, topography, and soil factors.

Of all site factors, climate exerts the most profound effect on the ability of a region or site to produce quality grapes. Over time the average climate structure of an area has proven to determine to a large degree the defining wine style, with variations in wine production and quality being chiefly controlled by husbandry decisions and short-term climate variability. Climate factors include solar radiation, heat accumulation, temperature extremes (including winter freezes and spring and fall frosts), precipitation during the principal growth stages, wind, and extreme weather events such as hail. Each climate factor should be assessed from the overriding macroscale down to the site or toposcale and even into the canopy (microscale).

The topography and soils of a site play important roles in grapevine growth and quality, and have interactive effects with climatic elements. Topographic factors that exert the greatest influence on a site's climate include elevation, slope, aspect, hill isolation and how it affects air drainage, and proximity to bodies of water. A marginal climate can be mitigated to some degree by locating the vineyard on an ideal site. From a soil standpoint, high-quality wines are made from grapes grown in many different types of soils with no single type considered ideal, however each soil imparts its own unique taste and mouth-feel to a given variety (Wilson, 1998). Grapevines will tolerate a wide range of soils, with nutrient imbalances and moisture issues being the principal factors to consider. Soil drainage is the most important physical property. Waterlogged soils will lead to a reduction in vine health and added difficulties in vineyard management. Adequate soil depth is next in importance. A minimum of 30-40 inches of permeable soil and no impeding layers (shallow bedrock, chemical or physical hardpans) is needed for optimum vine growth (Dry and Smart, 1988). Finally, soil fertility should be moderate for grapevines. Soil that is too fertile along with excessive vegetative growth can be problematic; in contrast, impoverished soils can require economically intensive applications of nutrients. Soil pH is a good indicator of the availability of nutrient cations and therefore fertility.

Grape Growing in Oregon

Oregon currently ranks as the fourth largest wine producer in the United States, growing grapes in six appellations (Federally designated grape growing and wine making regions, called American Viticultural Areas - AVAs): Applegate Valley, Columbia

Valley, Rogue Valley, Umpqua Valley, Walla Walla Valley, and Willamette Valley (Fig. 2). In 2002 there were 582 vineyards and 176 wineries growing more than 25 different varieties on over 10,000 harvestable acres with an economic benefit of over 250 million dollars to the state (Oregon Vineyard and Winery Reports, 2002). While Oregon is widely known for its Pinot Noir, other regions in Oregon are finding that warmer climate varieties can be ripened to produce quality wines (e.g., Merlot, Tempranillo, Syrah, and Malbec).

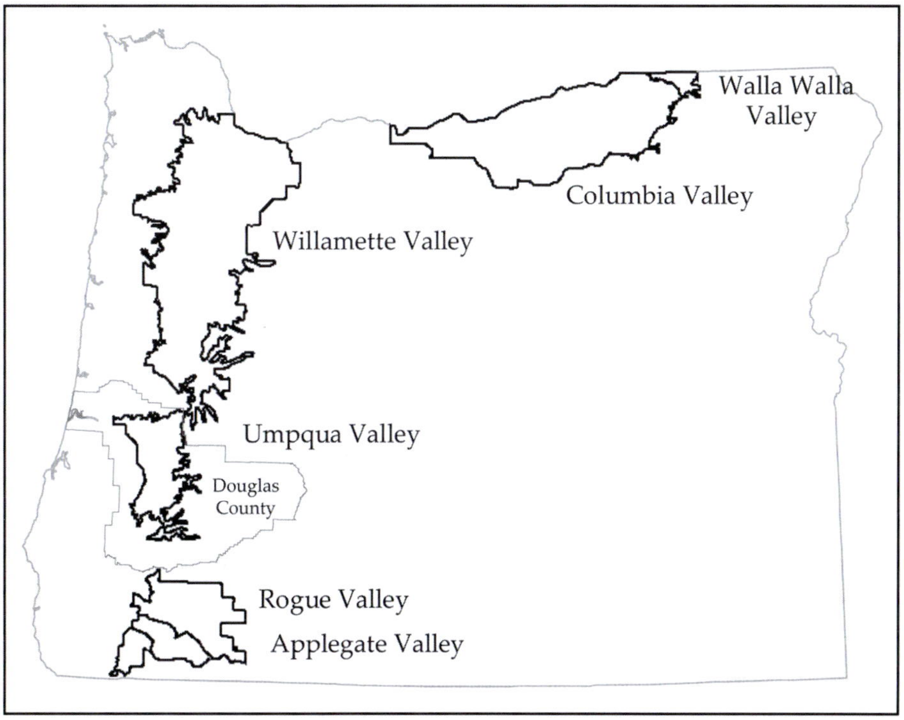

Figure 2 The spatial depiction of Oregon's American Viticultural Areas (AVAs). The Umpqua Valley AVA is completely contained in Douglas County.

Umpqua Valley AVA

The Umpqua Valley American Viticultural Area (AVA) is located entirely within Douglas County (Fig. 3) (Code of Federal Regulations, 2000). Grape growing and wine making started in the Umpqua Valley during the 1880s, but because of the region's distance to a larger market (i.e., Portland, San Francisco) it never established itself as a dominant industry in the region. In 1961 the region became the home for some of the earliest plantings of post-prohibition winegrapes in Oregon and in 1984 the Umpqua Valley AVA was established. Within the region, there is a diverse array of landscapes and climates that offer the various conditions needed to produce both cool and warm climate grape varieties. Today, the Umpqua Valley AVA has over 650,000 acres zoned for agriculture and, as of 2002, was planted with over 1,100 acres of vines on nearly 60 vineyards (Jones, 2003a). Production totaled over 1200 tons of grapes crushed at twelve wineries in 2002 (Oregon Vineyard and Winery Reports, 2002) with the dominant varieties grown being Pinot Noir, Chardonnay, Riesling, Cabernet Sauvignon, Pinot Gris, Merlot, and Gewurztraminer. Although the varieties listed above make up over 80 percent of the crop in the Umpqua Valley, the region is proving to be well suited for other varieties such as Baco Noir, Maréchal Foch, Tempranillo, Syrah, Malbec, and Dolcetto.

In comparison to the other grape growing regions in Oregon, the Umpqua Valley AVA has the most

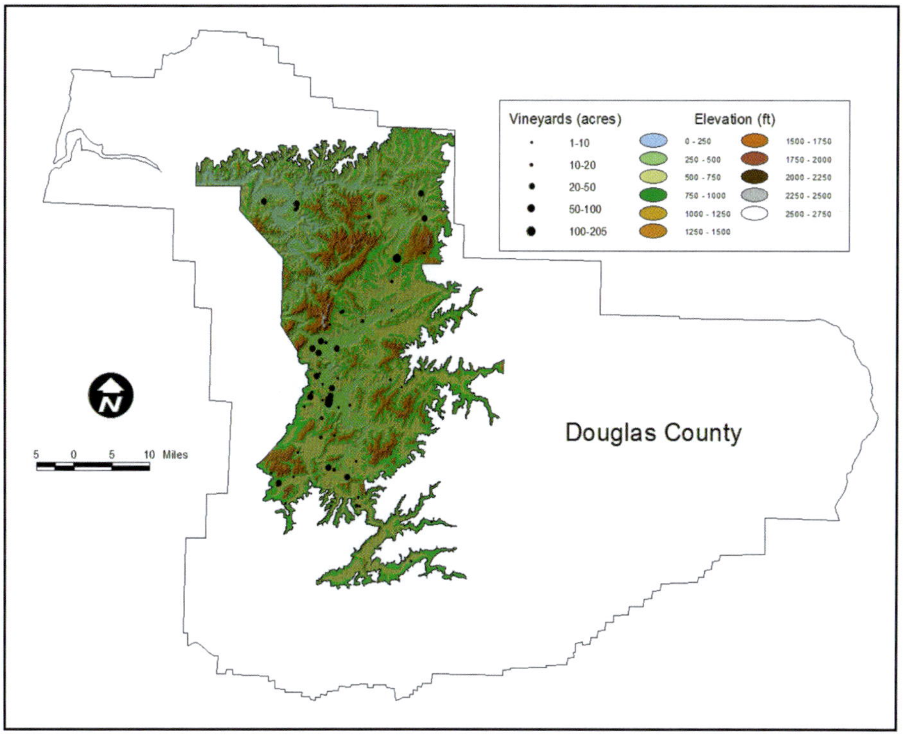

Figure 3 Location of the Umpqua Valley AVA relative to Douglas County, Oregon. Also depicted are elevation and existing vineyards as of summer, 2002.

beneficial climate structure in the state, with the longest growing season, the lowest risk of both spring and fall frosts, and low-ripening period rainfall and temperature extremes (Jones, 2003b). The region is warmer than the Willamette Valley to the north, but cooler than the Rogue Valley to the south (Fig. 2). Heat accumulation varies north to south in the AVA with growing degree days averaging 2200-2400 in the north and 2400-2700 in the south. Precipitation is also quite variable from north to south, averaging 30-60 inches annually (Oregon Climate Service, 2002). The growing season averages 200 days across the AVA, varying from 180 to over 215 days depending on elevation, while the median last and first frost (32°F) occurs on April 10 and October 31, respectively.

The topography of the region is extremely diverse, being derived from the joining of three mountain ranges of varying ages and structure: the Klamath Mountains, the Coastal Range, and the Cascades. The Klamath Mountains extend into the southwestern portion of the AVA and consist of complex folded and faulted igneous and metamorphic rocks that are the oldest in the region. The Cascade Mountains to the east consist of the younger High Cascades and the older, more deeply eroded Western Cascades that make up the eastern boundary of the AVA. The region is protected from the ocean largely by the Coastal Mountains, which are composed of mostly oceanic sedimentary rocks and volcanic islands that were accreted to the landscape over the last 50 million years. Often called the "Hundred Valleys of the Umpqua," the region is drained by the North and South Umpqua rivers, which join near the center of the AVA and empty into the Pacific Ocean. The landscape of the Umpqua Valley AVA is mostly valley lowlands with some isolated hills, stream terraces or benches, and footslopes of alluvial fans scattered by isolated hilltops and ridges. From this diverse geology comes a widely varying mix of metamorphic, sedimentary, and volcanic derived soils. The lower elevations of the valley are mostly deep alluvial material or heavy clays while the hillside and bench locations have mixed alluvial, silt, or clay structures. Complex faulting, especially in the southern part of the AVA, can produce large variations in soil types over areas the size of a vineyard. Drainage and moisture-holding capacity vary greatly by soil type, and while most soils in the region do retain water into the growing season, available water for irrigation during mid to late summer growth is generally needed. Soil fertility varies greatly over the region with issues generally related to either imbalances of nitrogen, calcium, potassium, phosphorous, magnesium, boron, or zinc. Soil pH also varies from region to region (roughly from 4.5 to 7.0) and is mostly due to differences in climate and parent rock material. In general, the soils in the northern portion of the AVA are slightly more acidic than those of the south as a result of more rainfall and greater leaching potential.

Even with over 40 years of post-prohibition grape growing experience, deciding to grow grapes in the Umpqua Valley can be a complex issue owing to the diverse geography and climate. For any existing and potential grape grower, vineyard site selection, or choosing the best terroir, is the single most important decision to be made. Therefore, the research described herein attempts to clarify the locational factors important to grape growing in the Umpqua Valley AVA through a multi-factor terroir-related analysis. The goal of the research is to model the best grape growing landscapes in the Umpqua Valley AVA, and to provide an inventory of sites to facilitate greater success in the grape growing to wine quality continuum.

DATA AND METHODS

To analyze the terroir of the Umpqua Valley, a multi-stage Geographic Information System (GIS) analysis was set up to incorporate factors related to the topography, soils, land zoning, and climate. The research started with the construction of a comprehensive GIS database of the individual vineyard sites in the Umpqua Valley AVA (Jones, 2003a). During the summer of 2002 each vineyard was surveyed using a Global Positioning System (GPS) to geographically locate the sites. Locational data from the existing vineyards, along with conversations with growers and winemakers, was then used to define the topographical and soil characteristics of the best sites for grape growing in the region.

Topographic Suitability

The topographic landscape was analyzed through the use of over 120 United States Geological Survey 10 meter digital elevation models (DEM). Each 1:24,000 scale quadrangle for Douglas County, Oregon was mosaiked into a complete county DEM grid (raster data). The entire landscape in Douglas County and the Umpqua Valley AVA was then categorized for the most advantageous elevations, slopes, and aspects for growing grapes. The categorization was constructed as a multilayer, topographically driven potential site analysis using ArcGIS (ESRI, 2003) with class rankings given each grid based on its potential. Elevations were categorized from 0-1200 feet into six classes (although grapes are grown at elevations greater than 1200 feet in the region, anything above that value has a much greater frost/freeze risk). Elevations from 400-799 feet were given the highest value and those greater than 1200 feet given a -1 or not viable (Table 1). Slopes were categorized into six classes from less than 1% (or basically flat with poor cold air drainage) to those over 30% being classed -1 or not viable (increasing slopes cause problems using vineyard equipment). The best slopes are considered to be those in the 5-15% range (Table 1). The aspect of the landscape was categorized into five classes with those from south-southeast to south-southwest given the highest ranking because of greater solar radiation receipt and ripening potential (Table 1). The three separate grids were then added together to produce a single topographical suitability grid with values ranging from not viable (anything with a negative score, coming from either being too high in elevation or on extremely steep slopes) to least suitable through most suitable landscapes. Owing to the spatial heterogeneity of most landscapes and realizing that potential sites need to be of adequate size to be profitable, the final grid was

Table 1 Topographic categorization of the landscape in Douglas County and the Umpqua Valley AVA using a 10 meter DEM (elevation, slope, and aspect). Class rankings represent a range of values related to the potential of the landscape for grape growing.

Elevation (ft.)	Class Ranking	Slope (%)	Class Ranking	Aspect (°)	Class Ranking
0-199	0	< 1 (flat)	0	0-89	0
200-39911-5	1	90-134	1	90-134	
400-799	2	5-15	2	135-224	2
800-999	1	15-20	1	225-269	1
1000-1199	0	20-30	0	270-360	0
>1200	-1	> 30	-1		

resampled using a nine acre neighborhood mean calculation. This size was used as smaller neighborhoods did not adequately produce homogeneous regions while larger neighborhoods created unrealistically smoothed regions.

Soil Suitability

To analyze the Umpqua Valley AVA's soils, data were obtained from the State Soil Geographic (STATSGO) Data Base for Douglas County Area, Oregon (NRCS, 1997). Site suitability relative to soils was analyzed in a similar manner to Margary et al. (1998) for New York, using the Soils Suitability Extension (SSE) with adjustments made for soils found in Oregon (Stulz, 2001). Four soil properties were used in the categorization of suitability: drainage, depth to bedrock, available water-holding capacity, and pH.

Drainage is thought to be the most important soil factor in establishing and maintaining a vineyard (Cass, 1999) and is influenced by many structural issues such as texture, depth, slope, and aspect. To assess soil drainage, the STATSGO database was analyzed by individual Hydrologic Soil Groups by map unit in the database. The four groups represent variations in drainage from good to poor (groups A to D). In dry growing regions like Oregon, depth to bedrock gives an indication of how well vines can cope with dry periods, with a minimum of 30-40 inches generally needed (Jordan et al., 1980; Dry and Smart, 1988). Mean depth to bedrock was calculated using the STATSGO database from the low to high bedrock depths for each component, and a weighted average was obtained for each map unit. While drainage is extremely important in vineyards, a soil's available water-holding capacity (AWC) is important as those soils with adequate water-holding capacity are at an advantage, giving vines the greatest ability to tolerate periods of moderate drought (Cass, 1999). AWC was calculated from the STATSGO database by computing the mean value for each soil layer, summed over the layers for each component, and then weighted by the percentage of each component per map unit (Margary et al., 1998). Soil pH gives an indication of fertility and nutrient balance with most ideal vineyard soils being found between 5.5-8.0. Outside this range, nutrients may become out of balance, with deficiencies or toxic levels affecting vine uptake or beneficial relationships with microorganisms. Soil pH was computed from the STATSGO database by computing the mean value for each soil component and then weighting by the percentage of each component per map unit.

The spatial polygon data was then converted to grids at the same nine acre resolution to match the landscape suitability grid. Each soil factor was then grouped into classes based upon their individual values; drainage from poor to excessive, available water holding capacity from 0.10 to 0.30 inches of water per inch of soil; depth to bedrock from 25 to 65 inches; and pH from values less than 5.0 to values over 6.0 (Table 2). All classes in each grid were then scaled and weighted with drainage given the greatest weight (40%) and each of the other factors weighted 20%. A final grid of soil suitability was then constructed from the weighting of the four soil factors.

Land-Use Suitability

To incorporate land-use issues relative to agricultural development, a statewide generalized zoning coverage was used in the analysis (OGDC, 2003). The zoning data used was initially digitized from data collected from 1983 through 1986 by the Oregon Department of Land Conservation and Development. Since its production, limited zoning changes have occurred and the dataset represents the best statewide zoning coverage available for the state of Oregon. For this analysis only lands zoned agriculture, farm/forest transition, and rural residential are considered as viable parcels on which to develop a vineyard. The farm/forest transition zoning was included, as this type of land tends to be on slopes that are currently used by vineyards and is easily converted to new vineyard sites. The rural residential zoning is set for both two and five acre parcels and is included because many vineyards in southern Oregon can be found on sites in this zoning. The coverage was converted to a grid with all agriculture and farm/forest transition zones given a value of 2, rural residential given a value of 1, and all others a -1 representing a zoning limit to viability.

Table 2 Criteria for developing soil suitability in Douglas County and the Umpqua Valley AVA using the STATSGO database. The soil factors were classed and then weighted to produce a final soil suitability grid.

Soil Factor	Lower Threshold	Upper Threshold	# of Classes	% Weighting
Drainage	Poor	Excessive	4	40
AWHC (inch H_2O/inch soil)	0.10	0.30	5	20
Depth to bedrock (inches)	25	65	3	20
pH	< 5.0	> 6.0	4	20

Climate Suitability

To assess the climate suitability of the Umpqua Valley AVA, a composite climate model was constructed based upon earlier work conducted for Oregon by Aney (1974) and Mastin (1983). These early analyses of climate suitability used available climate station data to create crude isolines of suitable areas for either cool or warm climate grape varieties. Climate factors used included winter minimum temperatures, growing season length, growing season potential evapotranspiration, heat summations, sunshine hours, and a precipitation index. To better depict the spatial variation in climate factors, this analysis uses the PRISM (Parameter-elevation Regressions on Independent Slopes Model) model, which is derived from a combination of point data, a digital elevation model, and other spatial data sets to create estimates of monthly and annual climate variables that are gridded at roughly 1.45 miles resolution (Daly et al., 2001). The composite climate model was developed for the western Oregon AVAs (Jones et al., 2002) using the most agriculturally and economically important variables.

Absolute climate viability was determined from extreme winter temperatures where areas that experience values lower than 5°F were deemed not viable for grape growing. The model then proceeds to build climate/maturity groupings for cool, intermediate, warm, and hot climates (Fig. 4) based initially on growing degree days (GrDDs; from April to October using a base of 50°F) using practical experience in Oregon. Each climate maturity group is classed to a 25 point range that represents a growing degree day potential within that grouping. The initial GrDD grids are then adjusted for the relative impact that each of the other climate variables has on production viability (frost-free period, last spring frost, first fall frost, rainfall frequency and amount during flowering and harvest, and ripening period temperature variability). For example, the length of the growing season, defined as the days between the first and last frost, is most advantageous when it is at least 180 days or greater (Dry and Smart, 1988). In the modeling, the frost free period was weighted as +2 if over 200 days; +1 if 180-200 days; 0 if 160-180 days; -1 if 140-160 days; and -2 if less than 140 days. The resulting grid was then added to the initial GrDD grid to adjust the grouping relative to how equitable the growing season length is to grape maturation potential. Each of the other climate variables are weighted individually, then added or subtracted to/from the initial GrDD grid. Frost timing is included by using the median last spring and first fall dates (based on 32°F) in reference to the known timing of phenological events for grape varieties grown in Oregon (Jones et al., 2002). Precipitation effects were assessed by weighting climate grids on the frequency and amount of rainfall that occurs during the critical growth stages of flowering and maturation. Finally, a ripening period temperature variability factor (extremes of both daily and monthly temperature), as defined by Gladstones (1992), was included. The temperature variability index accounts for how relatively constant, intermediate temperatures during the ripening period enhance the biochemical developmental processes of colour, flavour, and aroma that lead to high-quality wine. Grids are scored relative to a given ideal range as defined by Gladstones (1992). The final grid represents an adjusted climate/maturity value relative to the sum effects of multiple climate factors that attempt to define maturity potential by minimizing climate-related risks.

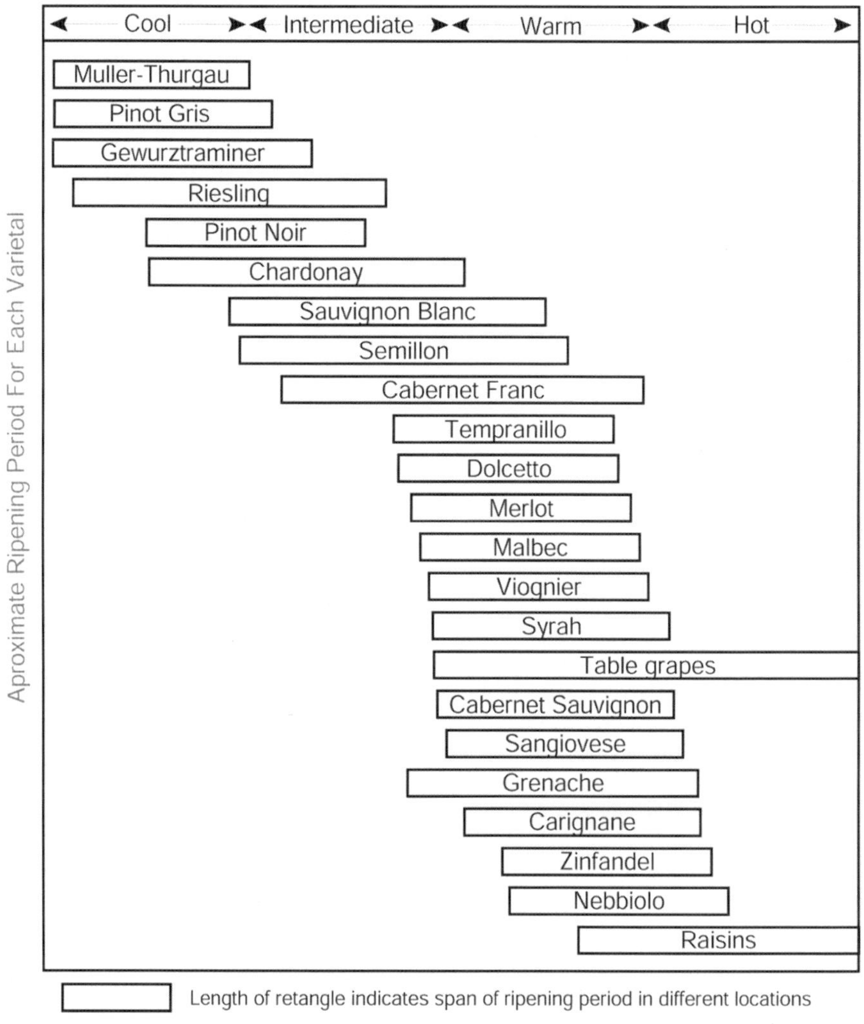

Figure 4 Climate/maturity groupings defined for Oregon based on the approximate ripening period potential for each variety in different locations: from Jones et al. (2002).

Composite Suitability

The three site factors – topography, soil, and land-use grids – after being internally scaled relative to their individual grape-growing influences (as defined above), were then combined to produce a composite suitability grid taking into account the combined landscape/land-use effect. The composite grid was then masked with the climate maturity group grid to remove all non-climatically viable areas and then overlaid with the climate maturity group grid to produce a spatial depiction of the best vineyard sites in each maturity group in the Douglas County and Umpqua Valley AVA areas.

RESULTS

The Umpqua Valley AVA encompasses a total of 689,154 acres in Douglas County, Oregon (Fig. 3). The GPS survey identified 58 vineyards geographically dispersed from north to south over the AVA, with the greatest concentration in the west-central portion of the AVA (Fig. 3). Overall, the 58 vineyards consist of a total of 1,106 acres and range from one acre or less to 205 acres, with a median vineyard size of 9.3 acres. Vineyards are planted at elevations ranging from 197-1296 ft., with an average elevation of 581 ft. and a majority found at between 400 and 800 ft. Vineyard slopes in the region range from flat to over 50%, averaging 8.3% with a standard deviation of 6.4%. Aspects vary over all cardinal directions with south-southeast to west most common, and an average of south-southwest.

The information derived from the GPS survey, along with a general understanding of the ideal topographic characteristics (Table 1), resulted in a grid depicting those landscapes most topographically suited to grape growing in the region (Fig. 5). The vast majority of the county is considered not suitable (60.8%) or least suitable (15.2%) for grape growing, owing to a combination of elevation, slope, and aspect that represents unsuitable landscapes (Table 3). Although the suitability grid reveals that no ideal landscapes (as scoring the highest on all three factors combined) exist in the region, a summation of all intermediate to most

Table 3 Topographic suitability acreage as derived from the combined elevation, slope, and aspect grids used in the analysis. Values given for both Douglas County and the Umpqua Valley AVA.

County-Wide Topographic Suitability Value	Acres	% of Total
Not suitable	1,983,028.6	60.8
Least suitable	495,742.2	15.2
	315,513.9	9.7
	247,147.7	7.6
Intermediate suitability	146,242.7	4.5
	58,634.9	1.8
	13,671.8	0.4
Most suitable	0.0	0.0
Total acres	3,259,981.8	

Umpqua AVA Topographic Suitability Value	Acres	% of Total	% of County
Not suitable	45,886.0	6.7	2.3
Least suitable	136,369.1	19.8	27.5
	152,175.1	22.1	48.2
	162,897.0	23.6	65.9
Intermediate suitability	124,835.0	18.1	85.4
	53,792.6	7.8	91.7
	12,954.7	1.9	94.8
Most suitable	0.0	0.0	0.0
Total acres	688,909.5		

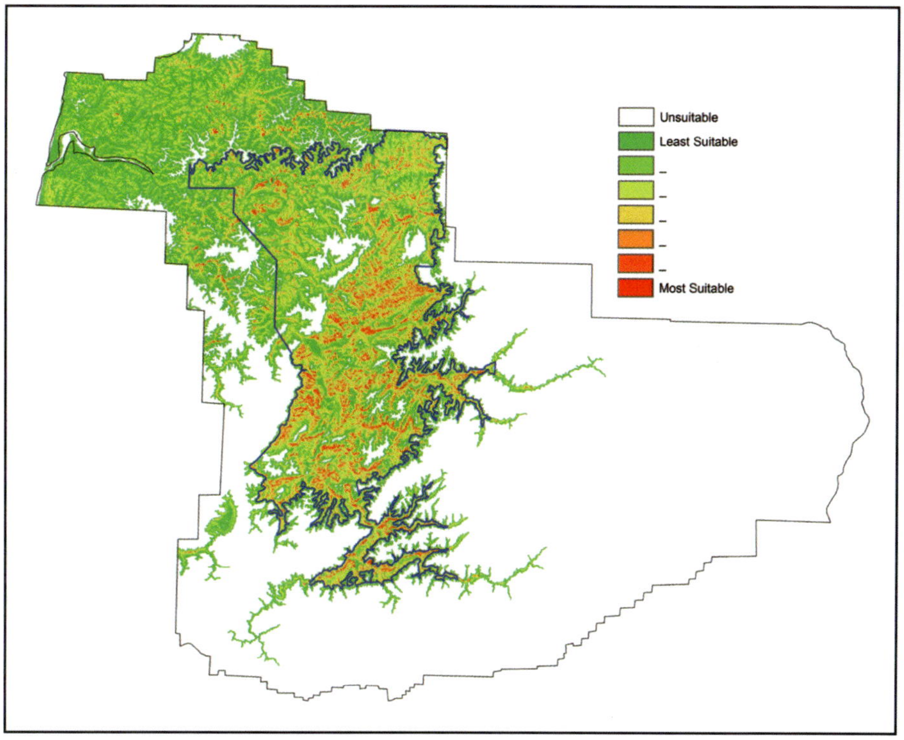

Figure 5 Topographic suitability as defined through the combination of elevation, slope, and aspect. The results are depicted for Douglas County as a whole and with the Umpqua Valley AVA outlined in blue.

suitable landscapes identifies over 218,000 acres in Douglas County. At the appellation level, the Umpqua AVA contains the vast majority of the region's intermediate to most suitable landscapes (88%), with the difference being the area in the northwest of the AVA where good sloping land can be found, but at lower elevations (Fig. 5). Using the highest topographic suitability class, nearly 13,000 acres can be found within the Umpqua Valley AVA (Table 3).

The soil analysis resulted in four individual grids that depict drainage, depth to bedrock, available water-holding capacity, and pH variations for individual map units for Douglas County (not shown individually). Spatial variations in hydrological groups indicate that the best drained soils are found near the coast, in sandy soils and along the alluvium-filled valleys and terraces of the Umpqua River. Poor drainage areas are most notable in the lands underlain by older metamorphic rocks of the Klamath Mountains to the south. Similar patterns are found with depth to bedrock, with the deepest soils found near the coast, in broad areas surrounding the main river drainages, in upland regions in the volcanic soils of the Cascades, and in isolated areas throughout the county. Depth to bedrock averages 44 inches over the county, with a minimum of just less than 20 inches and a maximum of just over 60 inches. Available water-holding capacity (AWC) varies from the coast, where higher values can be found in the more silty to clay-rich soils, to lower values in the soils of the Klamath Mountains and some of the more excessively drained soils in the area. AWC averages 0.15 inch of water per inch of soil in the county, with a minimum of under 0.01 to a maximum of just over 0.30 inches. Soil pH in Douglas County ranges from 4.7 to 6.6 with a mean of 5.6. Spatial variations are seen in the more acidic coastal soils, coastal mountains, and in much of the Cascades. More neutral pH soils are found along the main reaches of the middle portions of the South Umpqua River and throughout the Klamath Mountains.

The combined weighted soils-suitability categorization results in a single grid that depicts the best soils for grape growing with moderate to well drained soils, adequate depths to bedrock, moderate to good water-holding capacity, and more neutral soil pH (Fig. 6). The most suitable soils, those scoring more than the mean composite grid value, are found along the coastal zone, in a broad upland area in the northwestern portion of the county, along the main reaches of the middle portions of the North and South Umpqua Rivers, and along a broad upland region stretching from the southwest portion of the county into the western Cascades.

Land use in Douglas County is dominated by rural-designated zoning types where over 77% of the land is zoned for forestry, 19% is zoned for agriculture, and 2% for rural residential zoning categories (Fig. 7). For land zoned agriculture, the designations are further divided into three broad categories: agriculture (40%), farm/forest transition (56%), and rural residential (4%). In the Umpqua AVA, 67% of the land area is zoned agriculture with roughly equal amounts of exclusive agriculture and farm/forest transition (44% and 49%, respectively) along with 7% zoned rural residential.

Combining the topographic, soil, and land-use grids produced a point scale of suitability from which viable vineyard sites in Douglas County and the Umpqua Valley AVA are mapped (Fig. 8). Suitability scores range from a low of 26 to a maximum of 100, with an average of 62.1 and a 20.9 standard deviation. Using the unsuitable criteria in the topographic grid (too high in elevation or too steep slopes), along with the lands that are not zoned appropriately (other than agriculture or rural residential), reduces the suitable areas to lands mostly within the AVA (82%). Some suitable areas are found outside the AVA near the coast along the Umpqua River, in scattered valleys west and southwest of the AVA boundary, and along many of the stream reaches just outside the AVA boundary. Examining only those areas that scored greater than 1.5 standard deviations above the mean finds that there are nearly 3000 acres of the best sites clustered throughout the AVA boundary.

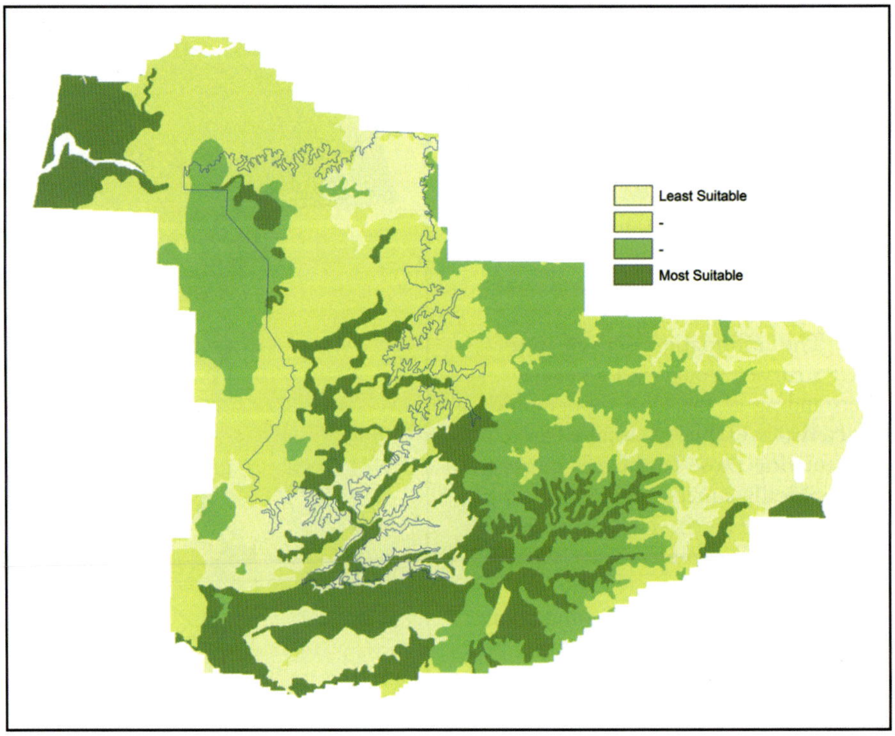

Figure 6 Soil suitability as defined through the combination of drainage, depth to bedrock, available water-holding capacity, and pH. The results are depicted for Douglas County as a whole and with the Umpqua Valley AVA outlined in blue.

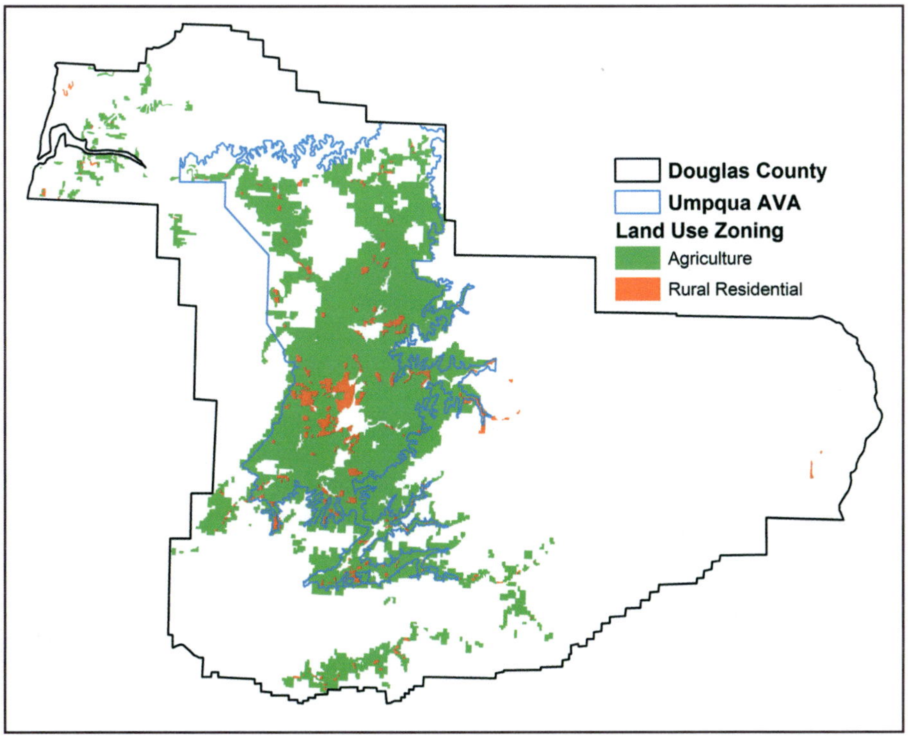

Figure 7 Land-use suitability as defined by all lands zoned either agriculture or rural residential. The results are depicted for Douglas County as a whole and with the Umpqua Valley AVA outlined in blue.

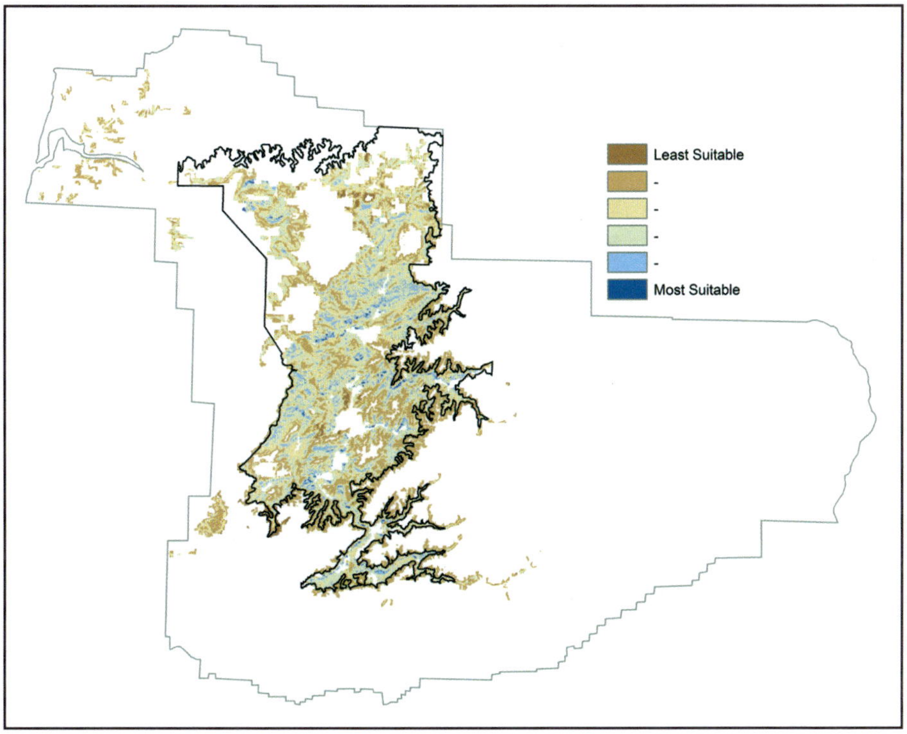

Figure 8 Composite suitability as defined through the combination of topographic, soil, and land-use suitability grids. The results are depicted for Douglas County as a whole and with the Umpqua Valley AVA outlined in black.

The climate maturity analysis produced a grid that spatially represents the potential for the region to ripen given varieties as depicted in Figure 4. The resulting map depicts the region as one that is largely viable with respect to its grape-growing climate (Fig. 9), with the majority of the landscape being classified as a cool climate maturity group (52%). The intermediate climate maturity group area (19%) is found mostly in the intermountain valleys of the Umpqua River drainage. The warm climate maturity group makes up the smallest area at 0.5% or roughly 5,300 acres. There are no areas deemed suitable to the hot climate maturity grouping in the AVA. Examining the climate suitability inside the Umpqua Valley AVA finds that the area is largely an intermediate (60%) and cool (38%) climate maturity region with less than 2% of the area deemed climatically not-viable (Fig. 9).

The combined topographic, soil, and land-use suitability grid was then masked with the climate maturity group grid's not viable area to further reduce the suitability to only those landscapes that can adequately ripen fruit across any of the climate maturity groups. The composite suitability grid is then depicted using the individual climate maturity groupings to show the spatial variation of terroir potential in the AVA (Fig. 10). Areas of non-suitable land are generally being limited through one of four main factors: elevation, slope, urban- or forest-zoned lands, or climate (shown in white in Fig. 10). Those lands deemed suitable that fall outside the AVA boundary include mostly cool climate groups with a few intermediate climate groups and can be found near the coast, to the west and southwest of the AVA in isolated valleys, and extending up some of the river valleys throughout the region. Cool-climate suitable land is found throughout the region with the greatest concentrations along the eastern, southern, and western AVA boundary area as elevations increase, and in a broad area in the northern portion of the AVA. Intermediate-climate suitable land is found covering much of the AVA, extending from the southern valleys, throughout the broader central portion

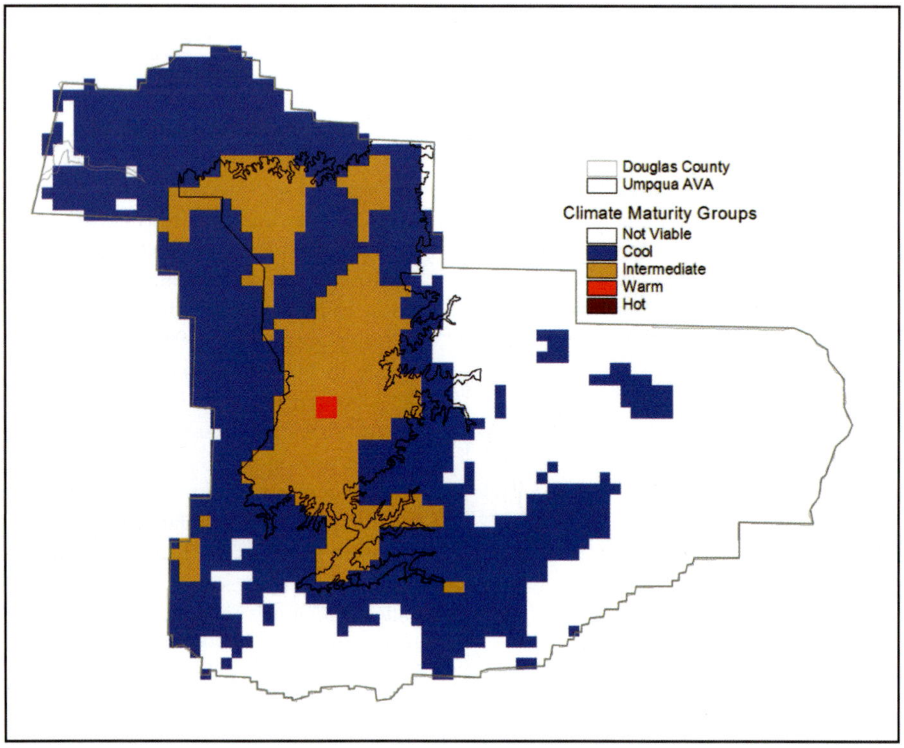

Figure 9 Climate maturity groups as derived through the combination of growing degree days, the frost free period, last and first frosts during the season, precipitation frequency and amount during flowering and harvest, and temperature variability during ripening. The results are depicted for Douglas County as a whole and with the Umpqua Valley AVA outlined in black.

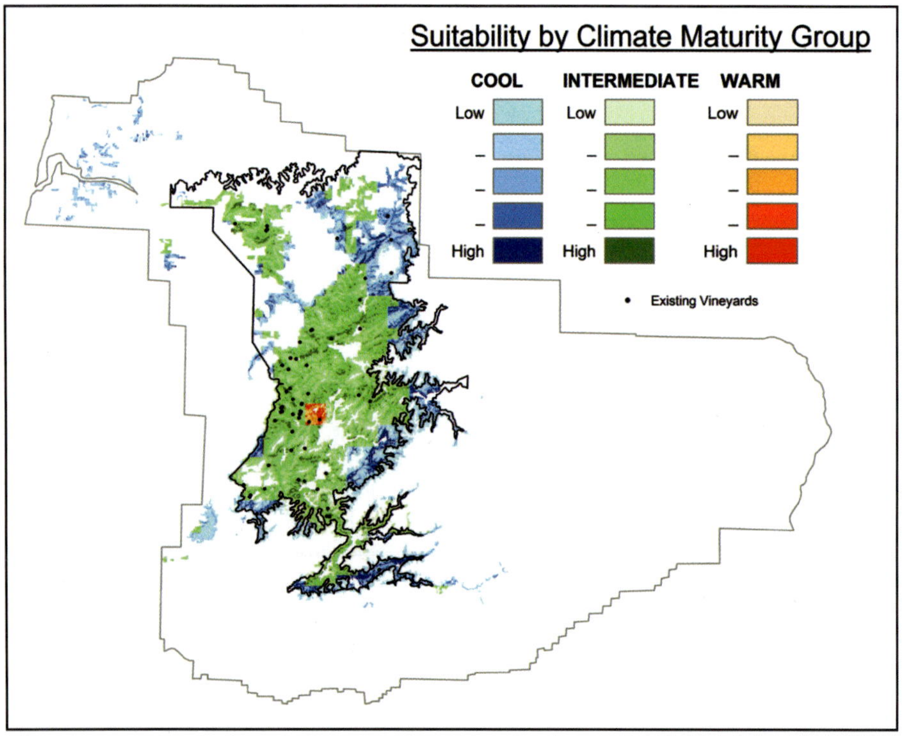

Figure 10 Composite landscape suitability from low to high (Fig. 8) depicted by climate maturity groups (Fig. 9). The results are depicted for Douglas County as a whole and with the Umpqua Valley AVA outlined in black.

of the AVA, and into a small area to the north (Fig. 10). Warm-climate suitable land is confined to an area in the west-central portion of the AVA. Examining only those landscapes that scored greater than 1.5 standard deviations above the mean landscape suitability results in 300, 2600, and 35 acres of the most suitable lands found in the cool, intermediate, and warm climate regions, respectively.

Since the original suitability criteria were established from a baseline of the existing vineyards, a comparison of the various suitability factors with the existing vineyards reveals how well they are situated in terms of the modeled terroir. From a topographic standpoint (elevation, slope, and aspect), the existing vineyards rank generally in the 5th and 6th deciles of topographic suitability (a decile divides a distribution of ranked scores into equal intervals where each interval contains one-tenth of the scores) with only six vineyards ranking in the 8th decile or higher (no existing vineyards are found in the topographically not viable regions). The main limiting factor related to topography that affects the existing vineyards is being planted to a flat site. Compared to the derived soil suitability (Fig. 6), the existing vineyards are found to be located on soils in the 6th and 7th deciles of the combine drainage, AWC, depth to bedrock, and ph suitability. Examining the relationship between the composite landscape suitability and climate maturity groups finds that 7, 49, and 2 of the existing vineyards are found in the cool, intermediate, and warm climate areas, respectively (Fig. 10). For the climate groupings, the seven vineyards in the cool region are mostly in the 6th and 7th decile of landscape suitability, the 49 vineyards in the intermediate region are in the 7th and 8th decile of the landscape suitability, and the two vineyards in the warm climate region are in the 6th decile of the landscape suitability. Potentially more important is that the existing vineyards represent only 24%, 13%, and 6% of the most suitable acreage that scored greater than 1.5 standard deviations above the mean landscape suitability in the model.

CONCLUSIONS

This research attempted to define and map terroir potential by examining the grape-growing landscape in the Umpqua Valley AVA of Oregon through the spatial analysis of topography, soil, land use, and climate. Using a locational survey of existing vineyards and grower knowledge of the best landscapes for grape growing, the topography of the region was classed into the most suitable elevations, slopes, and aspects for vineyards. The combined topographic suitability analysis shows that much of the area in the AVA, and some areas outside the AVA, provide fair to very good landscapes for grape growing. An assessment of soil suitability finds spatial variations in drainage, depth to bedrock, available water-holding capacity, and pH that are related to the underlying geology. The most advantageous soils with moderate to good drainage, adequate soil depths, fair to good water-holding capacity, and near neutral pH are found in a broad belt of the Umpqua Valley AVA, along the coastal zone, and in isolated areas throughout the county. Douglas County and the Umpqua Valley AVA have a large percentage of land zoned for agriculture, farm/forest transition, or rural residential that is more easily developed into vineyard sites.

Combining data on topography, soil, and land use reveals that the lands most suitable for vineyard development are located mostly within the Umpqua Valley AVA, with over 3000 acres of nearly ideal sites found spread throughout the region. Masking this composite suitability with the "not-viable" climate maturity regions reduces some of the locations both inside and outside the AVA. Overlaying climate and landscape factors results in a collection of suitable sites depicted by each of the cool, intermediate, and warm climate maturity groupings. Furthermore, since a good site can mitigate a marginal climate to some degree, and since the climate maturity groups are scored relative to a ripening potential with locations falling into a range of values for a climate maturity group, the best landscapes can increase the viability of ripening a given variety. For example, a site that scores at the high end of the intermediate climate maturity group and scores high on the landscape suitability, could potentially ripen warm climate maturity group varieties (i.e., Merlot could be ripened more consistently than on a marginal landscape in an intermediate climate). Comparing the terroir modeling to the existing vineyards finds that most vineyards in the region are sited on landscapes greater than average (6th–8th deciles of suitability) in terms of topography and soil. In addition, much of the best landscapes in each climate maturity grouping identified in the modeling (those greater than 1.5 standard deviations above the mean suitability) are not currently planted to vineyards. The results indicate that very good landscapes exist across all climate maturity types with strong potential for future development and production of quality fruit and wines.

While the results from this analysis are applied to a collection of land units (grids) that have varying levels of grape-growing suitability, many land units may not be available for vineyard development owing to various existing factors (e.g., already developed for other uses, not for sale, changes in zoning, etc.). Furthermore, the resulting grids do not depict availability of water for irrigation, which should be examined on a site-by-site basis. The results presented here are also subject to spatial variations that cannot be depicted from the resolution of the original digital data. For example, the climate grids have a resolution of roughly 1.45 miles, which from the toposcale to the microscale may not accurately model the climate structure when the landscape constricts to areas less than this resolution. In addition, spatial differences in soil types and properties at the site level can vary substantially and should be assessed through on-site soil analyses. Finally, the generalization of the topographic suitability to a homogeneous nine acre size will produce a potential vineyard site that can have small, less than ideal landscape features within it (e.g., swales, low spots, etc.) and should be analyzed on a site-by-site basis to define the vineyard size, shape, and block structure.

During its brief history, the Umpqua Valley AVA has proven to be a region in which cool- to warm-climate grape varieties can be grown. Rapid growth of the industry is likely to continue as a result of current successes and the availability of land. Through the use of GPS and GIS technologies, this research has helped to further define the terroir potential of grape growing in the Umpqua Valley AVA. The results provide existing and future growers with baseline knowledge of the region's grape-growing potential relative to its topography, soil, land use, and climate. While not specifically addressing the cultural aspects of terroir (e.g., style-directed viticultural and enological practices), which typically take many years to become dominant, the results presented here should serve to initiate better decisions in the site selection process, thus leading to fewer and/or more efficient trial and error procedures. In addition, for most potential growers, site selection will involve compromises, in that few sites will possess ideal characteristics in every respect. While compromise in many cases has been the rule, this body of research presents one of the best tools yet to enhance the site selection process for future growers in the Umpqua Valley AVA. Finally, the process developed here theoretically can be applied to any area where adequate spatial data resources are available.

ACKNOWLEDGEMENTS

This research is a compilation of various ongoing research projects that were funded by numerous sources: The Oregon Wine Advisory Board, the Umpqua Chapter of the Oregon Winegrowers' Association, and the Carpenter Foundation at Southern Oregon University. The authors would like to extend their appreciation to all of the grape growers in the Umpqua Valley AVA and H. Earl Jones for his insights into the industry.

REFERENCES

Aney, W.W., 1974, Oregon Climates Exhibiting Adaptation Potential for Vinifera: American Journal of Enology and Viticulture, v. 25, no. 4, p. 212-216.

Becker, H., 1988, Mapping of soils and climate in the Rheingau: Proceedings of the Second

Cool Climate Viticulture and Oenology symposium, Auckland, New Zealand.

Boyer, J. and Wolf, T., 2000, GIS and GPS aid the exploration of viticultural potential in Virginia: Vineyard and Winery Management, Nov/Dec 2000, p. 48-54.

Carey, V.A., 2001, Spatial characterisation of natural terroir units for viticulture in the Bottelaryberg-Simonsberg-Helderberg winegrowing area: M.Sc. Agriculture Thesis, University of Stellenbosch, South Africa.

Carey, V.A., Bonnardot, V., Schmidt, A. and Theron, J., 2001, The interaction between vintage, vineyard site (mesoclimate) and wine aroma of Vitis vinifera L. cvs. Sauvignon blanc, Chardonnay and Cabernet Sauvignon in the Stellenbosch-Klein Drakenstein wine growing area, South Africa (1996-2000): *in* Proceedings of the 26th World Congress & 81st General Assembly of the OIV, Section 1, p. 139-152.

Cass, A., 1999, What soil factors really determine water availability to vine?: The Australian Grapegrower and Winemaker, Annual Technical Issue, p. 95-97.

Code of Federal Regulations: Alcohol, Tobacco Products, and Firearms, 2000, (Title 27, Parts 1-199), Part 9 – American Viticultural Areas, p. 101-223.

Daly, C., Taylor, G.H., Gibson, W.P., Parzybok, T.W., Johnson, G.L. and Pasteris, P., 2001, High quality spatial climate data sets for the United States and beyond: Transactions of the American Society of Agricultural Engineers, v. 43, p. 1957-1962.

Davis, R., Chilton, R., Ottenbreit, L., Scheeler, M., Vielvoye, J., Williams, R. and Wittneben, U., 1984, Atlas of suitable grape growing locations in the Okanagan and Similkameen valleys of British Columbia: Agriculture Canada.

Department of Land Conservation and Development, 2002, Generalized Zoning Coverage for the State of Oregon: http://www.lcd.state.or.us/

De Villiers F. S., 1997, The use of Geographic Information System (GIS) in the selection of wine cultivars for specific areas by using temperature climatic models: *in* Proceedings, XXII Congrès de la Vigne et du Vin, December, 1997, Buenos Aires, Argentina.

Dry, P.R. and Smart, R.E., 1988, Vineyard site selection: *in* Coombe, B.G. and Dry, P.R. (eds.), Viticulture; Winetitles, Adelaide, Australia.

Environmental Systems Research Institute, Inc. (ESRI). Copyright © 1995-2003. All rights reserved. Published in the United States of America.

Gladstones, J., 1992, Viticulture and Environment: Winetitles, Adelaide, Australia. 310 p.

Haynes, S.J., 1999, Geology and Wine 1. Concept of terroir and the role of geology: Geoscience Canada, v. 26, p. 190-194.

Haynes, S.J., 2000, Geology and Wine 2. A geological foundation for terroirs and potential sub-appellations of Niagara Peninsula wines, Ontario, Canada: Geoscience Canada, v. 27, p. 67-87.

Jackson, D.I. and Cherry, N.J., 1988, Prediction of a district's grape-ripening capacity using a latitude-temperature index: American Journal of Enology and Viticulture, v. 39, n.1, p. 19-28.

Jones, G.V., 2001, Using GIS/GPS Technology to Assess Viticultural Status and Potential in the Rogue Valley AVA: Proceedings of the 115th Annual Meeting of the Oregon Horticultural Society, Portland, Oregon, p. 215-227.

Jones, G.V., Hille, P., Pine, J. and Jones, H.E., 2002, Mapping Varietal Potential for Quality Wines in Oregon using PRISM Spatial Climate Data: American Meteorological Society, 13th Conference on Applied Climatology, Portland, Oregon, May 2002 (in preparation, winter, 2004).

Jones, G.V., 2003, Trends in Frost Occurrence and the Frost-Free Growing Season Length in Oregon Grape-Growing Regions: Proceedings of the 117th Annual Meeting of the Oregon Horticultural Society, Portland, Oregon, p 223-231.

Jones, G.V. and Hellman, E., 2003, Site Assessment: *in* Hellman, E. (ed.), Oregon Viticulture: 5th Edition, Oregon State University Press, Corvallis, Oregon, p 44-50.

Jones, G.V. and Hellman, E., 2003, Oregon Winegrowing Regions: *in* Hellman, E. (ed.), Oregon Viticulture: 5th Edition, Oregon State University Press, Corvallis, Oregon, p 51-69.

Jordan, T.D., Pool, R.M., Zabadal, T.J. and Tompkins, J.P., 1980, Cultural practices for commercial vineyards: New York State College of Agriculture and Life Sciences, Miscellaneous Bulletin 111, 69 p.

Magarey, R.D., Seem, R. C. and DeGloria, S.D., 1998, Prediction of vineyard site suitability: Grape Research News, v. 9, no. 1, p. 1, 2.

Mastin, M. C., 1983, Site Selection of Vineyards in Oregon: Master's Thesis, Department of Geography, University of Oregon, June, 1983, 274 p.

Meinert, L.D. and Busacca, A.J., 2000, Geology and Wine 3. Terroirs of the Walla Walla Valley appellation, southeastern Washington State, USA: Geoscience Canada, v. 27, n. 4, p. 149-171.

Meinert, L.D. and Busacca, A.J., 2002, Geology and Wine 6. Terroir of the Red Mountain Appellation, Central Washington State, USA: Geoscience Canada, v. 29, n. 4, p. 149-168.

Natural Resources Conservation Service, 1997, State Soil Geographic (STATSGO) Data Base for Douglas County Area, Oregon: http://www.ftw.nrcs.usda.gov/ssur_data.html

Oregon Climate Service (OCS), 2002, Office of the State Climatologist, Oregon State University, Corvallis, Oregon: http://www.ocs.orst.edu/

Oregon Geospatial Data Clearinghouse (OGDC), 2003, Spatial Data Library – Generalized Zoning Coverage for the State of Oregon: Department of Administrative Services, State of Oregon: http://www.sscgis.state.or.us

Oregon Vineyard and Winery Reports (1987-2002), Oregon Agricultural Statistics Service, Salem, Oregon: http://www.oda.state.or.us/oass/vinewine.htm

Sayed, H.A., 1992, Vineyard site suitability in Ontario: Ontario Grape and Wine Adjustment Program, OMAFRA and Agriculture Canada Publication N.10.92, Ministry of Agriculture and Food, Agriculture Canada.

Shaulis, N. and Dethier, B., 1970, New York site selection for wine grapes: New York State Horticultural Society, Proceedings, v. 115, p. 288-294.

Smart, R. E. and Dry, P.R., 1980, A climatic classification for Australian viticultural regions: Australian Grapegrower and Winemaker, v. 17, p. 8-16.

Stulz, M. B., 2001, The Soils Suitability Extension (SSE): National Science and Technology Center, Resource Notes, no. 49, 4/18/01.

Tesic, D., Woolley, D.J., Hewett, E.W. and Martin, D.J., 2002a, Environmental effect on cv Cabernet Sauvignon (Vitis Vinifera L.) grown in Hawkes Bay, New Zealand, 1. Phenology and characterization of viticultural environments: Australian Journal of Grape and Wine Research, v. 8, p. 15-26.

Tesic, D., Woolley, D.J., Hewett, E.W. and Martin, D.J., 2002b, Environmental effect on cv Cabernet Sauvignon (Vitis Vinifera L.) grown in Hawkes Bay, New Zealand, 2. Development of a site index: Australian Journal of Grape and Wine Research, v. 8, p. 27-35.

Tukey, R.B., and Clore, W.J., 1972, Grapes - their characteristics and suitability for production in Washington: EB. 635, Cooperative Extension Service College of Agriculture, Washington State University, Pullman, Washington, 12 p.

United States Geological Survey (USGS), 2001, 1:24,000 Scale Digital Elevation Models (DEM): http://edcwww.cr.usgs.gov/doc/edchome/ndcdb/ndcdb.html

Vaudour, E., 2002, The quality of grapes and wine in relation to geography: Notions of terroir at various scales: Journal of Wine Research, v. 13, no. 2, p. 117-141.

Wilson, J.E., 1998, Terroir: the Role of Geology, Climate, and Culture in the Making of French Wines: Mitchell Beazley, London, UK, 336 p.

Wolf, T.K., 1997, Site Selection for Commercial Vineyards: Virginia Agricultural Experiment Station, Winchester, Virginia, Publication no. 463-016.

Accepted as revised 8 August 2004

SERIES

Geology and Wine 9: Regional Trace Element Fingerprinting of Canadian Wines

John D. Greenough[1], Leanne M. Mallory-Greenough[1], and Brian J. Fryer[2]

[1]*Department of Earth and Environmental Sciences, University of British Columbia - Okanagan, 3333 University Way, Kelowna, B.C., Canada, V1V 1V7, John.Greenough@ubc.ca*

[2]*Department of Earth Sciences, University of Windsor, Windsor, Ontario, Canada, N9B 3P4*

SUMMARY

New, and previously determined, trace-element analyses (20 elements) for 162 wines from five regions across Canada (Nova Scotia, Quebec, Ontario (Niagara and Pelee Island) and British Columbia (Okanagan and Vancouver Island)) test the hypothesis that wines can be regionally "fingerprinted", using routine ICP-MS analyses. Exploratory statistics show that wine trace-element composition is predominantly related to region of origin and wine colour. Compared to white wines (128 samples), red wines (34 samples) have higher Rb, Cs, Sr, Ba and Mo and lower Li, U and Th. This may be explained by extraction from grape skins (Rb, Cs, Sr, Ba, Mo) and adsorption onto grape skins (Li, U, Th) during fermentation of red wines, although other explanations (e.g., bisulfite addition) cannot be ruled out. Nearly all 20 trace elements are useful for creating discrimination diagrams that separate wines by region with considerable accuracy (≥90%). The geochemical behavior of trace elements in all wines suggests that a common mechanism, the effect of climate on trace element solubility, affects trace-element concentration patterns. Further, regional average concentrations of most trace elements correlate strongly with Degree Days, indicating that more heat results in increased evaporation, which in turn increases water uptake, thus yielding higher trace-element concentrations in grapes. High Sr and Ba in Okanagan wines may be derived from alkaline soils in this semi-arid environment. Thus, terroir-related climate and soil differences lead to quantifiable trace-element fingerprints for Canadian wines by region.

SUMMAIRE

Les résultats précédents et récents d'analyses d'éléments traces (20 éléments) sur 162 vins provenant de cinq régions du Canada (Nouvelle-Écosse, Québec, Ontario (Niagara et île Pelee) et de Colombie-Britannique (l'Okanagan et l'île de Vancouver)) permettent de tester l'hypothèse selon laquelle il serait possible de caractériser régionalement les vins à partir d'analyses de routine de SM/PIHF. Une analyse exploratoire statistique des résultats montre que la composition en éléments traces est principalement reliée à la région d'origine des vins et à leur couleur. Comparés aux vins blancs (128 échantillons), les vins rouges (34 échantillons) ont des teneurs plus élevés en Rb, Cs, Sr, Ba et Mo, et moins élevés en Li, U et Th. Cela pourrait s'expliquer par un phénomène d'extraction à partir des peaux de raisin (Rb, Cs, Sr, Ba, Mo) et d'adsorption sur les peaux de raisin (Li, U, Th) durant la fermentation des vins rouges, bien qu'on ne puisse exclure d'autres explications telle (l'ajout de bisulfite par ex.). Presque chacun des 20 éléments traces contribue à la réalisation de diagrammes de différenciation qui permettent de départager les vins par région avec une grande précision (=90 %). Le comportement des éléments traces dans tous les vins permettent de penser qu'un mécanisme commun, soit l'effet du climat sur la solubilité des éléments traces, affecte les profils de concentration en éléments traces. De plus, les concentrations régionales moyennes de la plupart des éléments traces montrent qu'il existe une forte corrélation avec les degrés-jours, ce qui indique qu'un surplus de chaleur amène un accroissement de l'évaporation, ce qui entraîne un accroissement de l'absorption d'eau, ce qui explique les plus fortes teneurs en éléments traces des vins. Les fortes teneurs en Sr et en Ba des vins de l'Okanagan pourraient s'expliquer par ses sols alcalins dans un environnement semi-aride. Et c'est ainsi que les facteurs climatiques et pédologiques permettent la caractériser de manière quantifiable les vins canadiens par région.

INTRODUCTION

In the early 1900s, the French began certifying the origin and authenticity of wines using the Appellation d'Origine Contrôlée (Johnson, 1991, p. 44). Underpinning this geographic certification was the idea that *terroir* (see Haynes, 1999, for a definition) determines the quality and character of wines. Today, quality wines from around the world bear labels testifying to where their grapes came from. In reality, it is difficult or impossible to quantitatively certify the origin of a bottle of wine. Given that individual wines can fetch thousands of dollars, and that organizations purchase millions of dollars-worth of wine assuming that the label actually reflects place-of-origin, the potential value of a simple method for verification is obvious.

During the mid-1990s, several studies indicated that trace elements in wines might be used to determine the origin of their grapes at vineyard and regional levels (Latorre et al., 1994; Stroh et al., 1994; Greenough et al., 1996, 1997). These studies left many unanswered questions. To what extent do processing methods affect the trace-element composition of wine? What happens to trace elements as wines age, and how much inter-year variability is there in wines from one vineyard or region? It is now known that these issues may confound the practical aspects of legal fingerprinting of wines, but the most important control on composition appears to be the geographic origin of the grapes (Taylor et al., 2002, 2003). Three other points have emerged: 1) fingerprinting requires a multi-element dataset; 2) multivariate statistical techniques are needed for data analysis; and 3) one of the most promising analytical methods for trace-element fingerprinting of wines is Inductively Coupled Plasma – Mass Spectrometry (ICP-MS), which is inexpensive, routine and accurate.

Canada's small wine industry has an international reputation for producing fine wines, primarily from the Niagara and Okanagan regions of Ontario and British Columbia, respectively. Fledgling products from Nova Scotia, Quebec and Vancouver Island enjoy loyal local support. Taylor et al. (2003) showed that Okanagan and Niagara wines are distinguishable using ICP-MS trace-element data. The present study has two objectives: 1) to determine if wines from the five Canadian wine-producing areas noted above can be "fingerprinted" with ICP-MS data; and 2) to re-examine why red and white wines appear to have different trace-element compositions.

SAMPLES, GEOLOGY AND CLIMATE

Seventy-six bottles of wine, made from grapes derived from 35 vineyards, were purchased from 28 wineries in Nova Scotia (30 bottles; 12 vineyards; 5 wineries), Quebec (9; 5; 5), Ontario (Pelee Island, 2; 1; 1), the Okanagan Valley, British Columbia (16; 13; 13) and Vancouver Island, British Columbia (19; 4; 4). The emphasis was placed on obtaining wines made from grapes from a large number of geographically distinct vineyards representing the five main regions noted above (Fig. 1). Many Nova Scotia and Vancouver Island vineyards are represented by more than one wine sample. The 76 new samples of wine are augmented by data for 86 wines from the Niagara area of Ontario (36) and the Okanagan Valley (50) in the Taylor et al. (2003) study. Ice wines and late harvest wines were excluded because their production involves distinct processing and harvesting practices that affect trace-element concentrations. Altogether there are 34 red wines and 128 white wines in this study. Most wines were made from a single grape variety with $\geq$ 42 varieties represented. Wines from Nova Scotia, Quebec and Vancouver Island are commonly made from hybrid grapes such as the popular Maréchal Foch (red) and Seyval Blanc (white). Okanagan and Niagara wines mostly use *Vitis vinifera* varietals with Chardonnay and Pinot Blanc (white wines), and Merlot and Pinot Noir (red wines) being important varieties; vintages range from 1992 to 2002.

Figure 1. Map showing areas of Canadian wine production represented in this study.

Canadian vineyards are planted on a wide variety of soil types underlain by diverse bedrock geology (Table 1). Each wine producing area has distinctive climate characteristics reflected in annual precipitation, average temperature and number of Degree Days (= the cumulative amount of time above the threshold temperature of 10°C for each day of the growing season. For example, if the mean temperature of a day is 25°C, this would be 1 x 15 or 15 Degree Days; Table 1). This latter variable, in various forms, is used extensively in agriculture, and in the wine industry for identifying suitable locations for viticulture (Davis et al., 1984, p. 31). In addition to geochemical studies, relationships between geology and terroir for the Okanagan and Niagara regions are discussed in Haynes (2000), Fulton (2003), Grant et al. (2004), Greenough et al. (2004) and Bowen et al. (2005).

ANALYTICAL TECHNIQUES AND STATISTICAL METHODS

The 76 "new" wines were analyzed for 20 trace elements by using the ICP-MS at the University of Windsor. Wine samples were diluted 10:1 using high-purity 2% HNO_3 that was spiked with internal standard elements (Be = 5, In = 0.5, Tl = 1 ppm) to correct for matrix effects and instrument drift during ICP-MS analysis. Elemental analysis was performed using a ThermoElemental X-7, ICP-MS unit coupled with a CETAC ASX 500 autosampler. ICP-MS calibration used 3 separate multi-element solutions (supplier = Inorganic Ventures Ltd.). Dilution blanks were measured from the internal standard spiked with 2% HNO_3. Data reduction and correction for matrix effects, instrument drift, and molecular ion interferences used commercial spreadsheet software. Duplicate samples (one for every 20 wines) were collected during opening of the wines and treated as unknowns. Precision, based on the duplicates, is +/- 1-2% except where close to the detection limit. Some of the 162 wine samples have Th (8 samples) or Ag (14 samples) below detection limits.

Data statistical analysis employed SYSTAT software. Multidimensional scaling (MDS; Wilkinson et al., 1992; Borg and Groenen, 1997, p. 1-14), an exploratory data analysis technique, was used to uncover natural patterns in the dataset. MDS yields "maps" of the distances between objects (trace-element compositions of wine samples in this study) based on a distance measure, in this case correlation coefficients. Two wines with identical concentrations of all trace elements would have a correlation coefficient of 1, whereas two wines with a perfect negative correlation would yield a correlation coefficient of -1. Using a matrix with all possible wine-versus-wine correlation coefficients, MDS yields diagrams in which chemically similar samples plot close together. Samples on opposite sides of an MDS diagram are most-different. MDS can also be used to compare the behaviour of elements by inverting the dataset (Fig. 2).

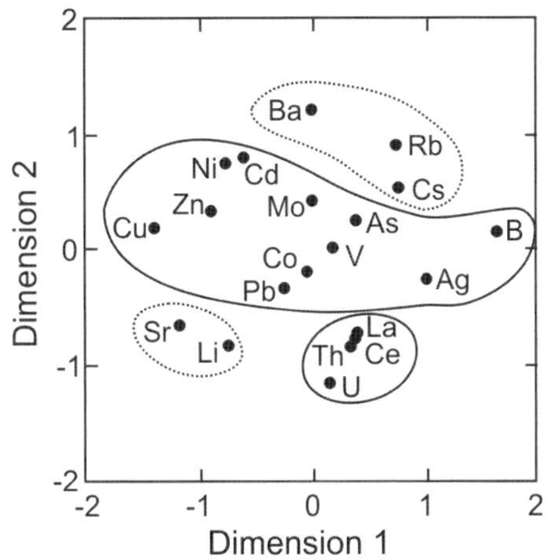

Figure 2. Multidimensional scaling diagram showing the tendency for trace elements in wine samples to associate according to geochemical character with chalcophile (e.g., Cu, Bi) and siderophile elements (Ni, Co) forming a group; highly soluble lithophile elements (Ba, Rb, Cs; and Sr and Li) forming two groups; and the insoluble high-field-strength elements (La, Ce, Th, and U) plotting together to form a third group. Field boundaries illustrating trace-element groupings are hand-drawn. See text for discussion.

After determining the predominant relationships among samples using MDS (Fig. 3), discriminant analysis can help explain the origin of the data patterns and put them to use (Tabachnick and Fidell, 2001, p. 456-464; Rollinson, 1993, p. 44-46). In this technique, samples are categorized (e.g., wine region) and the chemical data used to develop equations that place samples optimally within these categories. The equations allow new wine samples of unknown origin to be classified i.e., discrimination diagrams are created that classify wines according to origin using combinations of trace-element data (Fig. 4). This treatment of trace-element analyses uses Log_{10} transformed data to minimize the effect of outliers, and to make the analyses better fit the normal data distribution requisite of most inferential multivariate statistical methods (Rollinson, 1993, p. 44-46).

DATA

Trace-element behavior in the wine dataset is not random. First, the high-field strength lithophile elements, U, Th,

Table 1: Information on climate, soil and bedrock contributing to the terroir of Canadian viticultural regions.

Region	Ppt.	T	°Days	Soil Types	Bedrock
Nova Scotia	1211	6.9	1006	Brunisolic Gray Luvisols, Humo-Ferric Podzols.	Triassic siliciclastic redbeds (Annapolis Valley) and Pennsylvanian siliciclastics and Mississippian carbonates/evaporites.
Quebec	1134	5.3	1002	Dystric Brunisols, Ferro-Humic and Humo-Ferric Podzols, Gleysols.	Devonian marine siliciclastics and argillaceous limestones.
Niagara ON	944	9.0	1471	Gleysols and Gray Brown Luvisols.	Paleozoic dolostones, limestones and siliciclastic sediments.
Okanagan BC	333	9.2	1181	Brown & Dark Brown Chernozems; Eutric Brunisols.	Proterozoic to Cenozoic granite gneisses. Paleozoic arc rocks and sediments, Mesozoic granites, Cenozoic felsic volcanics/sediments.
Vancouver Is. BC	883	9.7	761	Dystric Brunisols.	Ealy Jurassic diorite and Jurassic arc volcanics & near-shore clastics.

Notes: Ppt = Precipitation (mm/year), T = average temperature (C°) and °Days = Degree Days (with a temperature above 10°C) are averages for the years 1971 to 2000 from the Environment Canada, Climate Normals and Averages web site, http://www.climate.weatheroffice.ec.gc.ca/Welcome_e.html. Weather stations (Kentville, NS; Magog, PQ; Niagara Falls, ON; Penticton, BC; Victoria, BC) were selected based on proximity to the majority of vineyards for the region and if possible were centrally located. Soil types are from the Agriculture Canada, Soil Landscapes of Canada v 2.2 website, http://sis.agr.gc.ca/cansis/nsdb/slc/intro.html.

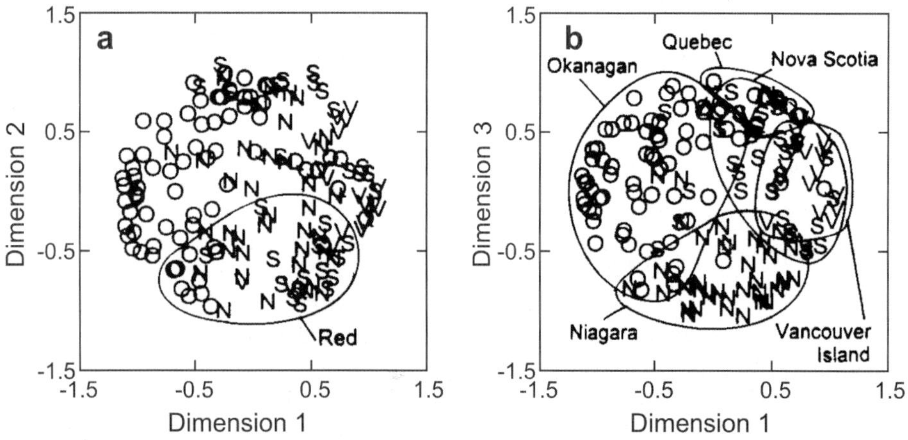

Figure 3. Multidimensional scaling diagrams showing relationships between wine samples in three dimensions. Symbols O = Okanagan, V = Vancouver Island, S = Nova Scotia, N = Niagara including two Pelee Island samples, Q = Quebec. The hand-drawn field in Figure 3a encompasses 90% of the red wine samples of this study. Dimension 2 predominantly reflects the impact of wine colour on composition. Hand-drawn fields in Figure 3b show that wine trace-element compositions reflect region of origin. Okanagan wines plot apart from most other wines along Dimension 1, and apparently represent the predominant source of chemical variation in the data set. Chemical differences between Niagara, Vancouver Island and Quebec samples are summarized along Dimension 3, but Nova Scotia samples appear to have less-distinct compositions.

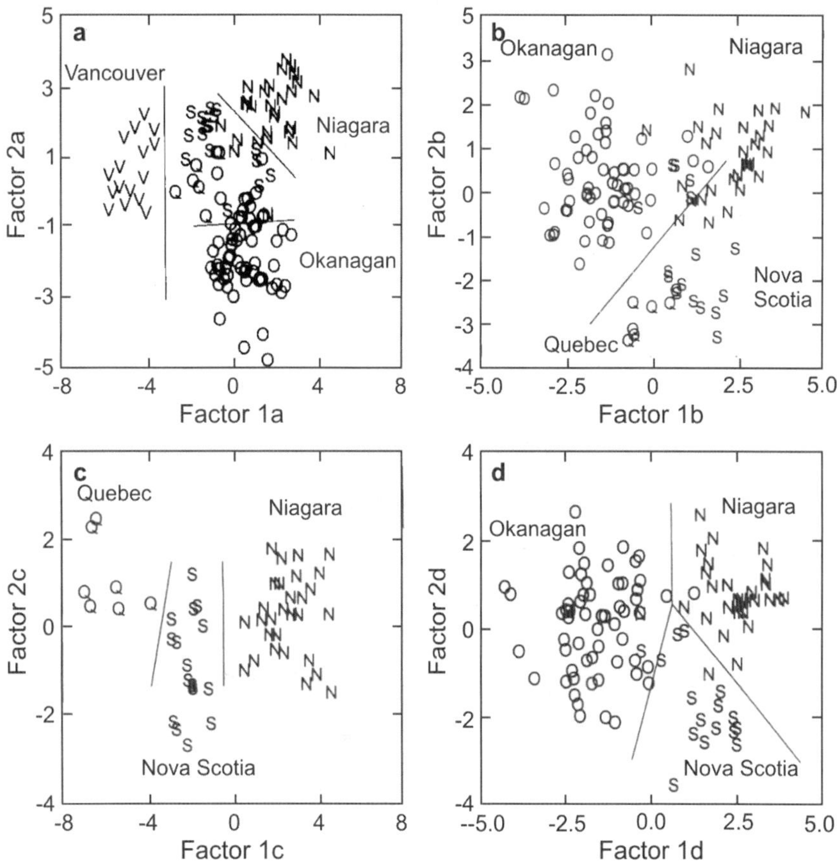

Figure 4. Discrimination diagrams for the regional fingerprinting of Canadian white wines. New samples are plotted on the diagrams using equations in Table 4. Symbols: O = Okanagan, V = Vancouver Island, S = Nova Scotia, Q = Quebec; N = Niagara. Field boundaries are hand-drawn. See text for discussion.

La and Ce plot close together on the MDS diagram (Fig. 2) indicating similar behavior. "Similar behavior" means that the concentrations of these trace elements are positively correlated in the wines and an X-Y plot (not shown) of the concentrations of any two elements, in all wines, would yield a straight line with a high correlation coefficient. Thus, when one of these high-field strength lithophile elements is present in large amounts (or small amounts) in a wine, all such elements tend to be present in large (or small) amounts. Second, the large ion lithophile elements form two groups on opposite sides of Figure 2, also suggesting coupled behavior: When Sr and Li are present in large amounts; Cs, Rb and Ba tend to be present in small amounts, and vice versa. Third, the chalcophile (Cu, Pb, Zn, Cd, Mo, As, Ag, Bi) and siderophile elements (Ni, Co) form a large group occupying the middle of the diagram.

On average, the red wines have significantly higher concentrations of Rb, Cs, Sr and Ba, than the white wines ($P \leq 0.05$, where P = t-test probability; see Table 2 for values and brief explanation of probability factor, P); whereas the white wines have higher Li, U and Th (bolded values, Table 2). Because wine colour and composition are related, Table 3 compares the regional average compositions of only the white wines, the largest dataset. P values (right column) indicate that the average concentration of nearly all trace elements is strongly correlated with region ($P \leq 0.05$). Niagara region white wines have the highest average concentrations for most trace elements. Exceptions are the high Sr and Ba concentrations in Okanagan wines. Okanagan wines also have the smallest amounts of Cs. In contrast with Niagara, Vancouver Island wines have either the lowest or second lowest average concentrations of nearly all trace elements. Bolded values in Table 3 point out the two elements from each region with mean concentrations that are statistically distinct.

Figures 3a and 3b provide 3-D, MDS "maps" of white and red wine trace-element sample relationships. Samples plotting close together are very similar in trace-element compositions. To understand how "similarity" is determined for wines, imagine a graph (not shown) with all trace-element concentrations in one wine plotted against all concentrations in a similar wine (i.e., two wines close together in Fig. 3). The two wines would produce a straight line graph with a high correlation coefficient. MDS arranges samples using a set (matrix) of correlation coefficients that measure (estimate) the distances between all possible pairs of wines. Stated another way, the MDS coordinates for a single wine (Fig. 3) are determined from the estimated distances to all other wines, a procedure something like locating the epicentre of an earthquake by using three or more seismic stations. Finally, instead of determining "distances" between samples in two dimensions, as done for elements in Figure 2, coordinates were determined in three dimensions, analogous to longitude, latitude and depth in the Earth. Because three-dimensional relationships are difficult to portray in two-dimensional diagrams on paper, sample coordinates are projected onto the Dimension 1 versus Dimension 2 plane (Fig. 3a), and Dimension 1 versus Dimension 3 plane (Fig. 3b).

The hand-drawn field in Figure 3a encompasses 90% of the trace-element compositional variability of all red wines of this study, but only 17% of the trace-element compositional variability of the white wines, indicating that Dimension 2 reflects the impact of wine colour on trace-element composition. The labeled fields in Figure 3b encompass ~ 90% of the trace-element compositional variability of wine samples from the Okanagan Valley, Quebec, Niagara and Vancouver Island with little overlap. Nova Scotia wines show more overlap with other regions, but most samples fall close to the Vancouver Island field. It is apparent that Dimensions 1 and 3 express differences in wine trace-element composition controlled by region-of-origin. In summary, Figure 3 diagrams indicate that wine colour and region-of-origin represent the two predominant factors controlling wine composition.

DISCUSSION

Trace Element Compositions of Red and White Wines

Our results confirm that red and white wines have distinct trace-element compositions (Greenough et al., 1997; Baxter et al., 1997). Elements that have significantly different ($P \leq 0.05$) average concentrations include Rb, Cs, Sr and Ba, which are higher in red wines; and Li, U and Th which are higher in white wines (Table 2). Greenough et al. (1997) reported similar results: Ba, Mo, Ca, P and Mg were significantly higher in red wines ($P \leq 0.05$) with Li higher in white wines. Average Rb, Cs and Sr concentrations were also reported higher in reds, and U lower, but the differences were not as large (significant) as for Ba, etc. Combining study results, red wines have higher Rb, Cs, Sr, Ba and Mo, and white wines higher Li, U and Th concentrations.

The differences among wine types (varietal or hybrid) could reflect trace-element concentrations in grapes. However, there are no other known relationships between grape variety and trace-element concentrations, and the trace-element differences between red and white wines apply to *Vitis vinifera* and non-*Vitis vinifera* (e.g., hybrid) varieties alike. That is, they apply to grapes with genetic differences arguably more fundamental than those relating to colour. One wine, a blush wine made from red wine grapes (Pinot Noir) and processed in a white wine style, plotted with white wines on the MDS diagram (Fig. 3). Thus trace-element differences between red and white wines do not appear to be related to grape variety.

Effects of Processing

The inorganic composition of wine undoubtedly is affected by how wines are produced. Bentonite, a fining (clarifying) agent, can increase rare-earth-element concentrations (Jakubowski et al., 1999), and storage of wine in stainless steel vats may raise Ni and Fe content in

Table 2: Mean concentrations (ppb), standard deviations (S.D.), and t-test probabilities (P) for differences between mean element concentrations in red and white wines.

Element	*Red*	*S.D.*	*White*	*S.D.*	*P*
Li	2.35	2.22	**4.9**	6.04	0
Rb	**856**	313	479	235	0
Cs	**2.92**	1.68	1.97	1.65	0
Sr	**610**	283	449	281	0
Ba	**193**	79	101	46	0
La	0.56	0.64	0.66	1	0.46
Ce	0.97	1.12	1.37	2.04	0.13
U	0.37	0.4	**0.58**	0.92	0.05
Th	0.057	0.064	**0.096**	0.136	0.02
V	29.1	46.1	21	37.1	0.35
Co	2.1	1.26	2.33	1.66	0.36
Ni	22.1	27.2	16.1	16.5	0.24
Cu	66.1	121.4	73.2	140.5	0.77
Zn	444	251	444	389	0.99
As	7.11	8.92	5	4.66	0.2
Mo	14.2	25.5	6.5	10.2	0.1
Ag	0.023	0.029	0.023	0.03	0.99
Cd	0.61	0.56	0.51	0.65	0.39
Pb	17.8	16.6	13.2	14.1	0.15
Bi	0.49	0.63	0.34	0.46	0.22

Sample numbers for red and white wines are 33 and 129, respectively, for all elements except Th (32, 122) and Ag (30, 118). Bold values highlight whether red or white wines have a higher average element concentration at $P < 0.05$. P values are t-test (= student's t-test) probabilities of the significance of analytical data, in this case trace elements in wines, as measured by the t-test. A P value of 0.05 represents the 95% confidence level, with analytical results at this level by convention regarded as "significant"; a P value of 0.01 represents the 99% confidence level, regarded as "highly significant". Thus the lower the value of P, the higher the significance of the results; the higher the value of P, the lower the significance of the results.

such wines (Eschnauer, 1982; Murányi and Papp, 1997). The addition of K or Na metabisulfite to wines to control oxidation and microbial activity (Cox, 1985), the natural precipitation of K bitartrate (wine diamonds), storage in oak casks (that add organic compounds; Eggers et al. *in press*), or contact with the bottle's cork may also affect trace-element concentrations (Taylor et al., 2002). Whether some of these factors can account for differences between red and white wine trace-element compositions is explored below.

Differences in alkalis (Na, K) might be caused by the precipitation of K bitartrate but this occurs in both red and white wines, and does not explain differences in high-field-strength elements such as U, Th, La, and Ce. Higher alkali and alkaline-earth-element concentrations in red wines could reflect the addition of Na or K metabisulfite. However, metabisulfite use is not ubiquitous; yet, as pointed out above, the same trace-element compositional differences between red and white wines turn up from one study to another. Also, metabisulfite addition cannot explain lower Li and high-field-strength elements in red wines. The Sr isotopic composition of wines closely reflects soil composition (Barbaste et al., 2002) indicating that little Sr is added to wine during processing. This argues against metabisulfite as a source for Sr. Finally, metabisulfite is used more extensively with white wines. Thus, higher alkali and alkaline earth-element concentrations in red wines appear to be unrelated to matabisulfite addition.

Bentonite is apparently more commonly used as a fining agent in white wines than in red wines. Its use raises rare-earth-element concentrations (Jakubowski et al., 1999) and might similarly affect other high-field-strength elements. If so, bentonite would also have to adsorb Rb, Cs, Sr, Ba and Mo, and release Li, to explain the other differences between red and white wines. However, this explanation has problems. Average Niagara and Vancouver Island red and white wines show the same relative pattern of Rb, Cs, Sr, Ba, Mo, Li, U and Th enrichment/depletion as is seen in the dataset (Table 2). Both red and white Niagara wines have higher concentrations of all trace elements than Vancouver Island wines. These regional differences in composition would be difficult if not impossible to preserve if bentonite addition has had a major impact on wine chemistry. Similar arguments can be made that less-common or non-ubiquitous addition of fining agents

Table 3: Mean concentrations and standard deviations, of elements in Canadian white wines by region.

Region	Niagara		Okanagan		Quebec		Nova Scotia		Vancouver Is.	
Samples	29	S.D.	59	S.D.	7	S.D.	16	S.D.	16	S.D.
Vineyards	18		29		4		6		5	
Wineries	17		27		4		4		5	
Li	7.55	7.23	6.35	6.07	**0.62**	0.42	1.9	2.24	**0.18**	0.12
Rb	647	304	358	173	384	82	475	139	648	110
Cs	3.34	2.1	**1.2**	1.03	2.53	1.39	2.33	1.77	1.73	0.7
Sr	369	162	**645**	281	263	46	**241**	118	189	53
Ba	105.3	47.1	112.4	51.9	99.5	17.4	94.9	26.6	63.3	23.9
La	1.42	1.7	0.39	0.5	0.36	0.19	0.72	0.62	0.37	0.4
Ce	2.75	3.43	0.82	0.99	0.73	0.36	1.8	1.76	0.78	0.88
U	0.75	0.66	0.62	1.21	0.26	0.14	0.74	0.63	0.14	0.13
Th	0.166	0.21	0.085	0.104	0.02	0.01	0.096	0.098	0.038	0.045
V	52.5	55.8	14.7	28.4	6	3.1	13.1	12.8	3.7	3.5
Co	**4.25**	1.89	1.94	1.26	1.99	0.33	2.14	0.46	0.84	0.41
Ni	24.5	11.5	15.9	21.4	12.6	2.7	13.8	4.1	6.6	4.7
Cu	135	233	44	60	38	20	40	26	118	193
Zn	**792**	526	412	321	267	95	287	108	182	69
As	8.24	6.64	3.91	3.53	3.62	1.26	**6.44**	4.08	2.55	1.29
Mo	12.9	16.1	6.32	7.54	1.33	0.53	4.26	5.83	**0.44**	0.14
Ag	0.027	0.032	0.016	0.016	0.014	0.007	0.049	0.057	0.018	0.018
Cd	0.66	0.55	0.49	0.87	0.53	0.19	0.46	0.14	0.35	0.23
Pb	27.3	20.9	10.7	8.3	6.16	2.62	11.6	6.5	2.75	1.12
Bi	0.17	0.16	0.31	0.46	**1.13**	0.34	0.37	0.23	0.21	0.07

Notes: Concentrations in ppb. Samples = number of bottles of wines though counts for Th are 29, 58, 6, 14 and 14 and for Ag are 27, 57, 7, 15 and 11 for regions from left to right, respectively. S.D. = standard deviation. P = probability of correlation with categorical variable "region" from F statistic. Bolded values represent the two most statistically distinct elements (univariant F statistic) for each region compared to all other wines grouped together. Niagara average does not include the Pelee Island sample.

such as egg albumin, isinglass (= gelatin; more commonly used with red wines) and casein (= milk protein; more commonly used with white wines), have minimal impact on wine trace-element chemistry. The same can be said for oaking and cork effects.

The simplest explanation for trace-element concentration differences between red and white wines is that contact between grape skins, seeds and must (juice) during red wine fermentation affects the extraction of highly soluble elements (e.g., Rb, Cs) that are metabolically concentrated in skins or seeds during growth. Under this hypothesis, Li and highly insoluble elements including U, Th, La and Ce are absorbed by the damaged skins. The end result is differences in element concentrations between red wines and white wines that also preserve or reflect regional compositional differences.

Regional Differences in White Wine Trace-Element Composition

As noted previously, only white wine trace-element concentrations are used here to discuss regional differences in composition. P values given in Table 3 show that nearly all trace elements in the white wines are highly correlated with "region of origin". Bolded values highlight two of the more distinct elements for each region, but regions should be more accurately distinguished using combinations of elements. Figure 4 shows how Canadian white wines could be regionally "fingerprinted". These discriminant analysis diagrams use multiple elements from the dataset to visually separate samples by region. The statistical output includes equations (Table 4) that allow new samples, of unknown origin, to be classified. All equations in Table 4 use $\leq$ 10 elements.

Trace-element concentrations for a new wine are converted to a Log_{10} format and inserted into Table 4 equations. The calculated Factor values (Table 4) are plotted on the diagrams of Figure 4a to d.

The hand-drawn vertical line in Figure 4a separates Vancouver Island white wines with 100% accuracy. Non-Vancouver Island white wines are plotted on Figure 4b, which divides samples into two groups for plotting on Figures 4c and 4d. The hand-drawn field lines in Figure 4c separate Quebec, Nova Scotia and Niagara wines in our dataset with 100% accuracy. The hand-drawn field lines shown in Figure 4d leave only 4 of 104 samples from the Okanagan, Niagara and Nova Scotia incorrectly classified.

The efficiency of separation shown in Figure 4 may be optimistic, however, owing to limited numbers of white wine/trace-element samples from some of the regions of this study. Niagara and Okanagan separations appear rigorous because they are based on many white wines from numerous vineyards. Quebec, in contrast, is represented by only seven samples from four vineyards (Table 3). Nevertheless, numerous possible diagrams and combinations of elements (not shown) yield similarly accurate classifications. Separations tend to be more effective - fewer samples are incorrectly classified – when larger numbers of trace elements are used. From a geological and analytical perspective this is reasonable, because more elements filter out variation of unknown origin (randomness) and ameliorate analytical error. However, with small numbers of samples (e.g., Quebec), large numbers of elements increase the likelihood that the classification procedure will only work with the original samples. New samples may be incorrectly classified.

Many studies suggest that wines can be fingerprinted with trace-element data at the vineyard level through to regional level (Latorre et al., 1994; Stroh et al., 1994; Day et al., 1995; Greenough et al., 1996; 1997; Baxter et al., 1997; Danzer et al., 1999; Taylor et al., 2002; 2003; Diaz et al., 2003). Although wine isotopic ratios reflect soil ratios (Barbaste et al., 2002), absolute element concentrations are difficult to relate to soil or bedrock compositions (Greenough et al., 1996; Taylor et al., 2002). As with plants, however (Brooks, 1972, p. 96-99), absolute concentrations of individual elements in wines correlate with solubility in water (predicted from ionic potential, = charge/radius ratios). Elements such as La and Ce show low concentrations in wines compared to soils and rocks because these elements are insoluble, whereas elements such as Rb and Sr are present in higher concentrations owing to higher solubility in water (Greenough et al., 1996). Also, studies of Okanagan wines have shown that trace-element concentrations correlate qualitatively with a climate gradient between the south (hot and dry) and north (more-temperate and wetter) ends of the valley (Greenough et al., 1997; Taylor et al., 2002).

Figure 2 provides clues to the causes of regional differences in the composition of Canadian wines. When data from several unrelated chemical sys-

Table 4: Discrimination equations for plotting new samples on Figure 4 diagrams.

Factor	Equation
1A	11.204 + 2.624*Li + 2.048*Zn - 4.473*Rb - 2.561*Sr + 1.399*Cs + 0.829*Pb + 0.789*Mo - 0.305*V + 0.275*Ce
2A	2.969 + 0.143*Li + 1.722*Zn + 1.774*Rb - 5.506*Sr + 1.066*Cs + 0.956*Pb - 0.378*Mo + 1.350*V - 0.135*Ce
1B	2.405 + 0.667*Li + 1.613*V + 2.047*Zn - 5.832*Sr - 0.312*Mo - 0.169*Bi + 2.219* Rb
2B	0.599 + 1.921*Li + 0.154*V + 0.614*Zn - 0.492*Sr + 0.628*Mo - 1.296*Bi - 1.281*Rb
1C	-29.479 + 2.940*V + 2.659*Zn - 1.970*As + 4.563*Rb - 4.318*Cd - 6.742*La + 7.343*Ce - 4.091*Bi + 1.572*Pb
2C	-2.446 + 1.127*V + 4.006*Zn - 2.235*As - 1.583*Rb - 1.202*Cd + 7.038*La - 7.464*Ce + 2.950*Bi + 0.093*Pb
1D	3.849 + 0.919*V + 1.099*Co + 1.835*Zn + 0.457*As + 0.951*Rb - 5.121*Sr - 0.231*Mo - 0.002*Cd + 0.953*Cs + 0.861*Pb
2D	-9.279 + 0.869*V - 1.317*Co + 3.071*Zn - 1.874*As - 1.893*Rb + 1.421*Sr + 1.575*Mo - 2.576*Cd + 1.806*Cs + 0.829*Pb

Notes: Concentrations of elements are entered as Log_{10} of concentration in ppb. Equations cannot classigy Pelee Island (ON) wines which appear to be distinct from Niagara wines.

tems (e.g., random analyses of unrelated rocks) are examined with multidimensional scaling, trace elements tend to plot randomly (e.g., Mallory-Greenough et al., 1998). Non-random behaviour (e.g., Fig. 2) implies that similar geochemical processes are controlling trace-element behaviour in the wines. Because many different rocks underlie vineyards, if bedrock had a predominant influence on wine chemistry, trace-element behaviour in Figure 2 should be more random. Further, the separation of soluble from insoluble lithophile elements in the dataset suggests that climate-related weathering and element solubility are important factors in determining trace-element concentrations in wines (Greenough et al., 1996). Bedrock composition may help to distinguish wines at the vineyard level, but for regional discrimination of wines it is likely of secondary importance. Note, however, that this conclusion does not apply to isotopic ratios.

Tables 1 and 3 and Figure 5 suggest further links between climate and wine composition. Vancouver Island wines show low average concentrations of most trace elements whereas the highest concentrations occur in Niagara wines. Except for Bi, all averaged element data (Table 3) have a positive correlation with climate-related Degree Days (Table 1). In fact, 14 of 20 trace elements show a correlation coefficient (r) with Degree Days of ≥ 0.74 ($P \leq 0.15$) and eight (Li, V, Co, Ni, Zn, Mo, Cd, Pb) have $r \geq 0.91$ ($P \leq 0.03$; and e.g., Fig. 5)! One explanation for this feature may be that higher temperatures lead to higher solubility of elements in soils. An alternative explanation may be that areas with higher Degree Day values - warmer temperatures - yield greater evaporative losses leading to more water uptake by grapes, thus increasing trace-element concentrations in grapes. This second hypothesis is supported by the linear, arithmetic relationships between trace-element concentrations and Degree Days, and by somewhat similar concentration factors [geometric mean and standard deviation = 8.5 ± 2.9 times; range = 41 (Li) to 1.9 (Cd) times] for these elements between Vancouver Island and Niagara (Table 3). Rb, Cs and Sr correlate poorly with Degree Days ($r < 0.55$; $P < 0.35$), and Okanagan wines have high Sr concentrations (Taylor et al., 2002). This may reflect the concentration of alkaline earth metals (Ca, Sr) in soil and groundwater as is typical in dry climates.

Regardless of the underlying causes, there are strong regional variations in the trace-element composition of Canadian wines. Our dataset includes 20 trace elements, but the ICP-MS is capable of routinely determining more than 40 elements. Considering that nearly all 20 trace elements reported here are highly correlated with "region", other trace elements should be similarly useful. With a larger sample base, assembled over several years, more trace elements could be "safely" used in discrimination. In addition, ICP-MS isotopic data promise to become routinely available, providing more discriminating power as indicated by experiments fingerprinting Italian and French wines (Day et al., 1995; Barbaste et al., 2002; Larcher et al., 2003). Still, there are unresolved issues. For example, ice wines and late harvest wines appear to have anomalous compositions compared to "normal" wines from the same vineyard.

Will trace-element fingerprinting become part of the Vintners Quality Alliance (VQA) certification procedure for wines? Undoubtedly ICP-MS analysis of trace elements can provide routine, accurate and inexpensive data and, as shown here, such analyses are capable of separating wines by region with a high degree of accuracy. There are, however, potential legal ramifications to incorrectly classifying a wine. Whether analytical and statistical methods can be refined, or regulations written with limitations in mind, remains to be seen. For now it is clear that there are academic and potential private uses for wine fingerprinting.

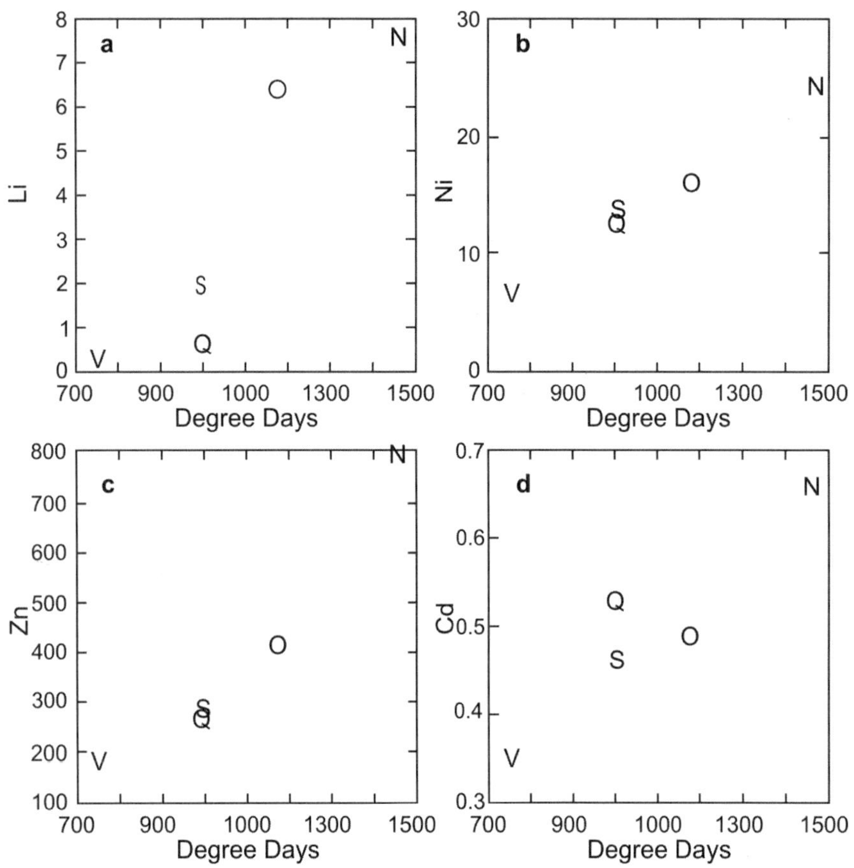

Figure 5. Plot of Degree Days versus the average concentrations of selected elements in wines from each of the Canadian wine producing areas. Concentrations are highly correlated with Degree Days. See text for discussion. Symbols as in Figure 4.

ACKNOWLEDGEMENTS

We acknowledge the many viticulturalists and wine makers across Canada who helped with sample selection. H. Soon (Sandhill), G. Stanley (Quails Gate), and D. Barnett (Lailey) advised on wine-making practices. H. Muggeridge and R. Corney prepared the diagrams. We are grateful for critical reading by Georgia Pe-Piper and Roger Macqueen; their comments improved presentation of the paper. This research was in part supported by NSERC research grants to JDG and BJF.

REFERENCES

Barbaste, M., Robinson, K., Guilfoyle, S., Medina, B., and Lobinski, R., 2002, Precise determination of the strontium isotope ratios in wine by inductively coupled plasma sector field multicollector mass spectrometry (ICP-SF-MC-MS): Journal of Analytical Atomic Spectrometry, v. 17, p. 135-137.

Baxter, M.J., Crews, H.M., Dennis, J.M., Goodall, I. and Anderson, D., 1997, The determination of the authenticity of wine from its trace element composition: Food Chemistry, v. 60, p. 443-450.

Borg, I. and Groenen, P., 1997, Modern Multidimensional Scaling, Theory and Applications: Springer-Verlag, New York, 471 p.

Bowen, P.A., Bogdanoff, C.P., Estergaard, B.F., Marsh, S.G., Usher, K.B., Smith, C.A.S., and Frank, G., 2005, Use of Geographical Information System technology to assess viticultural performance, Okanagan and Similkameen valleys, British Columbia: Geoscience Canada, v. 32, no.4, *In press*.

Brooks, R.R., 1972, Geobotany and Biogeochemistry in Mineral Exploration: Harper and Row, New York, 290 p.

Cox, J., 1985, From Vines to Wines: Harper and Row, New York, 253 p.

Danzer, K., De La Calle Garcia, D., Thiel, G. and Reichenbacher, M., 1999, Classification of wine samples according to origin and grape varieties on the basis of inorganic and organic trace analyses: American Laboratory, v. 31, p. 26-33.

Day, M.P., Zhang, B. and Martin, G.J., 1995, Determinatoin of the geographical origin of wine using joint analysis of elemental and isotopic composition. II Differentiation of the principle production zones in France for the 1990 vintage: Journal of the Science of Food and Agriculture, v. 67, p. 113-123.

Davis, R., Chilton, R., Ottenbreit, L., Scheeler, M., Vielvoye, J., Williams, R. and Wittneben, U., 1984, Atlas of Suitable Grape Growing Locations in the Okanagan and Similkameen Valleys of British Columbia: Agriculture Canada, Summerland, British Columbia, 141 p.

Diaz, C., Conde, J.E. and Estevez, D., 2003, Application of multivariate analysis and artificial neural networks for the differentiation of red wines from the Canary Islands according to the island of origin: Journal of Agricultural and Food Chemistry, v. 51, p. 4303-4307.

Eggers, N., Greenough, J.D. and Cernak, T., *in press*, Identifying the Terroir of Okanagan Valley, British Columbia, Chardonnay Wines Using Volatile Components: Geoscience Canada reprint volume, Geology and Wine.

Eschnauer, H., 1982, Trace elements in must and wine: primary and secondary contents: American Journal of Enology and Viticulture, v. 33, p. 226-230.

Fulton, R.J., 2003, Geology of Okanagan Valley wine region of British Columbia. [Abstract]: 2003 Seattle Annual Meeting of the Geological Society of America, Paper 74-2.

Grant, E.B., Ker, K. and Haynes, S., 2004, Niagara Geology and Wine: Fieldtrip Guidebook, Geological Association of Canada - Mineralogical Association of Canada, Joint Annual Meeting 2004, Brock Univ., St. Catharines, ON, 17 p.

Greenough, J.D., Longerich, H.P. and Jackson, S.E., 1997, Element fingerprinting of Okanagan Valley wines using ICP-MS: Relationships between wine composition, vineyard and wine colour: Australian Journal of Grape and Wine Research, v. 3, p. 75-83.

Greenough, J.D., Longerich, H.P. and Jackson, S.E., 1996, Trace element concentrations in wines by ICP-MS: Evidence for the role of solubility in determining uptake by plants: Canadian Journal of Applied Spectroscopy, v. 41, p. 76-80.

Greenough, J.D., Lutmerding, H., and Mallory-Greenough, L., 2004, Chapter 7, Geology and terroir of Okanagan Valley wine: *in* Roed, M. and Greenough, J.D., eds., Okanagan Geology, Kelowna Geology Committee, Kelowna, BC, p. 67-70.

Haynes, S.J., 1999, Geology and Wine 1. Concept of Terroir and the Role of Geology: Geoscience Canada, v. 26, p. 190-194.

Haynes, S.J., 2000, Geology and Wine 2, A geological foundation for *terroirs* and potential for sub-*appellations* of Niagara Peninsula wines, Ontario, Canada: Geoscience Canada, v. 27, p. 67-87.

Jakubowski, N., Brandt, R., Stuewer, D., Eschnauer, H.R. and Gortges, S., 1999, Analysis of wines by ICP-MS: Is the pattern of the rare earth elements a reliable fingerprint of provenance?: Fresenius Journal of Analytical Chemistry, v. 364, p. 424-428.

Johnson, H., 1991, Hugh Johnson's Modern Encyclopedia of Wine: Simon and Schuster, New York, 576 p.

Larcher, R., Nicolini, G. and Pangrazzi, P., 2003, Isotope ratios of lead in Italian wines by inductively coupled plasma mass spectrometry: Journal of Agricultural and Food Chemistry, v. 51, p. 5956-5961.

Latorre, M.J., Garcia-Jares, C., Medina, B. and Herrero, C., 1994, Pattern recognition analysis applied to classification of wines from Galicia (northwest Spain) with certified brand of origin: Journal of Agriculture and Food Chemistry, v. 42, p. 1451-1455.

Mallory-Greenough, L.M., Greenough, J.D. and Owen, J.V., 1998, New data for old pots: Trace element characterization of ancient Egyptian pottery using ICP-MS: Journal of Archaeological Science, v. 25, p. 85-97.

Murányi Z. and Papp L., 1997, ICP-AES metal content analysis of wines made with different technologies: ACH-models in Chemistry, v. 134 p. 529–537.

Rollinson, H., 1993, Using Geochemical Data: Evaluation, Presentation, Interpretation: Longman, Essex, England, 352 p.

Stroh, A., Brückner, P. and Völlkopf, U., 1994, Multielement analysis of wine samples using ICP-MS: Atomic Spectroscopy, v. 15, p. 100-106.

Tabachnick, B.G. and Fidell, L.S., 2001, Using Multivariate Statistics, 4th edition: Allyn and Bacon, Toronto, 966 p.

Taylor, V.F., Longerich, H.P. and Greenough, J.D., 2002, Geology and Wine 5. Provenance of Okanagan Valley wines, British Columbia, using trace elements: promise and limitations: Geoscience Canada, v. 29, p. 110-120.

Taylor, V.F., Longerich, H.P. and Greenough, J.D., 2003, Multielement analysis of Canadian wines by inductively coupled plasma mass spectrometry (ICP-MS) and multivariate statistics: Journal of Agricultural and Food Chemistry, v. 51, p. 856-860.

Wilkinson, L., Hill, M., Welna, J.P. and Birkenbeuel, G.K., 1992, SYSTAT for Windows: Statistics, Version 5 Edition: SYSTAT Inc., Evanston, IL, 750 p.

SERIES

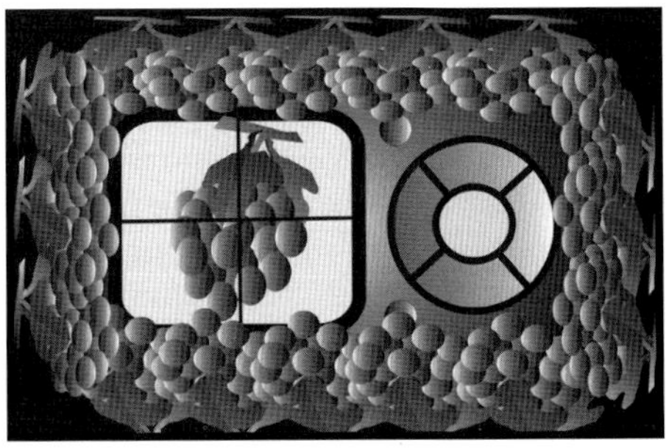

Geology and Wine 10: Use of Geographic Information System Technology to Assess Viticulture Performance in the Okanagan and Similkameen Valleys, British Columbia

P. A. Bowen, C.P. Bogdanoff, B.F. Estergaard, S.G. Marsh, K.B. Usher, C.A.S. Smith and G. Frank
Pacific Agri-Food Research Centre (PARC), Agriculture and Agri-Food Canada, Summerland, BC, V0H 1Z0 Corresponding author E-mail: bowenp@agr.gc.ca

SUMMARY
The complex geological history of the Okanagan and Similkameen valleys in British Columbia has created a wine growing region by way of diverse bedrock, soils, terrain and climate. Although wine grapes have been grown in the area for many decades, a recent conversion of vineyards to predominately *Vitis vinifera* varieties, and the subsequent world recognition of wine quality, suggests that there is further potential to improve wine quality through fine-tuning of grape varietal choices and vineyard management techniques. A geographic information system (GIS) application has been developed to study the relationships among site conditions, management practices and vineyard performance including fruit and wine quality. The production area was divided into six regions based on landform and climate: 1. Kelowna, 2. Penticton, 3. Vaseaux - Oliver, 4. Golden Mile, 5. Black Sage - Osoyoos, and 6. Similkameen. The complement of grape varieties planted varies among the regions. Comparisons of regional patterns of varieties planted, and medals received, have revealed significant regional differences in varietal suitability. Although most of the vineyards are sited on coarse-textured soils, comparisons of the distributions of all and medal-winning vineyard blocks among soil textural classes have revealed that quality wine grapes are grown on a broad range of soil types within, and among, the six regions studied. Loamy soils appear to be especially suitable for producing quality wine grapes in the Okanagan and Similkameen valleys.

SUMMAIRE
L'histoire géologique complexe des vallées d'Okanagan et de Similkameen en Colombie-Britannique a créé une région vinicole en réunissant divers socles rocheux, sols, terrains et climats. Bien que le raisin de cuve ait été cultivé dans la région depuis plusieurs décennies, une conversion récente aux variétés de *Vitis vinifera*, et la reconnaissance mondiale de la qualité de ses vins qui s'en est suivie, permet de penser que la qualité des vins pourrait encore être améliorer par le choix de variétés de vignes et l'amélioration des techniques de production. Une application d'un système d'information géographique (SIG) a été mise au point pour étudier les relations entre les caractéristiques des sites, les pratiques de gestion, et les performances du vignoble, dont la qualité des fruits et du vin. La zone de production a été subdivisée en six régions selon la forme du relief et le climat, soit : 1. Kelowna, 2. Penticton, 3. Vaseaux-Oliver, 4. Golden Mile, 5. Black Sage, et 6. Similkameen. Le complément des variétés de raisin plantées varie selon les régions. Les comparaisons d'arrangement des variétés plantées et des médailles reçues ont mis au jour d'importantes différences régionales quant aux variétés les mieux adaptées. Bien que la plupart des vignobles croissent dans des sols légers, les comparaisons des distributions des parcelles en fonction des types de texture des sols et des médailles reçues ont montré que des vignes de qualité sont cultivées sur une grande variété de types de sol à travers les six régions étudiées. Les sols loameux semblent particulièrement convenir pour la production de raisin de cuve dans les vallées d'Okanagan et de Similkameen.

INTRODUCTION
The Okanagan and neighbouring Similkameen valleys in south-central British Columbia (Fig. 1) are home to a young, expanding premium-wine industry. Premium-wine production in the two valleys has risen from 0.8 to 4.7 million litres in the last 12 years (British Columbia Wine Institute, 2004). Although *Vitis vinifera* varieties have been cultivated in the area for decades, prior to 1990, most of the vineyards were of French hybrid and *V. labrusca* varieties, which were found to be more productive, hardy and disease resistant than the enologically superior *V. vinifera* (Association of British Columbia Grape Growers, 1984). In the late 1980s, it became clear that development of an internationally competitive wine industry in the region must be based on premium *V. vinifera* wines. This spurred a govern-

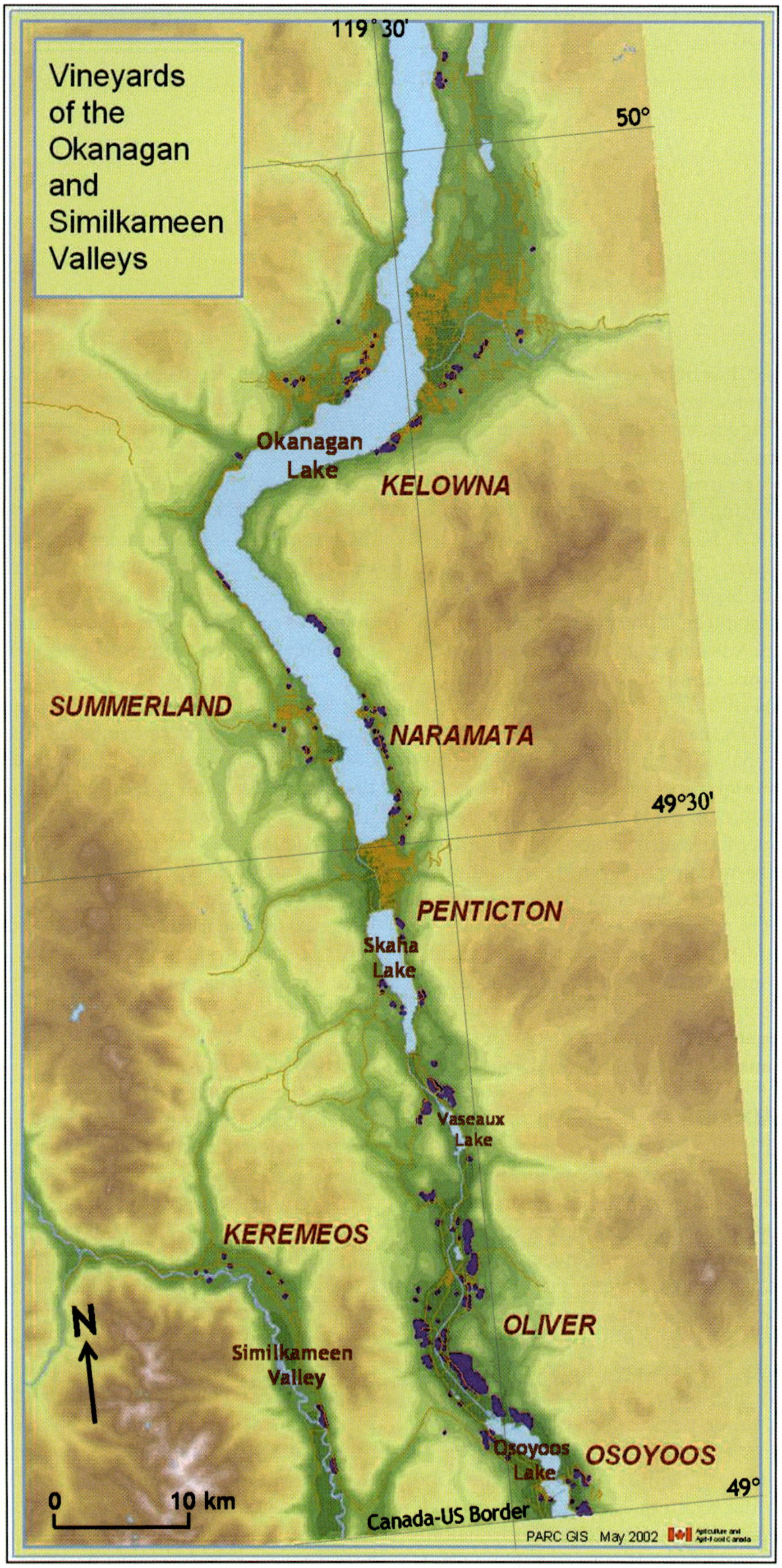

Figure 1. Distribution of vineyards (shown in purple) in the Canadian portions of the Similkameen and Okanagan valleys. The Okanagan Valley is occupied by Okanagan, Skaha, Vaseaux and Osoyoos lakes.

ment-supported vine pullout and replant program that resulted in the conversion of nearly all plantings to *V. vinifera*. In addition, the higher profit potential compared to some tree fruit production, and the opening of previously uncultivated vineyard-suitable lands, have resulted in further expansion of the vineyard area to its highest ever, totalling more than 2,000 ha. However, sites suitable for *V. vinifera* production are greatly limited by climate, terrain and the scarcity of arable soil. Thus individual vineyards are small, averaging *ca.* 10 ha.

The bedrock, soils and terrain in the two valleys are highly diverse owing to the complex geological history of the region. East of Okanagan Valley, the Okanagan Highlands are composed of the Monashee Gneiss, representing the ancient western margin of North America, which was uplifted during the Cretaceous (Roed and Greenough, 2004). West of the Okanagan Valley, the bedrock geology is dominated by Jurassic granites and Eocene volcanic sedimentary rocks, which slid downward and westward off the rising crustal massif of the Highlands during the Eocene. The separation plane between these two bedrock terrains crops out as a complex network of faults in and around the margins of the lake, collectively referred to as the Okanagan Fault (Templeman-Kluit and Parkinson, 1986). Eocene volcanics of the Marron Group are of andesitic composition and were extruded in association with large calderas, the remnants of which are still present at Vernon, Kelowna, Summerland and Penticton (Roed and Greenough, 2004). Eocene sedimentary rocks of the White Lake Formation, including conglomerate, sandstone, siltstone and minor coal, were deposited as fluvial sediments in the landscape surrounding the volcanic edifices; these bedrock units underlie vineyards in the Okanagan Valley.

The valleys have endured many glacial advances, the last occurring in the late Pleistocene (Nasmith, 1962). River valleys were clogged with large volumes of sand and gravel during glacial advances. Although glacial processes eroded bedrock locally, they mainly redistributed pre-existing unconsolidated materials. Thus the materials lining the Okanagan and Similkameen valleys are heterogeneous mixtures derived from many different bedrock sources (see detailed maps in Nasmith, 1962, and simplified map in Roed and Greenough, 2004)

The late glacial history of the Okanagan Valley has been described by Nasmith (1962). Most valley bottom landforms were created during deglaciation. The most influential activities were the melting of stagnant ice lobes in the valley, the periodic creation and breaching of ice dams south of Skaha Lake that created glacial Lake Penticton, and the deposition of thick glaciolacustrine silts which filled the valley. The draining of this lake and the erosional cutting led to benches and bluffs consisting of glacial lake sediments along the shores of Okanagan and Skaha lakes (Fig. 2), and fluvioglacial materials that gave rise to silty, sandy, gravelly and stony soils throughout the area but particularly south of Skaha Lake and to the east of Osoyoos Lake. The presence of ice lobes that filled the valley bottoms also allowed for the formation of alluvial fans that sit above the current valley floor.

Vineyards in the Okanagan and Similkameen valleys (Fig. 1) are established mostly on glaciofluvial and fluvial terraces (benches) and fans, glacial lake deposits, and slope deposit complexes that are reworked glacial deposits (Fulton, 2003). Soil parent materials are composed primarily of unconsolidated glacial deposits having a lithology reflecting regional geological inputs and producing a largely uniform, mixed mineralogy in most soils. What is variable is soil texture and thickness, characteristics that influence rooting depth and the water and nutrient holding capacities. Bedrock is rarely prominent in vineyards developed to date.

The geological events and processes, which led to the diverse bedrock, landforms and soils in the region, and the Okanagan Valley's wide latitude span (49° N to 50°30' N) create a wide range of vineyard edaphic and climatic conditions. Recognizing the challenge grape-growers face in selecting vineyard sites and varieties to cultivate in the valleys, government specialists in the early 1980s put significant effort into identifying and rating the viticultural suitability of agricultural lands. Lands were rated according to the suitability of their soil, climate and sun exposure, and maps showing these ratings were published in the widely used "Atlas of Suitable Grape Growing Locations" (Association of B.C. Grape Growers, 1984). At the time the atlas was published, French hybrid varieties were recommended, and dominated in plantings. Winter survival and high yields, rather than premium winemaking quality, were the main goals of viticulture in the Okanagan and Similkameen valleys. Since that publication, the conversion to *V. vinifera* and emphasis on quality for winemaking have created needs for more detailed ratings of vineyard site characteristics, and a better understanding of how these characteristics interact with management practices to affect the performance of *V. vinifera* varieties. While much has been learned from other areas about the response of *V. vinifera* varieties to climate, soils and management practices, information derived from local experience can be more relevant in developing strategies to improve crop performance while minimizing risk and cost. Indeed, within the Okanagan and Similkameen valleys the combinations of soils, landforms and climates - those characteristics that define *terroir* (Haynes, 1999; Wilson, 2001) - are unique.

Our approach in constructing our viticultural GIS application for the Okanagan and Similkameen valleys has been to focus on site-specific vineyard attributes. The GIS contains information on the conditions and activities that have resulted in the range of success growers have attained in producing high quality wine grapes from their sites and varieties planted. Its purpose is to provide a data model that growers and researchers can use to study relationships among the site characteristics, including physical, soil and climatic attributes, varieties grown, and viticultural practices that determine grape and wine quality. A similar model for France was described by Morlat (2001) in which a chain of factors contribute to the quality of the wine produced, beginning with site conditions, then influenced by planting material, annual climate, and human practices, including viticultural and winemaking methods.

Any grower or researcher who keeps and accesses records of field conditions, management practices and crop performance by block or site has, in effect, already created and is using a GIS. For large sets of location-based records, modern GIS software (i.e., ArcGIS, ESRI, Canada) facilitates data handling and analysis, and can be used to create maps displaying spatial relationships. In recent years the adoption of GIS analysis has become widespread in agriculture, both by single-farm operators and by regional commodity associations. For viticulture, GIS has been used successfully in the precision man-

Figure 2. A west-facing vineyard (foreground) on the benchlands north of Penticton, east of Okanagan Lake. The range of soil textures found on the benchlands results from the complex of medium to fine-textured glacial lacustrine materials that are overlain by a discontinuous sandy loam eolian veneer.

agement of nutrients and irrigation (e.g., Bramley and Williams, 2001; Lamb et al., 2004) and for modelling vineyard site suitability (e.g., Tesic et al., 2002a, b; Tesic, 2004; Wolf and Boyer, 2003). Application of GIS tools may be particularly useful in the Okanagan and Similkameen valleys where growers are challenged in managing vineyards sited on a limited land base characterized by variable bedrock, soils, climate and terrain.

Although *V. vinifera* culture in the Okanagan and Similkameen valleys is relatively recent, the potential for world-class wine production has already been verified by the increasing number of press commendations and gold medals received (British Columbia Wine Institute, 2004). Obviously some growers and wineries are already making good management choices and others may be able to learn from them. While France took centuries to develop its viticulture for each appellation, Okanagan and Similkameen growers have quickly adopted and modified techniques developed elsewhere (including France), and may be able to further fine-tune their varietal choices and management techniques for the specific conditions of their site. Since no two wine regions have the same growing conditions, the GIS should be able to provide growers with the most relevant information on what leads to the finest wines under the unique conditions in British Columbia's interior.

STRUCTURE OF THE GIS DATABASES

There are currently *ca.* 250 commercial vineyards in the Okanagan and Similkameen valleys. Vineyards operated by wineries average *ca.* 29 ha, whereas those operated by independent growers average *ca.* 4 ha. The average area of individual grape variety blocks is *ca.* 1 ha. As most vineyard managers keep records for cultural practices and performance on a variety block basis, information in our GIS is collected on the same basis and is organized in three modules: Site Conditions, Management Practices, and Vineyard Performance (Table 1). The Site Conditions module has information on soils, terrain and climate. The Management Practices module includes varieties planted, planting systems designs and management practices. The

Table 1: An example of data collected for an individual variety block for a single year. The list is incomplete, as the GIS database contains more detailed information on soil characteristics and cultural practices for this block.

SITE CONDITIONS	
Physical Attributes	
Region	5: Black Sage/Osoyoos
Block area (acres)	0.73
Average aspect (degrees)	265
Aspect range (degrees)	255-275
Aspect variability score (0 = not variable, 4 = highly variable)	1
Average slope (degrees)	15
Slope range (degrees)	0-20
Slope variability score (0 = not variable, 4 = highly variable)	3
Frost pocket(s) (% of block area)	0
Soil Attributes	
Soil series (verified by specialist)	Osoyoos
Soil (<2 mm) texture (1=coarse to 5=fine) (inspected)	1
Gravel (>2mm) content (0=none to 2=very gravelly) (inspected)	0
Surface stoniness (>25cm) (0=none to 5=excessive) (inspected)	0
Soil/land alteration (0=not altered to 3= alterations to 100% of area) (inspected)	0
Annual Frost Record	
Year	2001
Frost damage - spring, fall (%)	0, 0
Date of last spring frost	126
Date of first fall frost	278
MANAGEMENT PRACTICES	
Production System Attributes	
Variety	Chardonnay
Clone	76
Rootstock	SO4
Year planted	1997
Trellis system type	Bilateral cordon with vertical shoot positioning
Pruning style	mix (cane and spur)
Row direction (degrees)	265
In-row spacing (cm)	120
Between-row spacing (cm)	240

Production System Attributes (continued)	
Cordon wire height (cm)	80
Catch wire heights (cm)	104, 133, 163
Cover crop	Yes
Irrigation type	overhead sprinkler
Annual Cultural Practices	
Year	2001
50% bloom date	156
Shoot thinning (Y/N)	yes
Shoot thinning date	134
Shoots: number per vine	18
Hedged (shoots trimmed) (Y/N)	yes
Hedging date	185
Leaves removed in cluster zone (Y/N)	yes
Leaf removal date	220
Leaves removed (number/shoot)	3.5
Cluster thinning (Y/N)	no
Cluster thinning date	N/A
VINEYARD PERFORMANCE	
Annual Crop Performance	
Year	2001
Harvest date	262
Yield (tons/acre)	3.62
Soluble solids (tank) (°Brix)	23.5
pH (tank)	3.49
Titratable acidity (tank) (g/L)	8.9
Winemaker-expected quality score (1=poor, 5=excellent) (historical)	4
Winemaker-actual quality score (1=poor, 5=excellent)	4
Number of previous years the winemaker used grapes from this block	3
Annual Wine Awards	
Year (vintage)	2000
Competition	Okanagan Fall Wine Festival
Medal	silver
Percentage of wine (%)	3.5

Vineyard Performance module contains phenological dates, crop yield, basic juice composition, winemaking quality ratings and the vineyard sources of medal-winning wines. An outline of the procedures used for collecting data for each module follows.

Site Conditions
Block Locations
Perimeter coordinates for each grape-variety block were collected using a global positioning system (GPS); from these, the block area was calculated. Vineyard maps constructed by overlaying the block polygons onto base maps with elevation contours, roadways, and waterways using GIS software, were provided to growers. There are currently more than 240 mapped vineyards and more than 2,500 variety blocks in the GIS.

Terrain
Average slope and aspect of each grape-variety block were measured onsite, and terrain variation (roughness) was rated. More detailed terrain characterization, for determination of insolation (exposure to sunlight) and for mesoclimate modelling, is accomplished through analysis of elevation contours using GIS software (ArcGIS, Spatial Analyst extension, ESRI, Canada). Fruit ripening patterns can be affected by topography and terrain, which influence cold air drainage (Bowen et al., 2004) and insolation (Fialla et al., 2004).

Locations of vineyard areas frequently damaged by frost (frost pockets) were identified by the vineyard managers.

Soils
The classification of soils within each variety block was determined by first overlaying the block boundaries onto soil maps (Fig. 3) developed from a detailed (1:20,000) soil survey completed in 1981 (Wittneben, 1986). The soil series revealed for each block were verified by a specialist (Joe Fitzpatrick) who visited each vineyard and examined the soils and landforms. When necessary, he dug pits to observe the soil profile to confirm the series and the taxonomic classification. Subgroup names follow the Canadian system of soil classification (Soil Classification Working Group, 1998). For each soil series identified in vineyards, the physical and chemical properties as described in the soil survey report and observed by the specialist were entered into the relational database supporting the GIS application. Properties included are: depth to bedrock; surface stoniness; texture (resulting from the size distribution of mineral particles); perviousness class; drainage class; depth to root restriction; shear strength; permeability; pH; salinity class and cation exchange capacity. Soil maps, such as the one shown in Figure 3, are provided to growers and show the soil series and textural class in each variety block.

Climate
Climatic variation is known to be a major contributor to variation in fruit quality (i.e., Giomo et al., 1996). Climatic information is included in the GIS in two scales. Regional climate, representing areas spanning 50-60 km, is characterized using data gathered from grower- and government-owned weather stations. From these, indices such as growing degree days (base 10°C) (GDD), dates of last-spring and first-fall frosts and frost-free days are calculated. Annual variation in climate may be wide ranging (Fig. 4) and contributes to vintage differences in wines.

Mesoclimate, representing areas of $\leq$2 ha (i.e., on the scale of individual variety blocks), is characterized using data collected with fine-grained networks of climate monitoring devices such as radio-transmitting motes (Crossbow, USA); small sensor-loggers (Watchdog, Spectrum, USA; Smartbutton, ACR, Canada); or with a mobile weather station with attached GPS unit, which is moved through vineyards in a grid pattern. Owing to the irregular topography found in many vineyards, the range in ambient air temperature within an area as small as 1 ha can be as high as 8°C (Bowen et al., 2004). This spatial variation in temperature can also be dynamic, especially during the daytime, and can result in substantial spatial variation in the diurnal temperature range. Maps of such mesoclimate characteristics (Fig. 5) are produced using data from the fine-grained measurements.

Management Practices
Planting Systems
Vineyard planting systems and irrigation equipment types were determined onsite during the GPS survey. The planting system characteristics include training and trellis design, vine spacing within, and between, rows, and row direction. Irrigation equipment types include sprinkler, microjet and drip systems, and design features such as the spacing of emitters or risers.

Cultural Management
All management practices, including irrigation, pruning, thinning of shoots, leaves and clusters, crop protection (sprays and organic methods), nutrient applications and vineyard floor management (tillage methods and use of cover crops), are determined through an annual survey completed by vineyard managers.

Crop Performance
Crop Phenology
Phenological records including dates of budbreak, bloom, veraison (onset of berry ripening) and harvest are provided by growers in the annual survey.

Fruit Yield and Quality
Yield, and measures of basic fruit quality characteristics such as juice pH, titratable acidity (TA) and soluble solids (°Brix) are provided by growers in the annual survey. For selected vineyards and grape varieties, more detailed compositional information, e.g., levels of colour pigments, yeast assimilable nitrogen and aromatic volatiles, has been obtained from associated research (i.e., Usher et al., 2004). For grape-variety blocks, from which wine was made separately, winemakers provide a fruit quality rating that represents their experienced perception of winemaking quality. The ratings, 1 to 5, respectively, indicate quality levels of unacceptable, poor, fair (average), good and excellent or outstanding. A rating is provided for the block overall (over all years of winemaking experience) and for the current year. When the quality ratings are analyzed for relationships with cultural practices and crop characteristics such as yield or °Brix at harvest, the appropriateness of using current cultural practices for each variety or growing region can be assessed. For example, the effect of yield level on fruit

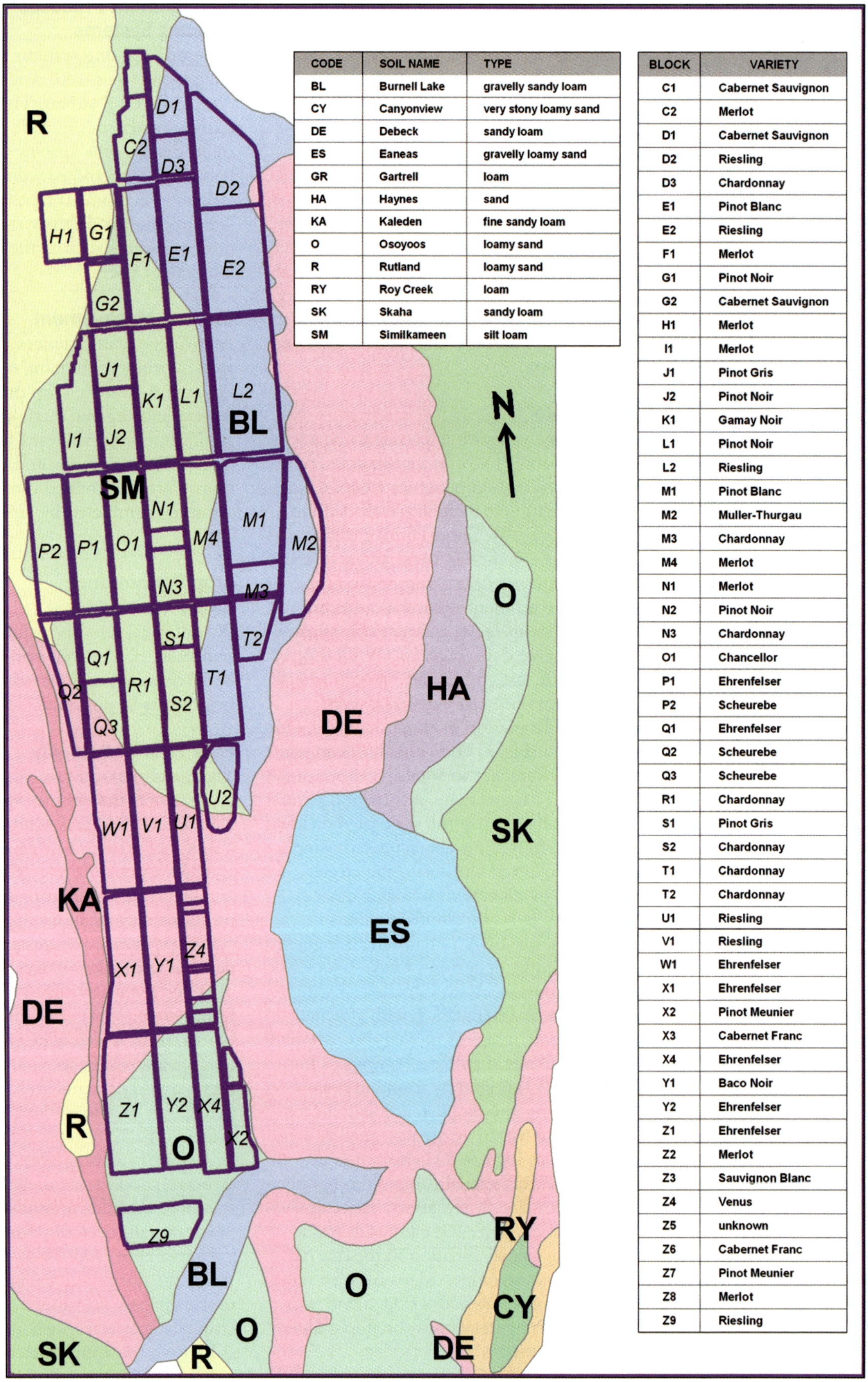

CODE	SOIL NAME	TYPE
BL	Burnell Lake	gravelly sandy loam
CY	Canyonview	very stony loamy sand
DE	Debeck	sandy loam
ES	Eaneas	gravelly loamy sand
GR	Gartrell	loam
HA	Haynes	sand
KA	Kaleden	fine sandy loam
O	Osoyoos	loamy sand
R	Rutland	loamy sand
RY	Roy Creek	loam
SK	Skaha	sandy loam
SM	Similkameen	silt loam

BLOCK	VARIETY
C1	Cabernet Sauvignon
C2	Merlot
D1	Cabernet Sauvignon
D2	Riesling
D3	Chardonnay
E1	Pinot Blanc
E2	Riesling
F1	Merlot
G1	Pinot Noir
G2	Cabernet Sauvignon
H1	Merlot
I1	Merlot
J1	Pinot Gris
J2	Pinot Noir
K1	Gamay Noir
L1	Pinot Noir
L2	Riesling
M1	Pinot Blanc
M2	Muller-Thurgau
M3	Chardonnay
M4	Merlot
N1	Merlot
N2	Pinot Noir
N3	Chardonnay
O1	Chancellor
P1	Ehrenfelser
P2	Scheurebe
Q1	Ehrenfelser
Q2	Scheurebe
Q3	Scheurebe
R1	Chardonnay
S1	Pinot Gris
S2	Chardonnay
T1	Chardonnay
T2	Chardonnay
U1	Riesling
V1	Riesling
W1	Ehrenfelser
X1	Ehrenfelser
X2	Pinot Meunier
X3	Cabernet Franc
X4	Ehrenfelser
Y1	Baco Noir
Y2	Ehrenfelser
Z1	Ehrenfelser
Z2	Merlot
Z3	Sauvignon Blanc
Z4	Venus
Z5	unknown
Z6	Cabernet Franc
Z7	Pinot Meunier
Z8	Merlot
Z9	Riesling

Figure 3. Grape variety block boundaries overlaid onto a soil map for Inkameep vineyard in Vaseaux – Oliver (Region 3). Type of soil (textural class) for each series is shown in the inset at top centre.

quality was analyzed to determine whether over-cropping (excess yield) has been a major quality-limiting factor. Results for Chardonnay, for example, indicate that although the overall relationship between yield and quality is not significant, excellent quality has been produced only in blocks yielding less than five tons of fruit per acre (Fig. 6); however, blocks producing low yields did not necessarily achieve high quality.

Wine quality from grape-variety blocks or vineyards is assessed by identifying the vineyard sources of fruit used in making wines that received medals in judged competitions. Both the fruit and wine quality assessment methods differentiate only levels of quality, and do not differentiate qualitative differences in fruit or wine sensory attributes or style which have already been shown to be influenced by site conditions in the Okanagan Valley (Reynolds et al., 1995, 1996). Such information is available, however, in data collected by the quality assurance program for wine in British Columbia, the Vintners Quality Alliance (VQA). Candidate wines for VQA status are evaluated by trained judges to ensure that these wines are free of defects and have varietal character. The judges also note wine quality characteristics including flavour and aroma notes, body, balance and other sensory attributes. With the winemakers' provision of the identity of the vineyard of origin, these data are being added to the GIS for studies of terroir effects on wine character and quality. Similar evaluations of the aging potential of medal-winning wines are being conducted in collaboration with researchers at the Wine Research Centre of the University of British Columbia.

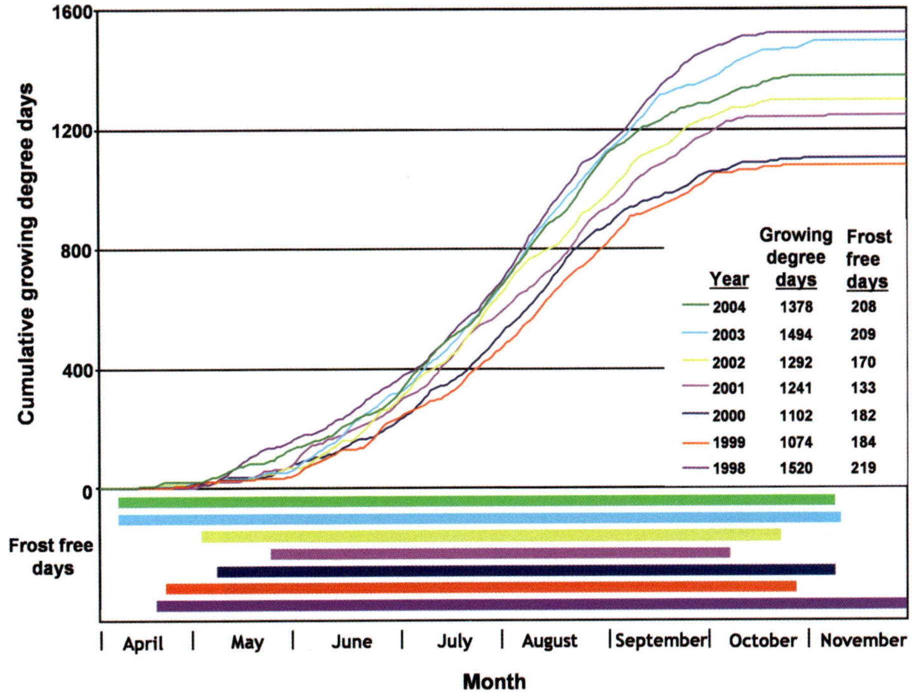

Figure 4. Cumulative growing degree days and frost free periods in Summerland, B.C. (Penticton, Region 2) from 1998 to 2004.

VITICULTURAL REGIONS

To help evaluate the effects of differences in regional conditions on viticultural performance, we have divided the Okanagan and Similkameen grape-growing area into six viticultural regions based primarily on landform and climate (Fig. 7). Differences in the complement of grape varieties planted among regions (Fig. 8) likely reflect general knowledge of the suitability of *V. vinifera* to climatic conditions, and to published climatic suitability ratings for grape-growing (Association of B.C. Grape Growers, 1984) available to growers when a high number of vineyards were planted in the early 1990s. White varieties dominated plantings in the northern regions and red varieties were planted more in the south. Experience from these *V. vinifera* plantings, and the value of varietal wines on the market, has likely affected varietal choices in more recent plantings as these contain relatively fewer varieties and are dominated by high-value noble varieties, including Chardonnay, Merlot and Cabernet Sauvignon. A general description of the site conditions and varieties grown in each region follow.

Kelowna: Region 1

The Kelowna region includes vineyards north of the bend in Okanagan Lake near Peachland (Fig. 7) which marks a major bedrock cross-fault. Many of the vineyards are located on terrain of Eocene sedimentary and volcanic bedrock. Kelowna is the coolest region, and is distinguished mainly by its relatively low total GDD accumulation of *ca.* 950 to 1,360 ($\bar{x} \pm$ SD, for 1998 to 2003). This cool climate reflects both the region's northern latitude and its proximity to Okanagan Lake, which leads to relatively low daily maximum and mean temperatures during the growing season. Most vineyards in the region are sited near Okanagan Lake on sloped benches or hillsides. These vineyards have excellent air drainage which, combined with the lake's moderating effect on nighttime temperatures, results in lower incidences of late-spring and early-fall frost episodes than in some regions to the south. Grape production occurs on a wide range of soils including very coarse glaciofluvial soils that have gravelly-sandy surface textures classified as Eluviated Eutric Brunisols (Gammil soils), and fine-textured soils formed on glaciolacustrine parent materials. These soils have silty clay loam surface textures and are classified as Orthic Grey Luvisols (Boucherie soils). Vineyard area is 12% (Fig. 9a) of the total in the two valleys, and the average age of plantings is 12 years.

The Kelowna region contains some of the oldest *V. vinifera* vineyards in the area, with blocks dating to the early 1960s. Varieties planted are predominantly whites, including a range of Germanic and French varieties suited to cool-climate viticulture, including Riesling, Chardonnay, Gerwurztraminer, and Ehrenfelser (Fig. 8). Most of the red varieties grown are also suited to cool climates and include substantial plantings of the Burgundian variety Pinot noir and the Beaujolais variety Gamay, as well as small amounts of Merlot and Marechal Foch.

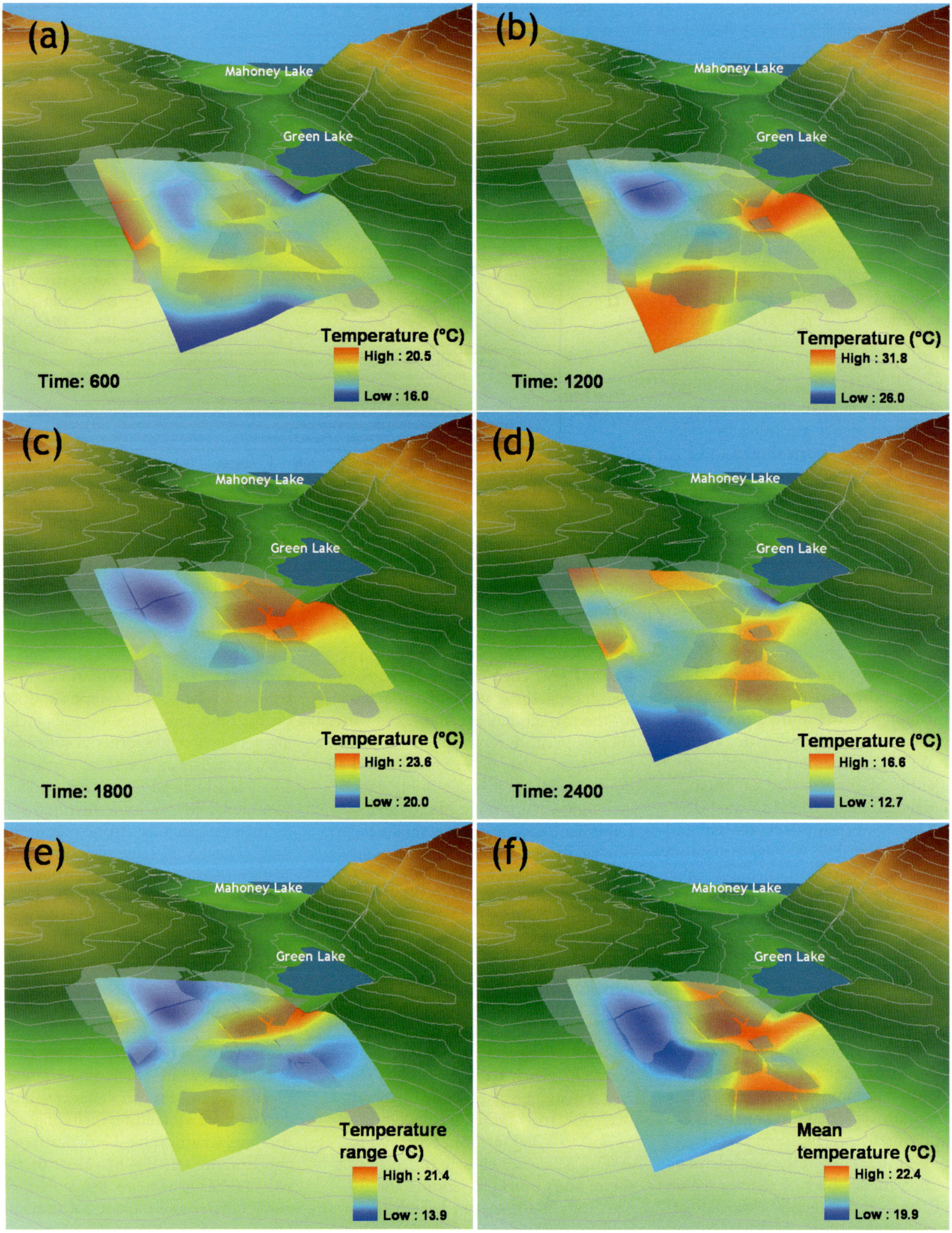

Figure 5. The dynamic spatial variation in ambient air temperature within Hawthorne Mountain Vineyard (Vaseaux – Oliver, Region 3) in the Okanagan Valley. Air temperature measurements were taken on August 9, 2004, using a 24-site sensor network installed in the *ca.* 40 ha vineyard. Maps show snapshots of air temperature at (a) 0600, (b) 1200, (c) 1800 and (d) 2400 hrs.; (e) the diurnal air temperature range and (f) mean air temperature. Vineyard blocks are shown in grey. Contour interval is 20 m.

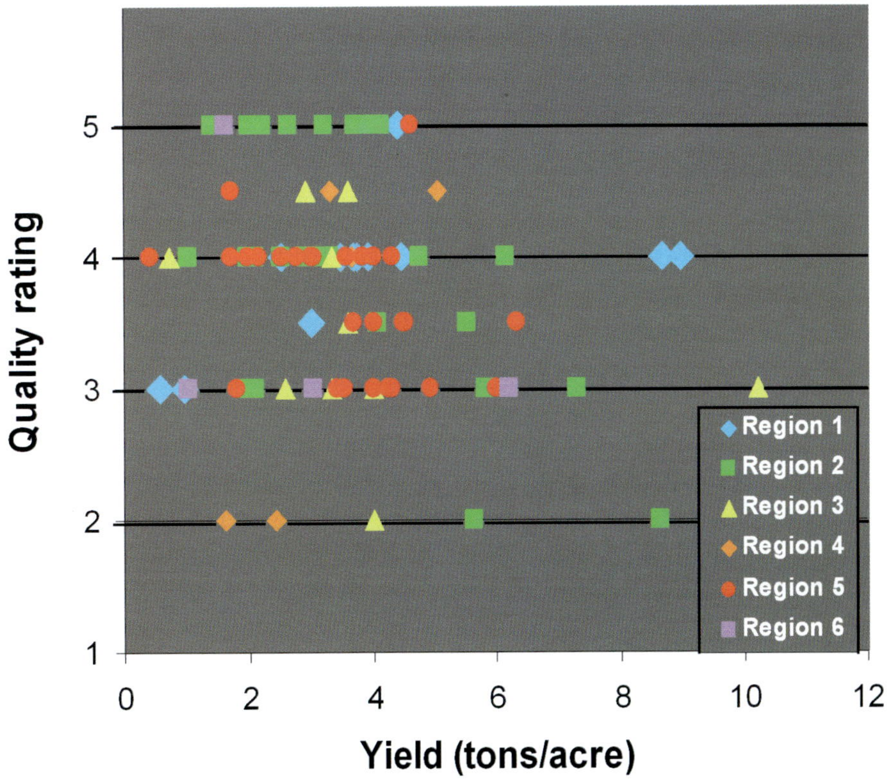

Figure 6. The relationship of Chardonnay fruit quality, as rated by winemakers, to yield (tons/acre). Quality ratings correspond to: 1=unacceptable, 2=poor, 3=fair (average), 4=good and 5=excellent or outstanding. Locations for Regions 1 to 6 are shown in Figure 7. A linear regression fit through the data was not significant ($P>0.05$, $R^2=0.02$, n=166).

Penticton: Region 2

Vineyards in the Penticton region are located along the southern-most part of Okanagan Lake and along Skaha Lake (Fig. 7). The region's total GDD range is *ca.* 1,140 to 1,500, which averages higher than that of the Kelowna region but significantly lower than in regions to the south. Vineyards in this region are concentrated along the glaciolacustrine benchlands on both the east and west sides of the southernmost reach of Okanagan Lake (Fig. 2) and along the benchlands above Skaha Lake. Those on the east sides of the lakes are underlain by Monashee Gneiss, whereas those on the west are underlain by Eocene volcanic and sedimentary substrates. The predominant vineyard soil of this region is composed of a sandy eolian veneer overlying clay loam to silty clay loam glaciolacustrine parent materials. The subsoils may be weakly saline. These soils are classified as Orthic Brown Chernozems (Olhausen soils). Where the sandy eolian veneer is missing and the fine-textured glaciolacustrine material is found at the surface, Orthic Brown Chernozems belonging to the Penticton series occur. In the Summerland area (Fig. 7), a sandy surface veneer overlies glacial till. These soils are also classified as Orthic Brown Chernozems, but are placed within the Giants Head soil series. In some cases, land levelling prior to vineyard planting has eradicated the native soil profile, and vines are growing in a mixture of unweathered parent materials. Most vineyards, particularly those on the east side of the lakes, slope gently to the west and have excellent air drainage which provides a relatively high number of frost free days. The region has 12% of the total vineyard area in the two valleys (Fig. 9a). The average age of plantings is 10 years.

Region 2 is planted predominantly with white varieties with less area in Germanic varieties than in Region 1, and more than a third of the total area planted with the whites Chardonnay, Pinot blanc and Pinot gris (Fig. 8). Bordeaux varieties dominate the red varieties, especially Merlot, but there is also significant area planted with Pinot noir.

Vaseaux - Oliver: Region 3

Vineyards in the Vaseaux-Oliver Region (Fig. 7) are sited on deep, sandy, glaciofluvial parent materials including kettled outwash deposits. Bedrock is dominated by gneisses and granite. The soils at higher elevations that were partially forested before clearing are classified as Eluviated Eutric Brunisols (Parkhill soils) while at lower elevations the soils formed under grassland conditions and are classified as Orthic Brown Chernozems (Osoyoos soils).

The rough topography of the region requires that vineyards are sited on a broad range of slopes at a variety of aspects. The total GDD range is *ca.* 1,320 to 1,490, with the lowest totals at the highest elevations and on north-facing slopes. Vineyard area in the region is 15% of the total in the two valleys (Fig. 9a), and the average age of plantings is 11 years.

Grape varieties planted are similar to those in the Kelowna region (Fig. 8) in that they include Germanic and French white varieties suited to cool climate viticulture, and the red varieties Pinot noir and Gamay. However, Vaseaux-Oliver has substantially more of the Bordeaux red varieties, Merlot and Cabernet Sauvignon, which better suit its warmer climate.

Golden Mile: Region 4

The viticultural region, which has been known locally for some time as the Golden Mile, is located on the west side of the Okanagan Valley south of the town of Oliver (Fig. 7). Vineyards are mostly sited on fluvial fan soils that were deposited against an ice lobe that filled the bottom of the valley during the last glacial episode, and overlie metamorphosed Paleozoic rocks. The soils are mostly dark brown, stony, gravelly sandy loams or gravelly loamy sands. These are soils that developed under grasslands and are classified as Orthic Dark Brown Chernozems belonging to the Ratnip soil series. The total GDD range of Golden Mile is *ca.* 1,340 to 1,630. Vineyard area in the region totals 6% of the total vineyard area in the two valleys (Fig. 9a). The average age of plantings is 10 years.

Although Golden Mile is one of the warmest regions in the Okanagan

Figure 7. Viticultural Regions 1-6 in the Okanagan and Similkameen valleys. Regions are: Kelowna (Region 1), Penticton (2), Vaseaux – Oliver (3), Golden Mile (4), Black Sage – Osoyoos (5), and Similkameen (6). Vineyards are shown in purple.

and Similkameen valleys growing area, a significant portion of its vineyard area is planted with white varieties, mainly Gewurztraminer and Chardonnay (Fig. 8). Other whites grown are Pinot gris, Riesling, Pinot blanc, and Chenin blanc, a Loire variety that is uncommon in the other regions. Bordeaux varieties dominate the red varieties grown and include Cabernet Sauvignon, Cabernet Franc and Merlot, but Pinot noir plantings are also significant.

Black Sage - Osoyoos: Region 5

The Black Sage - Osoyoos region includes a large vineyard-dominated area south of Oliver on the east side of the Okanagan Valley and in areas around Osoyoos Lake to the US border (Fig. 7). Vineyards in this region are nearly all sited on deep, stone-free, sandy glaciofluvial parent materials that dominate the benchlands to the east and north of Osoyoos Lake, and overlie granite bedrock. These soils formed under grassland conditions and are classified as Orthic Brown Chernozems belonging to the Osoyoos soil series. The soils are rapidly drained and have very low moisture holding capacity, but are well suited to grape production. The total GDD range is *ca.* 1,360 to 1,630. The region contains some of the largest vineyards in the area and the average age of plantings is 7 years.

Although Black Sage - Osoyoos is a relatively young growing region, it has more than half of the planted area in the Okanagan and Similkameen valleys (Fig. 9a). About 60% of the area is planted with red varieties, mainly the Bordeaux varieties Cabernet Sauvignon, Cabernet Franc and Merlot, but there is also significant Pinot noir and a small amount of the Rhone variety Syrah (Fig. 8). The major whites grown are Chardonnay, Sauvignon blanc, Pinot blanc and Pinot gris.

Similkameen: Region 6

The Similkameen region includes the viticultural area in the southern part of the Similkameen Valley to the US border (Fig. 7) Vineyards are underlain primarily by metamorphosed bedrock of Paleozoic age. Narrower than the Okanagan Valley, the Similkameen Valley is flanked by steep-sided mountains which significantly reduce sun exposure at some vineyard sites. However, most

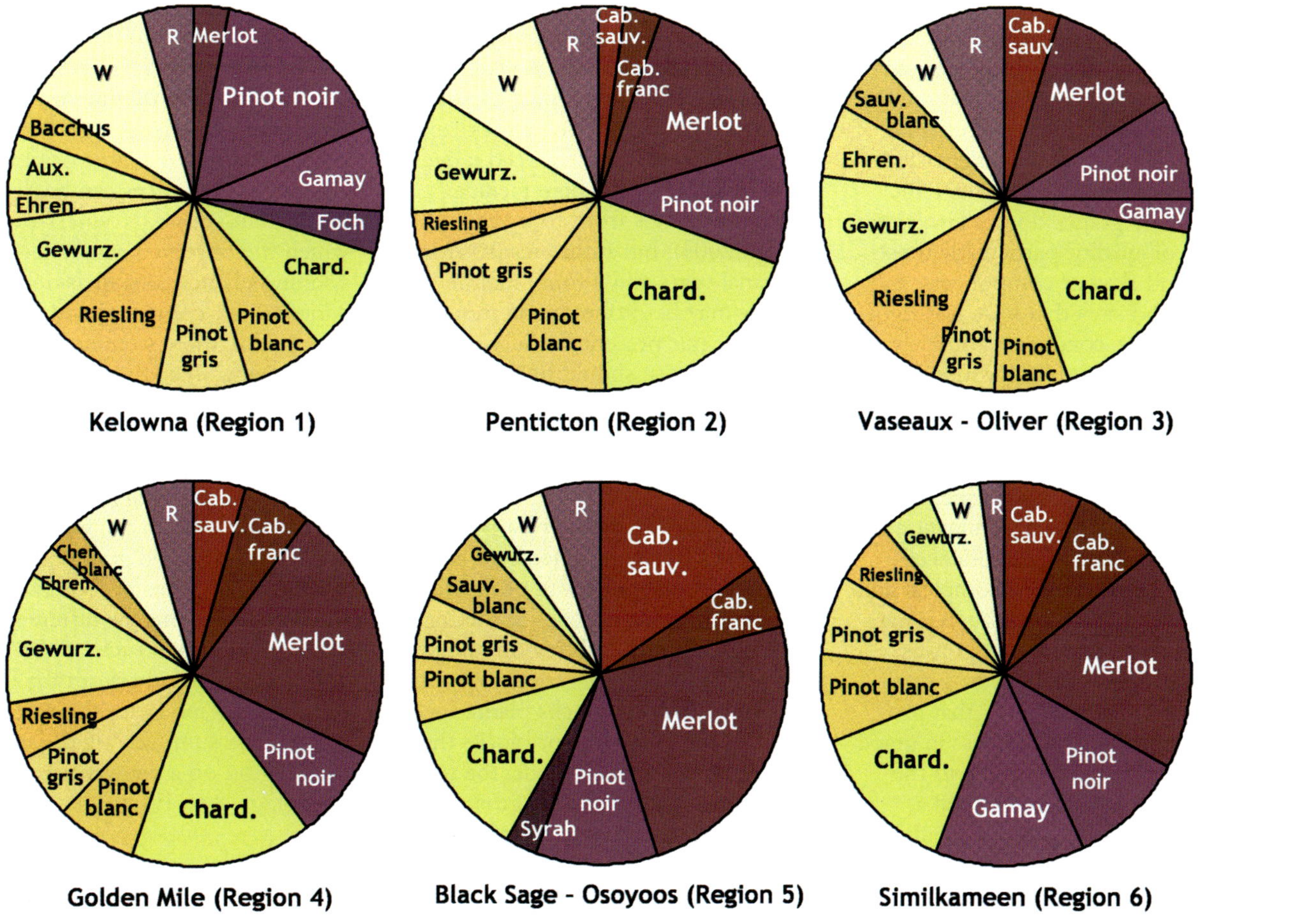

Figure 8. Division of planted area among grape varieties in each viticultural region in the Okanagan and Similkameen valleys. W and R, respectively, refer to total area planted to white and red varieties that individually total less than 2%, but collectively may total 15% (e.g. white varieties in Kelowna and Penticton regions).

vineyards have very favourable west-facing aspects and the total GDD range is *ca.* 1,180 to 1,540.

Vineyards are scattered throughout the region on gently to moderately sloping, gravelly, fluvial fan parent materials and in some locations, these soils have considerable surface stoniness. The soils formed under grassland conditions, and are classified as Orthic Dark Brown Chernozems belonging to the Stemwinder series.

Currently, Similkameen has 5% of the total vineyard area (Fig. 9a) and the age of plantings averages 5 years. More than half of the vineyard area is planted with red varieties including the Bordeaux varieties Merlot, Cabernet Sauvignon and Cabernet Franc, as well as Gamay and Pinot noir (Fig. 8). The major white varieties are Chardonnay, Pinot blanc, Pinot gris, Riesling and Gewurztraminer.

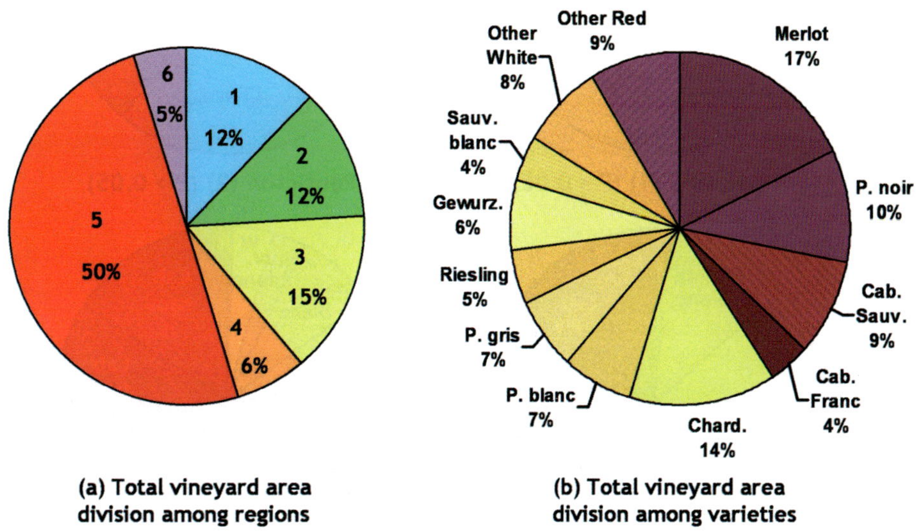

Figure 9. Total vineyard area in the Okanagan and Similkameen valleys. (9a) division among the six regions: note that Black Sage - Osoyoos (Region 5) has 50% of the total vineyard area; and (9b) division among grape varieties. "Other red" and "other white" varieties include the total areas of red and white grape varieties, respectively, that individually total less than 3%, but collectively total 9% (other red) and 8% (other white) as seen in 9b.

VARIETAL PERFORMANCE

Gauging the suitability of grape varieties to growing regions can be difficult, as differences in viticultural practices and winemaking styles may confound the effects of site conditions on fruit and wine quality. Unfortunately, chemical analyses used routinely to evaluate the composition of grapes do not give clear indications of quality, particularly those for aroma and flavour components, and it is rare to find sites that have had rigorous fruit analyses conducted routinely. A further complication in a young industry is the evolution of grape and wine quality resulting from adjustments made in grape-growing and winemaking methods while the industry gains experience. During that time sufficient evaluation of wines across vintages by experienced judges can be difficult to accomplish and may take several years. Although it may not be feasible to conduct a rigorous analysis of varietal suitability in a small and evolving industry, studies that provide early indications of success in grape varietal choices can be valuable to the industry. Thus we have developed a method for evaluating varietal suitability to regions in the Okanagan and Similkameen valleys based on outcomes of an annual contest for medal awards, the Fall Okanagan Wine Festival, in which wines from all regions are entered and judged blindly by well-recognized and internationally experienced judges. The analysis reported here includes judging outcomes from four years (2000 through 2003), but future results from additional years and a more mature industry may be different and more reliable. Thus our present findings on varietal performance should be considered as a preliminary evaluation.

Best-Performing Grape Varieties in Each Region

An analysis to determine best-performing varieties in a region needs to take into account differences in production from the different varieties in the region. By comparing, for each region, the proportions among grape varieties of total area or numbers of blocks planted with the proportions of area or blocks that produce medal-winning wine, the relative success of the varieties grown can be evaluated. If the performance of all varieties planted in a region is similar, the proportions planted and medal-winning are expected to be the same. If some varieties perform better than others, their portion of the medal-winning area or blocks will be larger than their portion of the total area or blocks planted. Such differences can be detected using a chi-square goodness-of-fit test in which the observed results (i.e., medal-winning planted area division among varieties) are compared with the results expected if there were no differences in performance (i.e., the total planted area division). The analysis determines the probability that the expected and observed proportions are the same and there are no differences in the performance of varieties in the region. We applied chi-squared goodness-of-fit tests, with a significance level of 5% (P=0.05), to detect performance differences among varieties in each of the six regions using results from the Fall Okanagan Wine Festival judging in 2000 to 2003. A summary of the results for each region on a vineyard area basis follows (Figs. 8 to 10).

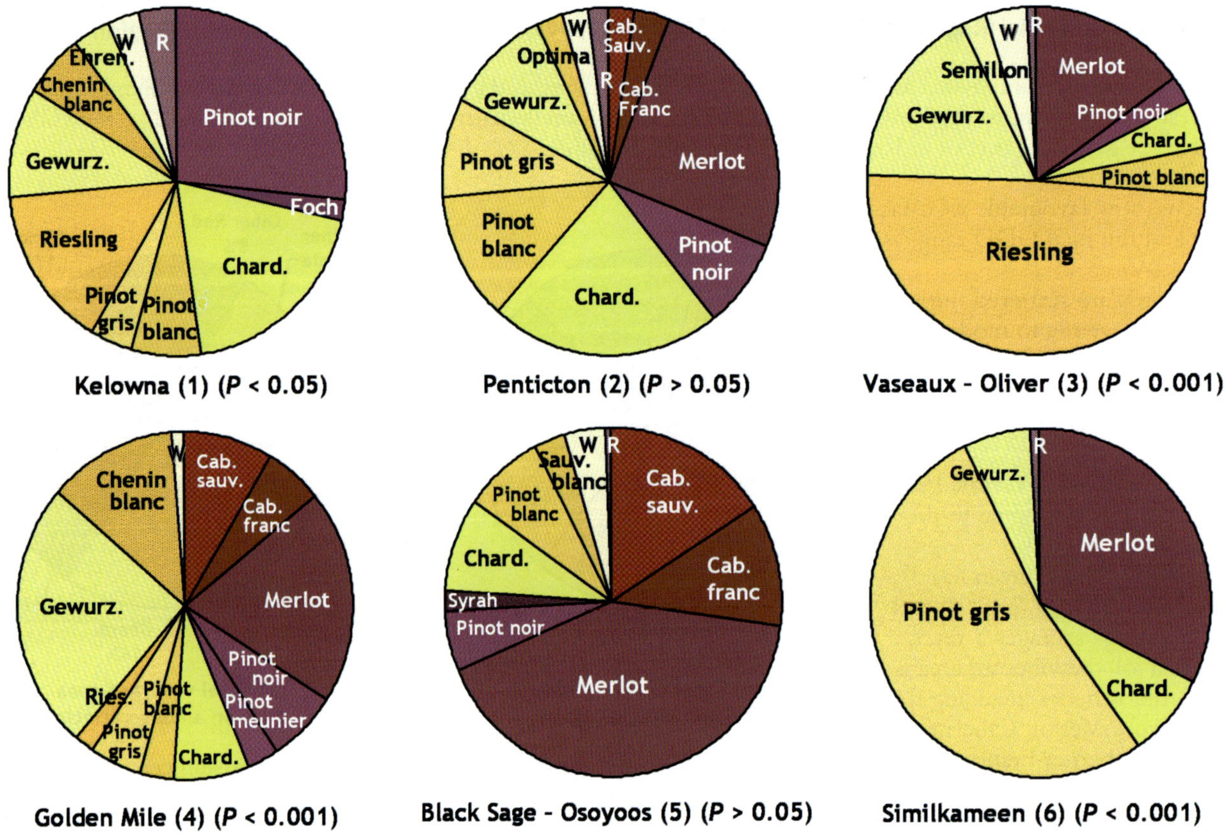

Figure 10. Division of medal-winning vineyard area among varieties in each viticultural region in the Okanagan and Similkameen valleys. P values are for the significance of chi-square goodness-of-fit tests comparing total and medal-winning area division among varieties.

Kelowna - Region 1:
The analysis detected differences in varietal performance. White varieties predominate in the total and medal-winning vineyard area, and Chardonnay is the strongest performer. Pinot noir is the most planted red variety and the strongest performing red.

Penticton - Region 2:
No differences in varietal performance were detected. White varieties predominate in the total and medal-winning vineyard areas. Merlot predominates in red plantings and medals.

Vaseaux-Oliver – Region 3:
Substantial differences in varietal performance were detected in the analysis. Riesling accounts for only *ca.* 10% of the total vineyard area, but about half of the medal-winning vineyard area. Gewurztraminer is another strong-performing white. Merlot is the strongest performing red variety.

Golden Mile – Region 4:
Substantial differences in varietal performance were detected. White varieties are planted in more than half of the total and medal-winning vineyard area, with Gewurztraminer and Chenin blanc showing the strongest performance. Bordeaux reds account for most of the total and medal-winning vineyard area in red varieties.

Black Sage - Osoyoos – Region 5:
No differences in varietal performance were detected. Chardonnay, Pinot blanc and Sauvignon blanc account for most of the medal-winning area in white varieties. Bordeaux reds account for most of the medal-winning area in red varieties.

Similkameen – Region 6
Substantial differences in varietal performance were detected. Pinot gris is planted in only 7% of the vineyard area, but accounts for more than half of the medal-winning area. Merlot is favoured in the medals for red varieties.

Best-performing Regions for Grape Varieties

The regions in which specific grape varieties have performed well can be identified using an approach similar to that used to identify best varieties in each region. For each variety, this is accomplished on a vineyard area basis by comparing the proportions among regions of total planted area and medal-winning area. If a variety performs similarly in all regions, the proportions are expected to be the same as the planted proportions: i.e., no region is favoured. Conversely, if a grape variety performs better in some regions than in others, its portion of the medal-winning planted area will be greater than its portion of the total planted area. Application of a chi-squared goodness-of-fit test determines the probability that the medal-winning proportions are not different from the planted proportions, indicating that no regions are favoured for the variety. We applied such chi-square tests to results from Fall Okanagan Wine Festival judging, from 2000 to 2003, for the eight most widely planted varieties in the Okanagan and Similkameen valleys. These included five white varieties: Chardonnay, Pinot blanc, Pinot gris,

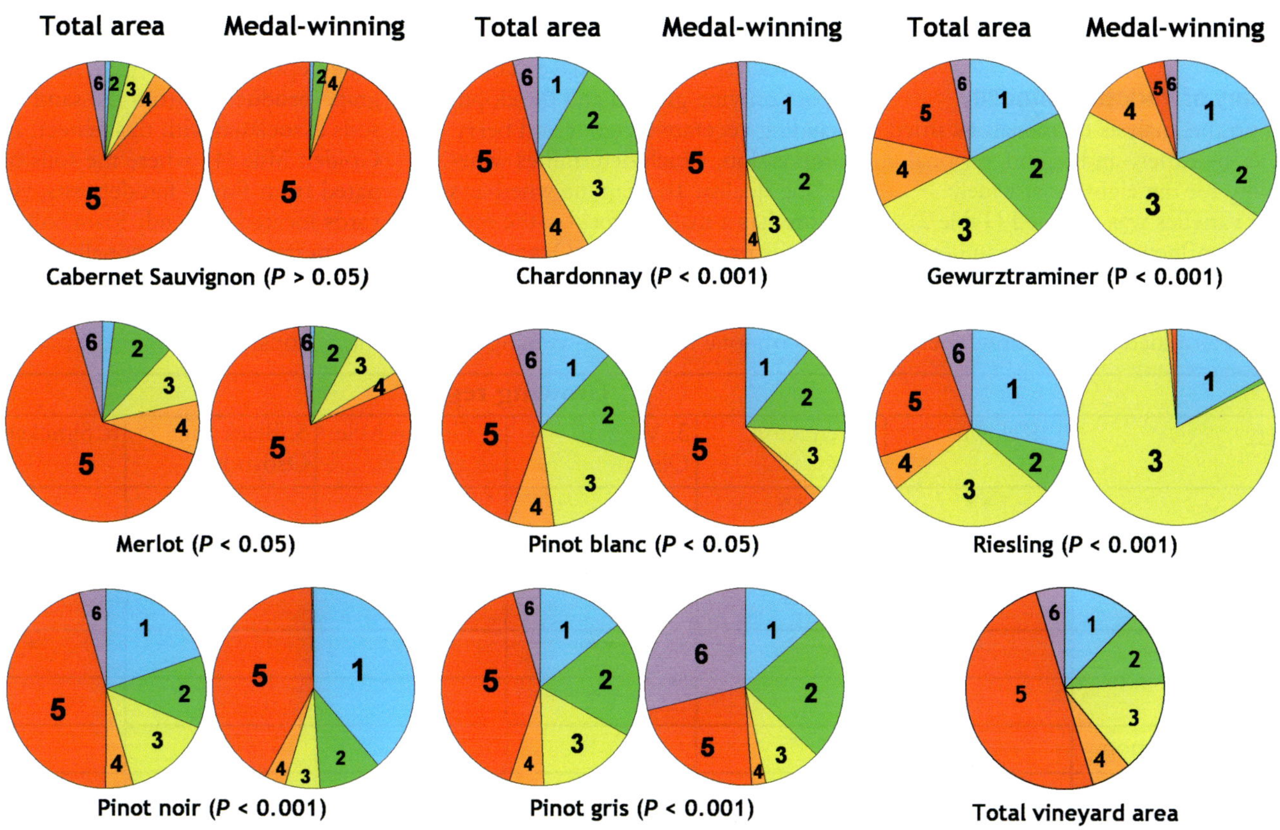

Figure 11. Total and medal-winning vineyard area divided among viticultural regions for each of the top eight varieties grown in the Okanagan and Similkameen valleys. P values are for the significance of chi-square goodness of fit tests comparing total and medal-winning area division among regions.

Gewurztraminer, and Riesling; and three red varieties: Cabernet Sauvignon, Merlot and Pinot noir (Fig. 9b). A summary of the results (Fig. 11) follows.

White Varieties

Although Chardonnay, Pinot blanc and Pinot gris are planted in similar proportions of the total vineyard area in all regions, analysis results indicate that the medals have favoured:

- Chardonnay from Kelowna and Penticton (Regions 1 and 2).
- Pinot blanc from Black Sage - Osoyoos (Region 5).
- Pinot gris from Penticton and Similkameen (Regions 2 and 6)

Gewurztraminer and Riesling are planted heavily in Vaseaux - Oliver (Region 3) which is favoured in the medals for both varieties, especially Riesling.

Red Varieties

Most of the total and medal-winning Cabernet Sauvignon and Merlot vineyard area is in Black Sage – Osoyoos (Region 5). The analysis results indicate that Merlot from Black Sage – Osoyoos is favoured. Pinot noir from Kelowna (Region 1) is strongly favoured.

Summary of Varietal Suitability

Based on the analysis of medals awarded to wines by variety and region, regional suitability for the eight most widely planted varieties was assessed (Table 2). While the influence of climate in determining high quality is apparent for red varieties (Bordeaux reds, including Merlot, Cabernet Sauvignon and Cabernet Franc, perform best in the southernmost regions where total growing degree days are highest, and Pinot noir, a Burgundian variety, performs best in Kelowna which has the coolest climate), climate does not appear to be a major quality-determining factor for white varieties. The influences on white wine quality of management practices and site conditions other than regional climate, such as soils or landforms, needs further study. It is notable that nearly all varieties, white and red, have performed well in Penticton (Region 2).

RESPONSE TO SOIL TEXTURE

One of the most variable characteristics of soils in the Okanagan and Similkameen vineyards is texture. Soil texture is determined by the size distribution of component mineral particles (Wittneben, 1986). Coarse-textured soils have sand-sized particles as their main component, whereas fine-textured soils consist mostly of silt- and clay-sized particles. Soil texture affects permeability for drainage, water holding capacity, and the tension with which water is held to soil particles. It can thus have substantial impact on vine growth and development including vegetative vigour and berry composition, which determines winemaking quality. To examine whether soil texture has a major effect on the quality of wines produced in the Okanagan and Similkameen valleys, the distribution among soil textural classes of all grape-variety blocks was compared with that of blocks from which medal-winning wines were produced (Fig. 12). If soil texture has no effect on wine quality the distributions are expected to be similar. Chi-square goodness-of-fit tests were used to detect differences in the distributions.

Although most of the vineyards in the Okanagan and Similkameen valleys are sited on coarse (i.e., loamy sand) or moderately coarse (i.e., sandy loam) soils, we found that medal-winning wines have been made from grapes originating from a range of soil textures, from loamy sands and sandy loams to finer textured clay loams and silty clay loams (Fig. 12). Nonetheless, based on the chi-square analyses, the distributions of all blocks and medal winning blocks among textural classes were different in Kelowna, Penticton, Golden Mile and Similkameen. Kelowna is the only region with significant grape production on moderately fine-textured soils (i.e., silty clay loams; Fig. 12), and these soils have produced a disproportionately high number of medal-winning wines. In Penticton (2) where a discontinuous eolian veneer overlies the glacial lacustrine deposits, medals have favoured grape production from the sandy loam soils associated with this veneer. In Golden Mile (4), where the soils originated from fluvial fan deposits, vineyards located on sandy loam soils have also performed exceptionally well. In

Table 2: The regional suitability of the eight most commonly grown wine grape varieties in the Okanagan and Similkameen valleys, determined by comparing the proportions of total and medal-winning areas among regions.

	Growing region					
Variety	1. Kelowna	2. Penticton	3. Vaseaux – Oliver	4. Golden Mile	5. Black Sage - Osoyoos	6. Similkameen
Chardonnay	**	**			*	
Pinot blanc	*	*			**	
Pinot gris	*	**				**
Gewurztraminer	*	*	**	*		*
Riesling			**			
Cabernet Sauvignon		*		*	*	
Merlot		*	*		**	
Pinot noir	**	*		*	*	

**,*: Performs better than (**) or as well as (*) expected from the area planted of the variety in each region. Blank cells indicate either performance was lower than expected or there was insufficient data to evaluate for that variety and region.

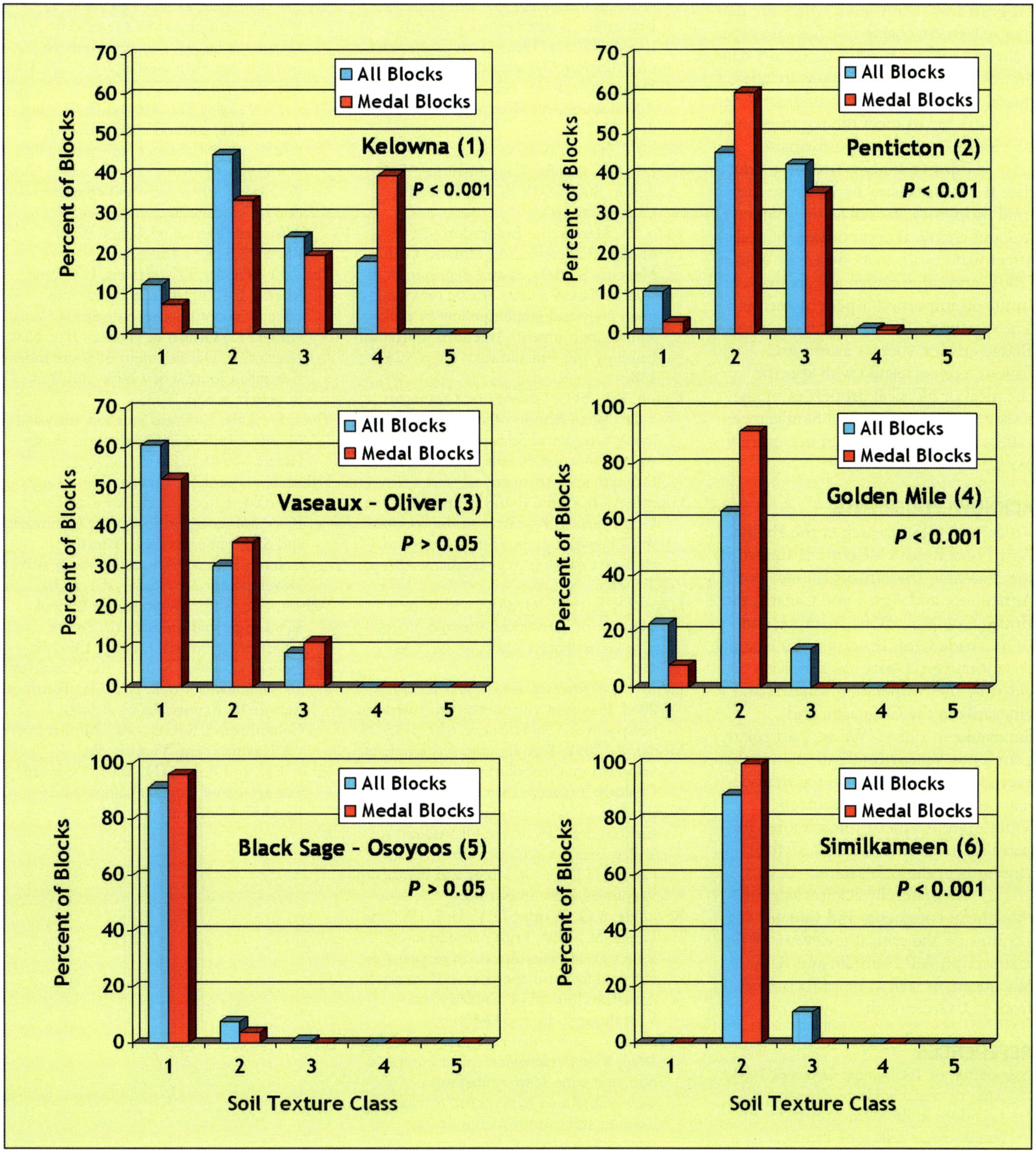

Figure 12. The division of all grape variety blocks and medal-winning grape variety blocks among soil textural classes in each viticultural region. Textural classes, ranging from coarse (1) to fine (5) texture, are: 1 = sand and loamy sand; 2 = sandy loam; 3 = loam, silt loam and silt; 4 = sandy clay loam, clay loam and silty clay loam; 5 = clay, heavy clay, silty clay and sandy clay. P values are for the significance of chi-square goodness-of-fit tests comparing the division of all and medal-winning blocks among soil textural classes.

Similkameen (6), medals also favour grapes produced on sandy loams. This suggests that loamy soils, which are neither extremely coarse nor fine, are especially suitable for producing quality wine grapes, perhaps because they are adequately drained yet hold sufficient water and nutrients to meet the requirements of vines for optimum performance.

Further study is needed to elucidate the mechanisms by which soil textural properties, as well as the mineralogical and chemical composition of soils, affect wine quality. As noted by Wilson (2001) and others, such affects may constitute an important aspect of terroir. The question of whether wine quality characteristics, such as aroma and flavour, are correlated with specific chemical or physical properties of the soils in the Okanagan and Similkameen valleys is being explored in our current work.

ACKNOWLEDGEMENTS

Viticulture GIS research at the Pacific Agri-Food Research Centre is funded by the Matching Investment Initiative of Agriculture and Agri-Food Canada, the British Columbia Wine Institute, and BC/Canada Crop Insurance. Substantial collaborative support has also been provided by the cooperating wineries and vineyards in the Okanagan and Similkameen valleys. We are particularly grateful to Vincor International for their excellent collaboration on the mesoclimate research, and to Inkameep Vineyard and Vincor International for permission to include maps of their vineyards in the publication.

We gratefully acknowledge the thoughtful comments and valuable suggestions of the critical reviewers of the manuscript, A.P Hamblin and R.W. Macqueen, as well as helpful comments from J.D. Greenough.

REFERENCES

Association of B.C. Grape Growers, 1984, Atlas of Suitable Grape Growing Locations in the Okanagan and Similkameen Valleys of British Columbia: Agriculture Canada, Research Station, Summerland, BC, 141 p.

Bowen, P., Estergaard, B., Bogdanoff, C., Marsh, S., Beckwith, R. and Tiebel, D., 2004 (abs.) Use of a geographic information system to study mesoclimate variation in Okanagan Valley vineyards: American Journal of Enology and Viticulture, v. 55, p. 301A.

Bramley, R.G.V. and Williams, S.K., 2001, A protocol for winegrape yield maps, *in* Blackmore, S. and Greisner, G., eds., ECPA 2001 - 3rd European Conference on Precision Agriculture, agro Montpellier, Ecole Nationale Superieure Agronomique de Montpellier, France, v. 2, p.773-778.

British Columbia Wine Institute, 2004, British Columbia Wine Institute Annual Report 2003-2004: BCWI, Kelowna, Canada, 26 pp.

Failla, O., Mariani, L., Brancadoro, L., Minelli, R., Scienza, A., Murada, G. and Mancini, S., 2004, Spatial distribution of solar radiation and its effects on vine phenology and grape ripening in an alpine environment: American Journal of Enology and Viticulture, v. 55, p. 128-138.

Fulton, R.J., 2003, Geology of Okanagan Valley Wine Region of British Columbia (abs.): Geological Society of America Annual Meeting, Seattle, Washington, Program and Abstracts.

Giomo, A., Borsetta, P. and Zironi, R., 1996, Grape quality: Research on the relationships between grape composition and climatic variables: Acta Hortic. v. 427, p. 277-285.

Haynes, S. J., 1999, Geology and Wine 1. Concept of terroir and the role of geology: Geoscience Canada, v. 26, p. 190-194.

Lamb, D.W., Bramley, R.G.V. and Hall, A., 2004, Precision viticulture - an Australian perspective: Acta Hortic. v. 640, p. 15-25.

Morlat, R., 2001, Terroirs viticoles: Étude et valorization: Oenoplurimédia Sarl, Chaintré, France, 118 p.

Nasmith, H., 1962, Late glacial history and surficial deposits of the Okanagan Valley, British Columbia: British Columbia Dept. of Mines and Petroleum Resources, Bulletin No. 46.

Reynolds, A.G., Wardle, D.A., Hall, J.W., and Dever, M., 1995, Fruit maturation of four *Vitis vinifera* cultivars in response to vineyard location and basal leaf removal: American Journal of Enology and Viticulture, v. 46, p. 542-558.

Reynolds, A.G., Wardle, D.A. and Dever, M., 1996, Vine performance, fruit compostition, and wine sensory attributes of Gewurztraminer in response to vineyard location and canopy manipulation: American Journal of Enology and Viticulture, v. 47, p. 77-92.

Roed, M.A., Greenough, J.D. (eds.), 2004, Okanagan Geology British Columbia: Kelowna Geology Committee, Kelowna, 220 pp.

Soil Classification Working Group, 1998, The Canadian System of Soil Classification: Agric. and Agri-Food Can. Publ. 1646 (revised), Ottawa, 187 pp.

Templeman-Kluit, D. and Parkinson, D., 1986, Extension across the Eocene Okanogan crustal shear in southern British Columbia: Geology, v. 14, p. 318-321.

Tesic, D., Woolley, D.J., Hewett, E.W., and Martin, D.J., 2002a, Environmental effects on cv Cabernet Sauvignon (Vitis vinifera L.) Grown in Hawke's Bay, New Zealand. 1. Phenology and characterization of viticultural environments. Australian Journal of Grape Wine Research, v. 8, p. 15-26.

Tesic, D., Woolley, D.J., Hewett, E.W., and Martin, D.J., 2002b, Environmental effects on cv Cabernet Sauvignon (Vitis vinifera L.) Grown in Hawke's Bay, New Zealand. 2. Development of a site index: Australian Journal of Grape and Wine Research, v. 8, p. 27-35.

Tesic, D., 2004, Vineyard location analysis for site selection and evaluation: Acta Hortic. v. 640, p. 27-35.

Usher, K., Girard, B., Bowen, P. and Eggers, N., 2004, Effects of site conditions on Pinot noir composition in the Okanagan and Similkameen valleys: British Columbia Wine Institute Research and Development Newsletter, v. 1, p. 10.

Wilson, J.E., 2001, Geology and Wine 4. The Origin and Odyssey of *Terroir*: Geoscience Canada, v. 28. p. 139-141.

Wittneben, U., 1986, Soils of the Okanagan and Similkameen valleys. MOE Technical Report 18. Report No.52, British Columbia Soil Survey. B.C. Ministry of the Environment, Victoria, B.C.

Wolf, T.K. and Boyer, J.D., 2003, Vineyard Site Selection: Virginia Cooperative Extension Publication 463-020.

SERIES

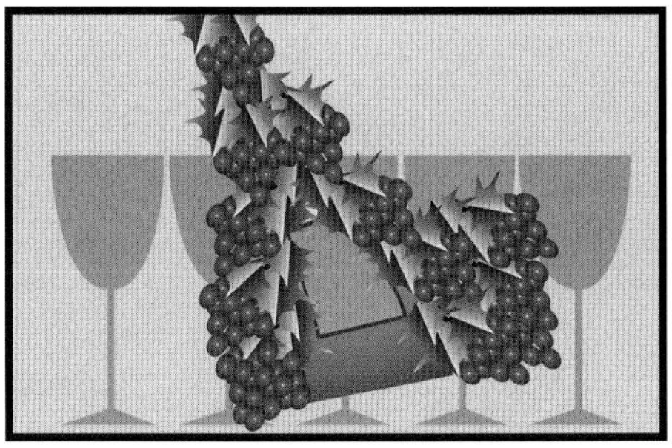

Geology and Wine 11. Terroir of the Western Snake River Plain, Idaho, USA

Virginia S. Gillerman[1], David Wilkins[2], Krista Shellie[3] and Ron Bitner[4]

[1]*Idaho Geological Survey, Boise State University, Boise, ID 83725-1535, vgillerm@boisestate.edu*
[2]*Department of Geosciences, Boise State University, Boise, ID 83725-1535, dwilkins@boisestate.edu*
[3]*USDA-ARS Horticultural Crops Research Laboratory, Parma, ID 83660, kshellie@uidaho.edu*
[4]*Bitner Vineyards, Caldwell, ID 83607, rmbitner@bitnervineyards.com*

SUMMARY

This article explores the unique combination of factors that shape the terroir of Idaho's principal wine grape-growing district. Most Idaho wine grape vineyards are located in the Western Snake River Plain (WSRP) rift basin (~43°N, ~117°W) on soils derived from lake, river, volcanic and wind-blown sediments. The underlying Tertiary and Quaternary rocks record the geologic history of ancient Lake Idaho, its interaction with basaltic volcanism, and subsequent Pleistocene fluvial processes and catastrophic floods. The arid to semi-arid, mid-latitude steppe climate of the WSRP provides fewer growing degree days than American Viticultural Areas (AVAs) in Walla Walla, Washington and Napa Valley, California, but still allows cultivation of *Vitis vinifera* grapes. Other differences include lower precipitation, higher solar radiation during the growing season, and greater threat of cold injury. Wine grapes grown in the WSRP require irrigation, and irrigation is used to manage canopy size and manipulate vine physiology. Wine grape acreage in Idaho has increased dramatically since 1993 and is estimated, in 2003, at about 500 ha with the white wine cultivars Riesling, Chardonnay, and Gewürztraminer comprising about 60% of production, and Cabernet Sauvignon, Merlot and Syrah as principal red wine cultivars.

RÉSUMÉ

Le présent article porte sur la combinaison particulière de facteurs qui définit le terroir du principal district viticole de l'État d'Idaho. La plupart des vignobles de l'Idaho sont situés dans le bassin de fossé tectonique (~43°N, ~117°O) de la Western Snake River Plain (WSRP), sur des sols formés de sédiments lacustres, fluviatiles, volcaniques et éoliens. Les couches tertiaires et quaternaires sous-jacentes témoignent des événements constitutifs de l'histoire géologique de l'ancien lac Idaho, de phénomènes interactifs dont il a été le théâtre, soit un volcanisme basaltique, ainsi que des processus fluviatiles et des inondations catastrophiques pléistocènes. Bien que le climat aride à semi-aride de steppe en altitude moyenne de la WSRP comporte moins de degrés-jours de croissance que les zone les zones viticoles étasuniennes (AVA) de Walla Walla de l'État de Washhington et de la vallée de Napa de l'État de Californie, il permet tout de même la culture des raisins de *Vitis vinifera*. De plus, cette région reçoit moins de précipitations, plus d'ensoleillement durant la saison de croissance, et elle est davantage exposée aux meurtrissures du froid. Les vignes de raisins de cuve cultivés dans la WSRP doivent être irriguées, l'irrigation permettant d'agir sur l'ampleur du feuillage et sur la physiologie du vin. La superficie de culture du raisin de cuve en Idaho s'est considérablement accrue depuis 1993 pour atteindre 500 ha environ en 2003, les cultivars à vin blanc de Riesling, Chardonnay, et Gewürztraminer constituant 60 % de la production, et ceux du Cabernet Sauvignon, du Merlot et du Syrah constituant les principaux cultivars à vin rouge.

INTRODUCTION

Best known for scenic beauty, whitewater rivers, and quality potatoes, the state of Idaho is receiving medals for premium red, white, and ice wines produced from Idaho-grown *Vitis vinifera* L. grapevines. Idaho normally is associated with high mountains and cold temperatures, but southwestern Idaho's low elevation and relatively moderate climate is suitable for growing European wine grapes. The Snake River Plain (SRP) is a crescent-shaped belt of sagebrush-covered volcanic rocks ranging in width from 65 to 100 km, and stretching roughly 600 km across southern Idaho; the principal wine grape-growing district is located in the Western part of the Snake River Plain (WSRP, Figs 1, 2). Geologically, the WSRP is distinguished from the Eastern Snake River Plain (ESRP, Fig. 1) by the much greater proportion of sedimentary rocks relative to basalts and a more fault-bounded, rift-

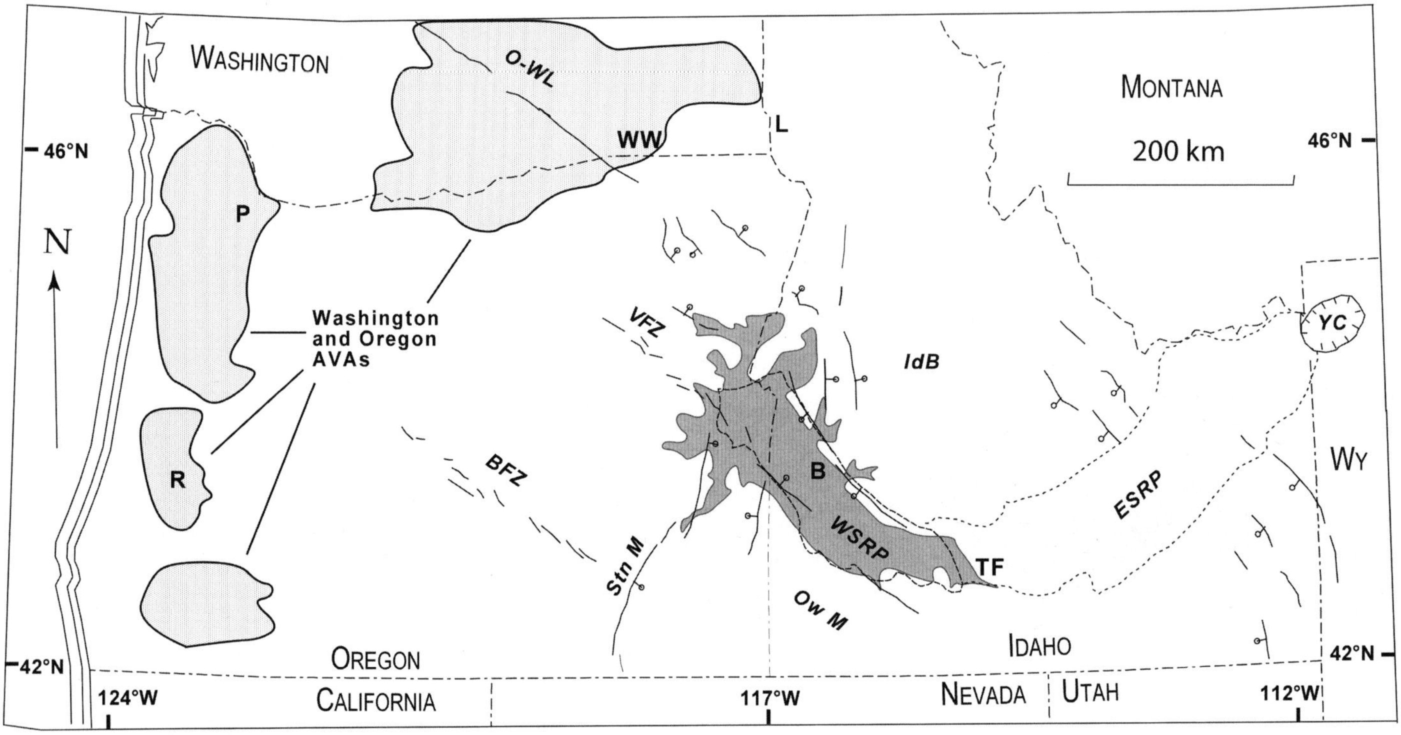

Figure 1. Map showing the location of the Western Snake River Plain (WSRP) relative to other major grape-growing regions and tectonic elements in the Pacific Northwest. Hells Canyon of the Snake River lies along the Idaho-Oregon border from the WSRP north to Lewiston. Physiographic and structural features (Ball and bar symbols indicate the downthrown side of the fault): BFZ, Brothers Fault Zone; ESRP, Eastern Snake River Plain; IdB, Idaho Batholith; O-WL, Olympic-Wallowa Lineament; Ow M, Owyhee Mountains; Stn M, Steens Mountains; VFZ, Vale Fault Zone; WSRP, Western Snake River Plain; YC, Yellowstone Caldera. Cities: B, Boise; L, Lewiston; R, Roseburg; P, Portland; TF, Twin Falls; WW, Walla Walla. AVA refers to American Viticultural Area, as designated by the Alcohol and Tobacco Tax and Trade Bureau (TTB) formerly called the Bureau of Alcohol, Tobacco and Firearms (BATF). Geology modified from Wood and Clemens (2002).

basin geometry. Yellowstone and Grand Teton National Parks lie at the eastern terminus of the SRP and contain the headwaters of the Snake River, which drains about 283,000 km^2 during its ~1120 km trek across the SRP and down Hells Canyon, before joining the Columbia River en route to the Pacific Ocean (Fig. 3). Irrigation from the Snake River has been instrumental for much of Idaho agriculture, including viticulture.

The WSRP includes parts of 10 Idaho counties and a small part of Oregon, but 75% of commercial wine grape acreage, including the largest vineyard at 150 ha, is located in Idaho's Canyon County [United States Department of Agriculture (USDA), Idaho Agricultural Statistics Service, 1999; Fig. 2]. Elevation in the WSRP ranges from about 660 to 1100 m a.s.l. (USDA, 1972) and vineyards are located at elevations ranging from 695 to 890 m. The WSRP is at a similar latitude (43°N to 44°N) as wine regions in France, Italy and Spain, and chapters of its geologic history are similar to the history of the neighbouring states of Washington and Oregon (Meinert and Busacca, 2000). In Idaho, European wine grape production north (~47°N) or east (~114°W) of the WSRP is limited by low winter minimum temperatures and limited length of growing season.

The history of wine production in Idaho is similar to that of neighbouring states and the province of British Columbia, Canada (Meinert and Busacca, 2000; Taylor et al., 2002), dating to the mid-1800s when French and German immigrants cultivated European grapes and produced wines near the confluence of the Snake and Columbia rivers. Native North American grape species that host pests detrimental to European vines, like the insect phylloxera [*Daktulosphaira vitifoliae* (Fitch)], were not present in this region, and own-rooted European vines were successfully cultivated. After the United States' prohibition of alcohol (1920-1933), wine grape production did not recover until around 1970. Idaho created a state commission in 1984 to promote growth and development of the grape and wine industry. Acreage and number of wineries increased steadily such that by 1998, wine grapes were Idaho's fourth largest fruit crop (USDA, Idaho Agricultural Statistics Service, 1999). Between 1993 and 1999, the latest year for which official statistics are available, acreage and number of vineyards doubled to 262 ha in 27 vineyards. An informal 2003 survey by the Idaho Grape Growers and Wine Producers Commission of 35 of its ~50 grower members, suggested further doubling of acreage to about 489 ha with > 50% of vineyards at ≤ 6 ha. Cultivar acreage, based on 65% of estimated total acreage, suggests a predominance of white wine cultivars (60%) including Riesling (32%), Chardonnay (18%), and Gewürztraminer (7%). Principal red wine cultivars by acreage are Cabernet Sauvignon (19%), Merlot (12%), and Syrah (5%). An extensive

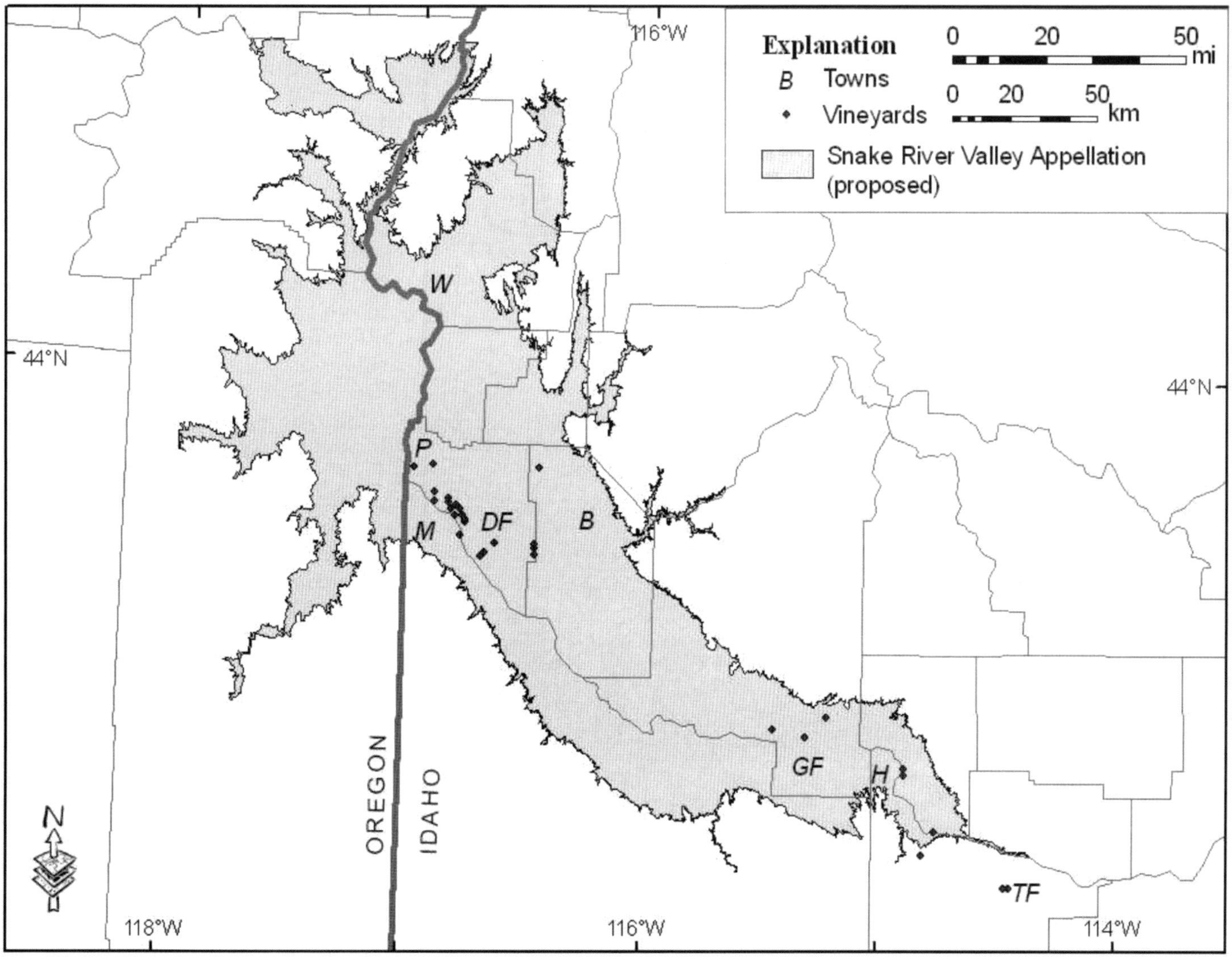

Figure 2. Map of the Western Snake River Plain grape-growing region, noting the locations of vineyards. Shaded area marks the extent (at an elevation of 1050 m a.s.l.) of Pliocene Lake Idaho, and is the boundary of the proposed Snake River Valley AVA. The Sunnyslope area referred to in the text is the cluster of vineyards west of Deer Flat (DF), which is located in Canyon County. Towns shown are Weiser (W), Boise (B), Parma (P), Marsing (M), Glenns Ferry (GF), Hagerman (H), and Twin Falls (TF).

amount of arable land is available for future plantings, limited more by access to water for irrigation than by land suitability. The Idaho Grape Growers and Wine Producers Commission currently is petitioning for the first designated American Viticultural Area (AVA) in Idaho, to be delimited by the 1050 m a.s.l. elevation boundary of ancient Lake Idaho (Fig. 2).

TERROIR OF IDAHO'S WESTERN SNAKE RIVER PLAIN

Physiography

The SRP is a large, arcuate-shaped, topographic depression, mostly filled with volcanic rocks, that crosses the entire state (Fig. 1). The westward-flowing Snake River lies near the southern boundary of the SRP and historically has provided water for much of the region's agriculture (Fig. 3). The river has formed either a steep-walled canyon where it incises thick piles of basaltic lava flows, or a more open valley where it cuts Tertiary and Quaternary lacustrine and fluvial sediments. The eastern part of the SRP is higher in elevation and too cold for *V. Vinifera*, so most Idaho wine originates from the WSRP, most notably the Sunnyslope area in Canyon County (Fig. 2), where the vineyards are located on the tops and flanks of a series of ridges between Homedale and Lake Lowell, just east of the town of Marsing (43°33'N, 116°48'W) on the Snake River (Fig. 4). Though the vineyards are located within a few kilometers of the Snake River, most slopes would support only native vegetation, such as sagebrush, rabbitbrush, and bunch grasses, were it not for widespread irrigation from a large number of irrigation canals. A second cluster of vineyards is located near Glenns Ferry (42°57'N, 115°18'W) in Elmore County (Fig. 2), and there also are vineyards near Lewiston (46°25'N, 117°01'W; Fig. 1) in northern Idaho.

Geologic Setting

The geologic history of southwestern Idaho resembles that of eastern Washington with its flood basalts, north-

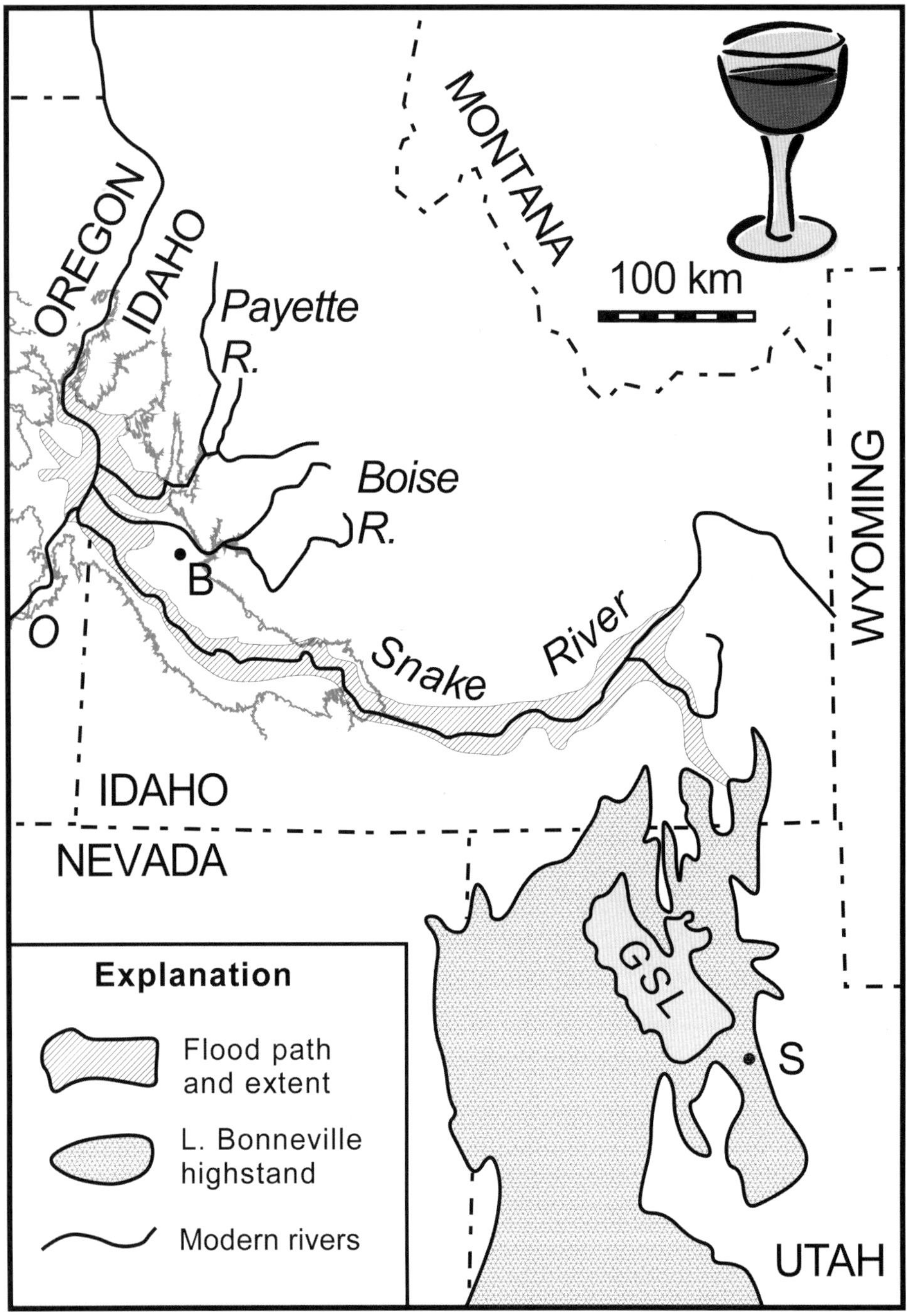

Figure 3. Regional map of the WSRP showing modern river drainages and outline of late Pleistocene Lake Bonneville with its associated outburst flood (flood path and extent), after O'Connor (1993). Abbreviations: O (Owyhee River), B (Boise), S (Salt Lake City), GSL (modern Great Salt Lake).

west-trending structures, extensive loess blankets, and glacial outburst floods (Meinert and Busacca, 2000, 2002). However, Idaho's geologic history includes Lake Idaho, a paleo-system of lakes and floodplains which, at its maximum, stretched 240 km northwest-to-southeast from what is now the Oregon-Idaho state line (117°W) to just west of Twin Falls (42°33.5'N, 114°28'W; Fig. 2).

North of the SRP are Cretaceous granites of the Idaho Batholith, along with assorted Eocene volcanic rocks, older sedimentary rocks, and the Miocene (14-17 Ma) Columbia River Basalts of the Weiser embayment (Figs. 1, 2). South of the WSRP are 12-15 Ma volcanic rocks of the Owyhee Mountains that overlie the southern extension of the granitic basement. The WSRP is a northwest-trending, 300-km long and 70-km wide intracontinental rift basin, whose margins are well-defined boundary faults that parallel other structural zones such as the Olympic-Wallowa lineament and Brothers fault zone (Fig. 1) in Oregon and Washington (Wood and Clemens, 2002). In contrast, the eastern SRP is a structural downwarp, associated with extension and magmatism along the track of the Yellowstone hot spot. This mantle plume helped generate the voluminous basalt-rhyolite volcanism of the SRP, and the Pleistocene-Recent Yellowstone caldera of Yellowstone National Park (Pierce et al., 2002).

The major faulting which downdropped the centre of the WSRP began about 12 Ma and ended by approximately 9 Ma, although minor warping and structural adjustment may have continued locally (Wood, 2004). Rhyolitic flow domes mark the margins; basaltic volcanic rocks interbedded with the earliest sediments show that volcanism and early basin formation were contemporaneous (Wood and Clemens, 2002). A generalized stratigraphic sequence is given in Table 1. Vineyards are planted on soils derived from many units, but most notably the sands and silts of the Pliocene Glenns Ferry Formation, local basalts, the Tenmile and Terrace gravels, and finer grained Holocene deposits of the Bonneville Flood (Fig. 4).

As the rift formed, water accumulated in a series of lakes, floodplains and wetlands, with marginal beaches and streams. Fish and terrestrial vertebrate fossils are abundant locally in the complex sequence of lacustrine and related floodplain-to-shoreline facies sedimentary rocks which make up the Tertiary Idaho Group deposits of paleo-Lake Idaho. Sediments deposited in Lake Idaho include sand, silt, and clay as well as local volcanic ash. Lake level rose and fell as the basin subsided; the maximum extent of Lake Idaho was about 4 million years ago, near what is currently the 3600-ft (1100-m) elevation contour (Wood and Clemens, 2002).

To the east, in the Hagerman Fossil National Monument (42°49'N, 114°57'W), basalts interbedded with several hundred feet of Glenns Ferry sediments have been dated at 3.4 Ma (Hart et al., 1999). By approximately 2 Ma, floodplain and marsh sediments of the later part of the Glenns Ferry Formation were deposited east of

Marsing on the Chalk Hills topographic ridge (Wood, 2004; Reppening et al., 1994) that underlies the Bitner vineyards. Lizard Butte (Fig. 4) is a phreatomagmatic basaltic volcano that erupted through wet lake and floodplain sediments. A short distance southeast of Marsing, near Pickles Butte and Idaho's largest vineyard, is a subaerial basalt flow that buries stream gravels of the ancestral Snake River. The age of the basalt flow, dated at 1.58 Ma by the Ar^{40}/Ar^{39} method, demonstrates that by early Pleistocene time, the WSRP had completely transformed from a filling rift basin to an incised lowland (Othberg, 1994). The draining of Lake Idaho was a consequence of headward erosion of ancestral Hells Canyon by the Snake-Salmon river system coupled with Lake Idaho overtopping a divide and draining northward through the ancestral Hells Canyon. Although timing of this event is poorly constrained, the initial spillover most likely occurred between 2 and 4

Table 1: General stratigraphy of the WSRP, modified from Wood and Clemens (2002). Q = Quaternary; T = Tertiary.

Epoch/Period	Age (Ma)	Group	Rock Unit
Holocene Q	14,500 yrs.	Bonneville Flood	Sediments
Pleistocene	0-2 Ma	Terrace Gravels	River Gravels w/ local basalt
	< 2 Ma	Snake River Group	Basalt flows, hydrovolcanoes
Pliocene T		Tenmile Gravel	Gravels
(2-5 Ma)	2-5 Ma?	Idaho Group	Glenns Ferry Formation, or
		~~	Terteling Springs Fm. (N. side)
Miocene T	7-10 Ma?	Idaho Group	Chalk Hills Formation
(5-24 Ma)		Older Basalts/Seds	inc. Banbury/Poison Creek Fm.
	11-12 Ma?	Margin Rhyolites	"Dry Rhyolites"
	12-15 Ma	Idavada Volcanics	Rhyolites, Volcanics of Owyhee Co.
	14-17 Ma	Columbia River Group	Basalts

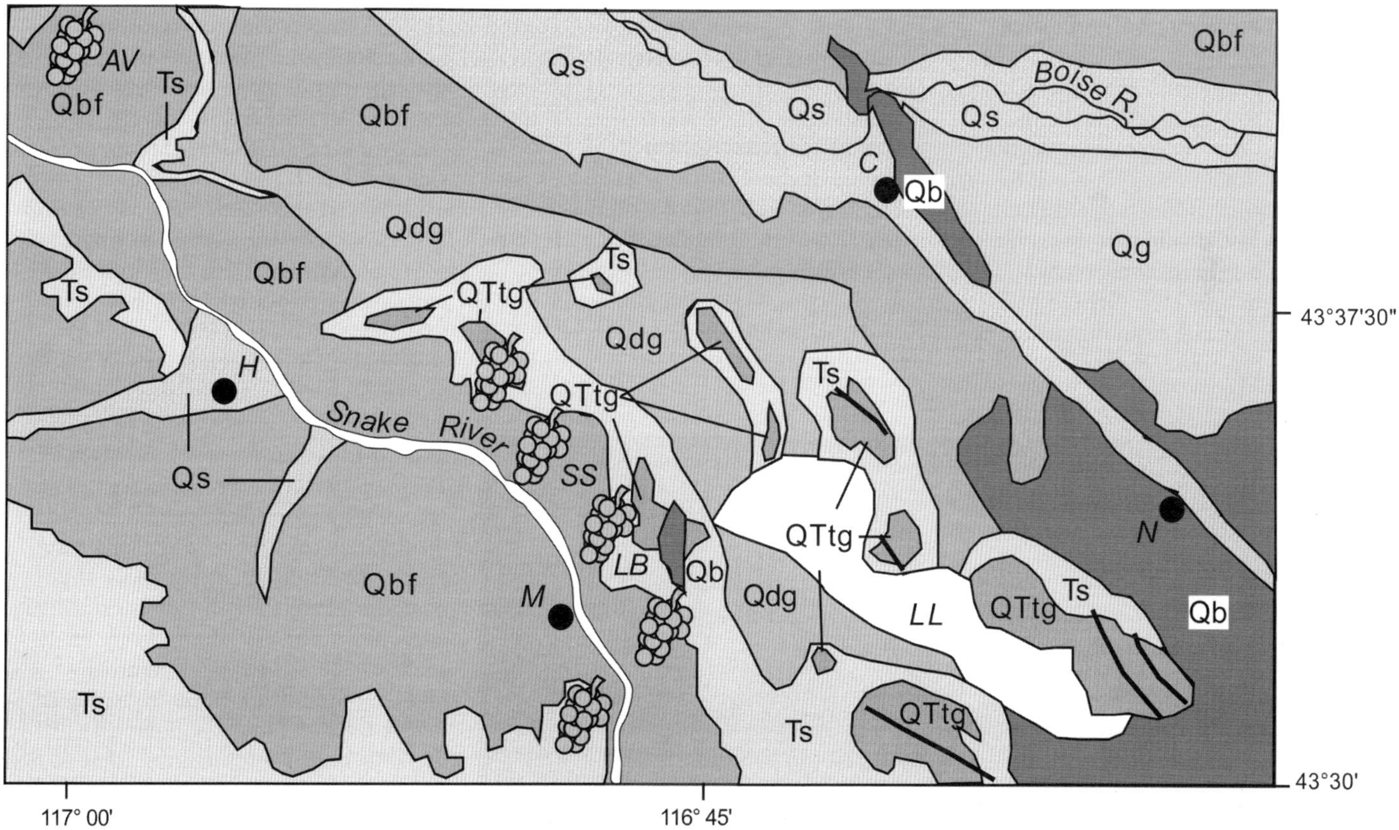

Figure 4. Simplified geologic map for major WSRP grape-growing region, modified from Othberg and Stanford (1992). Towns are Homedale (H), Marsing (M), Caldwell (C), Nampa (N), with other features of Sunnyslope (SS), Lake Lowell (LL), Arena Valley (AV), and Lizard Butte (LB). Wineries and general areas of vineyards shown schematically by grape clusters; others are east of this map. Geologic units: Qs, Quaternary alluvium/surficial deposits; Qbf, Bonneville Flood deposits undifferentiated; Qg, Quaternary terrace gravels, younger; Qdg, Deer Flat Terrace Gravel; Qb, Quaternary basalts; QTtg, Tenmile Gravel; and Ts, Tertiary lacustrine and fluvial sediments, Idaho Group. Heavy black lines are faults. Table 1 gives generalized stratigraphy.

Ma with subsequent slow downcutting of the divide (Wood and Clemens, 2002).

Draining of the lake and lowering base level allowed incision of the old lacustrine sediments by the ancestral Boise River, forming a stepped series of Quaternary stream terrace gravels that mark previous river base level stands (Othberg, 1994; Othberg and Stanford, 1992). The oldest, the Tenmile Gravel (Table 1), caps the northwest-trending ridge of Glenns Ferry sediments at Sunnyslope, where many vineyards are located (near 43°35'N, 116°47'W), and towards Chalk Hills where additional vineyards are located (Fig. 4). A younger, lower elevation terrace gravel, known as the Gravel of Deer Flat Terrace, overlies the Tertiary sediments and flanks the ridges. Several vineyards are planted on the sandy pebble gravel of the Deer Flat Terrace, which is locally overlain by loess. Terrace elevations were controlled by paleo-base levels, and in turn, those terrace elevations dictate the vineyard elevations and influence land use.

The Bonneville Flood, which occurred 14,500 years ago, is the most recent geologic event important to the vineyards of Idaho (Scott et al., 1982). This catastrophic flood resulted from erosion of a low divide that was overtopped by a northern arm of Lake Bonneville, the ancestral Great Salt Lake in Utah (Fig. 3). The resulting deluge of water down the Snake River lasted 6 months (D. Currey, personal communication, 2005) and the discharge peaked at about 1 million $m^3 \cdot s^{-1}$ at the Lake Bonneville outlet (O'Connor, 1993). This single flood event created the western Snake River Canyon of today. Because the discharge was variable over the duration of the flood, multi-metre-sized boulders to sand to silt-sized sediments were deposited (Fig. 5). During the highest flood discharges, water was hydraulically backed up by constricted flow through Hells Canyon, causing relatively quiet water to pond in the lower reaches of the Snake, Payette, Boise, and Owyhee river valleys (Othberg, 1994; Fig. 3). As a result, from Hells Canyon south to Marsing, fine-grained slackwater silt blankets the late Pleistocene terraces below an elevation of 747 m a.s.l. (2,450 feet), which is lower than the early Pleistocene Deer Flat Terrace (Othberg and Stanford, 1992). Vineyards in Arena Valley northwest of Marsing are located adjacent to a circular erosion feature where the late Pleistocene Whitney terrace was scoured by late stages of the Bonneville Flood (Fig. 4).

Pedological Description

Soil type in the WSRP varies somewhat according to the lithology of the parent material, and the timeframe and climate under which the soil developed. Surficial loess, sand, and Bonneville Flood slackwater silts are the predominant parent materials at the vineyards, and the soils normally contain abundant quartz and feldspar grains derived from the Tertiary units, though fields near the basaltic vents may contain more clay and mafic minerals.

Older soils generally tend to be more complex and show more extensive duripan (caliche) development and clay-rich B-horizons (Othberg, 1994). In some areas, soil has developed on a blanket of loess up to 4 m thick (Othberg and Stanford, 1992). The Bonneville Flood sediments are younger than the loess and typically show little soil development. Soils on the Deer Flat Terrace Gravel, which underlies some vineyards, contain more than 25% pedogenic clay and a buried duripan greater than one metre thick. Soils on the older Tenmile Gravel, which underlies ridge-top vineyards, may have 50% clay and a 2-m thick duripan (Othberg, 1994). The thicker, platy duripans promote alkaline soils and may impede subsurface drainage and root penetration by the vines.

Vineyard locations in the Sunnyslope area were spatially compared with their underlying soil characteristics listed in the NRCS Soil Survey Geographic (SSURGO) Database. The soil database for Canyon County lists over 50 soil series and eight associations. The GIS analysis revealed no single soil series common among the vineyards; 19 vineyards are located on 11 soil series that share many characteristics. Most soils underlying the vineyards in the Sunnyslope area are silty to sandy loams where silt percentages range from 58% to 67% in the upper horizons (USDA, 1972). The series are characterized as moderately to extremely well-drained, moderately calcareous and alkaline subgroups of aridisols and entisols; they have moderate to high cation exchange capacities. Most of the soils are fairly shallow (<1 m) with soil depth an inverse function of slope (i.e., steeper slopes have shallower soil depths). All the soils used for agriculture (that is, not range or urban) require surface irrigation (USDA, 1972). The combination of moderate to steep slopes, moderately to excessively drained soils, and easily eroded sediments places further limitations on land use in the region. The soils associated with the vineyard sites typically are characterized as not being prime farm land (i.e., having moderate to

Figure 5. Boulders in Bonneville Flood Melon Gravels, at Celebration Park along Snake River, south of Melba, Idaho.

severe limitations; USDA, 1972), with prescribed uses ranging from irrigated pasture to fruit orchards (where hard freezes are not a danger). Prime farm land in the area (i.e., having higher soil water availability, deeper soil depths, lower gradient slopes) is used primarily for large scale row crop operations such as sugar beets, alfalfa, corn, and onions. More recently, prime farmland on the urban fringes has been converted to housing subdivisions.

Climate

The climatic factors of precipitation (amount and seasonal pattern), growing season length, and growing degree days (e.g., Winkler et al., 1974) all affect grape and wine quality (van Leeuwen et al., 2004; Jones and Davis, 2000) and thus contribute strongly to terroir. The WSRP is located in the continental interior of the western U.S. approximately 500 km east of the Cascade Range (Fig. 1). Even with its continental interior location, the region is on a climatic hinge line and exhibits influences of both continental and marine climates. Winter months (November 1 through March 30) provide two-thirds of the precipitation for the region (Fig. 6). Winter precipitation is caused both by storms from the Gulf of Alaska tracking under the influence of the dominant westerlies at this latitude (Godfrey, 1999), and more tropical moisture originating near Hawaii tracking under the influence of the subtropical jet stream and producing what is colloquially referred to as the "Pineapple Express." Whereas winters may be cold and overcast, the summer growing season (April 1 through September 30) is characterized by warm, dry days with a possible average of 70% of sunshine [Western Regional Climate Center (WRCC), 2005; http://www.wrcc.dri.edu/].

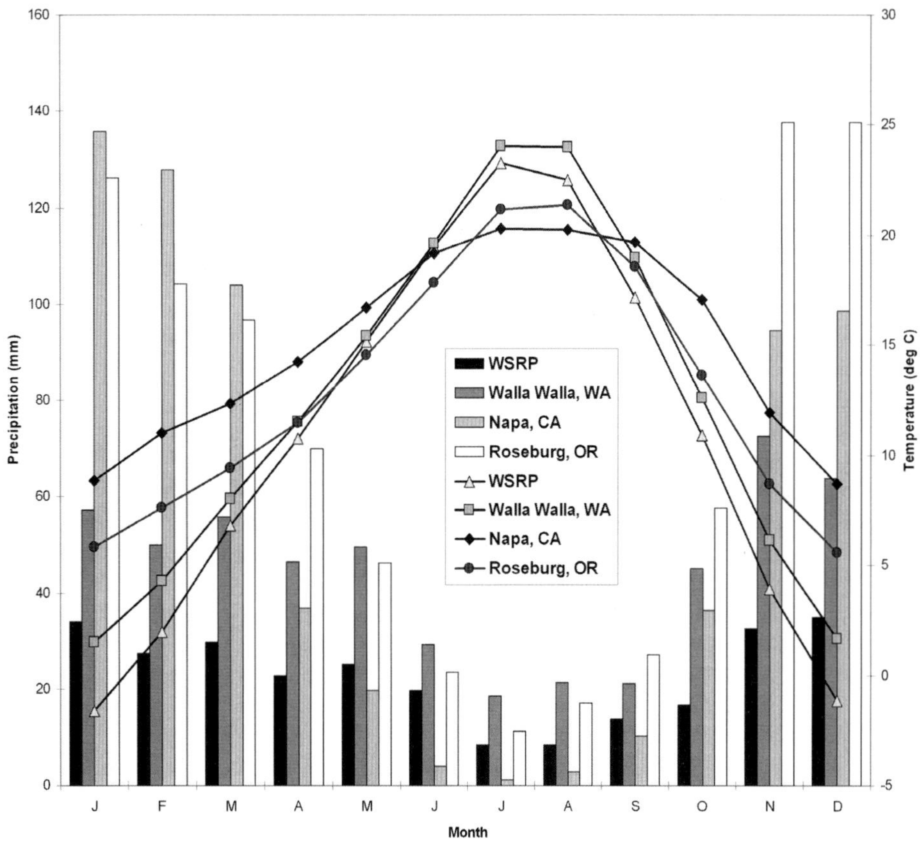

Figure 6. Comparison of climate variables, with mean precipitation as bars and mean temperature as lines, for the averaged WSRP stations; Walla Walla, Washington; Napa, California; and Roseburg, Oregon, in the western United States. Data were compiled from online archives available through the National Climate Data Center and the Western Regional Climate Center, Desert Research Institute, Reno, Nevada (http://www.wrcc.dri.edu/summary/).

Climatic Comparisons

Thirty-year monthly climate normals, covering the period from 1971-2000, were obtained from the National Climate Data Center's (NCDC) online archives (http://www.ncdc.noaa.gov/oa/ncdc.html) for four locations in the WSRP (Glenns Ferry, Weiser, Parma, and Deer Flat Dam, Idaho). These Idaho climate normals were compared to climate data from other grape-growing regions (Walla Walla, Washington; Roseburg, Oregon; and Napa, California) in the western United States (Table 2; Fig. 6). The NCDC climate normals summarize mean values for monthly maximum and minimum temperatures and precipitation.

Despite latitudinal and situational variation among the regions, seasonality of precipitation is strikingly similar (Fig. 6). All four regions experience a pronounced summer precipitation minimum, with the Napa climate station recording only a trace of July precipitation. Despite a similar seasonal precipitation pattern, the WSRP receives about half the annual precipitation of the other regions (Fig. 6). The lower annual precipitation in the WSRP may be partly attributed to the rain shadows created by the Cascade, Sierra Nevada, and more locally the Owyhee ranges.

Temperature is what most distinguishes the different regions, resulting at least in part from differences in elevation. The WSRP ranges from 640 to 765 m, compared to a low of 12 m at Napa and to 365 m at Walla Walla. The mean annual temperature in the WSRP of 10.8°C is the lowest in the regional comparison group, and is close to the 10°C isotherm described as the poleward temperature limit for cultivation of European grapes (Jones et al., 2004; De Blij, 1983). The WSRP also has significantly lower mid-winter (January) mean minimum temperatures than the other western US districts, and two months, December and January, have mean temperatures below 0°C (Fig. 6). Whereas these mean temperatures are not solely limiting, they provide evidence for the potential of damage from severe cold temperatures (i.e., <-18°C, Table 2; Winkler et al., 1974). This potential for deep freezing temperatures has implications for viticultural practices.

The temperature contrast between the WSRP and the other

regions also translates into differences in length of growing season (i.e., number of days during the growing season with less than 50% probability of reaching 0°C; Winkler et al., 1974). With its higher elevation and more interior location resulting in colder winters, the WSRP has the shortest growing season of the regions in Table 2, and this may be a limiting factor for some grape varieties.

Another climate factor, continentality, is defined by the annual range in temperature and reflects remoteness from moderating ocean influences; higher temperature ranges indicate a greater degree of continentality. Napa and Roseburg have the most moderate temperature ranges, reflecting their proximity to the ocean, whereas the WSRP has mean monthly temperatures that vary by almost 25°C (Fig. 6). This range is slightly greater than at Walla Walla and much greater than at Roseburg, which is at the same 43°N latitude but has a monthly mean temperature range of only 16°C.

Growing Degree Days

Growing degree days (GDD) is a summation of accumulated heat units as measured by days during the growing season (April 1 to October 31 in the northern hemisphere) with a mean temperature over an established base (10°C for grapes; Winkler et al., 1974); under the 10°C temperature base threshold, little growth or development of wine grapes occurs. Growing degree days also can be used as an indicator of the timing, independent of the calendar, of phenological events including dates of budbreak, bloom, veraison (onset of berry ripening) and harvest.

The climate data indicate that the Walla Walla Valley and Napa districts each fall within the Winkler Region III range (1666-1944 Base 10°C GDD), with higher growing degree days than the WSRP and Roseburg districts, which are within the Winkler Region II range (1389-1665 Base 10°C GDD; Table 2; Fig. 7). The higher growing degree days rating for Napa Valley reflects the relatively warm early growing season, whereas values for Walla Walla reflect high temperatures during June through August. In contrast, Roseburg has the lowest growing degree day rating, reflecting that district's marine influence and generally low summer temperatures. Seasonal temperatures rise quickly in the WSRP, with an average last and first day of 0°C frost on May 10 and September 29, respectively. Thus, the WSRP has a shorter growing season than the other three districts.

Table 2: Climate Data for the WSRP and other grape growing districts in the western United States. Column heading abbreviations: MAT, Mean Annual Temperature; MAP, Mean Annual Precipitation; GDD, Growing Degree Days; GSL, Growing Season Length; XMT, 30-year extreme minimum winter temperature (event year in parenthesis); CNT, Continentality (mean annual temperature range). Data compiled from the National Climate Data Center (http://www.ncdc.noaa.gov).

Station Name	Elev (m)	Location (lat/long)	MAT (°C)	MAP (mm)	GDD 10°C	GSL 0°C	XMT (°C)	CNT (°C)
Parma Exp. Station, Idaho	698	43.8°N 116.95°W	9.9	283	1342	140	-32 (1990)	25
Weiser, Idaho	646	43.67°N 116.68°W	11.0	307	1637	136	-34 (1990)	27
Deer Flat Dam, Idaho	765	43.59°N 116.75°W	11.6	258	1626	165	-30 (1989)	24
Glenns Ferry, Idaho	752	43.61°N 116.57°W	10.5	248	1413	125	-32 (1989)	24
WSRP	**715**		**10.8**	**274**	**1504**	**142**		**25**
Roseburg, Oregon	130	43.2°N 123.36°W	13.0	855	1484	218	-16 (1989)	16
Walla Walla Airport, Washington	355	46.05°N 118.28°W	12.3	530	1715	206	-11 (1985)	23
Napa State Hospital, California	11	38.28°N 122.27°W	15.0	672	1753	259	-10 (1990)	12

Viticultural Practices

Successful production of wine grapes in the WSRP requires careful consideration of site and cultivar selection, number of frost-free days, and availability of supplemental water. Risk of frost damage is minimized by locating vineyards on slopes with good air drainage towards river or valley bottoms (Fig. 8a). Vineyards tend to be located on hillsides with a southern or southwestern aspect to maximize heat unit accumulation and to avoid direct exposure to prevailing northwesterly winds. Vines are generally planted in north-south rows to facilitate cold air drainage and to provide equal sunlight exposure on both sides of the vine canopy (Fig. 8b). The predominant vine training system is cordon-trained, spur-pruned, with vertical shoot positioning on a six wire trellis. Common in-row spacing is 2.1 m with 2.7 m between rows. Target shoot length is about 1.5 m. Dormant vines typically are pruned manually to two (red cultivars) or three (white cultivars) bud spurs, yielding 24 to 28 (red) or 36-42 (white) buds per vine. Some growers mechanically pre-prune dormant vines, and many mechanically harvest their grapes.

An estimated 60% of total producing wine grape acreage in the WSRP is composed of fairly cold hardy, white wine cultivars (Riesling, Chardonnay, and Gewürztraminer) that have a low heat unit requirement to reach maturity (Wolfe, 1998). The red wine cultivars Cabernet Sauvignon, a late maturing cultivar with a relatively high heat unit requirement (Wolfe, 1998), and Merlot, a less cold hardy cultivar (Wolfe, 1998), comprise about a third of producing acreage, and many new plantings include red wine cultivars with similar temperature requirements.

The large range (about 280) in GDD among WSRP weather station sites as depicted in Table 2, as well as the successful commercial production of Cabernet Sauvignon and Merlot, highlights the importance of the vineyard

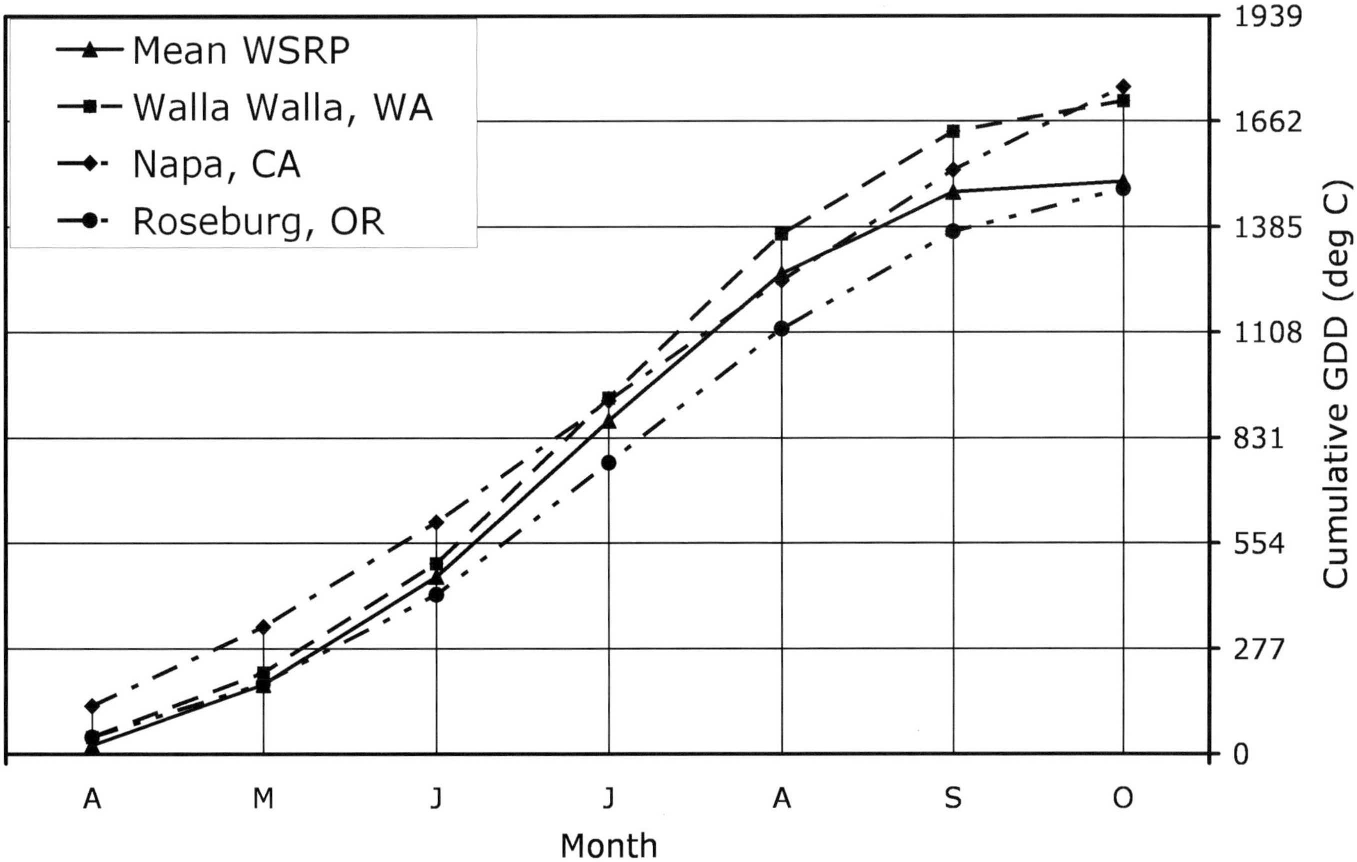

Figure 7. Mean cumulative growing degree days (GDD) from 1971-2000 for the four averaged WSRP stations; Walla Walla, Washington; Napa, California; and Roseburg, Oregon, calculated from National Climate Data Center from average monthly maximum and minimum temperatures between April 1 and October 31 using a base temperature of 10°C.

mesoclimate. For example, a 140-day growing season with 1442 growing degree days is reported for the Idaho Parma Experimental weather station in Table 2 (Fig. 2). However, temperature data collected in a hillside vineyard planted in 1997 with a northern aspect and north-south row orientation recorded 1581, 1851, and 1644 growing degree days at the Parma Experiment Station (simple average, base 10°C, daily temperature), respectively over three vintages (2002-2004). The average number of frost-free days needed in this vineyard for fruit to reach maturity was 147-150 for Merlot, 150-154 for Chardonnay, and 160 or more for Cabernet Sauvignon and Syrah (Table 3). Fruit composition of each cultivar at harvest reached optimum brix (~23%; this is a measure of sugar concentration in the grapes), pH (~3.6), and titratable acidity (0.6 g/dl), suggesting adequate season duration and temperature accumulation. The growing season and heat unit accumulation data in Table 2 suggest that the climate in

Table 3: For own-rooted, *Vitis vinifera* L. vines (wine grapes of European origin) planted in 1997 at the Parma Experimental Station, Idaho, table shows: Average number of days from budburst to harvest (DTM); Harvest percent brix (a measure of sugar concentration in the grapes); pH; titratable acidity (TA); and berry weight of fruit harvested in 2002, 2003, and 2004 from the cultivars listed.

Cultivar (clone)	DTM (days)	Brix (%)	pH	TA (g/dl)	100 berry weight (g)
Merlot (01)	147	23.6	3.56	0.52	104.5
Merlot (08)	149	24.0	3.70	0.51	105.5
Chardonnay (29)	154	22.9	3.38	0.75	97.9
Chardonnay (49)	150	23.1	3.38	0.74	102.5
Cabernet Sauvignon (02)	162	24.1	3.65	0.45	95.9
Cabernet Sauvignon (04)	160	23.3	3.62	0.46	95.0
Cabernet Sauvignon (11)	171	23.7	3.51	0.59	104.0
Syrah (07)	164	24.3	3.73	0.50	113.7

Parma is marginal for cultivation of Merlot and Chardonnay and not suitable for production of Cabernet Sauvignon or Syrah, yet vineyard heat unit accumulation and number of frost-free days suggest a suitable vineyard mesoclimate for production of these red wine cultivars.

Vine cold-hardiness is not well understood because of its complex interaction with environmental conditions, including temperature and photoperiod (Howell, 2000), tissue maturity (Goffinet, 2000), and vine water status (Wample et al., 2000), but growers in the WSRP minimize yield loss from cold injury by adopting preventative cultivation practices. For example, own-rooted cuttings are planted at a depth of 30 to 36 cm to facilitate root survival in the event of above ground vine loss from prolonged low temperature. The vines are trained to two trunks (Fig. 8b) with each trunk forming a unilateral cordon that extends to one side or other of the vine. In the event of cold damage, the second trunk can be used to replenish damaged wood. The absence of phylloxera in the region permits cultivation of own-rooted rather than grafted vines, enabling trunk re-establishment without replanting. Despite these cultivation practices, injury from prolonged minimum mid-winter low temperature was anecdotally reported in the very cold years of 1989 and 1990. Another common type of cold injury observed in the WSRP occurs during winter dormancy when several days of high solar radiation and warm ambient temperature precede an abrupt return to freezing or near freezing temperature.

Figure 8a. Cliffside Vineyard planted on gently tilted Lake Idaho sediments of the Glenns Ferry Formation (Pliocene), facing west. The Snake River lies at the base of the steep slope, along the line of green trees and houses.

Figure 8b. North-south rows of grape vines at Skyline Vineyard with a basaltic cinder pit in distance adjacent to Sawtooth Vineyard, facing north. Vines are planted with double trunks and require drip irrigation.

The low annual and growing season precipitation and the shallow soils in the WSRP facilitate manipulation of vine physiology through irrigation management. Most irrigation water is obtained from annually recharged snowpack in mountain ranges surrounding the WSRP and is delivered through an extensive network of reservoirs and canals. The majority of vineyards are irrigated with above ground drip lines, although some vineyards utilize overhead sprinklers or furrows. Irrigation scheduling is used to prepare for the first fall frost by imposing water stress on the vine to encourage periderm formation on green shoots. Periderm is a visible indicator of tissue maturity and has been associated with bud and cane cold hardiness (Goffinet, 2000). Irrigation is then normally applied prior to the first fall frost to bring soil moisture up to field capacity in an effort to protect roots during the winter. Growers also manipulate vine water stress during the growing season to shift growth from vegetative to reproductive structures (Greenspan, 2005) and to control canopy as well as berry size. Regulated deficit irrigation is used to control plant water status and to optimize fruit quality during the growing season. Ongoing research is being conducted by researchers at the USDA-Agricultural Research Service in Parma, Idaho, to understand how vine water status influences grape components that contribute to wine quality.

CONCLUSIONS

The Western Snake River Plain is the principal wine grape-growing district in

Idaho. Formation of the WSRP began between 9 and 12 Ma with major extensional faulting and volcanism resulting in a down-dropped topographic basin. Over approximately the next 7 million years, Lake Idaho and associated streams and floodplains deposited a succession of fine-grained sand, silt, and ash, with the lake high stand reaching up to near the present-day 1100-metre topographic contour of the WSRP. After Lake Idaho drained about 2 Ma, the ancestral Boise River incised these Tertiary sediments, forming a stepped series of Quaternary stream terrace gravels surrounding remnant highlands of the earlier sediments. The final major geologic event to shape the region occurred about 14,500 years ago when Lake Bonneville discharged catastrophically into the Snake River canyon, enlarging it and leaving behind widespread flood deposits from conglomerates to silt over-bank deposits. Vineyards are planted on all of these Tertiary to Quaternary units where the landscape allows cold air drainage.

Low temperatures and a short growing season limit the range of European wine grape cultivars that are suitable for production in the WSRP. However, late maturing, less cold hardy cultivars like Cabernet Sauvignon and Merlot currently comprise an estimated 31% of producing acreage in the WSRP, suggesting that vineyard mesoclimate is critical for production success. Furthermore, controlled irrigation is a critical tool for managing vine cold hardiness.

The geologic and physiographic diversity within the WSRP suggests that subregions, such as the Marsing Valley, the Hagerman Valley, Glenns Ferry region, and Boise Foothills, may emerge as future viticultural areas. Much work remains to document the key elements of Idaho terroir that may enhance future vintages.

ACKNOWLEDGEMENTS

We thank Larry Meinert, Roger Macqueen, and an anonymous reviewer for comments which greatly improved the original manuscript. We also thank members of the Idaho Grape Growers and Wine Producers Commission for introducing us to the subject and providing access to the vineyards.

REFERENCES

De Blij, H., 1983, Wine, A Geographic Appreciation: Rowmand and Allanheld Publishers, New Jersey, 239p.

Godfrey, B., 1999, Delineation of agroclimate zones in Idaho: M.S. thesis, University of Idaho, Moscow, Idaho, 202p.

Greenspan, M., 2005, Integrated irrigation of California winegrapes: Practical Winery and Vineyard, v. 17, p. 21-79.

Goffinet, M., 2000, The anatomy of low-temperature injury of grapevines: Proceedings of the ASEV 50th Anniversary Annual Meeting, Seattle WA, American Society for Enology and Viticulture, p.94-100.

Hart, W.K., Brueske, M.E., Renne, P.R., and McDonald, H.G., 1999, Chronostratigraphy of the Pliocene Glenns Ferry Formation, Hagerman Fossil Beds National Monument, Idaho [abstract]: Geological Society of America Abstracts with Programs, v. 31: No 4, p. A15.

Howell, G. S., 2000, Grapevine cold hardiness: mechanisms of cold acclimation, mid-winter hardiness maintenance, and spring deacclimation: Proceedings of the ASEV 50th Anniversary Annual Meeting, Seattle WA, American Society for Enology and Viticulture, p. 35-47.

Jones, G., and Davis, R., 2000, Climate influences on grapevine phenology, grape composition, and wine production and quality for Bordeaux, France: American Journal of Enology and Viticulture, v. 51, p. 249-261.

Jones, G., Snead, N., and Nelson, P., 2004, Geology and Wine 8. Modelling viticultural landscapes: a GIS analysis of the *terroir* potential in the Umpqua Valley of Oregon: Geoscience Canada, v. 31, p. 167-178.

Meinert, L.D., and Busacca, A.J., 2000, Geology and Wine 3. *Terroirs* of the Walla Walla Valley *appellation*, southeastern Washington State, USA: Geoscience Canada, v. 27, p. 149-171.

Meinert, L.D., and Busacca, A.J., 2002, Geology and Wine 6. *Terroir* of the Red Mountain Appellation, central Washington State, U.S.A.: Geoscience Canada, v. 29, p. 149-168.

O'Connor, J.E., 1993, Hydrology, hydraulics, and geomorphology of the Bonneville flood: Geological Society of America Special Paper 274, 83p.

Othberg, K.L., 1994, Geology and geomorphology of the Boise Valley and adjoining areas, western Snake River Plain, Idaho: Idaho Geological Survey Bulletin 29, 54p.

Othberg, K.L., and Stanford, L.R., 1992, Geologic map of the Boise Valley and adjoining area, Western Snake River Plain, Idaho: Idaho Geological Survey GM-18, 1:100,000 scale, 1 sheet.

Pierce, K.L., Morgan, L.A., and Saltus, R.W., 2002, Yellowstone plume head: Postulated tectonic relations to the Vancouver slab, continental boundaries, and climate, *in* Bonnichsen, B., White, C.M., and McCurry, M., eds., Tectonic and Magmatic Evolution of the Snake River Plain Volcanic Province: Idaho Geological Survey Bulletin 30, p. 5-33.

Reppening, C.A., Weasma, T.R., and Scott, G.R., 1994, The early Pleistocene (latest Blancan-earliest Irvingtonian) Froman Ferry fauna and history of the Glenns Ferry Formation, southwestern Idaho: U.S. Geological Survey Bulletin 2105, 86p.

Scott, W.E., Pierce, K.L., Bradbury, J.P., and Forester, R.M., 1982, Revised Quaternary stratigraphy and chronology in the American Falls area, Southeastern Idaho, *in* Bonnichsen, B., and Breckenridge, R.M., eds., Cenozoic Geology of Idaho: Idaho Bureau of Mines and Geology Bulletin 26, p. 581-595.

Taylor, V., Longerich, H., and Greenough, J., 2002, Geology and Wine 5. Provenance of Okanagan Valley wines, British Columbia, using trace elements: promise and limitations: Geoscience Canada, v. 29, p. 110-120.

U.S. Department of Agriculture, Idaho Agricultural Statistics Service, 1999, Idaho fruit tree census 1999, 21p. (http://www.nass.usda.gov/id/).

U.S. Department of Agriculture, Soil Conservation Service, 1972, Soil Survey of Canyon Area, Idaho: Superintendent of Documents, U.S. Government Printing Office Washington, D.C., 118p.

van Leeuwen, C., Friant, P., Chone, X., Tregoat, O., Koundouras, S., and Dubourdieu, D., 2004, Influence of climate, soil, and cultivar on *terroir*: American Journal of Enology and Viticulture, v. 55, p. 207-217.

Wample, R.L., Hartley, S., and Mills, L., 2000, Dynamics of grapevine cold hardiness: Proceedings of the ASEV 50th Anniversary Annual Meeting, Seattle WA, American Society for Enology and Viticulture, p. 81-93.

Winkler, A.J., Cook, J.A., Kliewer, W.M., and Lider, L.A., 1974, General Viticulture: University of California Press, 710p.

Wolfe, W., 1998, Site selection in eastern Washington: optimizing site and variety choices, in Watson, J., ed., Growing Grapes in Eastern Washington: Good Fruit Grower, Yakima, WA, USA, p.27-30.

Wood, S.H., 2004, Geology across and under the Western Snake River Plain, Idaho:

Owyhee Mountains to the Boise Foothills, in Haller, K.M., and Wood, S.H., eds., Geological Fieldtrips in southern Idaho, eastern Oregon, and northern Nevada: Dept. of Geosciences, Boise State University, pp. 82-105 (USGS OFR2004-1222).

Wood, S.H., and Clemens, D.L., 2002, Geologic and tectonic history of the western Snake River Plain, Idaho and Oregon, in Bonnichsen, B.,White, C.M., and McCurry, M., eds., 2004, Tectonic and magmatic evolution of the Snake River Plain Volcanic Province: Idaho Geological Survey Bulletin 30, p. 69-103.

Accepted as revised 17 February, 2006

PAPER 1 FROM THE "TERROIR, GEOLOGY AND WINE: A TRIBUTE TO SIMON J. HAYNES" SESSION HELD AT THE GEOLOGICAL SOCIETY OF AMERICAN ANNUAL MEETING, SEATTLE, WASHINGTON, NOVEMBER 3, 2003

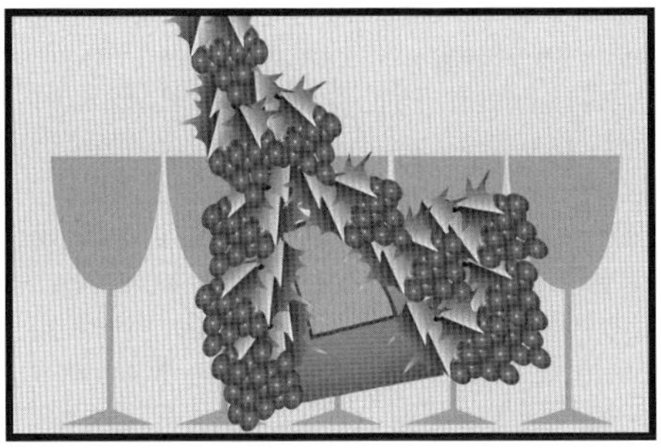

Grape Growing and Wine Making in Two Unique Italian Terroirs: Teroldego Rotaliano and Aglianico del Vulture

Maria Bianca Cita[1] and Antonello Fiore[2]

[1]*Dipartimento di Scienze della Terra Ardito Desio, Universita Degli Studi Milano, via Mangiagalli 34, 20133, Italy, maria.cita@unimi.it*

[2]*Consulting geologist, via Scanderberg 28, Ginestra (PZ), 85020, Italy, infogaia@libero.it*

SUMMARY

Two terroirs, Teroldego Rotaliano in the Adige River Valley of the Southern Alps at approximately 46° N; and Aglianico del Vulture from the lower slopes of a volcano in the Southern Apennines at approximately 40° N, are fine wine production areas of Italy, the world's largest producer at ~60 million hectolitres per year. Both terroirs produce single varietal red wines that are prize winners and both terroirs have long historical traditions. Most wine in these regions comes from vines derived from indigenous grapevines of local or limited areal distribution. The training (trellising) methods in the two regions are very different: large pergolas for Teroldego, and goblet-trained or short rows for Aglianico, and are well suited to the local climatic and microclimatic conditions. Teroldego vineyards are planted on post-glacial coarse-grained calcareous alluvial sediments of the Noce River alluvial fan; a limiting factor on further development of Teroldego vineyards is the availability of land, since the DOC (appellation) is entirely occupied by vineyards. In contrast, Aglianico del Vulture vineyards are planted on fertile, potassium-rich volcaniclastic terrain. The limiting factor in this area seems to be water, owing to the long, hot, dry summers. In terms of land value, the Teroldego vineyards are by far the most expensive in Italy, but this does not affect the price of the wine.

RÉSUMÉ

Deux terroirs, le Teroldego Rotaliano dans la vallée de la rivière Adige dans les Alpes méridionales situé à environ 46° N de latitude, et l'Agliano del Vulture sur les pentes basses d'un volcan du sud des Apennins situé à environ 40° N de latitude, sont deux zones de production de grands vins de l'Italie, le plus grand producteur au monde, soit environ 60 millions d'hectolitres par année. Ces deux terroirs qui donnent des vins rouges d'une même variété et qui ont gagné de nombreux prix, découlent tous deux de traditions qui remontent loin dans l'histoire. La plupart des vignobles de ces deux régions proviennent de vignes issues de cépages indigènes locales ou à distribution géographique limitée. Les méthodes de tuteurage par treillis de ces deux régions diffèrent considérablement; on a recours à de grandes pergolas dans le cas du Teroldego, et à un alignement sur des rangées de faibles hauteurs dans le cas de l'Aglianico, lesquels sont bien adaptées aux conditions climatiques et micro-climatiques locales. Les vignes du Teroldego croissent dans les sédiments alluvionnaires post-glaciaires calcareux grossiers du cône alluvial de la Rivière Noce; le manque de terres disponibles est l'un des facteurs limitatifs du développement des vignobles du Teroldego, puisque tout le DOC (appellation) est déjà occupé par des vignobles. Par contre, les vignobles de l'Agliano del Vulture croissent sur des terrains volcanoclastiques fertiles riches en potassium. Il semble que le facteur limitatif dans cette région soit le manque d'eau, étant donné ses étés longs, chauds et secs. En terme de valeurs, les vignobles du Teroldego sont de loin les plus dispendieux d'Italie, mais cela n'affecte pas le prix du vin.

INTRODUCTION

Today an increasing interest in wine is arising in Italy, where grape growing and wine making have an incredibly long and uninterrupted history dating back to almost 2000 years before Christ, when Italy was called "Oenotria" or "Land of Wine" by the colonizing Greeks, because wine grapes grow and wine production occurs over the length and breadth of the country (Johnson and Robinson, 2001). Thus grape growing and wine making is a deeply rooted tradition in Italy, in part because there is a great variety of local, indigenous grapevines. Also, wine is, and has been almost since time immemorial, a part of the Mediterranean diet, and has played an important role in nutrition even in times of economic stress. Wine has been a continuing success story in Italy in the last decade, not only for the large quantity produced, but also for the increasingly appreciated quality - prize-winning, top quality products - as well as for the great variety of typical products. For these reasons we were pleased to attend the Geological Society of America symposium dedicated to honour the memory of Simon J. Haynes, a Canadian geologist who has been instrumental in developing and popularizing the scientific aspects of viticulture related to geological factors (Haynes, 1999, 2000; Meinert and Busacca, 2000, 2002). In Italy, the interest of the geological community in viticulture has just started to develop (Cita *et al.*, 2001; Cita *et al.*, 2003; Colacicchi and Parotto, in press; Cita *et al*, 2004), but is now growing quickly, as seen by the large number of participants in a Topical Symposium dedicated to the subject of Geology and Wine during the 32nd International Geological Congress held in Italy in August, 2004. Well known Italian personalities that have had and still have fundamental importance in the great progress that has been made in the quality of Italian wines include Veronelli (1983), the reference point for the definition and qualification of sensory characters of wine; Fregoni (1985) for the scientific aspects of viticulture and ampelography (a term referring to the genealogy of grape vines, these days done largely by DNA analysis) and for recognition of the importance of grape varietals; and Scienza (Scienza *et al*, 1991, 1996; Failla *et al*, 1992) for the application of modern techniques of molecular biology to successful viticulture and selection of the best grape clones.

Terroir is a French term strongly rooted in France, where it has been used by geologists with reference to viticulture at least since the early 1980s (e.g. Wallace, 1972; Sittler and Marocke, 1981; Pomerol, 1984; Sittler, 1998; Marre, 2003; Bourrouillh, 2004) inter alias. A special Working Group on Terroirs Viticoles has been appointed by the International Association of Geomorphology: this group recently published a report arising from three international meetings (Biancotti *et al.*, 2003). The term terroir has not been used in Italy until now. We do not want to discuss here the conceptual problems that underlie this term and/or the various interpretations of the term as given by geologists, pedologists, geomorphologists, biologists, and geneticists that can be found in the literature (eg. Wilson, 1998, 2001). We point out that, although there is a wealth of scientific data available on terroir in published form, only a minor part of it is available in widely distributed scientific journals, which makes bibliographic research on terroir surprisingly difficult.

Terroir is concerned with landscape (as is the German term, "Landschaft"), including the physical, natural environment. Terroir is thus unlike the Italian word "paesaggio" (as well as the French term "paysage" that derives from paese, = village). In other words, "non c'è paesaggio senza paese" (literally, "there is no landscape without village") which means that the work of man, making a living in the fields, contributes to reshaping the natural environment plantings. This is why history and tradition seem to play a more important role in the old world of wine, mainly Europe, than in the new world including North America, Australia, New Zealand, and South America. The two Italian examples that we discuss in this paper are strongly supportive of this old world approach. Both terroirs have a long history and a very strong tradition clearly expressed by the unity of the grapevines, by their indigenous nature, and by the use of both traditional and unusual training (trellising) methods. For the Teroldego Rotaliano terroir in particular we could well use the words of Luca Benciolini (2004), stating that "the vineyards appear to be more of a cultural heritage than the result of a scientific or strategic choice".

On the legal/political level in Italy, the term "DOC" stands for "Denominazione di Origine Controllata" and defines the denomination of origin by setting up laws for producers to follow, and thus is similar in concept to the French AOC laws ("Appellation d'Origine Contrôlée") which govern the area, name, vines, yields and production practices. At present there are about 233 DOCs in Italy.

TEROLDEGO ROTALIANO DOC – SOUTHERN ALPS

Teroldego Rotaliano is considered to produce the best red wine of the Trentino-Alto Adige region of northern Italy (Fig. 1). The name comes from "Teroldego", the indigenous grapevine used for making this monogrape (= varietal) wine, and "Rotaliano" which refers to the locality (Campo Rotaliano or Piana Rotaliana), that is the flat area at the junction of the Noce River coming from Val di Non and the N-S trending major Adige River valley (Fig.1). Campo Rotaliano derives its name from King Rotari of the Longobards, a warrior who reigned from 636 to 652 in Northern Italy and won an important battle in this area. His famous edict, promulgated in 641, contains five articles dedicated to the protection of vineyards (Andreotti, 1999).

History of the Geographic Area

Gorgeous vineyards occupy the entire plain that the famous writer and hero Cesare Battisti called "il più bel giardino vitato d'Europa"-"Europe's best vineyard garden" (Fig. 2). Campo Rotaliano lies very close to the border between the German-speaking community to the north and the Italian-speaking community to the south. Relationships between the two have always been strongly competitive, and sometimes difficult and antagonistic. As a consequence, it is not easy to trace the history of the specific Teroldego vine, whereas maps and historical documents help in reconstructing the history of the terroir. During the long and important "Concilio di Trento" (1545-1563) when the Roman Catholic Church reconsidered the Controriforma, Teroldego wine was one of the most appreciated of the "Council wines", along with Marzemino. The oldest map of this area was published when Maria Theresa was the Koningin (~king) of Austria (starting from 1740) and was particularly interested in the Italian territory. The boundary between Austria and what was then called the "Venedisches Gebiet" crossed the Adige River Valley in the Piana Rotaliana area along the River Noce, which at that time joined the Adige River at Grumo: a bridge crossing the Noce connected the villages of Mezzolombardo to the south, on the right bank of the river, and Mezatedesca to the north, on the left bank. The names of the two villages were modified later, and Mezatedesca is now Mezzocorona (Fig. 1), both in the province of Trento. The powerful Prince-Bishop of Trento took over after the Republic of Venice, but Austrian domination persisted until World War I. Owing to the marshy nature of the ground near Grumo, the River Noce was diverted to the south, and the associated hydraulic works were terminated in 1842. An accurate map created by Baron Ignaz von Nowack in 1805 is reproduced here in Figure 3 as an important document for the past history of the terroir.

Origin of Teroldego Rotaliano DOC

The origin of the Teroldego vine is less clear owing to the lack of precise information in this area, which is marginal to both Italy (that did not exist as a unified state until 1870) and Austria. There are two possibilities: one is that Teroldego vines arrived in the vicinity of Trento from the south, from the province of Verona, where there was a red vine called "Terodola" or "Tirodola" (Scienza *et al.*, 1991). The other possibility is that "Teroldeghe" was the name of a locality near the southern limit of the Campo Rotaliano, and that the grapevine is indigenous to this area. According to the first director of the School of Agriculture of San Michele all'Adige, Sorni is the type-locality for the Teroldego varietal. Ampelographic and chromatographic analysis showed recently that the greatest affinity of Teroldego is with two indigenous grapevines: Marzemino, from the province of Trento, and Lagrein, typically from the vicinity of the abbey of Gries, on the outskirts of Bolzano (Scienza *et al.*, 1991; M. Bertamini, pers. comm., 2003).

Teroldego Rotaliano is a DOC that has a very limited and compact distribution in the Campo Rotaliano, including the municipalities of Mezzolombardo, Mezzocorona and a small part of the municipality of San Michele all'Adige (frazione Grumo) on the right bank of the Adige River but excluding all the surrounding slopes, which are entirely covered by specialized vineyards (Fig. 1). Teroldego vines are also grown in other DOCs from the province of Trento, and their juices blended with those of other grapes, to reinforce them (see below). A Consorzio (= consortium) del Teroldego was created by the local producers in 1948 and later modified, in 1954, to Consorzio del

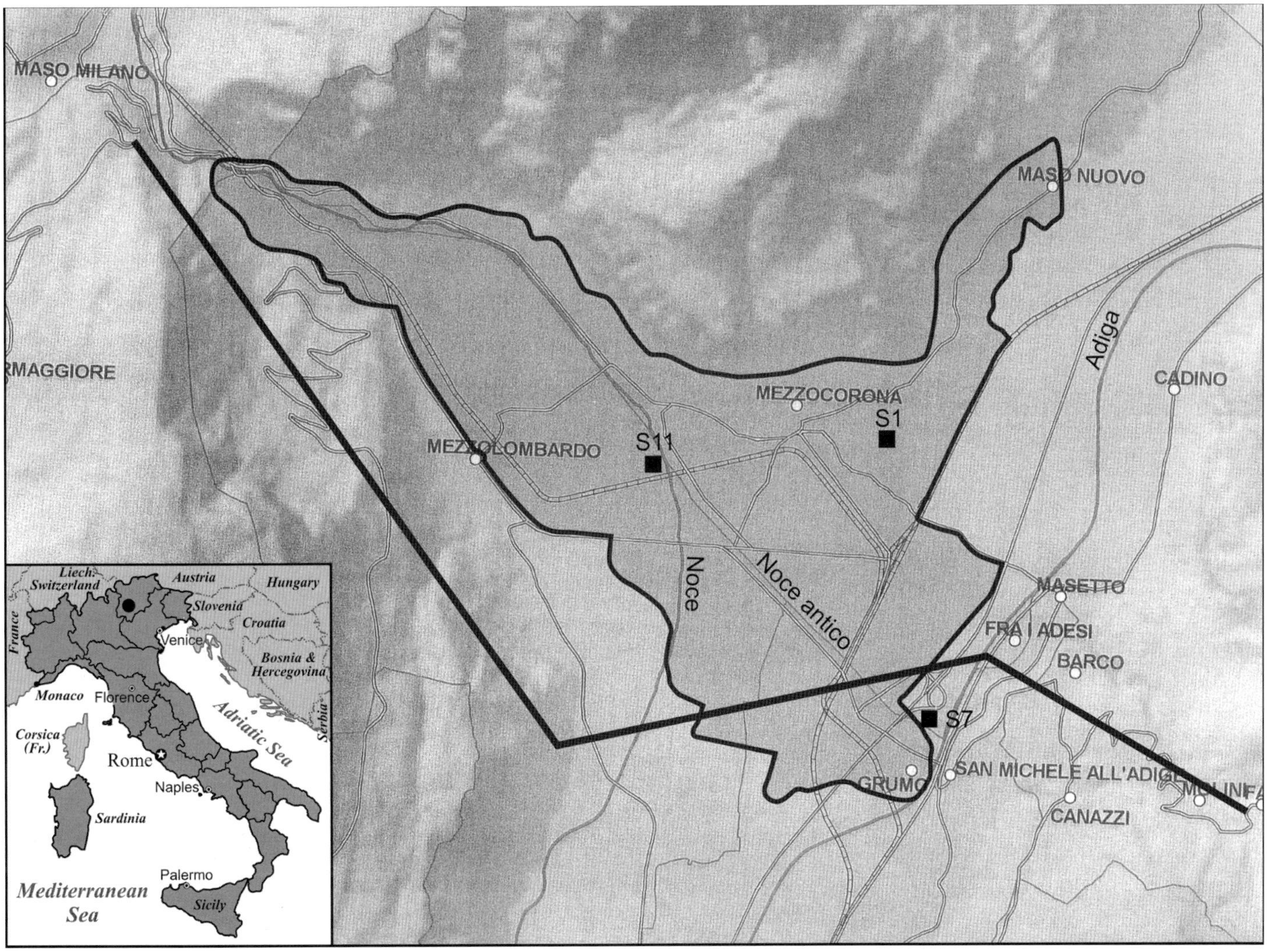

Figure 1 Location map of the Teroldego Rotaliano DOC (Denominazine di Origine Controllata) outlined by solid black line, in the region of Trentino-Alto Adige, province of Trento, Northern Italy. The Appellation is located in three municipalities: Mezzocorona, Mezzolombardo and the portion of San Michele all'Adige on the right bank of the Adige River (Frazione Grumo). Sites S1, S7 and S11, the locations of drill sites to sample lithologic compositions, are discussed in the text (also see Tables 2-4). Inset map at lower left shows the 20 regions of Italy, and indicates the location of this DOC (solid circle) in the region of Trentino-Alto Adige, Southern Alps. Segmented solid line indicates the approximate location of the geological cross-section of Figure 5.

Figure 2 The vineyards of the Campo Rotaliano. (A) Looking north from the bridge crossing the Highway A24 on the right bank of the Adige River; note the white, calcareous cliffs, several hundred metres high, that define the Adige River valley and concentrate solar heat. (B) Aerial view of part of the Campo Rotaliano, 100% covered by large double pergolas of Teroldego vines, showing the various orientations of the rows (modified after Andreotti, 1999). (C) single pergolas at the foot of the calcareous cliffs north of Mezzocorona. (D) Single pergolas charged with grapes in early summer.

Teroldego Rotaliano to make it more specific. The aims of the consortium were basically to provide a precise definition of the area based on geo-pedologic and climatic criteria, a selection of the clones to guarantee a homogeneous quality of grapevines, a rational study to optimize a consistent and scientifically controlled use of fertilizers, and specific treatments and agricultural procedures in order to improve the quality of the vineyards and of the wines produced.

The DOC designation was obtained in 1971, the first in the region. Wine production in the DOC increased in approximately thirty years from the initial 16,045 hectolitres (hl) to the 46,038 hl produced in 2001 (see Table 1). Since the area covered by the DOC remains constant, and no new plantings were established (only re-planting of old vines), the increase in wine production documents the success of the actions taken, and may account for the incredibly high price of the vineyards, over half a million Euros per hectare, according to the newspapers and to local experts. This is the highest price reported for DOC vineyards in Italy, but fortunately does not affect the price of the wine.

Geology of Teroldego Rotaliano Terroir

Campo Rotaliano is part of the Adige River Valley that runs N-S from Bolzano to Trento, deeply incised in the Permo-Triassic succession of the Southern Alps, with evidence of thrust tectonics (see Fig. 4, after Prosser and Selli, 1991). The U-shaped Adige River valley illustrates the important glacial erosion that took place during the ice ages. This major alpine valley is far more depressed than others south of the Alps, including Val d'Aosta (elevation around 500 m above sea level, a.s.l., at Aosta) and Valtellina (around 300-400 m a.s.l. in the wine-producing area): indeed, the elevation of the Adige Valley is only ~200 m a.s.l. at San Michele all'Adige and this accounts for the unusual local microclimate that is highly favourable for growing grapes, apples and pears. The entire flat bottom of the valley is covered by highly specialized cultures, not just the best exposed sunny slopes as is the case for the other major valleys. A graben, bounded by normal faults with offsets of up to one km, is present near Trento, where fossiliferous calcarenites of Eocene age at Doss di Trento are in contact with early Jurassic Calcari Grigi and/or with Triassic dolomites. The alluvial deposits are very thick, and the actual depth of the bedrock is unknown (Fig. 5).

The Campo Rotaliano DOC itself, however, is largely developed on the cone-shaped post-glacial Noce River alluvial fan which is made up of coarse-grained, calcareous alluvial sediments, and has very gentle slopes from 248 m a.s.l. at the apex of the fan to 205-206 m a.s.l. at the southern limit of the DOC. The Adige alluvial plain, in contrast, consists of fine-grained sediments: silts and mud with minor sand and gravel. Fuganti and Defrancesco (1992) investigated 11 eight-metre long drill sites specifically designed to study the lithologic composition of the sediments of the Noce River alluvial fan, including grain-size analysis, mineralogical and chemical composition of selected sam-

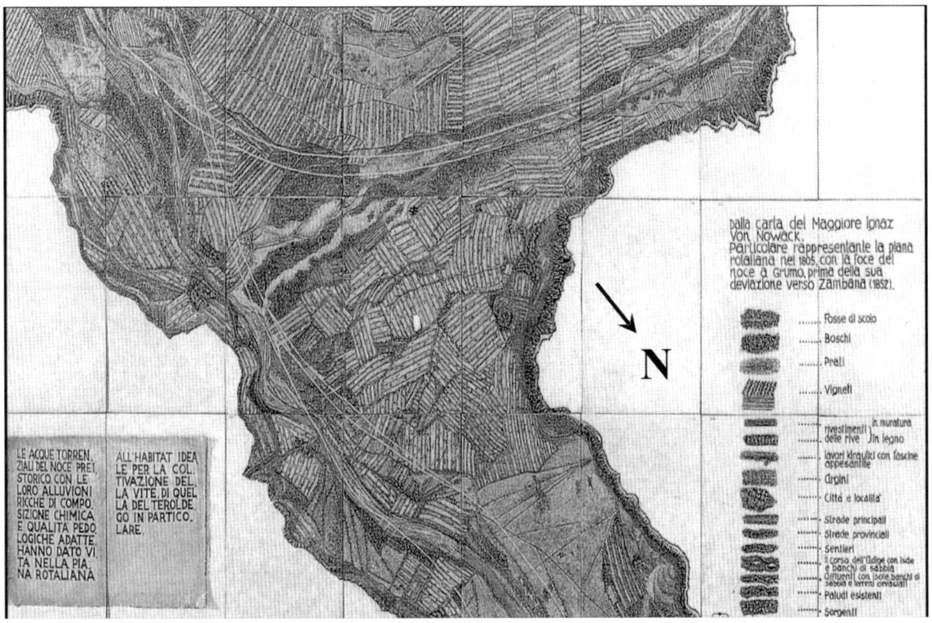

Figure 3 Ancient map of the Campo Rotaliano made by Baron Ignaz von Nowack in 1805 (after Facchini and Falcetti, 1991) showing the various orientations of the large double pergolas, practically identical with those of today, some two centuries later. Note that North is downward to the right.

Table 1: Record of the annual production of Teroldego Rotaliano wine, 1971-2001. (DOC official recognition took place in 1971).

Year	Wine produced (hl)
2001	46,038
2000	43,897
1999	35,766
1998	42,006
1997	35,448
1996	37,807
1995	31,277
1994	33,232
1993	32,923
1992	31,476
1991	27,614
1990	34,515
1989	27,229
1988	33,746
1987	20,599
1986	21,348
1985	25,463
1984	17,892
1983	20,444
1982	20,063
1981	16,222
1980	22,543
1979	21,823
1978	22,686
1977	17,076
1976	13,657
1975	9,372
1974	14,324
1973	15,857
1972	16,246
1971	16,045

From: C.C.I.A.A. Trento

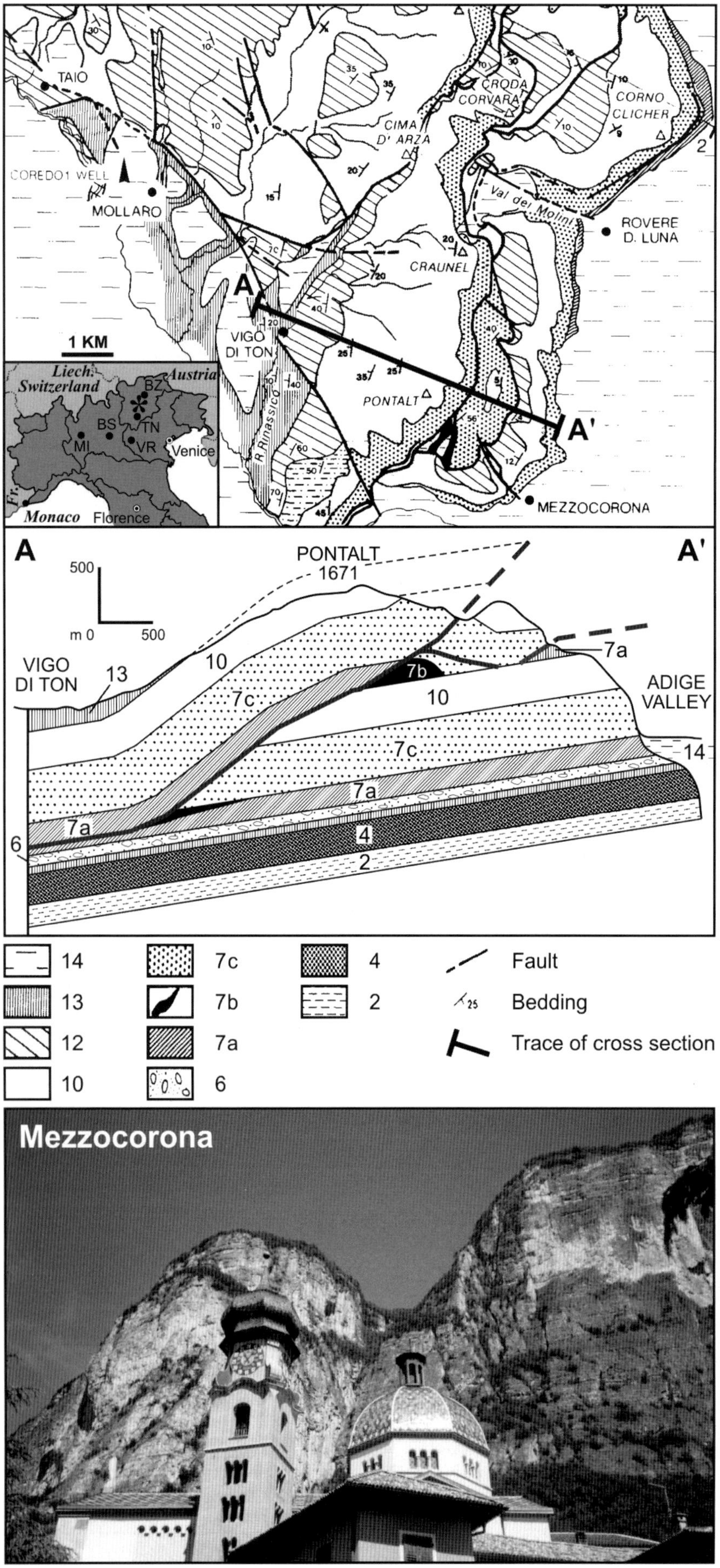

ples, and analysis of water samples. Tables 2 through 4 report their data relative to sites S11, Mezzolombardo; S1, Mezzocorona; and S7, Grumo, whose locations are indicated in Figure 1. The water table was reconstructed from existing wells and shows very clearly a progressive shallowing from 22 m below the ground level near the apex of the fan to 2 m in its distal, downslope part, at the limit of the DOC area.

On visiting this area in the summer of 2003 we observed a large excavation near the crossing of the railway in locality Ai Pini that shows very clearly the coarse, lenticular nature of the gravelly sediments that constitute the alluvial fan (Fig. 6). The bottom of the excavation, at approximately 4 m beneath ground level, was wet, in agreement with the map published by Fuganti and Defrancesco in 1992. It should be point-

Figure 4 Geological sketch map and cross-section of the western side of the Adige River Valley northwest of Mezzocorona (after Prosser and Selli, 1991). Location of cross-section A-A' shown by solid black line near base of geological sketch map; cross-section faces north from Vigo di Ton (A) to the Adige River Valley (A'). Lower part of figure shows a photograph of the church of Mezzocorona looking WNW towards the white calcareous cliffs of Pontalt (shown on cross-section A-A') in the background. Map and cross-section units from the base up are: 2 – Val Gardena sandstone Fm (Formation) (Upper Permian); and the overlying units (4-10, all Triassic): 4 – Werfen Fm (Scythian); 6 – Richthofen Conglomerate (Upper Anisian); 7a – Contrin Fm (Upper Anisian); 7b – Zwischenbildungen marls, limestones (Upper Anisian – Lower Ladinian); 7c – Sciliar Dolomite (Ladinian); 10 – Dolomia Principale (Norian – Rhaetian); 12 – Rosso Ammonitico, Biancone, Scaglia Rossa Fms (Middle Jurassic – Upper Cretaceous); 13 – Scalia Grigia (Paleocene-Eocene); 14 – Quaternary units. City and town symbols on the inset map (lower left corner of geological sketch map, top of figure) are: MI, Milano; BS, Brescia; VR, Verona; TN, Trento; BZ, Bolzano. Large asterisk on the inset map shows approximate location of Teroldego Rotaliano DOC.

ed out that the attitude of the isopieze reflects the old course of the Noce River, predating the artificial relocation of the river that took place in 1850. This illustrates the action of man as a geologic agent in a region with such a long history as Italy.

The catchment basin of the Noce River is largely incised in carbonate rocks of Mesozoic age (limestones and dolostones; Fig. 5) and thus the alluvial sediments are dominated by carbonates, with minor contributions of crystalline rocks from the Hercynian basement. In contrast, the Adige River sediments are poorer in carbonates and contain a large quantity of porphyries and related pyroclastic rocks of Permian age. According to some authors, the Noce River was captured by the deeply incised Adige River valley in post-glacial times, and this is why the Val di Non is not located above the level of the Adige, as is the case at Valsugana and Val di Cembra. The soils are poorly developed and the two main soil horizons have a combined thickness of < 150 cm.

Climate

The climatic conditions of the Campo Rotaliano are very well known, because at the Istituto Agrario of San Michele all'Adige (Fig. 1; one of the three municipalities of the DOC) founded in 1887 by the kings of Austria, there is the well equipped Centro Agrometeorologico Provinciale, where all the pertinent parameters have been recorded for more than a century. Figure 7 shows the annual variation of mean rainfall from 1959 to 1996: the strong variability shown is typical of the Mediterranean climate. Other climatic data are presented in a later section comparing the two terroirs.

Teroldego Characteristics and Production

Teroldego Rotaliano, as noted above, is strictly a monogrape (varietal), but according to the "Disciplinare di produzione" updated to 2003, a maximum of 6.5 % grape juice from outside the DOC area is permitted. There are three types of Teroldego Rotaliano: Rubino (red), Kretzer (rosé), and Superior. The maximum production of the grape is 17,000 kilograms per hectare of vineyard in specialized culture. This is indeed a very high production, rarely obtained in other DOC areas, and derives from the peculiar, traditional type of training (trellising) called pergola doppia trentina, pergola a doppio sbraccio, or tendone rotaliano (see Fig. 8). Pergola is a term to describe an elevated system of training of grape vines, usually up posts and over wire to form arbours (Fig. 8). The distance between the rows is in excess of 5 m, and the inclination of the wings allows a good insolation (exposure to sunlight) and ventilation of the grapes. The soil is permanently covered by grass. The orientation of the large pergolas is variable, from N-S to E-W, to NE-SW, as shown in Figures 2B and 3.

Pressing of grapes cannot exceed more wine than 70% of the initial grape weight. The potential alcohol as deduced from the initial grape sugar has to be no less than 10.5° for Rubino and for Kretzer and not less than 11.5° for Superior. Teroldego is not a rapidly maturing wine. When bottled at least one year after vintage, in bottles no more than 750 cc with cork lid, the wine has to meet the following requirements: Rubino, alcohol 11.5°, acidity total 4.5 per mil, dry extract 21 per mil; Kretzer, alcohol 11.5°, acidity total 4.5 per mil, dry extract 19 per mil; Superior, alcohol 12°, acidity total 4.5 per mil, dry extract 22 per mil. The Superior type may be called Riserva (reserve) after aging at least two years. Any addition of alcohol is forbidden. Watering of the vineyards also is forbidden, except for emergencies

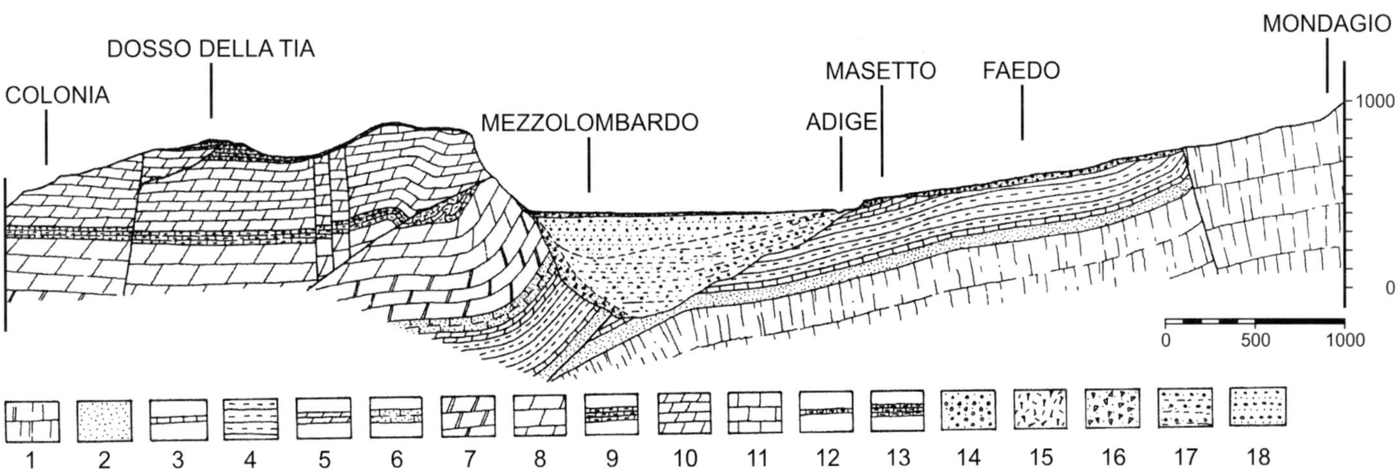

Figure 5 Geological cross-section of the Mezzolombardo - S. Michele all'Adige area facing northeast, from Colonia in the northwest to Mondagio in the southeast; approximate location shown in Figure 1. Lithological units of the cross-section from 1 – 18 are: **Permian:** 1 – volcanic pyroclastic flows (ignimbrites); 2 – Val Gardena sandstones; 3 – Bellerophon Fm (dolomite and gypsum). **Triassic:** 4 – Werfen Fm (clastics); 5 – Serla inferiore (dolomite); 6 – Gruppo di Braies (marls and limestones); 7 – Contrin Fm (limestones); 8 – Sciliar Fm (dolomites); 9 – Gruppo di Raibl (marls); 10 – Dolomia Principale (dolomites). **Jurassic:** 11 – Calcari Grigi (limestones); 12 – Ammonitico Rosso (limestones). **Cretaceous:** 13 – Scaglia Rossa. **Quaternary and Holocene** units of the Adige River Valley, centre of cross-section: 14 – alluvial fan deposits; 15 – glacial deposits; 16 – slope deposits; 17 – alluvial and fluvioglacial deposits of the last glacial maximum; 18 – alluvial and fluvio-glacial deposits post-dating last glacial maximum, which are the location of most wine grape growing of the Teroldego Rotaliano (see text). Cross-section unpublished, provided courtesy of Marco Avanzini, February, 2004.

Table 2: Grain size distribution and main mineralogical composition of an 8 metre long core taken from the Noce River alluvial fan at Site S-11 near Mezzolombardo, 220 m a.s.l. Location shown in Fig. 1.

cm	Mud %	Sand %	Gravel %	Carbonates %	Hornblende %	Biotite %	Garnet %
0 - 150	soil						
150 - 200	26	22	52	54.48	13.10	6.21	14.48
200 - 250	13	31	56	62.68	4.23	4.23	21.83
250 - 300	13	33	54	71.28	5.32	4.26	7.45
300 - 400	8	31	61				
400 - 500	6	28	66				
500 - 600	10	27	63	70.06	5.10	3.17	12.10
600 - 700	12	36	52				
700 - 800	7	28	65				

Data after Fuganti e Defrancesco (1992)

Table 3: Grain-size distribution and main mineralogical composition of an 8 metre long core taken from the Noce River alluvial fan at site S-1 near Mezzocorona, 210 m a.s.l. Location shown in Fig. 1.

cm	Mud %	Sand %	Gravel %	Carbonates %	Hornblende %	Biotite %	Garnet %
0 - 81	soil with roots						
81 - 161	80	20		63.91	6.77	8.27	7.52
161 - 210	46	22	32				
210 - 230	24	27	49				
230 - 250	11	28	61				
250 - 320	17	25	58				
320 - 450	18	27	58	80.89	3.66	6.91	5.28
450 - 490	11	19	70				
515 - 620	12	22	66	60.00	5.00	5.71	15.01
620 - 640	11	35	37	57.41	14.18	1.94	13.54
640 - 700	11	36	53				
700 - 800	17	28	55	56.64	7.96	7.09	15.93

Table 4: Grain-size distribution and main mineralogical composition of an 8 metre long core taken from the Noce River alluvial fan at site S-7 near Grumo, 206 m a.s.l. Location shown in Fig. 1.

cm	Mud %	Sand %	Gravel %	Carbonates %	Hornblende %	Biotite %	Garnet %
0 - 130	soil						
130 - 300	12	18	70	72.83	4.05	2.31	14.45
300 - 400	12	27	61	68.45	6.92	3.08	12.31
400 - 500	7	28	60	60.46	6.97	4.65	16.28
500 - 600	6	27	67				
650 - 720	5	19	76	72.72	3.79	6.82	11.36
720 - 770	11	50	39	77.31	3.61		9.27
770 - 800	5	28	67				

Mineralogical composition derived from analytical data

(young plantings or re-plantings, or hot and dry seasons such as the summer of 2003). Teroldego was largely used in the past to reinforce weaker wines, owing to its high alcohol content, dark hue, and tannic nature. In the last few years it has been successfully used to produce "vino novello", similar to Beaujolais Nouveau, that has been well received by the market.

The producers of Teroldego Rotaliano DOC number a few dozen, including private companies, noble families and large cooperative consortia or social cellars (Table 5). It is worthwhile mentioning here that the concept of cooperative efforts to improve the quality of the wine and to make grape growing a successful business originated in this part of Italy in the early days of the twentieth century, after the phylloxera plague, at a time when the local economy was particularly depressed. The first social cellar was created in the lower Val di Non, upstream from Campo Rotaliano in the village of Revò, where nowadays there is no vineyard left, only very successful, highly specialized agricultural activity, including the production of "Melinda" apples. The ratio of social cellars to private producers is anomalously high in the province of Trento: some 60 – 70 % of producers are social cellars. For additional information, see Bona et al. (2003), Colugani and Crespan (2000), Roncador and Falcetti (1991). In conclusion, tradition combined with innovative techniques, well controlled monitoring, and quality production in an unusual geological setting that benefits from very favourable climatic conditions, make Teroldego Rotaliano a unique terroir in the heart of the Southern Alps of northern Italy.

Figure 6 Temporary excavation at the center of the Campo Rotaliano (loc. Ai Pini) showing the coarse, laterally variable grain size typical of the alluvial fan of the Noce River. The photo was taken in the very hot and dry summer of 2003: the presence of water at the bottom of the excavation confirms the interpretation provided by Fuganti and Defrancesco (1992) that the water table is shallow, and follows the old course of the Noce River.

AGLIANICO DEL VULTURE DOC – SOUTHERN APPENNINES

History

This southern Apennines DOC is located in the region of Basilicata in Southern Italy, approximately due east of Naples and about halfway between the cities of Foggia to the north and Potenza to the south (inset map, Fig. 9). The Aglianico vine was imported into Southern Italy by the Greeks around the 6th-7th Century B.C. The origin of the name "Aglianico" is controversial. The name derives from Hellenica, modified in the 15th Century, during the domination of Aragon in the Kingdom of Naples, when the double "l" of the Hellenic took on the Spanish phonetic substitute of "gli". Ancient historical testimony concerning the cultivation of Aglianico in the northern part of the region of Basilicata comes from the 4th Century B.C., at the beginning of the Roman era, when the mythic wine "falernum" was created and bronze coins of the agricultural God Dionysius were minted in the city of Venosa (Fig. 9). In the following centuries the cult of Bacchus, the wine god, spread in the area. Venosa is the native land of the great Latin poet Quinto Orazio Flacco (Horace) who wrote about the beauty of his home and of its precious nectars. In the Medieval period, great attention was paid towards the cultivation of good wine by Frederick II of Swabia who visited these lands regularly. The land was inhabited primarily by shepherds and farmers, who used wine as an integral part of their diet to overcome difficulties of sustenance. A victim of frequent earthquakes and natural disasters (especially in the years 1456, 1857 and 1930), the Basilicata region suffered long periods of abandonment and decay. After Italy became unified as one country in 1870, conditions of poverty facing the

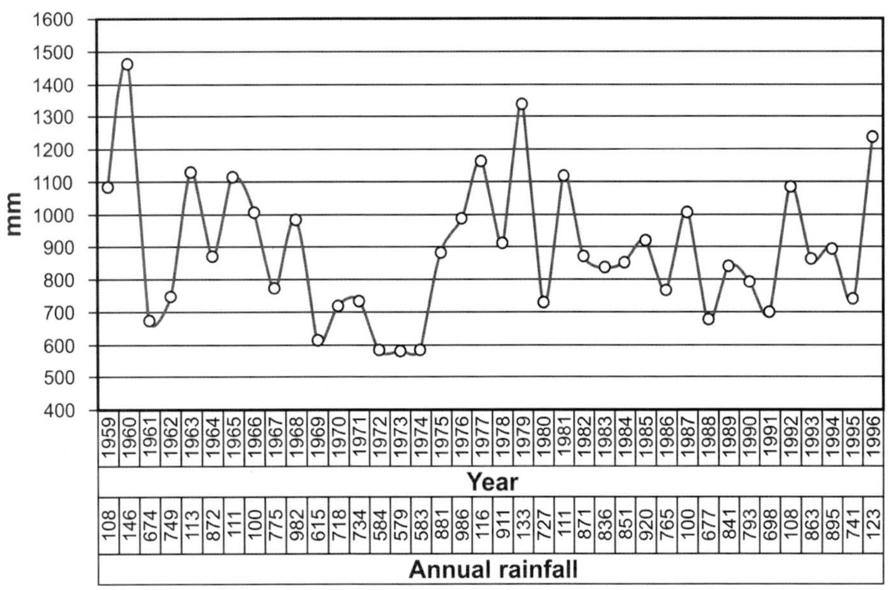

Figure 7 Mean annual rainfall recorded at the S. Michele all'Adige observation station (Fig. 1) from 1959 (left) to 1996 (right). The strong variability, from less than 600 mm to almost 1500 mm of annual rainfall, is typical of the Mediterranean climate.

population of Basilicata led to a mass emigration to the Americas and Australia, and it was only with the land reform of 1950 that the poor were given a small portion of the large agricultural lands previously owned by the rich. Presently, the farms are small and are characterized by mixed cultivation of vines and olive trees.

Origin of Aglianico del Vulture DOC

The year 1971 is a historical date for the study of wine in Basilicata. In that year a decree was published by the President of the Republic: "Denomination of the controlled origin of Aglianico wine of Vulture – Discipline and Production" that defined, in detail, the procedures for the production of Aglianico wine in Vulture. Today, the Aglianico vine produces a particular type of grape that can be used to make a vigorous and perfumed wine, which derives its characteristics from the natural volcanic soil of Monte Vulture. Until 2003 it was the only DOC wine from the Basilicata region. In May 2003, however, the DOC status was also assigned to "Terre dell'Alta Val d'Agri" wine produced in the municipalities of Viggiano and Grumento Nova e Moliterno, in southern Basilicata.

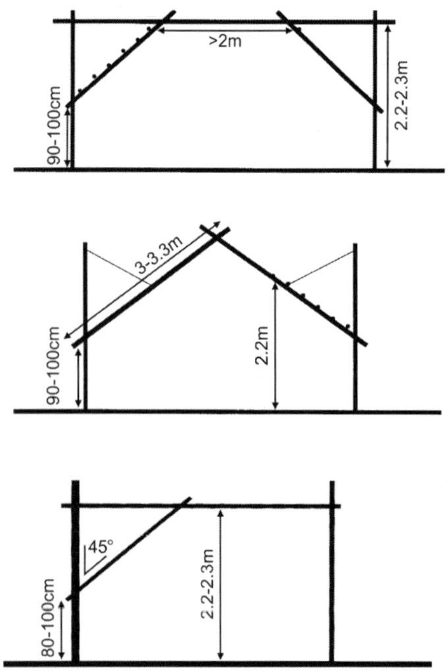

Figure 8 Scheme of the large pergolas typical of the topographically flat Campo Rotaliano. From top to bottom: open double pergolas; closed double pergolas; single pergolas; dimensions shown in metres.

Geology of Aglianico del Vulture Terroir

Monte Vulture (= Mount Vulture) is one of the volcanic complexes of the Quaternary potassic alkaline volcanism in Italy, the only one located on the eastern border of the Southern Apennines at the inner margin of the Bradanic foredeep (Fig. 9; De Fino *et al.*, 1986). Its volcanism is linked to a phase of Pliocene-aged extensional tectonics which followed the compressive tectonics responsible for the formation of the Apennine Orogen. This volcanic complex consists mainly of a composite volcano; its products represent the result of explosive and episodic effusive activity, occurring from the Middle Pleistocene (~730 ky B.P.) until Late Pleistocene (~132 ky B.P.). Monte Vulture volcanic rocks consist of a succession of pyroclastic flows, fall and surge deposits and, less commonly, lava flows. The chemical composition of the deposits is typical of potassic volcanism, ranging from tephrites to foidites (Caggianelli *et al.*, 1990). Volcanic activity ended with hydromagmatic explosions which formed two explosion craters, now occupied by two crater lakes, the large and small lakes of Monticchio. The vol-

Table 5: Producers of Teroldego Rotaliano wine (after Andreotti, 1999)

Cantina Sociale Roverè del Luna. Founded in 1919 with 14 original partners, it presently includes 260 farmers, with an annual production of 40,000 corked bottles, plus 100,000 litre bottles, including Teroldego Rotaliano.

Azienda Vinicola Gayerhof, Roverè della Luna. Founded in 1976, it includes some 200 farmers, and has an average yearly production of 500,000 bottles, including Teroldego Rotaliano, Novello di Teroldego, and Teroldego Rotaliano superiore.

Azienda Agricola Luigino Botta, Mezzocorona. Founded in 1974, this is a small family winery that produces some 8000 bottles a year; these are considered one of the best examples of Teroldego wine.

Cantine Mezzacorona SCARL. Founded in 1904, this group has changed significantly over the years. At present it represents 1200 farmers, owners of 2000 ha of vineyards (29% of Trento Province) with a production of 12 million bottles/year, including Teroldego and Teroldego Riserva.

Azienda Agricola Marco Donati, Mezzocorona. Founded in 1863, it has an annual production of 80,000 bottles, including a famous SANGUE DI DRAGO Teroldego Rotaliano.

Azienda Agricola Elio Endrizzi e F.lli, Mezzocorona. Founded in 1945, produces annually some 20,000 bottles, including Teroldego Rotaliano.

Cantina Rotaliana di Mezzolombardo SCARL. Founded in 1931. Presently 263 farmers supply to the social cellar 35,000 litres of selected grapes to produce 930,000 bottles yearly, mainly Teroldego Rotaliano classico and Riserva.

Azienda Agricola Foradori, Mezzolombardo. Founded in 1930 as a private company, this is one of the best producers from 15 ha of vineyards. The best known is Teroldego Foradori and Granato, with an annual production of 200,000 bottles (90% Teroldego).

Azienda Agricola Barone de Cles, Mezzolombardo. This private company originated from the fusion of two noble families that owned feuds (= lands) in the lower part of Val di Non. Annual production is 100,000 bottles, mostly Teroldego Rotaliano (Maso Scari) and Primizia clesiana (Novello di Teroldego).

Villa de Varda srl, Mezzolombardo. From its beginning in the XVII century, the de Varda family has produced remarkable wines from their vineyards. The present production is approximately 80,000 bottles, including Teroldego Rotaliano.

Casa Tait, Mezzolombardo. Founded in 1995, this private company produces some 30,000 bottles from two of the best crues of the Piana Rotaliana.

Azienda Agricola Luigi Zanini, Mezzolombardo. Founded in 1968, this private company produces some 20,000 bottles annually from their own vineyards, including Teroldego Rotaliano.

Azienda Agricola Cipriano Fedrizzi, Mezzolombardo. Founded in 1870, this small private company produces some 15,000 bottles a year (mostly Teroldego Rotaliano) from their own vineyards, called "Due Vigneti".

canic deposits are spread over an area of ~184 sq km, overlying a sedimentary basement represented mainly by clayey terrains, limestones, and sandstones of pre-Pliocene units at structural highs and by lower-middle Pliocene conglomerate, sand and clay sediments which fill the structural lows as seen in the cross-section of Figure 9 (La Volpe *et al.*, 1984; Ricchetti and Pieri, 1999).

Monte Vulture consists of a composite central cone of altitude 1326 m with some secondary eruptive centers mostly located in the peripheral areas. The summit of the central volcano is complex, showing a lobate, sickle-shaped rim whose western and southwestern parts were destroyed either by erosion or by later caldera collapses. The morphology of the area shows a volcanic edifice which is surrounded by numerous rivers (Ofanto River, Atella Creek, Arcidiaconata Creek, Melfi Creek) characterized by the occurrence of alluvial terraces bordered by sharp scarps of fluvial erosion and by smooth hills 400-700 m high, and represented by sedimentary terrains.

Monte Vulture is surrounded by chestnut, ash and hornbeam trees, that thrive in the loose and deep volcanic soils. At the lowest altitude, starting from ~700 meters a.s.l., the soils on the sides of the volcano and the nearby sedimentary hills to the east are used for food crops (Fig. 10A). Higher elevations include small plots of cultivated olive trees and vineyards, most of which are plantings of the Aglianico varietal.

Most Aglianico del Vulture vineyard soils are rich in nutrients because the pyroclastic substratum, in the form of residual lenses from 1-2 metres to 8-10 metres deep, easily releases minerals into the upper soil levels (Lulli and Bidini, 1978). These brown acidic soils are well drained andosoils. Pyroclastic soils occur in the Venosa, Maschito, Palazzo S.G., Genzano di L, and Banzi vineyards owing to the transportation of pyroclastic material from Monte Vulture via surficial runoff to the ancient lakes of Venosa, Melfi and Atella, which are presently infilled and covered by alluvial sediments.

Climate

The area has a typical Mediterranean climate with winter rainfall and summer drought (41 mm average rainfall). For

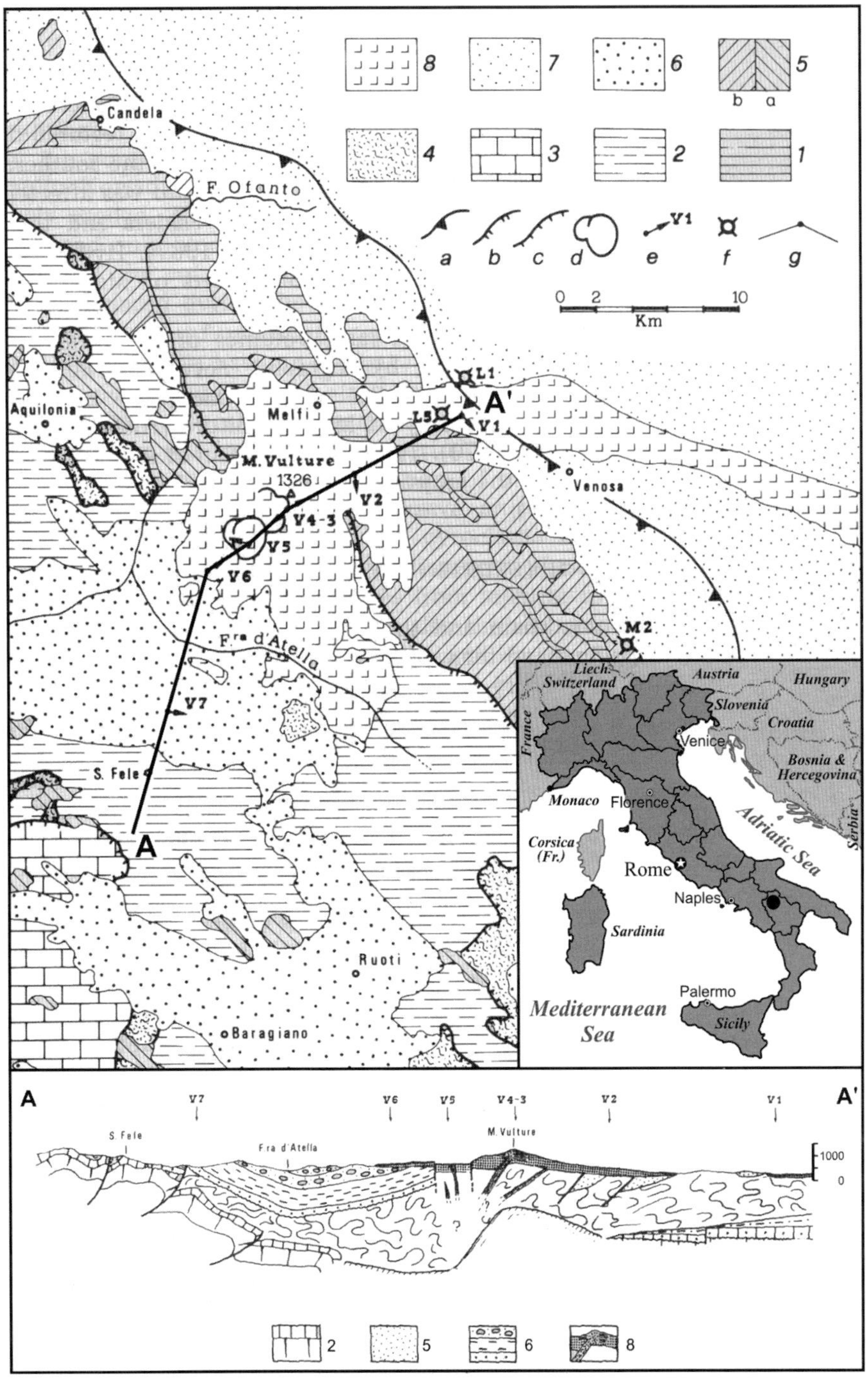

Figure 9 Schematic geological map (top of figure) and cross-section A-A' (bottom of figure). Inset map at lower right side of geological map shows the 20 regions of Italy, and indicates the location of this DOC (solid circle) in the region of Basilicata. Geological sketch map units are: **Oligocene-Miocene:** 1 - Sannitic flysch units; **Triassic:** 2 – Lagonegro cherty limestones and radiolarite units; **Cretaceous (mainly):** 3 - Campania-Lucania carbonate units; **Cretaceous-Paleocene:** 4 – Sicilide units, clastic; **Miocene:** 5 – Irpinian flysch units; **Plio-Pleistocene:** 6, 7 – Ariano units and fore-trough terrains, consisting of sands, clays and conglomerates; **Quaternary:** 8 – Monte Vulture volcanic rocks, tephrites and pyroclastic flow deposits. Cross-section A-A' faces NW, from S. Fele (south) through Monte Vulture volcano and environs to the north (after La Volpe et al, 1984). Major units shown in cross-section are: 2 – Lagonegro units; 5 – Irpinian units; 6 – Ariano units; and 8 – Monte Vulture volcanics. In cross-section A-A' the unidentified unit with curved line symbols below Monte Vulture represents undivided Sannitic and Lagonegro rocks with high electrical conductivity.

Figure 10 Aglianico del Vulture terroir vineyards and soils. (A) Panoramic view of Monte Vulture from the south. (B) New plantings with short, regular rows, with ground kept clear of grass. (C) Aglianico vineyards, with Monte Vulture in the background. (D) Typical Alberello type (=goblet-trained) vine. (E) Soil profiles overlying Pliocene conglomerates of the Bradanic foredeep. (F) allochthonous flysch deposits. (G) Monte Vulture volcaniclastic tephra. (H) Old, small cave-cellars dug in the tephra at Barile. (I, J) modern, large cellars in volcanic tephra of Monte Vulture.

Table 6: Basic climatic data from MELFI station

MELFI (PZ) 1960-1998 climate data. Latitude 40°59' N; Longitude 15° 37' W., Elevation = 531 m. (Servizio Idrografico Ministero Lavori Pubblici)													
	Jan	Feb	Mar	Apr	May	Jun	Jul	Aug	Sep	Oct	Nov	Dec	Annual
Mean max Temp. °C	8.8	9.6	12.2	15.8	21.2	25.7	28.9	28.9	24.6	19.1	14.0	10.1	18.2
Mean avg temp °C	5.8	6.1	8.3	11.4	16.1	20.1	23.0	23.0	19.4	14.7	10.2	6.9	13.7
Mean min temp. °C	2.7	2.6	4.3	6.9	11.0	14.5	17.0	17.1	14.2	10.3	6.5	3.8	9.3
Precipitation (mm)	80	80	76	76	56	45	39	39	54	86	93	86	811

the observational period of 1960-1998 rainfall averaged 811 mm with 89 mm in October, November and December and 41 mm in June, July, and August (Table 6). The summer rainfall is concentrated in storms of short duration. The average annual temperature is 13.7°C, averaging 6.8°C in winter (December-March) and 23°C in summer (June-September). Strong evapo-transpiration results in dry soil for most of the year with a hydraulic deficit in the months of May-September.

Table 7: Ratio between vineyard area and total area in the 15 municipalities of the Aglianico del Vulture DOC. See text for discussion.

COMUNE	SUPERFICIE investita a vigneto (ha)	SUPERFICIE amministrativa (ha)	Percentuale sup. Vitata a vino DOC
BARILE	189.63	2464	7.70%
RIONERO IN V.	34.09	5319	0.64%
RIPACANDIDA	117.96	3222	3.66%
GINESTRA	42.63	1321	3.23%
RAPOLLA	81.53	2905	2.81%
VENOSA	917.78	16934	5.42%
MASCHITO	283.85	4549	6.24%
LAVELLO	113.10	13292	0.85%
MELFI	60.44	7713	0.78%
ATELLA	30.00	11560	0.26%
ACERENZA	104.77	20515	0.51%
FORENZA	85.64	8828	0.97%
GENZANO IN L.	34.48	20704	0.17%
BANZI	21.34	8235	0.26%
PALAZZO S. G.	18.38	6226	0.30%
TOTALE superfici nei comuni della Doc Aglianico del Vulture	2135.62	133787	33.8%

Aglianico del Vulture Characteristics and Production

Aglianico vines are widespread in the regions of Basilicata and neighbouring Campania to the west, but when grown on the Monte Vulture soil, the Aglianico grape produces a particular type of delicate and perfumed wine that led to the area being relabeled as a DOC in 1971, and named "Aglianico del Vulture DOC". The production zone of the Aglianico del Vulture DOC includes the municipalities of Barile, Rionero in Vulture, Ripacandida, Ginestra, Rapolla, Venosa, Maschito, Lavello, Melfi, Atella, Acerenza, Forenza, Genzano in Lucania, Banzi, and Palazzo San Gervasio, of the province of Potenza in the northwestern part of Basilicata. The Aglianico del Vulture DOC area has definite altitudinal limits of 200 to 700 meters above sea level. In other words, the cultivating land for Aglianico is limited to two areas: that restricted to the horizontal border of the volcanic edifice covering the territory of Barile, Rapolla, Rionero in Vulture, Ginestra and Ripacandida; and an external area where river and lake sediments are abundant, in the territories of Venosa and Maschito. The rest of the DOC territory is covered by forest, grazing pastures, cornfields and olive groves. The lands dedicated to the cultivation of vines total 2,136 hectares, whereas the extent of vines recorded in the DOC list of the Chamber of Commerce in Potenza in 2002 was only 722 hectares (Table 7).

According to the "discipline of production" (1971) for the Aglianico del Vulture DOC, the maximum weight of grapes produced from one hectare of land should not be more than 10,000 kilograms; no more than 70% of the weight of the grapes harvested can go into the wines. Wine making, including the obligatory ageing process of one year, should be completed in the region in which the grapes were grown and harvested. The grapes have to reach an alcoholic strength of no less than 11.5 degrees, with a minimum acidity of 5 per mil and dry residual of 22 per mil. The colour is a ruby red, with orange scintillation after ageing. The smell or "nose" is vinous with a delicate perfume that improves with ageing. There are three types of Aglianico del Vulture wine: Vecchio (Old), Riserva (Reserve) and Spumante (Sparkling). The qualifications for a Vecchio wine are a minimum of 12.5% grade alcohol and at least three years of ageing, two of which must be done in wooden barrels. If the period of ageing is five years, two of which are completed in wooden barrels, the wine is qualified as Riserva. Following local tradition, the wine is

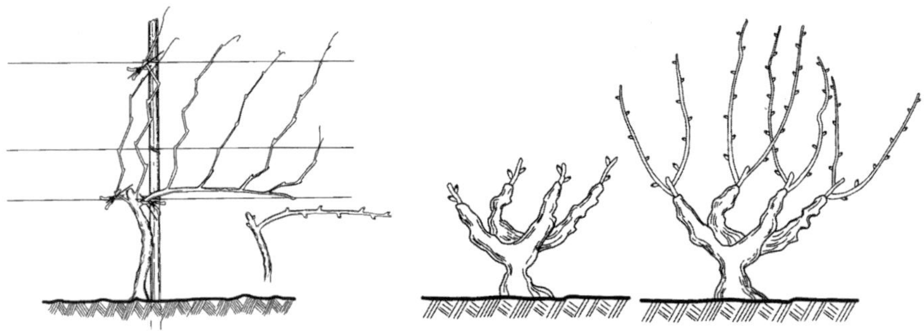

Figure 11 Schemes of the modern (Guyot type, at left) versus traditional (Alberello type, at right) training (trellising) methods used in the Aglianico del Vulture DOC; see text for discussion.

preserved in cave-like cellars dug into the volcanic rock where stable values of temperature and humidity conditions are ideal for ageing wine. Both old cellars (see Fig. 10H) and new cellars in volcanic rock (Fig. 10I, 10J) are located in the historic zone of Barile (Fig. 10H), where some cellars are up to 10 metres deep.

Most grapevines in the DOC are 25-30 years old, and consequently production is low with respect to the area occupied by the vineyards. Small farms still use cultural practices with short stumps held, when young, by three bamboo canes; this method of free standing, bush-style training (trellising) is termed "Alberello" and is in widespread use in Southern Italy (see Figs. 10D, 11). Alberello farmers make their own wine for family use or for local markets, or sell the grapes. Grape growing is slowly evolving to modern cultural practices, with long, regular rows pruned according to the Guyot system, a low style of training along wires, used in order to control the plant vigour in the volcanic, potash-rich soils of Monte Vulture and in the well structured soils of the Bradanic foredeep (Figs. 10B, 10C). Watering is forbidden by the DOC regulations, and mechanical cultural practices are limited to the use of mowers in the intra-row areas, that are kept clear in order to prevent water loss due to evapotranspiration. New farms are much larger than the old family-owned farms, where the direction of the vine rows is sometimes quite independent of the gradient of the slope, but parallels the boundary of the property. Wine production is strongly dependent on climate including rain, temperature, and even wind. Grape harvesting takes place in the second half of October.

Thanks to the national and international success of the Aglianico del Vulture DOC, grape growing and wine making represent a thriving sector of agriculture in the region of Basilicata. The DOC production in 2002 was 12,185 hl (data from Catasto Vinicolo Regionale), and is relatively low compared to the potential of the area. The number of producers is growing rapidly (Table 8) and includes 23 consortia plus private companies and three important wine-makers cooperative stores, those of Barile, Rionero in Vulture, and Venosa. The price of cork bottled wines (750 cc) ranges from 3 to 8 Euros, although some producers sell the best aged wines at as much as 30 Euros per bottle. Because Aglianico del Vulture consistently has been recognized by a number of wine connoisseurs as one of the great Italian red wines, the present orientation is towards the production of top quality wines. Initiatives aimed at promotion of the DOC include the "via del vino e dell'olio" that is a series of tourist-oriented routes that connect typical productive villages, and the creation of a "centro della civiltà dell'Aglianico" and an associated museum of local peasant culture. For additional information, see AA.VV. (1997), Calò and Costacurta (2003), Cassi et al. (1999).

Table 8: Producers of the Aglianico del Vulture DOC.

1	CASA MASCHITO
2	TERRE DEL VULTURE
3	VITIVINICOLA PELLEGRINO
4	TERRE DEGLI SVEVI
5	LA LUCE MICHELE
6	BASILIUM SCARL
7	AZIENDA AGRICOLA ALLEGRETTI
8	AZIENDA AGRICOLA BASILISCO
9,10	AZIENDA AGRICOLA ELENA FUCCI
11	AZIENDA AGRICOLA EUBEA FAM.SASSO
12	AZIENDA VINICOLA PATERNOSTER
13	CANTINA DI VENOSA SCARL
14	CANTINA GAMMONE
15	CANTINA SOCIALE DEL VULTURE COOP A.R.L.
16	CANTINE DEL NOTAIO di GERARDO GIURATRABOCCHETTI
17	CANTINE DI PALMA S.R.L.
18	CANTINE SASSO S.R.L.
19	CASA VINICOLA ARMANDO MARTINO
20	CASA VINICOLA D'ANGELO
21	CONSIGLIO DONATO MARIA
22	CONSORZIO VITICOLTORI ASSOCIATI DEL VULTURE SCARL
23	FRATELLI NAPOLITANO S.A.S.
24	IL MERCANTE DEL SUD
25	MERUM s.c.a.r.l.
26	TENUTA DEL PORTALE CA.VI.DA.
27	TENUTA LE QUERCIE SRL GRUPPO PIETRAFESA
28	TERRA DEIRE

DISCUSSION

The geology of the two terroirs, Teroldego Rotaliano and Aglianico del Vulture, is quite different. Coarse- to fine-grained, calcareous-dominated terrains characterized by low fertility, low content of organic matter and good to moderate permeability prevail in the Teroldego Rotaliano terroir, where the thickness of continental sediments is unknown, but likely approximates several hundred metres as demonstrated by water wells and seismic reflection profiles (G.A.Venzo, pers. comm., 1999; M. Avanzini, pers. comm., 2004). Glacial or post-glacial deposits play an essential role in the Teroldego Rotaliano terroir. In contrast, siliceous, potash-rich, fertile, well-drained terrains are typical of the Aglianico del Vulture terroir, with special reference to the lower slopes of the Monte Vulture volcano, where volcaniclastic sediments are subcropping, covered by a thin veneer of alluvium. In the DOC area outside the volcanic edifice of Monte Vulture the substrate consists either of conglomerates of the Apenninic foredeep, or of allochthonous flysch terrains (Fig. 10), partly covered by a thin veneer of volcanic ash. Unlike the Teroldego Rotaliano terroir, no glacial sediments are present.

The well controlled climatic data available for Teroldego Rotaliano and Aglianico del Vulture terroirs over time intervals of thirty years or more show that the thermal variation, both daily and seasonal, is much stronger in the heart of the Southern Alps than in the Apennines of Southern Italy (see Figs. 12a, 12b). The maximum summer temperature is higher in the northern site, notwithstanding a difference in latitude of almost 5°. This is accounted for by the very peculiar microclimate that results from the presence of white calcareous cliffs, several hundred metres high and with southern or southeastern exposure, that surround the Adige River Valley all around the Piana Rotaliana and concentrate solar heat (Fig. 2A). A comparison of the ratios between monthly temperature and calculated evapotranspiration (Fig. 13) shows that the moisture deficit area is much more pronounced and has a longer duration in the southern site, where the soil is more fertile but the production is very low. The availability of water is a limiting factor here (Fig. 13). Training methods are strongly differentiated in the two terroirs, and strongly affect the cultivated landscape. Large pergolas with a distance between the rows of more than five metres are typical of Campo Rotaliano, where all the poles (vertical, horizontal and oblique) are wooden (see Fig. 2A and 2C, and Fig. 8). The ground is permanently covered by grass that apparently does not compete with the vines. No zonation is noticed in this terroir, which is limited in area, and has no vertical relief: less than 50 metres from the apex of the Noce River alluvial fan at Rocchetta and the Adige River bank. The best wines are produced in the municipality of Mezzolombardo, on the southwestern part of the alluvial fan, where the grain size of the alluvial sediments is coarse, and the water table is relatively deep.

The training methods in the old, family-owned vineyards of the Aglianico del Vulture DOC are those of the Alberello method: free standing bush-style or goblets, in regular rows kept short by repeated pruning in modern plantings (Figs. 10B, 10C and 10D) and thus entirely different from the trellising methods of the Teroldego Rotaliano DOC. Grapevines are cut according to the Guyot (or Cordone speronato) method. Here the ground is kept mostly free of grass, to avoid competition with

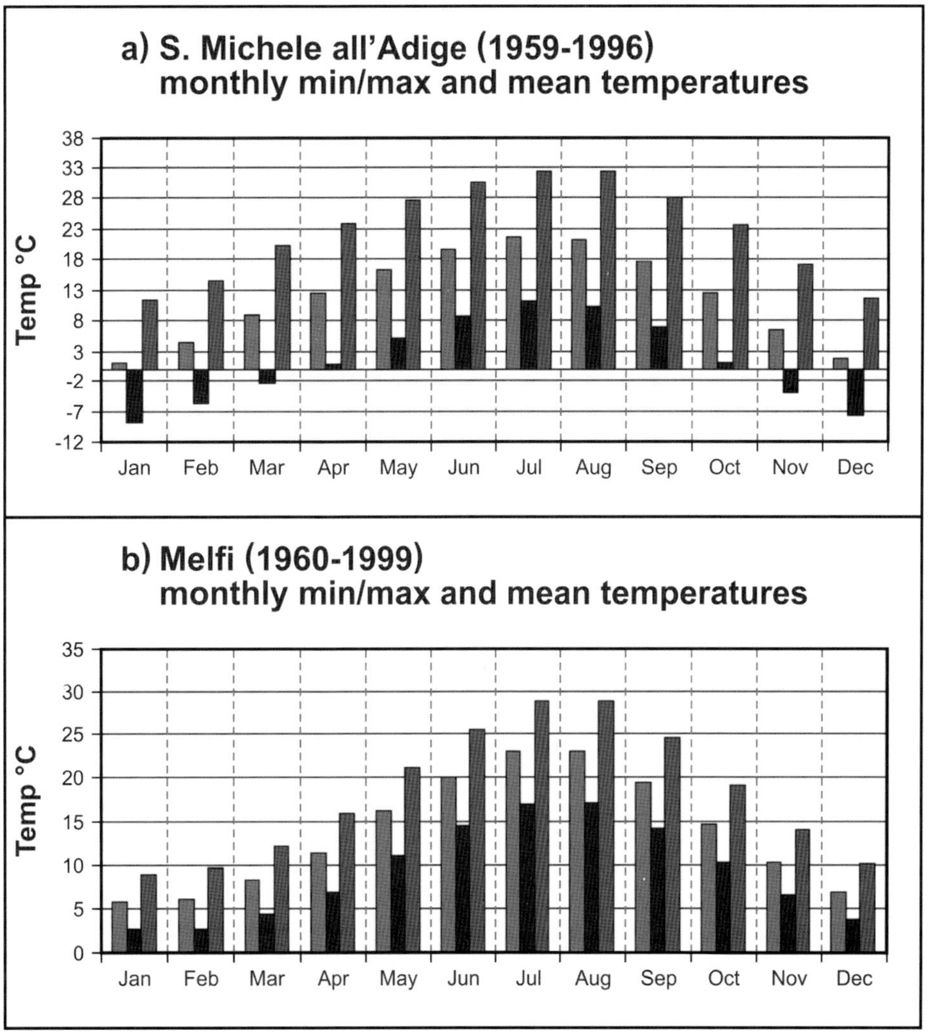

Figure 12 Comparison of temperature data from stations representative of the Teroldego Rotaliano DOC of the Southern Alps of Northern Italy (a, S. Michele all'Adige, 1959-1996, location shown on Fig. 1), and Aglianico del Vulture DOC of the Southern Appenines of Southern Italy (b, Melfi, 1960-1999, location shown on Fig. 9). For each month shown in a and b, left bar indicates average temperature, centre bar indicates minimum temperature, and right bar indicates maximum temperature. As can be seen, the thermal excursion, both daily and seasonal, is much larger in the northern (Alpine) location than in the southern (Appenines) location.

the vine in the long dry summer season. An altitudinal zonation is noticed on the flanks of the volcano (see Fig. 10A), where vineyards occupy the lower slopes, and olive groves and chestnut forests occupy the higher slopes.

Table 1 shows the historical record of wine produced in the Teroldego Rotaliano DOC since the official recognition of the DOC in 1971. Production is increasing, with some irregularities related to unfavourable climatic conditions, and this increase is accounted for by improvements in cultural methods or, in other words, documents the success of the DOC policy. The extremely high prices of the vineyards are related to the availability of land, and are independent of the price of the wine. For Aglianico del Vulture, the historical record of the DOC production is incomplete and difficult to interpret, since a large part of the DOC area is only partly covered by vineyards, and the data may be misleading. The soil is certainly fertile and the physical environment, quality of the vine and long tradition underlie the great success of this prize-winning red wine, considered the best of Southern Italy.

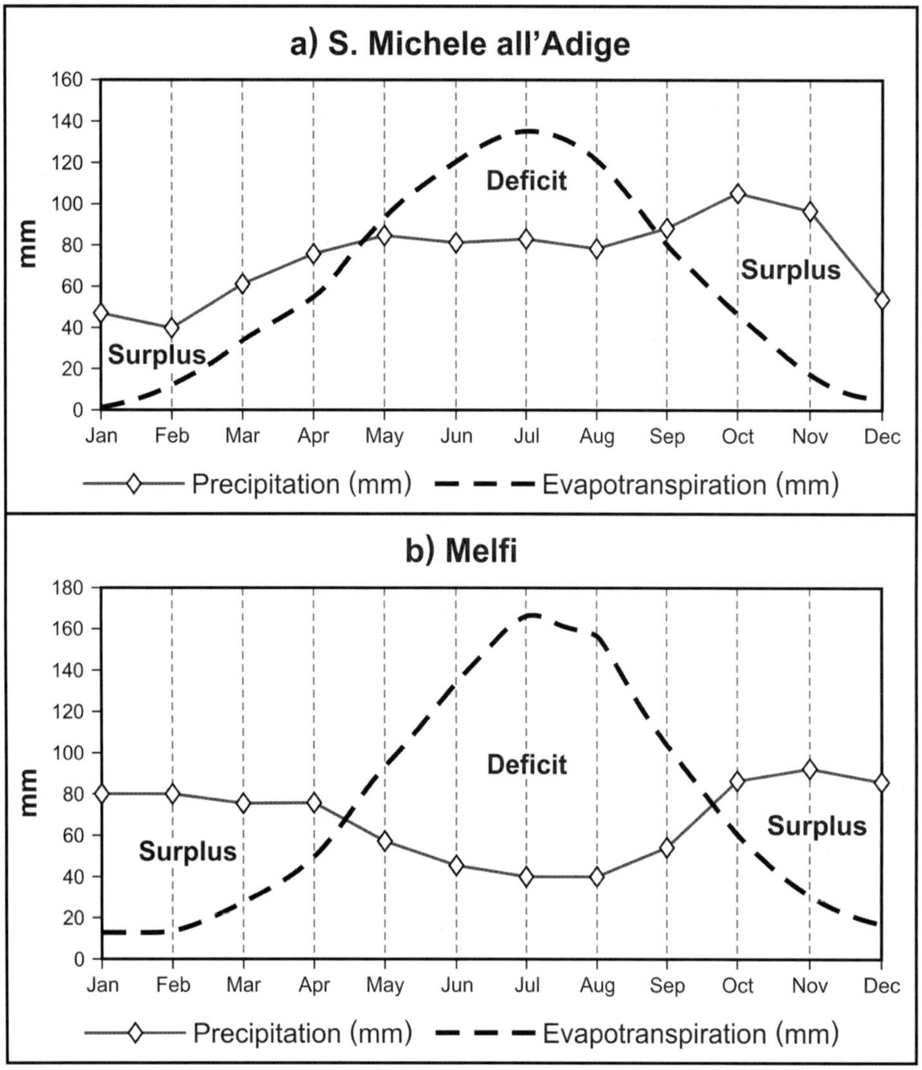

Figure 13 Comparison of the precipitation/calculated evapotranspiration diagrams obtained for long (>30 years) and comparable time periods at the observation stations of San Michele all'Adige, northern Italy (Fig. 1) and Melfi, southern Italy (Fig. 9; Table 6). The summer moisture deficit is much larger, and lasts longer, in the southern location.

CONCLUDING REMARKS

The success of Italian wines is not just the result of effective marketing. It derives from a combination of several factors, including latitude, climate, geomorphology, geological constitution and soils favourable to viticulture; and an extremely long tradition in grape growing and wine making that dates back to the prehistory of mankind. The dispersal of Vitis vinifera originated in the eastern Mediterranean region and its cultivation has been uninterrupted through millennia. Other factors include substantial improvements in viticultural methods and in wine making, resulting from a long campaign dominated by a few well known personalities that insisted on quality and qualification of Italian wines, and promulgation of a Law in 1967 for the recognition of certified appellations, the DOCs. Although some DOCs are poorly defined (too large, too small, too comprehensive, etc.), the existence of DOCs certainly contributes to high quality and increased visibility for Italian wines. From a few dozen in the early 1970s, DOCs now number about 233, and each year new areas apply for DOC recognition. More than 1500 varieties of grapevines are known in Italy, with a limited areal distribution. They are indigenous, or were imported in the distant past; and their quality is protected by a movement "per la salvaguardia e la tutela dei vitigni autoctoni italiani" (for the safeguard and the protection of autochthonous Italian grapevines) (Fregoni, 2003).

We have presented two examples of terroirs from Northern and Southern Italy, producing excellent, single-varietal "niche" wines. There are many more niche wines in Italy, however. Accordingly, the message is: be sure to consider more Italian wines than Chianti, Brunello or Barolo.

ACKNOWLEDGMENTS

Many people, including scientists, relatives, friends and wine makers deserve our gratitude for helping us to put together this paper. First of all we thank Symposium convener Larry Meinert for the invitation and encouragement, and volume co-editors Roger Macqueen and Larry Meinert for careful revisions. Our relatives Rosanna Caputo, and Pepe, Marco and Franco Sironi contributed their support in the field and in data collecting. Technical help was provided by Gabriele Pezzi and Barbara Aldighieri. Massimo Bertamini and Marco Avanzini were instrumental in putting together the Teroldego part of the paper, as was Cassandra Quave for the Aglianico del Vulture part of the paper. Cantine del

Notaio Gerardo Giuratrabocchetti, Tenuta Le Querce SRL gruppo Pietrafesa, Istituto Agrario Provinciale S. Michele all' Adige, and Cantina Sociale Mezzacorona provided photographs and access to rare publications not available in the open literature. We thank David Sargent of the Geological Survey of Canada, Calgary, who was instrumental in revision of many of the figures.

REFERENCES

Andreotti, A.,1999, Vino Trentino, Cantine, Tradizione, Benessere, Artimedia: Trento, 270 p.

Benciolini, L., 2004, Geology and Wine in the Isarco/Eisacktal Valley, Italian Eastern Alps: 32nd International Geological Congress, Florence, Italy, Abstracts, Topical Symposium TS13.

Biancotti, A., Pambianchi, G. and Pioletti, A.M. (eds.), 2003, Spaces, environments and landscapes of terroirs: BDEM, Milano, Italy, 166 p.

Bollettino Ufficiale Regione Basilicata, 1997: AA. VV. V. 56, p. 4729 – 4739.

Bona, E., Dagostin, F., Dorigati, G., Porro D., Roncadori I., Tonon, C., 2003, La tutela dei vini trentini: Grafiche Futura, Mattarello, Trento, Italy, 242 p.

Bourrouilh, R., 2004, Geology and Terroirs of Bordeaux Wines: 32nd International Geological Congress, Florence, Italy, Abstracts, Topical Symposium TS13.

Caggianelli, A., De Fino, M., La Volpe, L. and Picarreta, G, 1990, Mineral Chemistry of Monte Vulture volcanics: Petrological implications: Mineralogy and Petrology, v. 41, p. 215-227.

Calò, A., Costacurta, A., 2003, Aglianico re del Sud: Civiltà del bere, p. 93-102.

Cassi, F., Quaglino, E., and Rusco, E., 1999, Studio per la realizzazione di una carta pedologica di sintesi e di carte derivate applicative per il territorio della Provincia di Bari: Assessorato Agricoltura, Provincia di Bari.

Cita, M.B., Chiesa, S., Colacicchi, R., Crisci, G., Massiotta, P., and Parotto, M., 2004, Italian Wines and Geology: BE-MA editrice, Milano, Italy, 148 p.

Cita, M.B., Chiesa, S. and Crisci, G.M., 2003, Geologia dei vini italiani, Italia meridionale e insulare: BE-MA Editrice, Milano, 128 p.

Cita, M.B., Chiesa, S., and Massiotta, P., 2001, Geologia dei vini italiani, Italia settentrionale: BE-MA Editrice, Milano, 128 p.

Colacicchi, R., and Parotto, M. (in press), Geologia dei vini italiani. Italia centrale: BE-MA Editrice, Milano.

Colugnati, G., and Crespan, G., 2000, I rapporti tra vitigno e ambiente: L'Informatore agrario, v. 25, p. 75-83.

De Fino, M., La Volpe, L., Peccerillo, A., Picarreta, G., and Poli, G., 1986, Petrogenesis of Monte Vulture volcano (Italy): inferences from chemistry, major and trace element data: Contributions to Mineralogy and Petrology, v. 92, p.135-145.

Eccel, E., and Toller, G., 1997, Annale Meteorologico e climatologico della rete automatica SPARTAC, Ist. Agr. S. Michele all'Adige, p. 1-84.

Facchini, P., and Falcetti, M. (eds.), 1991, Atti del Convegno Il Teroldego Rotaliano (S.Michele all'Adige, Trento, 1-2 settembre, 1989) p. 1-100, Bollettino ISMA, suppl. 4a, 2.

Failla, O., Anzani, R., Scienza, A., 1992, La vite selvatica in Italia: diffusione, caratteristiche e conservazione del germoplasme: Vigne Vini N. 1/2, p. 37-46.

Fregoni, M., 2003, Rari e antichi vitigni da salvare: Viaggi e sapori, p. 122-125.

Fregoni, M., 1985, Viticoltura generale: REDA Edizioni, Roma, p. 1-300.

Fuganti, A., and Defrancesco, E.,1992, Le Rocce da Vino della Piana Rotaliana:. Centro di Studi Rotaliani Mezzocorona, p. 1-97, Trento, Italy.

Haynes, S.J., 1999, Geology and Wine 1. Concept of Terroir and the role of geology: Geoscience Canada, v. 26, p. 190-194.

Haynes, S.J., 2000, Geology and Wine 2. A geological foundation for terroirs and potential sub-appellations of Niagara peninsula wines, Ontario, Canada: Geoscience Canada, v. 27, p. 67-87.

Johnson, Hugh, and Robinson, Jancis, 2001,The World Atlas of Wine, 5th Edition: Mitchell Beazley, London, U.K., 352 p.

La Volpe, L., Patella, D., Rapisardi, L., and Tramacere, A., 1984, The evolution of the Monte Vulture volcano (Southern Italy): inferences from volcanological, geological and deep dipole electrical soundings data: Journal of Volcanology and Geothermal Research, v. 20, p. 617-638.

Lulli, L., and Bidini, D., 1978, Guida all'escursione-dibattito sui suoli dei vulcani Roccamonfina e Vulture, 27-30 settembre 1978: C.N.R. Centro di Studio per la Genesi, Classificazione e Cartografia del suolo, Pubbl. 51.

Marre, A., 2003, Les terroirs vitivinicoles: concepts generaux et application à la Champagne. *in*: Biancotti *et al.* (eds.), p .55-64.

Meinert, L.D., and Busacca, A.J., 2000, Geology and Wine 3. Terroirs of the Walla Walla Valley appellation, southeastern Washington State, USA: Geoscience Canada, v. 27, p. 149-171.

Meinert, L.D., and Busacca, A. J., 2002, Geology and Wine 6. Terroir of the Red Mountain Appellation, Central Washington State, USA: Geoscience Canada, v. 29, p.149-168.

Pomerol, C., 1984, Terroirs et Vins de France B.R.G.M., Orléans and TOTAL Edition-Presse, Paris, 370 p.

Prosser, G., and Selli, L., 1991, Thrusts of the Mezzocorona-Mendola Pass area (southern Alps, Italy); structural analysis and kinematic reconstruction: Bollettino Società Geologica Italiana, v. 110, p. 805-821.

Ricchetti, G., and Pieri, P., 1999, Puglia e Monte Vulture, Guide Geologiche regionali della Società Geologica Italiana: BE-MA Editrice, Milano, 285 p.

Roncador, I., and Falcetti, M., 1991 (eds.), Storia e tradizione del vitigno Teroldego: *in*: Facchini P. and Falcetti, M., 1991 (eds.) p. 13-21.

Scienza, A., and Failla, O., 1996, La circolazione dei vitigni in ambito podanoveneto e atesino: le fonti storio lettrarie e l'affrosccio biologico-molecolare: In "2500 anni di cultura della vite nell'ambito alpino e usolpino" Trento, Italia, p. 1-8.

Scienza, A., Mattini ,F., Villa, P., Gianazza, E, Tedeschi, G. and Failla, O., 1991, Rapporti filogenetici tra il Teroldego ed alcuni vitigni appartenenti a zone geografiche diverse: *in*: Facchini, P., and Falcetti, M., 1991 (eds.), p. 23-32.

Sittler, C. and Marocke, R., 1981, Géologie et oenologie en Alsace: Sols et terroirs géologiques, cépages et specificité des vins: Sci. Géol. Bull. Strasbourg, v. 34, p. 147-182.

Sittler, C., 1998, Le vignoble et ses terroirs géologiques. Rélations éco-physiologiques entre sol, vignes et vin: in: Bibliothèque du Naturaliste: l'Alsace et les Vosges. Délacheaux et Niestlè Edit. Lausanne, p. 1-452.

Veronelli, L.,1983, Catalogo dei Vini d'Italia:. Giorgio Mondadori Editore, Milano, 562 p.

Wallace, P., 1972, Geology of Wine: 24th International Geological Congress, Montreal, Section 6, Stratigraphy and Sedimentology, p. 359-365.

Wilson, J.E., 1998, Terroir. The role of Geology, Climate and Culture in the Making of French Wines. Mitchell Beazley, London, UK, 336 p.

Wilson, J.L., 2001, Geology and Wine 4. The origin and Odyssey of Terroir: Geoscience Canada, v. 28, p.139-142.

PAPER 2 FROM THE "TERROIR, GEOLOGY AND WINE: A TRIBUTE TO SIMON J. HAYNES" SESSION HELD AT THE GEOLOGICAL SOCIETY OF AMERICAN ANNUAL MEETING, SEATTLE, WASHINGTON, NOVEMBER 3, 2003

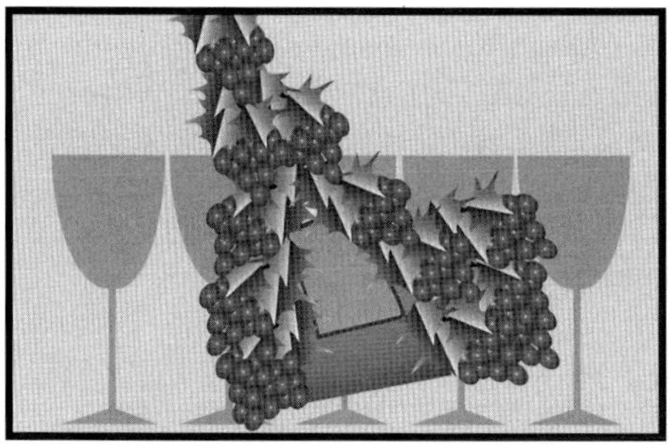

Classification of British Columbia's Okanagan Chardonnay Wines by Origin Using Volatile Components

Nigel J. Eggers*[1], John D. Greenough[2] and Tim Cernak[1]
Department of Chemistry, and the Department of Earth and Environmental Sciences, University of British Columbia Okanagan, Kelowna, BC, V1V 1V7.
**Corresponding author nigel.eggers@ubc.ca*
[1]Department of Chemistry
[2]Department of Earth and Environmental Sciences

SUMMARY

The relative concentrations of 31 organic volatile compounds present in 12 Okanagan Chardonnay wines from the year 2000 vintage have been determined using gas chromatography/mass spectrometry. Wines came from Mission Hill and Quails' Gate commercial wineries. Grapes used in the wines originated from the north (Kelowna, Westbank), central (Naramata) and southern Okanagan Valley (Oliver, Osoyoos), reflecting climatological and geographical variations in the valley. Exploratory statistics (two dimensional, and multidimensional scaling) employing all possible organic volatile compound-versus-compound correlation coefficients show that the organic volatiles, specifically fermentation esters, alcohols, and acids, tend to behave differently. Using the same statistical methods as used to compare compounds to compare samples, shows that the dominant factor controlling the organic volatiles is the winery that produced the wines. This suggests that processing factors such as fermentation organisms and conditions, or the use of oak barrels, have a dominant effect on wine organic volatile chemistry. In a third statistical dimension, however, wines separate according to geography, apparently predominantly reflecting important variations in ester concentrations. Discriminant analysis shows that wines can be separated according to geographic origin - the location of growth of grapes - using many different combinations of organic volatiles with as few as three organic volatile compounds. This study reveals that at least 16 of the 31 organic volatiles identified reflect biogeographic controls on grape composition and, ultimately, wine composition. It is apparent that factors contributing to terroir, e.g. climatic and soil-related aspects, have a measurable impact on the organic volatile composition of Chardonnay wines of the Okanagan Valley, thus providing a closer link between the aromatic and palate appeal of wines and place of grape origin.

RÉSUMÉ

On a mesuré, par chromatographie en phase gazeuse et par spectrographie de masse, la concentration relative de 31 composés volatiles présents dans 12 vins Chardonnay de la vallée de l'Okanagan du millésime 2000. Les vins provenaient des vignobles commerciaux de Mission Hill et de Quail's Gate. Les raisins de ces vins provenaient des régions du nord (Kelowna, Westbank), du centre (Naramat) et du sud de la vallée de l'Okanagan (Oliver, Osoyoos), reflétant ainsi les variations climatologiques et géographiques de la vallée. Les statistiques exploratoires recueillies (échelles bidimensionnelles et multidimensionnelles) basées sur tout type de combinaisons possibles de composés organiques volatiles par rapport aux coefficients de corrélation des composés montrent que les composés organiques volatiles, particulièrement les esters, les alcools, et les acides, semblent se comporter différemment. Les méthodes statistiques employées pour comparer les composés, qui ont été appliquées pour comparer les échantillons, montrent que le vignoble producteur est le principal facteur déterminant les composés organiques volatiles. On peut en conclure que les facteurs de vinification comme les organismes et les conditions de fermentation, et l'emploi de barils de chêne, ont un effet déterminant sur la chimie des composés organiques volatiles du vin. Cependant, selon une troisième dimension statistique, les vins se répartissent selon la provenance géographique, qui semble se refléter surtout dans des variations de la concentration en ester. L'analyse discriminante montre que les vins peuvent être distingués selon l'origine géographique - lieux de culture des vignes - en ayant recours à différentes combinaisons de composés organiques volatiles différents, dont aussi peu que trois. Cette étude montre qu'au moins 16 des 31 composés organiques volatiles définis sont le reflet des contrôles biogéographiques sur la composition du raisin, et finalement sur la composition du vin. Il semble que les facteurs déterminant le terroir, par exemple les aspects climatiques et pédologiques, ont un effet mesurable sur la composition des composés organiques volatiles des vins Chardonnay de la vallée de l'Okanagan, ce qui renforce le lien entre l'arôme et le goût des vins, et l'origine du raisin.

INTRODUCTION

For centuries anecdotal evidence has indicated that many factors impact the sensory character of wines. In studies to determine these factors, much recent effort has gone into quantitatively identifying the organic volatile composition of wines – those compounds in wine with a low vapour pressure and consequently that are present in the vapour above the wine in relatively high concentrations - and correlating it with other factors that affect the sensory character of wines. The fermentation process has a major impact on the aroma of wines. Naturally occurring yeast cells, by metabolizing carbohydrates in the absence of oxygen during the fermentation process, generate organic volatiles, liberate volatiles from nonvolatile precursors, and may produce what are commonly known as "off-notes", such as thio compounds or vinylphenols. Among the often cited "impact volatiles" - those volatiles which are important to the aroma of wine - are: 3-methylbutyl acetate (E3, E = ester; see Table 1), 3-methylbutan-1-ol (O5, O = alcohol; see Table 1), and the ethyl esters of acetic, hexanoic (E6), and octanoic acids (E11) (Moio *et al.*, 1994; Berger, 1995). Wines with similar higher-alcohol and ester contents may smell differently, however, indicating a specific role for minor and trace impact components. A further complicating factor is that winery practices, such as barrel aging, may alter the organic volatile compound profiles of wines by adding γ-lactones, furans and volatile phenolic compounds (Towey and Waterhouse, 1996; Cutzach *et al.*, 1997). These winery-related factors overprint and act on what may be thought of as naturally-occurring volatiles, those related to climatic and soil-related (edaphic) factors commonly referred to as terroir, and supplied initially by the grapes. Aroma compounds arising from the grapes are often present as odourless precursor compounds in the grapes, and are released by the acid conditions that prevail during fermentation. Francis *et al.* (1992) found that acid treatment of the glycosides (= non-volatile aroma precursor compounds) of Chardonnay juices produced a solution with aroma attributes characteristic of Chardonnay wines. This indicates that grape aroma precursors, on acid hydrolysis, contribute to and enhance key aromas in Chardonnay wines.

There has been considerable success in discriminating wines on the basis of origin, maturity, or variety using elemental analyses (e.g. Kwan *et al.*, 1979; Greenough *et al.*, 1997; 2005), volatile analyses or sensory properties. The organic volatiles providing the best classification by region, maturity or variety are apparently fermentation products (Kwan *et al.*, 1980; Noble *et al.*, 1980; Marais *et al.*, 1981a,b; Moret *et al.*, 1984; Garcia-Jares *et al.*, 1995; Marengo et al., 2001: and Sabon *et al.*, 2002). For example Marais *et al.*, (1981) successfully classified Colombar wines according to geographic origin using the organic volatiles isoamyl acetate (E3), isobutyl alcohol (O2) and hexyl acetate (E7) and, for Chenin Blanc wines, hexanol (O10) and 2-phenylethanol (O23). In addition, nitrogen concentrations in wine juices during the wine-making process can impact the amounts of higher alcohols, fatty acids, esters and carbonyl compounds (e.g. aldehydes and diacetyl) produced by yeasts (Fleet, 1997). Because nitrogen concentrations in wine juices can be an indicator of vine health, this implies a further possible tie between region and wine organic volatile composition.

Chardonnay wines are produced worldwide and their aromatic and palate components are specific, defined, and well known. These grapes have little aroma, yet their wines often develop complex olfactory profiles. Following the extensive replacement of hybrid grapes with *Vitis vinifera* in the late 1980s (Schreiner, 1994; 1998), Chardonnay has become one of the principal grape varieties grown in British Columbia. The major grape growing region in British Columbia is the Okanagan Valley (Bowen *et al.*, 2005; Greenough *et al.*, 2004). To date relatively few data are available on terroir traits for this grape in this valley, despite significant climatic changes between the northern (Kelowna) and southern (Osoyoos) portions of the valley (Fig. 1). This paper addresses the two-part question: (a) what aspects of commercial Okanagan Chardonnay wine organic volatile composition can be related to terroir factors and thus may be diagnostic of geographic place of grape origin, and (b) what aspects can be related to winery processing?

The geology of the vineyard areas has been summarized by Fulton (2003), Greenough *et al.* (2004) and Bowen *et al.* (2005). Bedrock for the Kelowna/Westbank vineyards typically consists of Paleozoic to Mesozoic metamorphic and sedimentary rocks (sandstone, limestone), locally intruded by Jurassic-aged granites and overlain by Tertiary-aged volcanic rocks and sediments. The soils formed from mixtures of glacial lake silt and clay, glacial river deposits, and till. Naramata and Oliver/Osoyoos vineyards tend to be underlain by "granite" gneiss of the Monashee complex, but there is considerable local variation (e.g. greenstones, amphibolites) and bedrock exposure where there are vineyards commonly is poor. Thick (tens to hundreds of metres) "benches' of fine-grained sand and silt sediments of glacial lake origin overlie bedrock at Naramata, whereas Oliver/Osoyoos vineyards tend to be planted on post-glacial river-formed flood plain and alluvial fan deposits. Soils in the Oliver/Osoyoos area tend to be Brown Chernozems reflecting hot and dry conditions. Cooler and more moist conditions near Kelowna/Westbank yielded Dark Brown Chernozems or, in previously forested areas, Eluviated Eutric Brunisols.

MATERIALS AND METHODS

Wine Samples

The wines used in this study came from bulk commercial wines made from specific grape blocks (= fields) of Chardonnay grapes, and were sampled before blending. Twelve, year 2000 vintage Okanagan Valley Chardonnays were supplied by two commercial wineries, Mission Hill (winery B), and Quails' Gate (winery C). Grapes originated from Kelowna/Westbank in the north, Naramata in the middle, and Oliver/Osoyoos vineyards in the southern Okanagan Valley (Fig. 1). The Mission Hill (B) wines were sampled from stainless steel tanks on 6 June 2001, and the Quails' Gate (C) wines from oak barrels on 22 June 2001. One of the samples (C1) was made from 2 blocks of Naramata grapes and two

Table 1: Average relative concentrations of the volatile compounds in 12 Chardonnay wines[a].

Winery		C1	C2	C3	C4	B5	B6	B7	B8	B9	B10	B11	B12
Geographic origin[b]		N1	K2	K3	K4	O5	O6	O7	N8	N9	O10	O11	N12
Name of compound[c]													
Ethyl butanoate	E1	0.12	0.19	0.18	0.19	0.18	0.31	0.3	0.11	0.4	0.22	0.19	0.07
Isobutyl alcohol (2-methylpropan-1-ol)	O2	1.5	2.16	1.11	0.85	0.58	0.46	1.01	0.88	1.58	0.53	0.82	0.93
Isoamyl acetate (3-methylbutyl acetate)	E3	0.22	0.6	1.43	1.3	1.5	2.47	2.85	1.15	2.63	1.91	2.64	1.8
Butan-1-ol	O4	0.15	0.11	0.06	0.06	0.04	0.05	0.07	0.1	0.14	0.05	0.12	0.04
Isoamyl alcohol (3-methylbutan-1-ol)	O5	29.91	39.67	33.12	33.15	28.01	28.07	35.63	27.54	37.27	31.3	30.46	25.47
Ethyl hexanoate	E6	0.23	0.44	0.4	0.57	0.37	0.89	0.58	0.58	0.91	0.85	0.68	0.66
Hexyl acetate	E7	0.01	0.04	0.05	0.09	0.23	0.43	0.57	0.71	0.77	0.47	0.41	0.38
Acetoine (3-hydroxybutan-2-one)	O8	0.36	0.07	0.01	0.18	0.08	0.1	1.72	0.57	0.39	0.015	0.37	0.25
Ethyl lactate (ethyl 2-hydroxypropanoate)	E9	8.98	10.49	1.07	6.58	6.57	8.25	10.43	4.94	3.09	0.17	13.66	2.5
Hexan-1-ol	O10	0.9	1.32	0.75	1.09	0.92	1.02	0.7	2.36	2.42	1.57	-	1.39
Ethyl octanoate	E11	0.33	0.37	0.53	0.72	0.61	1.32	0.68	0.93	0.98	1.06	1.12	1.59
Furfural (furan-2-carboxaldehyde)	M12	0.05	0.02	0.59	0.04	-	-	-	-	-	-	-	-
Ethyl 3-hydoxybutanoate	E13	0.03	0.16	0.03	0.05	0.05	0.05	0.08	0.08	0.04	0.04	0.11	0.08
5-Methylfurfural (5-methylfuran-2-carboxaldehyde)	M14	0.03	-	0.24	-	-	-	-	-	-	-	-	-
Butanoic acid	A15	0.1	0.18	0.24	0.25	0.34	0.24	1.04	0.32	0.31	0.77	0.42	0.19
γ-Butyrolactone (dihydro-2(3H)-furan-2-one)	E16	0.14	0.25	0.22	0.21	0.43	0.27	0.5	0.13	0.08	0.16	0.82	0.27
Furfuryl alcohol (furan-2-methanol)	M17	0.08	0.07	0.27	0.03	-	-	-	-	-	-	-	-
3-Methylbutanoic acid	A18	0.36	0.18	0.4	0.31	0.32	0.21	0.35	0.32	0.27	0.36	0.36	0.28
Diethyl succinate (diethyl butanedioate)	E19	2.24	5.2	2.17	0.94	0.32	0.41	0.25	0.14	0.13	0.55	0.27	0.29
2-Phenylethyl acetate	E20	0.29	0.3	1.52	0.31	0.44	0.37	0.38	0.55	0.39	0.57	0.44	0.61
Hexanoic acid	A21	3.82	3.25	5.2	6.66	6.62	6.79	4.62	9.38	6.99	7.72	4	3.84
Phenylmethanol	O22	0.21	0.18	0.09	0.1	0.05	0.08	0.16	0.14	0.07	0.11	0.12	0.2
2-Phenylethanol	O23	31.98	16.02	33.01	16.85	13.91	9.98	10.47	12.42	9.93	16.61	9.25	10.43
Oak lactone (4-hydroxy-3-methyloctanoic acid γ-lactone)	M24	0.23	0.18	0.2	-	-	-	-	-	-	-	-	-
Diethyl malate (diethyl 2-hydroxybutanedioate)	E25	1.24	1.82	1.81	2.46	1.1	0.86	0.67	0.26	0.17	2.53	1.07	0.07
Octanoic acid	A26	7.15	5.58	7.59	15.1	15.25	15.93	8.54	19.71	13.67	14.99	7.18	6.6
Eugenol (2-methoxy-4-(2-propenyl)phenol)	M27	0.03	0.03	0.04	-	-	-	-	-	-	-	-	-
Unknown	E28	0.08	0.25	0.18	0.18	0.39	0.24	0.27	0.18	0.11	0.38	0.18	0.07
Decanoic acid	A29	0.76	0.55	0.63	2.55	6.58	7	2.7	7.06	5.94	5.46	3.34	2.72
Vanillin (4-hydroxy-3-methoxybenzaldehyde)	M30	0.14	0.06	0.09	0.09	-	-	-	-	-	-	-	-
Acetovanillone (1-(4-hydroxy-3-methoxyphenyl)ethanone)	M31	0.03	0.04	0.02	-	-	-	-	-	-	-	-	-

[a] All results are percentage of total area.
[b] Geographic origin: K=Kelowna/Westbank, N=Naramata, O=Oliver/Oosoyos.
[c] Functional group: esters=E, acids=A, alcohols=O, oak compounds=M. The numbers given for specific compounds are percentages of total volatile compounds.

blocks of Kelowna grapes. It is labeled as a Naramata sample (N1, Table 1).

Isolation of Volatiles

Volatile isolation followed the procedure of Gunata *et al.*, (1985). At least two replicate extractions were performed for each wine. Wine samples were stored as received at 4°C until analyzed by gas chromatography/mass spectrometry.

Chromatographic Analyses

Analyses were carried out on a Hewlett Packard 5890 Series II gas chromatograph (GC) coupled to a Hewlett Packard quadrupole mass spectrometer, at the Pacific Agri-Food Research Centre in Summerland, B.C. A 60 m fused silica capillary column was employed with a Supelwax stationary phase and helium carrier gas. Following a 10:1 split, 1μL of the liquid sample was manually injected. The injector was held at 250°C and the oven ramped from 35°C to 220°C at 3°C/min and held at 220°C for 25 minutes.

Statistical Analyses

All statistical calculations were carried out using SYSTAT statistical software. Multidimensional scaling (MDS), an exploratory data analysis technique, was used to uncover patterns in the data set (Figs. 2, 3a,b, and 4). In this technique the unknown relationships between samples (or compounds) reflect natural patterns in the data. The statistical package does not "know" one sample (or compound) from another, and does not "go looking" for correlations supporting preconceived models for relationships between samples or compounds. Simplistically, these diagrams (Figs. 2, 3a,b, and 4) can be thought of as "maps" showing the two-dimensional (Figs. 2, 4) and three-dimensional average "distances" between compounds (Fig. 2) and between samples (Figs. 3a, 3b and 4). Wine samples or organic volatile compounds that are similar, or that behaved similarly, plot close together. Points on opposite sides of a diagram represent wine samples or organic volatile compounds that are dissimilar or behaved in an opposite manner. Multidimensional scaling determines the distances between samples or compounds (Figs. 3a, b and 4) from sample-versus-sample matrices of correlation coefficients calculated by plotting chemical data for each sample against data for all other samples. Similarly, the "map" of distances between compounds was determined from a matrix of correlation coefficients calculated by plotting chemical data for each compound against the concentrations for all other compounds (Fig. 2).

As a first step in creating Figures 2, 3a,b, and 4, organic volatile com-

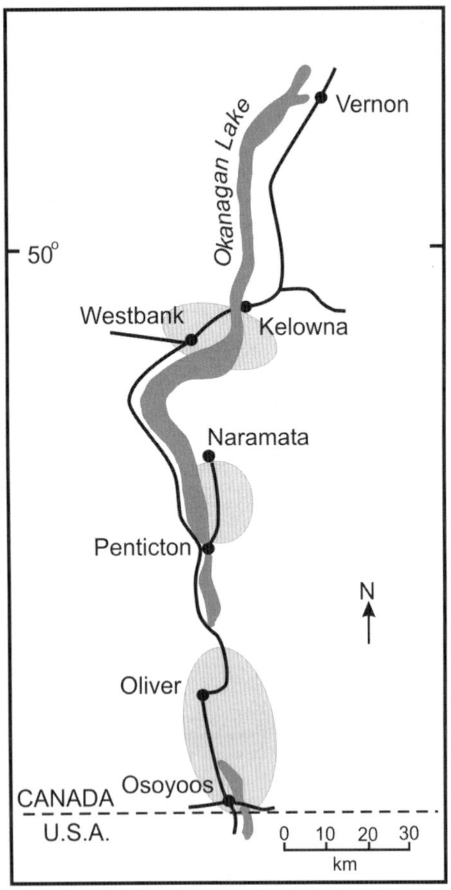

Figure 1. Sketch map of the Okanagan Valley, British Columbia, showing the regions (stippled) from which grapes were obtained for the Chardonnay wines used in this study.

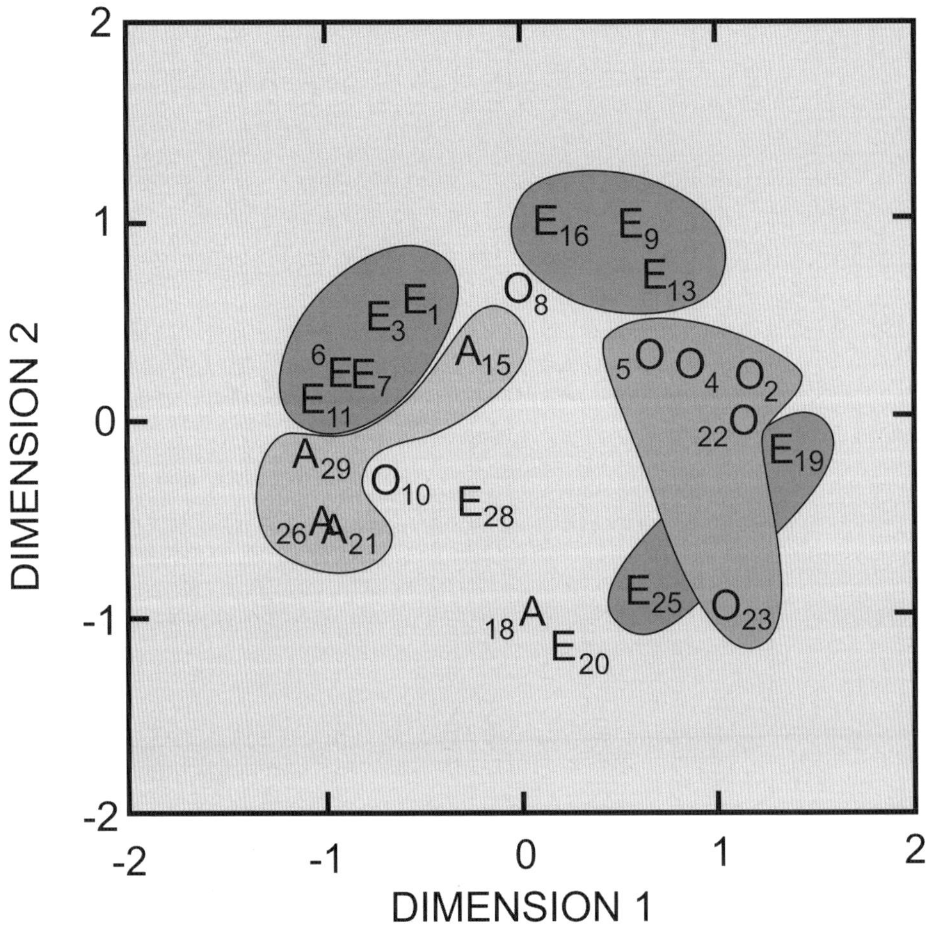

Figure 2. Multidimensional scaling (MDS) diagram summarizing relationships between organic volatile compounds occurring in essentially all wines (see Materials and Methods, and Table I for compound names and symbols). See text for discussion.

pounds were identified which occurred in all wines with one exception: compound O10 was missing in wine B11 (Table 1). The concentrations of these compounds were standardized (z-scored) to put them on the same scale. This ensures, for example, that no one compound, owing to high average concentration, has more impact than any other compound, in assessing how similar or dissimilar samples are. For other examples of the application of linear multidimensional scaling to chemical data also from wine, see Greenough *et al.*, (1997; 2005).

To identify compounds and combinations of compounds useful for separating wines based on geography, an inferential statistical technique, discriminant analysis, was employed (Fig. 5a,b,c,d). Discriminant analysis separates samples (wines) into pre-assigned categorical groupings, in this case the geographic origin of the grapes, using combinations of analytical data. The discriminant functions for all four diagrams yielded 100% correct classification of samples according to geographic origin. Not surprisingly the multivariate statistical tests related to classification are all highly significant. Owing to high standard deviations on the mean concentrations of most organic volatile compounds (variables) in each wine group, and the small number of samples, removing even one sample can strongly impact the efficiency of the classification.

RESULTS

Table 1 presents organic volatile compound relative concentrations and classifies these volatiles into esters (E), higher alcohols (O), carboxylic acids (A), and miscellaneous compounds, e.g. oak compounds (M). Multidimensional scaling (MDS; Borg and Groen, 1997) was used to examine relationships between volatile compounds and samples. Figure 2, constructed using multidimensional scaling, shows the relative behavior of 24 identified organic volatile compounds detected in at least 11 of the 12 Chardonnay wine samples (hexan-1-ol, O10, could not be measured in sample B11). Hand-drawn fields in Figure 2 emphasize that esters, alcohols and acids tend to plot in distinct areas on this two-dimensional MDS diagram. The esters are separated into three groups (Fig. 2). Using the

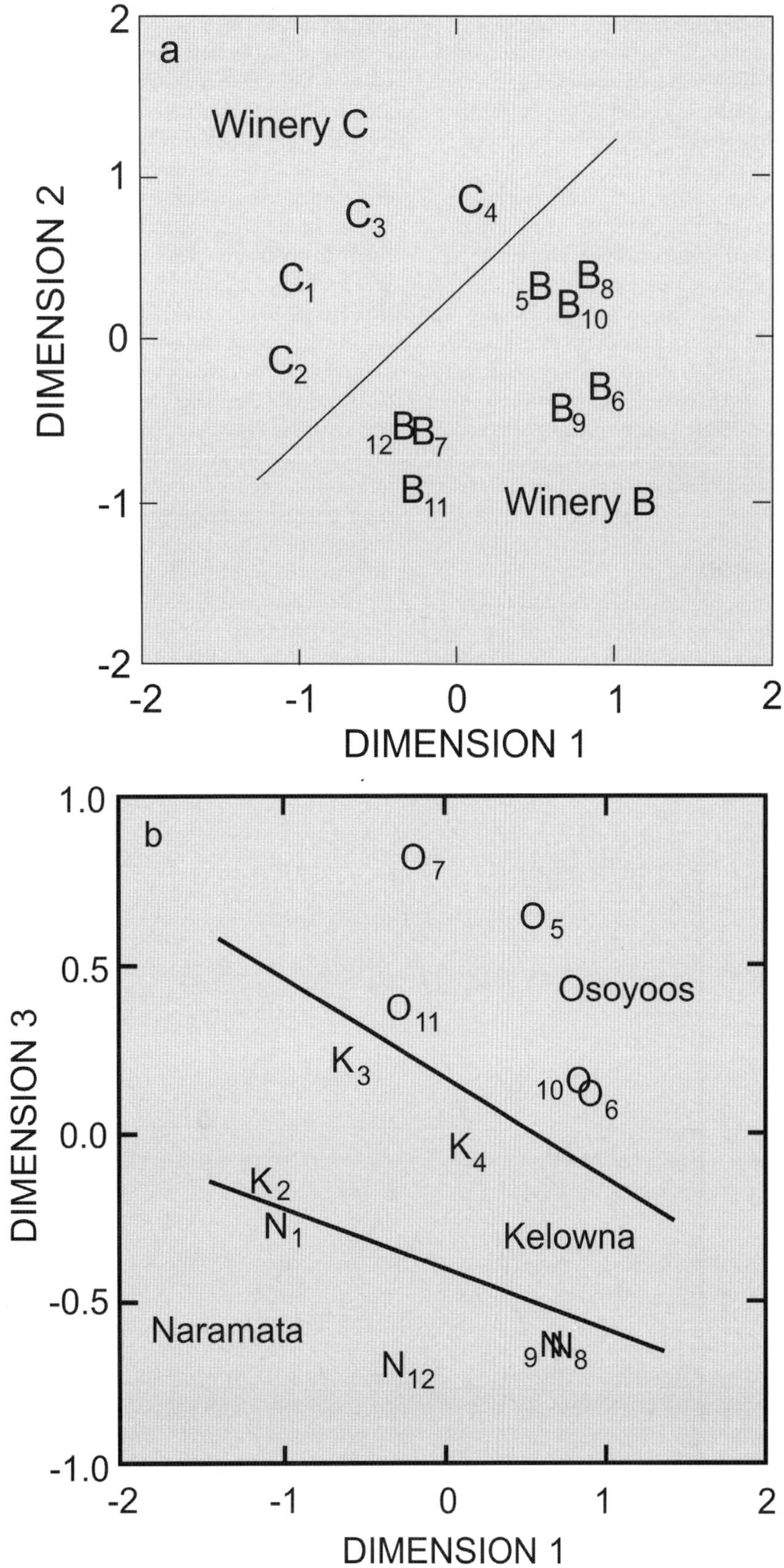

Figure 3. Multidimensional scaling diagrams showing relationships between samples based on organic volatile compounds occurring in essentially all samples. 3a) Plot of the Dimension 1 versus Dimension 2 plane showing the separation of winery samples. 3b) Plot emphasizing the importance of Dimension 3 for separating samples according grape geographic origin; see text for discussion.

same data set transposed to compare samples, multidimensional scaling shows that samples primarily separate according to which winery produced each wine (Fig. 3a), although they also separate according to the geographic origin of grapes used in the wines along Dimension 3 (Fig. 3b). MDS experiments comparing samples using esters, alcohols and acids revealed that the esters yield the best organization of samples according to geography (Fig. 4), and acids the poorest. The esters also efficiently separate samples according to winery. As discussed below, several other organic volatile compounds suspected to be related to the oak barrels used in Chardonnay fermentation were not used in Figures 2-4.

DISCUSSION

The esters, higher alcohols, and fatty acids are the most important groups of the yeast-synthesized aroma substances in the fermentation bouquet of wine.

Higher alcohols

Higher alcohols detected in all 12 samples of the Okanagan Chardonnays include 2-methylpropan-1-ol (O2), butan-1-ol (O4), 3-methylbutan-1-ol (O5), and 2-phenylethanol (O23). Their association on the right side of Figure 2 is consistent with a common origin; they form from amino acids by a mechanism known as the Ehrlich reaction (Ribéreau-Gayon *et al.*, 2001). Deamination of the amino acid to the α-keto acid is followed by decarboxylation to the aldehyde, and finally reduction to the alcohol. However, the production of higher alcohols by yeasts during fermentation also seems to be associated with their corresponding ketonic acids, and butan-1-ol (O4) appears to be formed in this way (Ribéreau-Gayon *et al.*, 2001). Figure 2 suggests that phenylmethanol (O22) is formed in a similar fashion, as it plots in the middle of the alcohol group. If as suggested these organic volatile compounds owe their formation to fermentation, their prominence along Dimension 1 (Fig. 2) suggests that this axis reflects wine production processes.

Another alcohol, hexan-1-ol (O10), plots on the left side of Figure 2 opposite the main group, suggesting that this compound is formed by a different metabolic process. Hexanol contributes to the green-leaf or herbaceous odour produced when grapes are crushed.

Acetoin, (O8) although classified as an alcohol, is in fact a ketol, formed from pyruvate via α-acetolactic acid and is found in the upper part of Figure 2 between the hydroxyacid esters (E9, E13, E16) and the fatty acid esters (E1, E5, E11; see next section).

Esters

Ethyl esters of straight-chain fatty acids [ethyl butanoate (E1), ethyl hexanoate (E6), ethyl octanoate (E11); acetates of higher alcohols (3-methylbutyl acetate (E3); and 2-phenylethyl acetate (E20)] are the dominant esters in wine: they form during fermentation. These esters, excepting 2-phenylethyl acetate (E20), appear in the upper left-hand corner of Figure 2.

The ethyl esters of the hydroxyacids, lactic acid (E9) and 3-hydroxybutanoic acid (E13), together with the internal lactone of 4-hydroxybutanoic acid, γ-butyrolactone (E16), are grouped together in the upper right-hand corner of Figure 2.

Chemical esterification by ethanol of succinic acid, formed via yeast metabolism (Schreier, 1979), and malic acid, originating in the grape, takes place slowly over many years to yield diethyl succinate (E19) and diethyl malate (E25) (Rapp, 1988). These two esters are found grouped together at the lower right side of Figure 2.

The unknown ester (E28), which we think may be a γ-lactone of a C6-C10 fatty acid, has an important role to play in determining geographic distribution, and so has been classified with the esters. It appears by itself in the center of Figure 2.

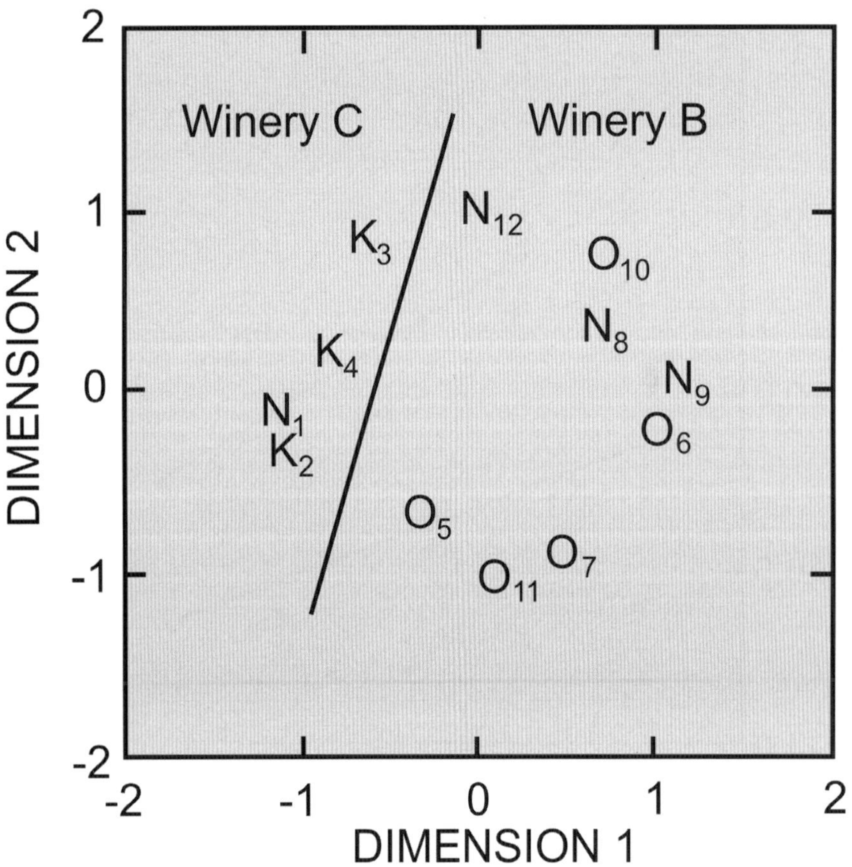

Figure 4. Multidimensional scaling diagram showing relationships between wine samples based on esters. Winery B is Mission Hill; Winery C is Quail's Gate. Geographic areas: K = Kelowna/Westbank; N = Naramata; O = Oliver/Osoyoos; locations shown in Figure 1.

Acids

The fatty acids [butanoic acid (A15), hexanoic acid (A21), octanoic acid (A26), and decanoic acid (A29)] appear on the left-hand-side of Figure 2, close to the ethyl esters of the fatty acids (E1,

E3, E6, E11, E20; see above). The yeast synthesizes the acetates in the same way as the fatty acids. The fatty acids originate from the fatty acid synthetase complex where the products of this enzyme complex are hydrolysed with water (Schreier, 1979). Hydrolysis of the products with ethanol leads to the formation of the ethyl esters (Rapp, 1988).

Fatty acids are formed in large quantities during fermentation at an earlier stage than the respective ethyl esters. Only when substantial amounts of ethanol have been formed do the fatty acid ethyl esters increase in amounts. The final quantities of fatty acids have already been achieved before maximum ethanol content has developed.

Relationships with Winery

Figure 3a shows that in terms of organic volatile composition the wines separate according to the winery they came from. Table 1 shows, however, that there are

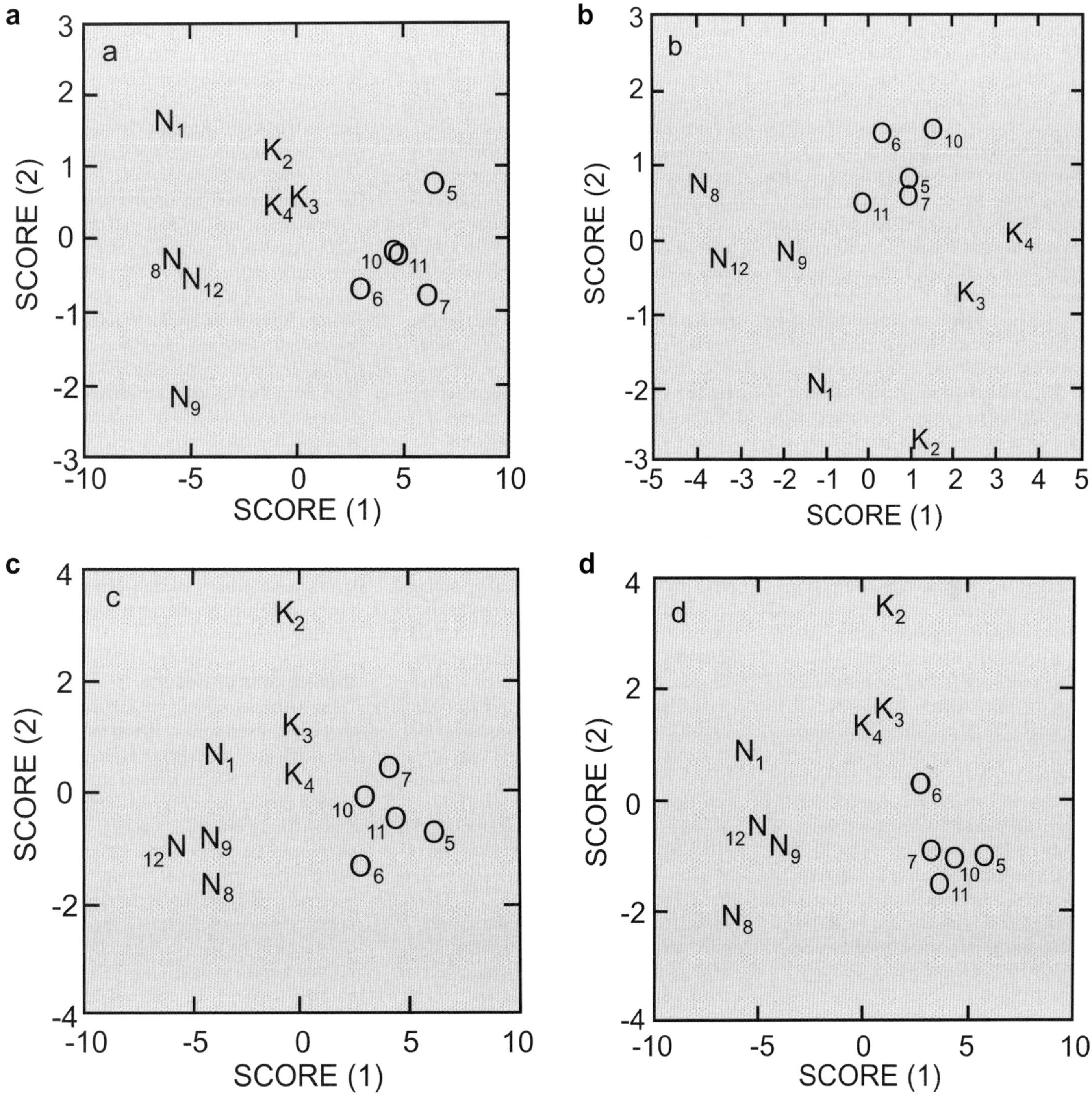

Figure 5. Discrimination diagrams illustrating the separation of wines based on geography; geographic symbos K, N, O as in Figure 4. 5a) Plot showing geographic separations based on the organic volatile compounds 3E, 10O, and 28E. 5b) Geographic separations based on compounds 2O, 5O, 7E, and 29A. 5c) Geographic separations using compounds 5O, 10O, 19E, 28E, 29A. 5d) Geographic separation based on compounds 3E, 7E, 13E, 16E, 25E and 28E. See text for discussion.

some organic volatile compounds that are present only in Quails' Gate (winery C) wines. These compounds are known to be extracted from oak barrels by wine (Schreier, 1979; Towey and Waterhouse, 1996), and include furfural (M12), 5-methylfurfural (M14), furfuryl alcohol (M17), oak lactone (M24), eugenol (M27), vanillin (M30), and acetovanillone (M31). Use of oak barrels in production of Chardonnay wines is a common industry practice, as such aging tends to give Chardonnay wines more complexity. Only the Quails' Gate wines had been fermented in oak barrels: Mission Hill wines had not – they were sampled from stainless steel tanks. Because of the apparent effect of oaking on the organic volatile composition of Quails' Gate wines, the above oak-related compounds were not used in Figure 3a, and thus the separation according to winery appears unrelated to the use of oak barrels.

There are many compounds for which concentrations are similar in the two or are, on average, higher in Mission Hill (B) wines than in Quails' Gate (C) wines. Different groups of compounds such as the esters (Fig. 4) and alcohols separate samples according to winery. Arguments have already been made that behavior of these compounds reflects fermentation processes that apparently vary between wineries. Univariant f-tests indicate that the following compounds are highly correlated with winery of origin in the following order: hexyl acetate (E7; $p = 0.001$), decanoic acid (A29), diethyl succinate (E19), 2-phenylethanol (O23), isoamyl acetate (E3), ethyl octanoate (E11), and ethyl hexanoate (E6; $p = 0.024$). Not surprisingly there are many compound-versus-compound plots that effectively separate samples by winery. These results resemble those of a study by Delteil and Jarry (1992) that showed that Chardonnay musts (crushed grapes at the beginning of fermentation) from different wineries can be distinguished using methanol, total butane-2,3-diols, (Z)-hex-3-en-1-ol, and ethyl lactate (E9).

Relationships with Geography

Exploration of sample relationships using all compounds (except the oak-related compounds noted above) and multidimensional scaling shows that wine samples also organize themselves according to the geographic origin of their grapes in the Okanagan valley (Dimension 3, Fig. 3b). The British Columbia grape growing region of the Okanagan valley extends from the warmer, open benches of the Oliver/Osoyoos area in the south (Fig. 1) to the Westbank/Kelowna area in the north, where the valley has narrowed and the climate is cooler and wetter. Note that the "Naramata" sample N1 (Table 1) was made from both Kelowna and Naramata grapes and plots between other Kelowna and Naramata samples in Figure 3b. Of the three major groups of organic volatiles, esters, alcohols and acids, the esters yield the best MDS organization of samples according to geography (Fig. 4), implying geoclimatic controls on grapes and hence on wine composition. Univariant f-statistics suggest that five esters, two alcohols and two acids are correlated with geographic origin. From strongest ($p = 0.013$) to weakest ($p = 0.114$) correlation, these are the unknown ester (E28), diethyl succinate (E19), hexyl acetate (E7), isobutyl alcohol (O2), γ-butyrolactone (E16), butanoic acid (A15), decanoic acid (A29), hexan-1-ol (O10) and isoamyl acetate (E3).

The potential for correlations in multidimensional chemical space was examined using discriminant analysis. This technique separates samples (wines) into pre-assigned categorical groupings, in this case the geographic origin of the grapes, using combinations of analytical data. Selected results in Figure 5, and similar results not shown herein, illustrate that in addition to the above list of useful discriminating compounds - the five esters, two alcohols and two acids noted above - the following compounds can be added: ethyl butanoate (E1), butan-1-ol (O4), isoamyl alcohol (O5), ethyl octanoate (E11), ethyl 3-hydroxybutanoate (E13), 2-phenylethyl acetate (E20), and diethyl malate (E25). Thus some 16 organic volatile compounds have proved useful in discriminating wine samples by place of origin: ten of these 16 compounds are esters. Subjectively, the unknown ester (E28), hexan-1-ol (O10), hexyl acetate (E7), isoamyl acetate (E3) and decanoic acid (A29), in that order, appear especially important for effective discrimination of wine samples by geography. Figure 5a illustrates that as few as any three of the esters, alcohols or acids can effectively separate samples.

Other Studies

These results confirm those of Marais *et al.*, (1981a) showing that small numbers of organic volatile compounds can separate wines according to origin. Further, three compounds found useful for separating their Colombar wines [isoamyl acetate (E3) and hexyl acetate (E7)], and Chenin Blanc wines [hexan-1-ol (O10)], by geographic origin proved especially useful for separating the Okanagan Chardonnay wines as discussed above. Marais *et al.*, (1981b) also successfully separated South African Pinotage wines into three regions, and Cabernet Sauvignon wines into two regions, using discriminant analysis on the solvent extracted organic volatiles. Moret *et al.*, (1984) also used solvent extraction of organic volatiles to resolve commercial white wines into three Venetian regions using multiple discriminant analysis, and two of the volatiles used [hexan-1-ol (O10) and ethyl octanoate (E11)] were also found to be useful in this study. Shimoda *et al.*, (1993) used discriminant analysis to classify commercial Cabernet Sauvignon wines into six regions of the Napa valley of California using chromatography of headspace organic volatiles, and Garcia-Jares *et al.*, (1995) distinguished Spanish Rias Baixas white wines from other Galician wines using purge and trap gas chromatography to identify organic volatiles.

Implications of Results

As noted, the organic volatile compounds chosen for statistical analysis in this study are considered to be fermentation products with the exception of diethyl malate, which probably arises from abiotic esterification of malic acid present in the grape juice. Interestingly, the chosen compounds group together by structural type, the esters on the upper left side of Figure 2, the alcohols on the right-hand-side, the acids at the middle left-hand-side, and the esters of hydroxyacids on the upper right side, with the esters of diacids on the lower right side. Most of these groupings appear to be related to their fermentation biosynthetic pathway. However, diethyl malate (E25) and diethyl succinate (E19) arise from the slow abiotic reaction with ethanol, independent of metabolization by yeast. Interestingly these two compounds plot on the right side of Figure 2, opposite from the esters considered to be formed by yeast; these appear on the left side of Figure 2. Because the two wineries do not use similar wine-making practices, it is not surprising that the samples primarily separate according to winery (Fig. 3a). More interestingly, the samples also separate according to the geographic origin of grapes used in the wines along Dimension 3 (Fig. 3b). Although the geographic separation appears to be related to climatic and soil-related factors, it also may be influenced by different vineyard cultural practices.

CONCLUSIONS

This study has shown that it is possible to analyze the organic volatile compounds in Chardonnay wine samples produced from specific blocks of grapes, in commercial wineries using different winery practices, and use these compounds to distinguish the wines into the region of the Okanagan valley from which the grapes were obtained. Presumably the success of using fermentation volatiles in resolving the wines into different regions is dependant on the nutrient status of the grapes affected by different terroirs. British Columbia has a very young industry based on *Vitis vinifera* and consequently has not established appellations as yet. The information presented in this study may well prove useful in classifying the wines of the Okanagan into regions according to their terroir.

This study shows that regional differences in wines are expressed in the organic volatile compounds believed to be produced by yeast fermentation of grapes. Fermentation overlays its unique pattern of aroma compounds on those provided initially by the grapes, giving the wine of each geographic region its own distinctive aroma. Although this study deals only with Chardonnay grapes, it is hoped that this technique eventually may be used to match other grape varieties to their optimal microclimate.

ACKNOWLEDGMENTS

We thank T. Cottrell and T. Kopp for their technical support and the Pacific Agri-Food Research Centre for use of their GC/MS. We also thank the Mission Hill and Quails' Gate wineries for sampling assistance. Diagrams were skillfully prepared by H. Muggeridge. J.D. Greenough acknowledges an NSERC discovery grant and operating grant. T. Cernak received support from an NSERC USRA scholarship. We also thank R.W. Macqueen, L.D. Meinert, and two anonymous reviewers for constructive comments in revision of the manuscript.

REFERENCES

Berger, R. G., 1995, Aroma compounds in Food: in Aroma Biotechnology; Springer-Verlag: Berlin, Germany, p. 14-21

Borg, I., and Groenen, P., 1997, Modern multidimensional scaling, theory and applications; Springer-Verlag, New York, 471 p.

Bowen, P., Bogdanoff, C.P., Estergaard, B.F., March, S.G., Usher, K.B., Smith, C.A.S., and Frank, G. Geoscience Canada, v. 32, p.161-176. Geology and Wine 10. Use of Geographic Information System Technology to Assess Viticultrual Performance, Okanagan and Similkameen Valleys, British Columbia: Geoscience Canada, v. 32, p. 161-176.

Cutzach, I., Chatonnet, P., Henry, R., and Dubourdieu, D., 1997, Identification of volatile compounds with a "toasty" aroma in heated oak used in barrelmaking: Journal of Agriculture and Food Chemistry, v. 45, p. 2217-2224.

Delteil, D., and Jarry, J.M., 1992, Characteristic effects of two commercial yeast strains on Chardonnay wine volatiles and polysaccharide composition: Australian and New Zealand Wine Industry Journal, v. 7, p. 29-33.

Fleet, G. H., 1997, Wine: *in*: Doyle, M. P., Beuchat, L. R., and Montville, T. J., (eds.), Food Microbiology, Fundamentals and Frontiers, ASM Press,: Washington, D.C., p 671-694.

Francis, I. L., Sefton, M. A., and Williams, P. J., 1992, Sensory descriptive analysis of the aroma of hydrolysed precursor fractions from Semillon, Chardonnay and Sauvignon Blanc grape juices: Journal of the Science of Food and Agriculture, v.59, p. 511-520.

Fulton, R.J., 2003, Geology of the Okanagan Valley wine region of British Columbia (abstr): Geological Society of America Annual Meeting, Seattle, Washington, Abstract 64-2.

García-Jares, C., García-Martín, S., and Cela-Torrijos, R., 1995, Analysis of some highly volatile compounds of wine by means of purge and cold trapping injector capillary gas chromatography. Application to the differentiation of Rias Baixas Spanish white wines: Journal of Agriculture and Food Chemistry, v. 43, p. 764-768.

Greenough, J. D., Longerich, H. P., and Jackson, S. E., 1997, Elemental fingerprinting of Okanagan valley wines using ICP-MS: relationships between wine composition, vineyard and wine colour: Australian Journal of Grape and Wine Research, v. 3, p. 75-83.

Greenough, J.D., Lutmerding, H., and Mallory-Greenough, L. 2004, Chapter 7, Geology and terroir of Okanagan Valley wine: *in* Roed, M. and Greenough, J.D., (eds.), Okanagan Geology, Kelowna Geology Committee, Kelowna, BC, p. 67-70.

Greenough, J.D., Mallory-Greenough, L.M., and Fryer, B.J., 2005, Geology and Wine 9: Regional Trace Element Fingerprinting of Canadian Wines: Geoscience Canada, v. 32, p. 129-137.

Gunata, Y. Z., Bayonove, C. L., Baumes, R. L., and Cordonnier, R. E., 1985, The aroma of grapes. I. Extraction and determination of free and glycosidically bound fractions of some grape aroma components: Journal of Chromatography, v. 331, p. 83-90.

Kwan, W., Kowalski, B. R., and Skogerboe, B., 1979, Pattern recognition analysis of elemental data. Wines of *Vitis vinifera* cv. Pinot Noir from France and the United States: Journal of Agriculture and Food Chemistry, v. 27, p. 1321-1326.

Kwan, W.-O., and Kowalski, B. R., 1980, Pattern recognition of gas chromatographic data. Geographic classification of wines of *Vitis vinifera* cv. Pinot Noir from France and the United States: Journal of Agriculture and Food Chemistry, v. 28, p. 356-359.

Marais, J., Van Rooyen, P. C., and Du Plessis, C.S., 1981, Differentiation between wines originating from different red wine cultivars and wine regions by the application of stepwise discriminant analysis to gas chromatographic data: South African Journal of Enology and Viticulture, v. 2, p. 19-24.

Marais, J., Van Rooyen, P. C., and Du Plessis, C. S, 1981, Classification of white cultivar wines by origin using volatile aroma components: South African Journal of Enology and Viticulture, v. 2, p. 45-49.

Marengo, E., Aceto, M., Maurino, V., 2001, Classification of Nebbiolo-based wines from Piedmont (Italy) by means of solid-phase microextraction-gas chromatography-mass spectrometry of volatile compounds: Journal of Chromatography A, v. 943, p. 123-137.

Moio, L., Schlich, P., and Etiévant, P., 1994, Acquisition et analyse d'arômagrammes de vins de Bourgogne issus de cépage

Chardonnay: Science Elements, v. 14, p. 601-608.
Moret, I., Scarponi, G., and Cescon, P., 1984, Aroma components as discriminating parameters in the chemometric classification of Venetian white wines: Journal of the Science of Food and Agriculture, v. 35, p. 1004-1011.
Noble, A. C., Flath, R. A., and Forrey, R. R., 1980, Wine headspace analysis. Reproducibility and application to varietal classification.: Journal of Agriculture and Food Chemistry, v. 28, p. 346-353.
Rapp, A., 1988, Wine aroma substances from gas chromatographic analysis: *in* Linskens, H. F., and Jackson, J. F. (eds.), Wine Analysis; Springer-Verlag, Berlin, Germany, p. 29-66.
Ribéreau-Gayon, P., Dubourdieu, D., Donèche, B., and Lonvaud, A., 2001, in Handbook of Enology, Volume 1, The Microbiology of Wine and Vinifications: John Wiley & Sons Ltd: Chichester, England, p. 71-73.
Sabon, I., De Revel, G., Kotseridis, Y., and Bertrand, A., 2002, Determination of volatile compounds in Grenache wines in relation with different terroirs in the Rhone Valley: Journal of Agriculture and Food Chemistry, v. 50, p. 6341-6345.
Schreier, P., 1979, Flavor composition of wines: A review: in C.R.C. Critical Reviews in Food Science and Nutrition, v. 12, p. 59-111.
Schreiner, J., 1994, The Wineries of British Columbia: Orca Book Publishers, Victoria, B.C., 218 p.
Schreiner, J., 1998, Chardonnay and Friends: Orca Book Publishers, Victoria, B.C., 218 p.
Shimoda, M., Shibamoto, T., and Noble, A. C., 1993, Evaluation of headspace volatiles of Cabernet sauvignon wines sampled by an on-column method: Journal of Agriculture and Food Chemistry, v. 41, p. 1664-1668.
Towey, J. P., and Waterhouse, A. L., 1996, Barrel-to-barrel variation of volatile oak extractives in barrel-fermented Chardonnay: American Journal of Enology and Viticulture, v. 47, p. 17-20.

PAPER 3 FROM THE "TERROIR, GEOLOGY AND WINE: A TRIBUTE TO SIMON J. HAYNES" SESSION HELD AT THE GEOLOGICAL SOCIETY OF AMERICAN ANNUAL MEETING, SEATTLE, WASHINGTON, NOVEMBER 3, 2003

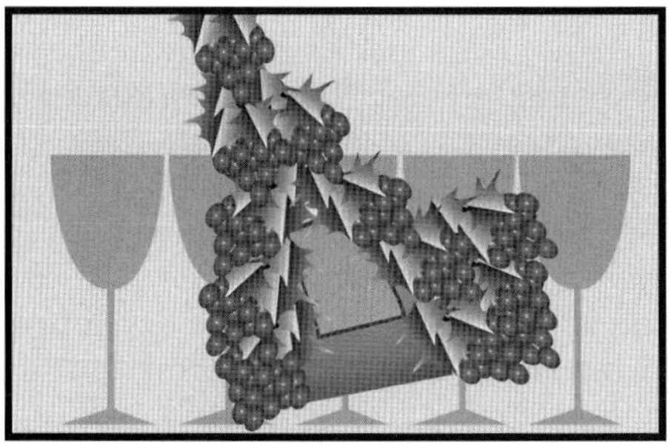

Vineyard Soil Water Content: Mapping Small-scale Variability using Ground Penetrating Radar

Susan Hubbard[1,2], Ian Lunt[3], Katherine Grote[4] and Yoram Rubin[1]
[1]*University of California at Berkeley, Department of Civil and Environmental Engineering, Berkeley, California 94720 USA*
[2]*Lawrence Berkeley National Laboratory, 1 Cyclotron Road, Mail Stop 90-1116, Berkeley, California 94720 USA*
[3]*Department of Earth Sciences, University of Leeds, Leeds LS2 9JT, United Kingdom.*
[4]*University of Wisconsin Eau-Claire, Department of Geology, 105 Garfield Avenue, Eau Claire, Wisconsin 54702 USA*

SUMMARY
Development of precision viticulture strategies, focused on promoting uniformly high wine quality throughout vineyard blocks, necessitates an ability to assess the factors that influence winegrape quality over various spatial and temporal scales. Although micrometeorological factors are often considered when developing an irrigation approach or assessing fruit quality, the influence of small-scale soil variations and their control on water use and associated fruit quality is poorly understood. We have investigated the use of surface GPR (Ground Penetrating Radar) data, using both 900 MHz groundwave and 100 MHz reflected wave techniques, to map shallow soil water content in high resolution with acceptable accuracy and in a non-invasive manner. By combining the different GPR techniques and by using different frequency antennas, it is possible to construct a 3-D soil water content cube and to monitor moisture distribution over time. Techniques such as these open the way for development of precision viticulture approaches that account for natural soil variabilities, such as precision irrigation. Additionally, the high resolution characterization methods could be very helpful for guiding vineyard development. For example, such detailed soil information can be used to lay out vineyards according to the natural geologic variations or to develop variable vine spacing in an attempt to encourage uniform development across a vineyard block.

RÉSUMÉ
L'élaboration de stratégies viticoles de précision, visant à produire de façon uniforme des vins de haute qualité par parcellisation des vignobles, impliquent que l'on puisse évaluer les facteurs qui déterminent les qualités vinifères du raisin à diverses échelles spatiales et temporelles. Bien que des facteurs micro-météorologiques soient souvent pris en compte au moment de déterminer une stratégie d'irrigation ou lors de l'évaluation de la qualité du fruit, l'influence des variations à petites échelles des sols et leur influence sur la disponibilité de l'eau et sa relation sur la qualité du fruit sont encore mal connues. Nous avons étudié des données de levés par géoradar de surface, à partir de techniques utilisant à des ondes de sol à 900 MHz et des ondes réfléchis à 100 MHz, afin d'obtenir une cartographie à haute résolution de la teneur des sols en eau à faible profondeur, avec une précision suffisante et cela de façon non-invasive. En combinant les différentes techniques de levé par géoradar et en ayant recours à des antennes à différentes fréquences, il est possible d'en arriver à un cube unitaire de la teneur hydrique du sol et ainsi à pouvoir suivre la distribution de la teneur hydrique en fonction du temps. De telles techniques ouvrent la voie à la mise au point d'approches viticoles de précision qui tiendront comptes de la variabilité des sols, comme l'irrigation de précision. Par ailleurs, les méthodes de caractérisation à haute résolution pourraient être fort utiles pour orienter le développement des vignobles. Ainsi, de telles informations pédologiques détaillées pourraient permettre d'établir les vignobles en fonction des variations géologiques des sols, ou encore, d'adapter l'espacement des vignes de manière à favoriser un développement uniforme sur toute la parcelle.

INTRODUCTION
Enhancing wine quality is critical to the financial success of the wine industry. Quality in turn is linked to the physiological status of grapevines, and specifically to the plant's water status. Through trial and error over hundreds of years, European winemakers have recognized that certain soils, and their associated water holding capacity, produce finer wines than do others, and that soils play a role in '*terroir*' (Wilson, 1998). Due to natural geological processes, soil properties can vary laterally over distances as small as several metres to tens of metres. Recognition and incorporation of information about small scale soil variability into crop management practices could lead to winegrapes that better

reflect the microterroir of the area. However, because viticulture within areas such as California and Australia is relatively new and human intervention there tends to play a significant role (e.g., Bohmrich, 2006), the perspective that has helped to define terroir in older winegrape growing areas is still evolving in newer winegrape growing regions. Geospatial information about properties that influence grape quality could be useful in these newer areas for guiding precision viticulture.

Soil parameters (such as soil texture, volume of rock fragments, and soil moisture) play an important role in winegrape growing. Soil texture, or the proportion of sand, silt, and clay, influences the depth to which the roots grow, the amount of water that is available for the roots, water movement through soil (hydraulic conductivity), the extent of soil compaction, and the amount of aeration (e.g., Wright, 2001; Rice, 2002; White, 2003). Sandy soils generally have a low water-holding capacity, and thus require frequent irrigation in regions that do not receive sufficient rainfall. Loams and clay loams have finer soil particles and a higher plant-available water-holding capacity, and therefore generally require less irrigation. Large-scale soil surveys, such as those provided by the United States Department of Agriculture (USDA) Soil Conservation Service, provide general information about the spatial variations in soil properties but do not provide the level of detail required to make management decisions at the vineyard scale. The "industry standard" approach to vineyard soil characterization is to dig backhoe pits for soil sample collection at points located on a 75 m grid. Measurement spacings such as this are also typical for collection of conventional 'point' soil water content measurements. A number of studies have shown, however, that soil properties may vary significantly *over lateral distances of less than a metre to tens of metres* (e.g., Saddiq *et al.*, 1985; Hendrickx *et al.*, 1990; Loague and Gardner, 1990; Or and Rubin, 1993; Western and Grayson, 1998; Huisman *et al.*, 2003; Grote *et al.*, 2003). Under these natural conditions, conventional approaches for characterizing soil properties using point sampling techniques may not be sufficient for capturing the variations in soil properties needed to guide precision agriculture (e.g., Lamb and Bramley, 2001).

A considerable amount of viticultural research has identified strategies that can be used to optimize winegrape quality, including canopy management cropping levels, and irrigation (e.g., Kennedy, 2002). Although the impact on microclimate is often considered when developing large-scale irrigation strategies or when assessing vintage quality, a clear and consistent message emerging from many viticulture workshops is that the wine industry has an inadequate understanding of the influence of small- and intermediate-scale soil variabilities on winegrape variability and irrigation needs. Below we briefly describe different approaches that are often used to guide winegrape irrigation. We then present two studies that describe how we have used a surface geophysical method to map soil moisture in very high spatial resolution, and suggest that such data can be used to guide precision viticulture strategies.

WINEGRAPE IRRIGATION APPROACHES

Because water is a key control on the physiological status of grapevines, good water management can be used to optimize crop quality. It is generally accepted that moderate water stress on grapevines at early stages in the growing season controls canopy density and has a positive impact on the fruit characteristics (e.g., Hardie and Considine, 1976; Freeman and Kliewer, 1983; Matthews *et al.*, 1987; Wample *et al.*, 1989; Loveys *et al.*, 1998; Dry *et al.*, 2000; Williams, 2001). In areas where there is little rainfall during this time period, such as in the Napa Valley of California, water management typically entails irrigation.

To optimize water stress, it is critical to understand how much irrigation water to apply and when to apply it. Because the amount of irrigation water that a plant requires is a function of the plant and soil characteristics as well as the microclimate, irrigation approaches are guided by measurements of those properties. Plant-based irrigation approaches often include using visual indicators or measurements at one location in the vineyard. Visual indicators of plant stress include assessing shoot-tip growth and leaf colour (e.g., Caldwell-Ewart, 2003). Plant based measurement techniques include collecting measurements, such as leaf water potential (LWP), stem water potential (SWP) (e.g., Matthews *et al.*, 1987; Naor 1998; Chone *et al.*, 2001), and canopy temperature (Van Zyl, 1986). Typically these measurements are collected at single plant locations within a subset of vineyard blocks during times when the vineyard manager suspects that water stress is critical.

Popular climate-based methods for guiding irrigation rely on weather data, empirical crop coefficients, and meteorological models to calculate the evapotranspiration (ETo) and irrigation requirements (e.g., Snyder *et al.*, 1989; Grattan *et al.*, 1998, White, 2003). Because moderate water stress in the early growing season is desirable, deficit irrigation strategies have been developed that strive to give the plant some fraction of the full ET_0, where ETc = ETo * k_c , where k_c is the crop coefficient. The crop coefficient is a function of canopy density and winegrape variety and is spatially and temporally variable (e.g., Williams, 2001). Deficit irrigation approaches require accurate monitoring of leaf water status or soil moisture content (Stamp, 2003) to ensure that the plant does not reach a critical state of water stress. Such monitoring can be challenging, given the natural variations that can occur over small scales.

Soil-based characterization methods often involve taking soil water content measurements at single locations within a vineyard by retrieving a soil sample and assessing the moisture, or by using an indirect geophysical probe or borehole approach. These 'point' soil measurement techniques often include gravimetric, time-domain reflectometry, and neutron probe approaches. Because soil properties can vary laterally on the order of metres, however, conventional sampling at a location or two within a vineyard block is commonly insufficient to assess the soil moisture distribution throughout a vineyard. Additionally, without destructive analysis of the soil system, it is often difficult to assess the vertical soil moisture variations. In some cases, such as under irrigated conditions, roots may be concentrated in the very shallow wetted zones. In other cases, 'feeder roots' may penetrate many metres below the ground surface in search of moist soil layers or groundwater. Because of the uncertainty associated with the root zone depth range, it is commonly difficult to determine at which depth or over which depth range conventional soil moisture measurements should be collected.

HIGH RESOLUTION CHARACTERIZATION OF SOIL WATER CONTENT USING GROUND-PENETRATING RADAR

Because surface geophysical tools can probe the subsurface with high spatial resolution and in a non-invasive manner, they offer great potential as a shallow soil investigation tool. Electrical methods have been used in recent years to guide precision agricultural studies (e.g., Lund *et al.*, 2001; Johnson, C.K. *et al.*, 2003; Bramley, 2005). The ability to conduct electrical current through the soil is a function of a variety of factors, such as the soil clay content, temperature, water content, and soil water salinity. As such, unique interpretation of electrical conductivity (or its inverse, electrical resistivity) data in terms of a single soil property can be challenging (e.g., Rhoades *et al.*, 1976; Harstock *et al.*, 2000), although it has been successfully performed using site-specific controls (e.g., Sheets and Hendrickx, 1995; Michot *et al.*, 2003; Reedy and Scanlon, 2003). Typically, electrical measurements are used to provide spatial patterns of electrical conductivity, rather than to provide unique estimates of soil properties.

Ground Penetrating Radar Background

Surface Ground Penetrating Radar (GPR) approaches have been used recently to map the dielectric properties of the soil and subsequently to estimate soil water content within vineyard sites in a non-invasive manner and with high spatial resolution (Hubbard *et al.*, 2002; Grote *et al.*, 2003; Lunt *et al.*, 2005). GPR uses high frequency electromagnetic waves (~50MHz to 1200 MHz) to probe the subsurface. Radar signal penetration varies as a function of several parameters, including the GPR system performance, the attenuation of the media, and the antenna frequency (Davis and Annan, 1989). GPR signal attenuation increases with both increasing frequency and increasing electrical conductivity of the subsurface material. For example, using a conventional radar system with lower frequency (50 or 100 MHz) antennas, signal penetration depths of ten to twenty metres below ground surface have been realized (e.g., Fisher *et al.*, 1992; van Overmeeren *et al.*, 1997; Hubbard *et al.*, 2001), especially in sandy areas or when using common midpoint (CMP) acquisition geometries to collect the data (which increases the signal to noise ratio; Annan, 2005). However, acquisition using higher frequency signals (such as 900 MHz) in clay rich sediments may permit GPR imaging of reflections located only a metre or less below the ground surface. Resolution refers to the ability to distinguish two signals that arrive very close in time. Because the maximum vertical resolution is generally considered to be a quarter of the predominant wavelength of the system, and wavelength decreases with frequency, higher resolution can be obtained using higher frequency signals. For example, resolutions on the order of 0.25 m are common using 100MHz data, whereas resolutions on the order of 0.05-0.1 m are possible using 900MHz data. There is a trade-off between achievable spatial resolution and penetration depth: higher frequency signals tend to yield higher resolution information than lower frequency signals, but because they attenuate more rapidly they cannot be used for probing deeper soil layers. In practice, the antenna frequency is chosen depending on target depth of the investigation and considerations of the electrical characteristics of the study site. In general, GPR methods work best in soils that have a low electrical conductivity, such as coarser grained soils with low-salinity soil water, as soils having a high electrical conductivity tend to attenuate the GPR signal rapidly.

The GPR concept of estimating water content is similar to that of the Time Domain Reflectometry (TDR) method, which is commonly used in agricultural applications to measure water content at a single location. With TDR applications, metal prongs are inserted into the ground, and a high frequency (~1-3GHz) electromagnetic pulse is sent out, which travels through the soil along the probes. The travel time (or the velocity, v) of the electromagnetic wave is used to estimate the dielectric constant (k) of the soils through which the wave travels using the approximate relationship $k=(c/v)^2$ where c is the electromagnetic speed of light in a vacuum ($3x10^8$ m/s). With the GPR technique, the instrument is pulled along the ground surface as is illustrated in Figure 1 or is attached to farm equipment, and an electromagnetic pulse is sent from the transmitting antenna (Tx) into the soil. Energy from the pulse travels directly through the air (A in Figure 1) to the receiving antenna (Rx), along the ground surface (the groundwave; G in Figure 1), and through the soil. Energy penetrating the soil is reflected at soil interfaces having differing dielectric constants (R in Figure 1). The travel-time of the wave is measured, and if the length of the travel path is known, the electromagnetic wave velocity can be estimated and converted into dielectric constant values. A detailed discussion of the use of GPR for hydrogeological investigations is given by Annan (2005).

The dielectric constant of soils is most sensitive to soil water content, which renders both TDR and GPR useful tools for measuring vineyard soil water content. The dielectric constant of air is 1, of water is 80; these two values represent the approximate end members of the dielectric constant range. The dielectric constant of dry soils is ~4-8, and as the pore spaces become filled with soil water, the effective dielectric constant increases. A relationship commonly used to relate dielectric constant measurements to volumetric soil water content was developed empirically by Topp *et al.* (1980) using TDR laboratory estimates of dielectric constant and various soil samples. Other researchers have used dielectric mixing models, where the effective dielectric constant is represented as a function of the volume fractions and the dielectric constant of each soil constituent, such as soil grains, air, and water (e.g. Roth *et al.*, 1990; Friedman, 1998). To obtain quantitative and accurate estimates of water content using TDR or GPR methods, other researchers have developed site- and frequency-specific calibration equations. Further discussion of these relationships, as well as the use of GPR methods to estimate water content, is given in Huisman *et al.*, (2003).

We have used both GPR groundwave and reflection techniques to provide spatially extensive estimates of soil water content distribution over time within two different California vineyards. Different GPR antenna frequencies (100 MHz and 900 MHz) were used so that we could map soil moisture in shallow as well as in deeper soil layers. The results of a vineyard GPR groundwave and a vineyard GPR reflection wave study are briefly described below.

GPR Groundwave Study at the Robert Mondavi Vineyard

We tested the concept of using the GPR groundwave technique to estimate water content within a Robert Mondavi Vineyard, located in Napa Valley, California. This study vineyard is located in Oakville, which is about 9 miles northwest of the city of Napa. Sediments at the Robert Mondavi study site were deposited within alluvial fans, flood plains, and low terrace settings during the Holocene Epoch. Soils at the study site are generally categorized as USDA Bale Loams. The water table is approximately 3 to 4 m below ground surface, and the topography is fairly level. The approximately 4 acre study site is planted with Cabernet Sauvignon grapes having row and vine spacing of 1.2 m. The mean annual precipitation in this valley is from 64 to 89 cm. Summers are hot and dry, and the winters are cool and moist with a mean annual air temperature of ~60 degrees F. All vines in the study area are subject to the same volume and frequency of irrigation water via a drip system, with an average of $0.02m^3$/vine/week during the warmest months (typically May through October).

Soil texture, soil moisture, remote sensing, and GPR data were collected at the study site. Approximately 30 soil texture measurements were obtained by hand-auguring soil samples and assessing the gravel, sand, silt, and clay content. Water content measurements were obtained using conventional sampling techniques (TDR, gravimetric, and neutron probe techniques) to compare with estimates of volumetric water content obtained from the GPR data. Remote sensing data sets were acquired using the airborne ADAR Multispectral System 5500 (Positive Systems) in the blue, green, red and near-infrared portions of the spectrum from a flight altitude of 4300 m above ground level and with a high spatial resolution of 2m x 2m (Johnson *et al.*, 2001, 2003). These data were used to estimate canopy vegetation density variations within the block; Figure 2a illustrates the normalized density vegetation index (NDVI) map created from the multispectral data for the site, where the brown tones illustrate areas of low vegetation density and the green tones illustrate areas of higher vegetation density, or higher vigor (Johnson, L.F. *et al.*, 2003). Note that although this vineyard block is planted with the same grape variety, has been subject the same farming practices, irrigation, and microclimate, and has relatively uniform topography, there are variations in the vegetation that likely reflect soil property variations.

GPR data were collected several times over a period of 18 months within the area shown in Figure 2a using a 900 MHz PulseEKKO 1000 System. The GPR data were collected with measurement spacing in the lateral direction of 0.1-0.2 m along the middle of every 5th row with a fixed antenna separation distance of 0.17 m (see Fig. 1). Because the high frequency groundwave only travels a short distance from the transmitting to the receiving antenna, the strength of the groundwave signal was significant enough to be detected at all locations, despite the presence of some finer textured soils that tend to attenuate the GPR signals. Using the GPR groundwave arrival travel time and the distance between the transmitting and receiving antenna, we estimated the electromagnetic velocity and subsequently the dielectric constant at each sampled location. As described by Grote *et al.* (2003), we used a site-specific relationship to portray the GPR-obtained dielectric constant values in the form of maps of estimated water content. Figures 2b and 2c illustrate the estimated water content obtained at two different times during the year using the GPR groundwave technique. Independent comparison with conventional 'point' soil moisture measurements, obtained using time domain reflectometry (TDR) and gravimetric techniques at the locations indicated in Figure 2a, revealed that the estimates of

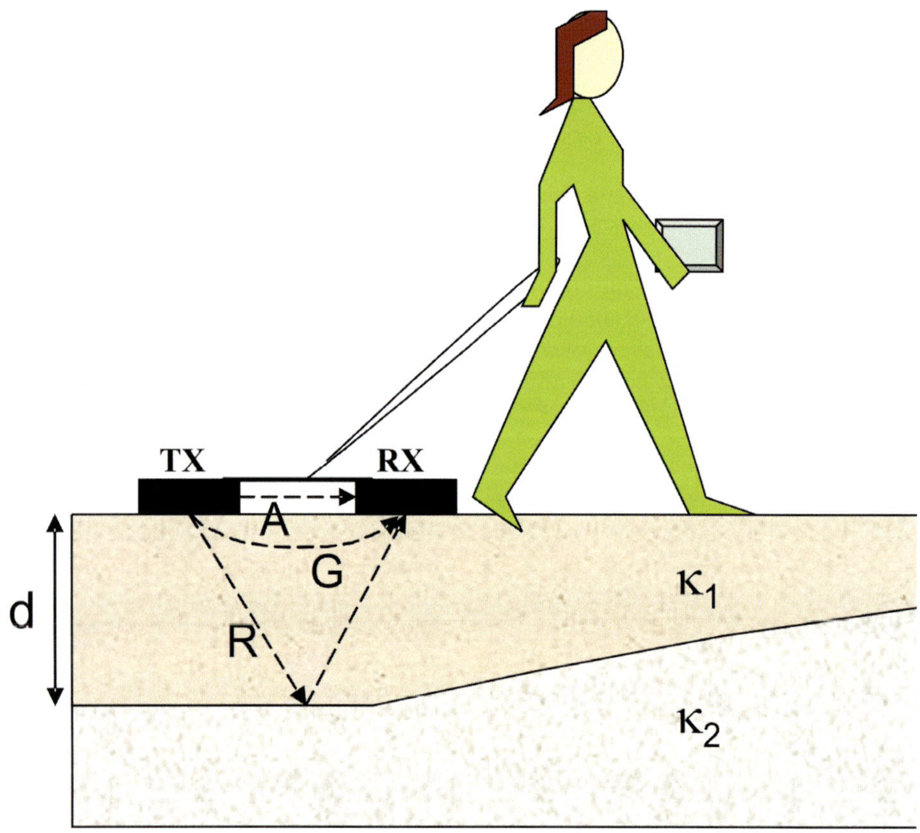

Figure 1. Surface GPR acquisition geometry. Waves travel from the transmitting antenna (TX) to the receiving antenna (RX) through the air (A), and through the soil as groundwaves (G) and as reflected waves (R). The signal is reflected at boundaries that have different dielectric constants (κ).

volumetric water content obtained using the 900 MHz GPR groundwave data were accurate to within 1% by volume, and that the 900 MHz groundwave sampled approximately the upper 15 cm of the soil layer. The spacing of the GPR-obtained water content estimates was extremely dense; each dataset in Figures 2b and 2c contains over 20,000 measurements, which is perhaps the highest density of shallow moisture measurements obtained to date in vineyards. We found that although the mean value of the water content estimates throughout the study block changed during the year, the spatial pattern remained fairly consistent. A detailed discussion of the GPR Mondavi groundwave study is given by Grote *et al.* (2003).

Figure 2d shows the spatial variations in sand content obtained from interpolating between shallow soil samples at the same study site. Comparison of the GPR-obtained water content estimates (Figs. 2b, 2c), and the soil textures (Fig. 2d) reveal the control of soil texture on shallow soil moisture; the sandy areas are consistently dryer while the finer textured soils, which retain moisture better, are consistently wetter. Comparison of these figures with the remote sensing vigor estimates (Fig. 2a) suggest that at this location, the sandy,

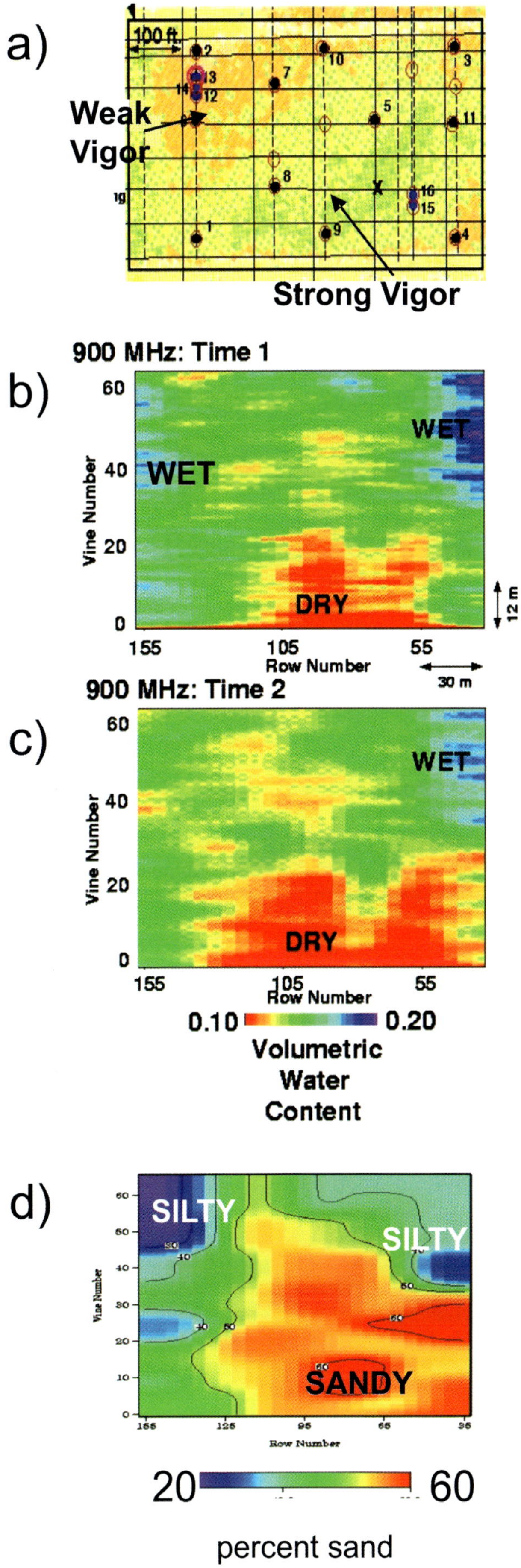

Figure 2. Datasets at the Mondavi Study Site. a) NDVI map obtained from multispectral data (e.g., Johnson, L.D. et al., 2003) indicating areas of weaker vegetation (brown) and stronger vegetation, or vigor (green). Solid black circles indicate locations of neutron probe access tubes; red circles indicate detailed 'study areas' where validation of the GPR moisture content estimates was performed; note 100 ft scale, top left corner; b) Estimates of volumetric water content (m^3/m^3) in the top ~0.15 m obtained using surface 900 MHz GPR groundwave data at one point in time. This map encompasses ~20,000 data points; note 12 and 30 m scales, lower right; c) estimates of volumetric water content (m^3/m^3) during another point in time obtained using surface GPR groundwave data; scale as in (b); d) soil texture maps obtained from interpolation of soil texture data collected in the near surface soil layer throughout the site.

dryer areas have a higher vegetation vigor than the more silt-rich, wet soils. A similar but better correlation was found by comparing soil textures over a thicker depth interval with NDVI responses (Hubbard *et al.*, 2003). The relationships between soil texture, soil moisture, and NDVI suggests that at this site, the silt-rich soils may be too compacted for proper root functioning, or that the uniform water that is applied to the block through irrigation, which is designed to satisfy the needs in the more sand-rich soils, results in over-watering of the finer textured soils. The consistent moisture patterns over time and the apparent linkages between the soil water content, soil texture, and vegetation density are in line with the findings of Bramley (2001), who found that, although annual winegrape yields per acre could vary as much as by a factor of ten within a vineyard, the patterns of spatial variation in yield typically were constant over time. Bramley (2001) found that the variations remained consistent, regardless of year-to-year variations in winegrape yield, which were driven largely by climate.

GPR Reflection Study at the Dehlinger Vineyard

Although the GPR groundwave approach provided high resolution and accurate information about water content at the Mondavi site, the zone of influence of the high frequency groundwave was quite shallow. In this section, we describe the use of GPR reflected waves (refer to Fig. 1) to obtain information about moisture content variations in deeper soil layers. We tested the concept of using GPR reflection arrivals to map deeper water content at the Dehlinger vineyard in Sonoma County. As illustrated in Figure 1, a GPR reflection occurs where there is a dielectric contrast between two subsurface units, such as between a soil layer and bedrock, unsaturated and saturated soils, or possibly between two different soil layers. However, in some soils, GPR reflections may be difficult to detect because soil properties can vary gradually with depth rather than as a series of distinct layers with detectable interfaces. Also, in-row tilling of vineyard soils may disrupt the natural geological variations that can create GPR reflections. Additionally, if the electrical conductivity of the soils is high, i.e. if significant volumes of clay or saline pore fluids are present, the signal attenuation may be too large to permit the GPR signal to travel down to the reflecting interface and back up to the receiving antenna (see Fig. 1). Importantly for GPR reflection methods, the difference in dielectric constant between different soil types alone typically is not large enough to give rise to a strong GPR reflection. However, different soil layers commonly retain different amounts of soil moisture, and these differences in soil moisture content may give rise to a GPR reflection (Martinez and Byrnes, 2001).

In order to estimate the velocity of the reflected event from the recorded two-way travel time of the signal, information about the depth of the reflector ('d' in Fig. 1) must be independently determined, for example, from borehole measurements. If the subsurface horizon that causes the GPR reflection varies greatly in depth over short distances, it may be difficult to determine the depth of the reflecting horizon in all locations. The absence of reasonably accurate depth information will reduce the accuracy of water content estimates. In spite of these difficulties, when a smoothly varying reflecting subsurface horizon can be identified using GPR and borehole data, estimates of water content can be made over time using the GPR reflection method.

We investigated the use of GPR reflection data for providing information about deeper soil layer water content at the Dehlinger Winery in Sonoma County, California. This vineyard is located on Guerneville Road of Sebastopol, a town situated about 8 miles southwest of the city of Santa Rosa. Vineyard soils at this site are composed of a 1.5 to 1.8 m thick red, fine-loamy soil (Sebastopol series) and a buff-coloured, clayey soil (Goldridge series) (Miller, 1972). We used conventional soil textural analysis techniques to assess soil variations within this 80 by 180 m study area. We found that the soil textures vary between sandy loam and clay loam, but are generally composed of sandy clay loam. Up to 30 weight percent gravel is present in the Sebastopol series soil, but gravel is absent in the Goldridge series soil. Topography varies by up to 3.5 m over the study area, with the highest elevations located in the north-western corner of the study site. The water table is approximately 4 m below the ground surface, as measured in nearby wells. Average air temperatures vary between 7°C between November and March and 16°C for the remainder of the year. Mean annual precipitation is around 100 cm, with the majority of the rain falling between November and March. Vines are spaced at 2.44 m (8 feet) intervals and are arranged in rows that are 3.48 m (10 feet) apart. Metal trellises are located at every fourth vine. The crops in this study area are 20 year-old Chardonnay vines, all of which are derived from the same rootstock. No in-row tilling has been performed at this site, and thus the natural soil horizons are 'intact'. An above-trellis spray system is used occasionally to irrigate the winegrapes during hot weather.

GPR reflection travel time data were used to estimate changes in soil water content under a range of soil saturation conditions throughout the growing season at the Dehlinger site. Data were collected during several data acquisition campaigns using 100 MHz and 900 MHz surface GPR antennas. GPR reflections, imaged using 100 MHz antennas, were associated with a thin (~0.1 m thick), low permeability clay layer located 0.8 to 1.3 m below the ground surface and identified from borehole information. GPR profiles indicated that the subsurface reflection is channel-shaped. Field infiltration tests and neutron probe logs suggested that the thin clay layer inhibited vertical water flow, and was coincident with high volumetric water content values. The GPR reflection two-way travel time and the depth of the reflector at the borehole locations were used to calculate an average dielectric constant for soils above the reflector. A site-specific relationship between the dielectric constant and volumetric water content was then used to estimate the depth-averaged volumetric water content of the soils above the reflector. Compared to average water content measurements from calibrated neutron probe logs over the same depth interval, the estimates obtained from GPR reflections at the borehole locations had an error of 1.8%. Details of the Dehlinger GPR reflection study are provided by Lunt *et al.* (2005).

Measurements of travel time to the GPR reflector (obtained using the

100 MHz antennas) and estimates of the depth of the reflecting clay layer (obtained using borehole measurements) were used within a Bayesian estimation approach (following Chen *et al.*, 2001) to estimate deeper, depth-averaged volumetric water content. Figure 3 illustrates the estimated volumetric water content for the zone located above the reflecting clay layer (i.e., at depths less than 1.5m below ground surface) at different times during the year. As was observed at the Mondavi study site, there is a consistent spatial pattern of water content variation over time at the Dehlinger site that might be difficult to detect using conventional measurement approaches. Figure 3 shows that a channel-shaped feature trends NE to SW through the study site, and that this feature influences water content distribution: within this area, the soils are consistently wetter than the surrounding soils. The Dehlinger Winery vineyard manager has recognized that the area outlined by this channel-shaped feature consistently has lower grapevine vegetative growth (= lower vigor) than that of the surrounding area, based on annual measurements of fruit yield and pruning weight. The close correspondence of the soil moisture distribution and the vegetation variations at the Dehlinger Winery site indicate how soil variations may influence grapevine characteristics.

These results suggest that the two-way travel time to a GPR reflection associated with a geological surface can be used under some natural conditions to obtain estimates of average water content when GPR reflectors are present and detectable over time, and when borehole control is available. By using different GPR arrivals and different frequency GPR events, we can sample different effective soil layers and thus resolve soil water content variations vertically as well as laterally. Figure 4 illustrates water content estimates obtained from 900 MHz groundwave data, 100 MHz groundwave data, and 100 MHz reflection data over the Dehlinger site during a single acquisition campaign, which are interpreted to have sampled approximately the top 0.15m, 0.8m, and 1.2m of the subsurface soil, respectively. This example indicates the potential of developing 3D "moisture cubes", using different GPR signal components and frequencies.

SUMMARY AND OUTLOOK

Development of targeted wine grape irrigation strategies, focused on promoting uniformly high wine quality throughout vineyard blocks, necessitates an ability to assess the soil water content across a range of spatial and temporal scales. Although micrometeorological factors are often considered when developing an irrigation approach or assessing fruit quality, the influence of small-scale soil variations and their control on water use and associated fruit quality is poorly understood within the industry. However, ground-based geophysical methods now make it possible to characterize variations in soil properties.

We have investigated the use of surface GPR data, using both groundwave and reflected wave techniques, to map shallow soil water content in high resolution, with acceptable accuracy, and in a non-invasive manner. We found that GPR groundwave techniques provide rapid and accurate estimates of soil water content in shallow soil layers. We found that the GPR reflection technique is less straightforward to perform, as it requires a-priori information about the depth of a reflector, and the application is limited to areas that have good GPR reflectors and low electrical conductivity environments. However, in areas such as the Dehlinger vineyard, which satisfied those criteria, the GPR reflection approach can be used to map water content in deeper soil layers. Use of multiple GPR methods and frequencies offers the potential to characterize the soil in 3-D space. The depth of penetration of the GPR energy is a function of the site-specific conditions. Our GPR reflection example at the Dehlinger site extended to only 1.5 m below the ground surface. However, in areas that are less electrically conductive and that have a deeper GPR reflector, such as a deeper water table or a deep interface between soils and bedrock, the same technique could be used to map soil water content at deeper levels. Both studies suggest that soil water content, soil texture, and vegetation vigor are correlated. Our experience suggests that there is often a temporal persistence of soil moisture patterns within vineyard blocks. This suggests that once soil patterns are identified, point measurements at 'hot spots' within a block may suffice

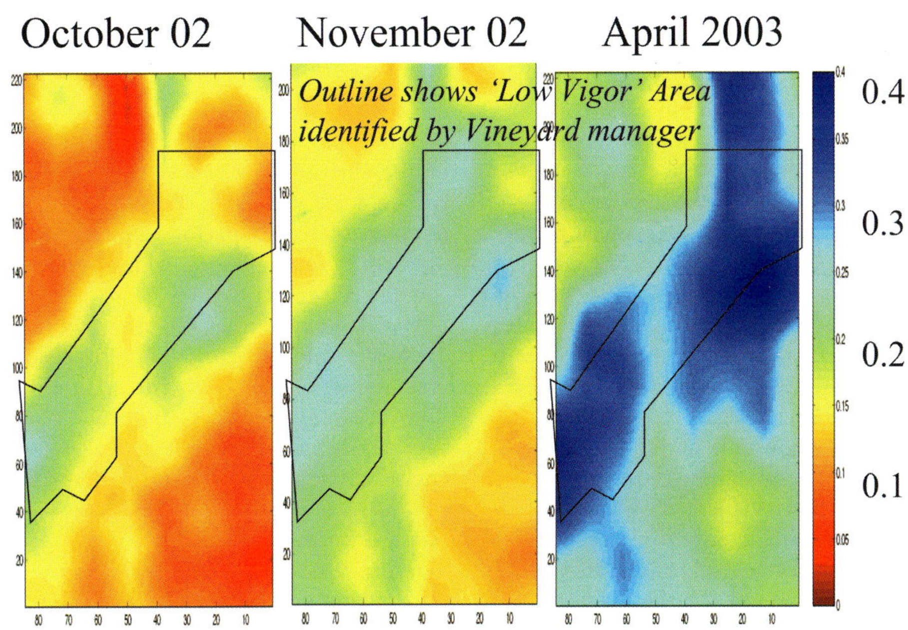

Figure 3 Average volumetric water content of the top soil layer (<1.5m below ground surface), estimated using 100 MHz GPR reflection travel time data at the Dehlinger Vineyards. Black line indicates the boundary of a low vigor area identified by the vineyard manager, which is coincident with the consistently wetter area identified using GPR reflection data. Color key at right indicates relative volumetric water content in units of m^3/m^3, from red (drier) to blue (wetter).

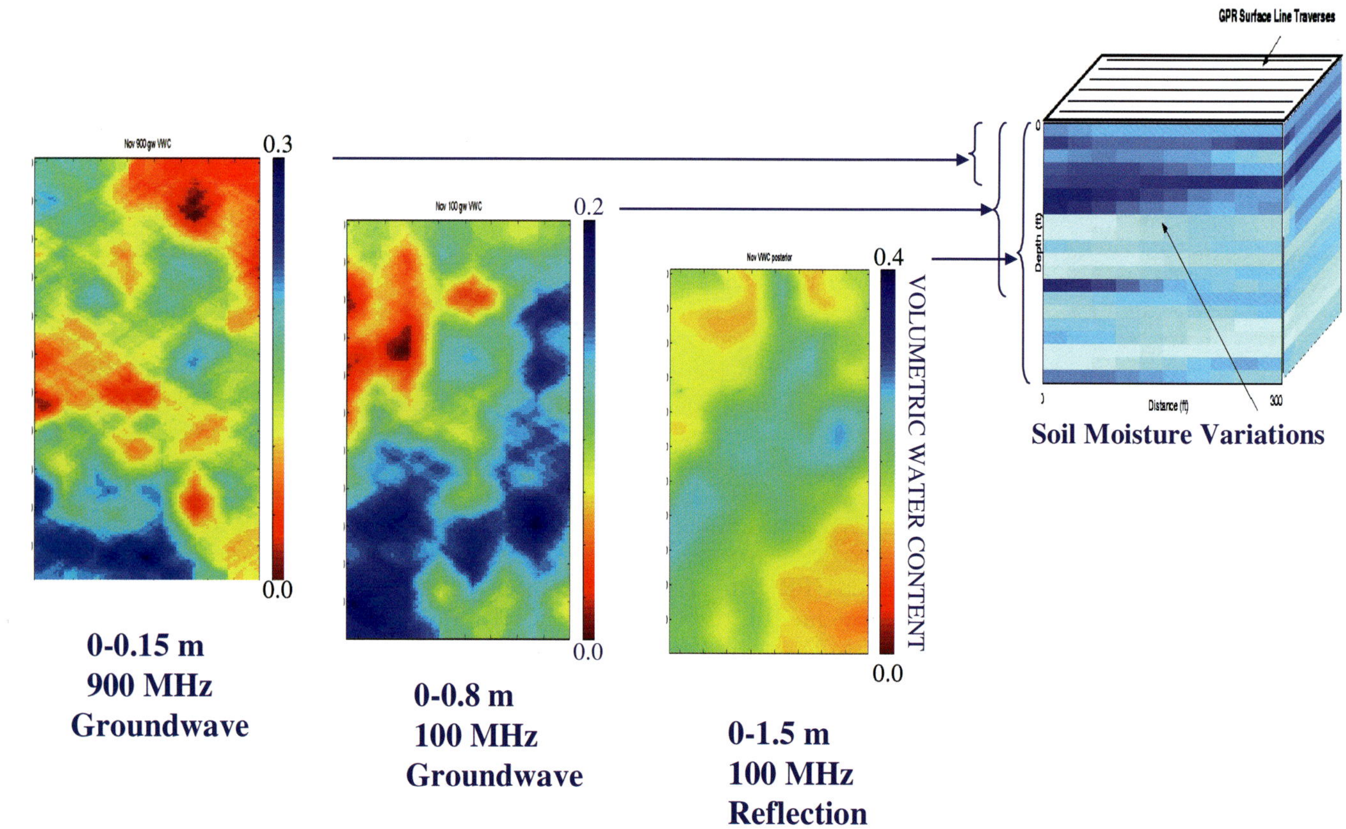

Figure 4. Estimates of volumetric water content (m^3/m^3) obtained over different soil layers using different GPR arrivals and antenna frequencies at the Dehlinger vineyard site, illustrating how different components of the GPR signal and different frequency antennas can potentially be used to construct a 3D soil water content cube. Color keys indicate relative moisture contents.

for management purposes. Geophysical techniques, such as the GPR method presented here, open the way for development of precision viticulture approaches that account for natural soil variabilities.

Because of the high gross revenue per acre and large financial premiums associated with improving wine quality, developing such abilities to map soil properties and use these as a guide for precision irrigation could prove economically feasible for optimizing winegrape production as well as reducing overall water use. In addition to using such data to guide precision farming (such as precision irrigation), the high resolution characterization methods could be very helpful for guiding vineyard development. We are currently in the process of working with Dr. Phillip Freese of WineGrow on development of the Dichiro Vineyards, located approximately five miles north of Healdsburg in Alexander Valley of Sonoma County, California. At this 20 acre site, we are using a variety of measurement methods, including GPR, electrical resistivity, cone penetrometer, neutron probe, TDR, and remote sensing, to delineate vineyard management zones. We are using the maps to guide lay-out of the vineyard blocks according to natural geologic variations and to choose the (variable) vine spacing to promote uniform development of winegrapes across management zones.

We expect that the use of geophysical techniques for guiding precision viticulture will increase as the competition for high fruit quality and the demand for water supplies increases. We envision that a combination of dense soil information, obtained using geophysical techniques, could be used together with remotely-sensed canopy information and microclimate climate data to develop a better understanding of the relationships between soil, vegetation, meteorological variables, and winegrape quality, or of the micro-terroir.

ACKNOLWEDGEMENTS

We would like to thank Lee Johnson (NASA Ames) for providing access to the remote sensing imagery at the Robert Mondavi Site, Daniel Bosch (Robert Mondavi), Marty Hedlund (Dehlinger Winery), Tom Dehlinger (Dehlinger Winery), Phillip Freese (WineGrow) and the DiChiros for providing vineyard access and viticulture expertise. We also thank Larry Meinert, Roger Macqueen, and Jim Hunter for useful comments and suggestions. The studies described herein were funded by USDA 2001-351-2-09866 and by UC Discovery itl04-10144 to Yoram Rubin.

REFERENCES

Annan, A.P., 2005, Ground Penetrating Radar for hydrogeological investigations: Chapter 7 *in* Rubin, Y., and Hubbard, S. (eds.), Hydrogeophysics, Springer, The Netherlands, p. 185-214.

Bramley, R.G.V. 2001, Progress in the development of precision viticulture - Variation in yield, quality and soil properties in contrasting Australian vineyards: *in* Currie, L.D., and Loganathan, P. (eds.), Precision tools for improving land management; Occasional report n. 14,

Fertilizer and Lime Research Centre, Massey University, Palmerston North, Australia, p. 25-43.

Bramley, R.G.V., 2005, Understanding variability in winegrape production systems, 2. Within vineyard variation in quality over several vintages: Australian Journal of Grape and Wine Research, v. 11, p. 33-42.

Bohmrich, R.C., 2006, The next chapter in the terroir debate: Wine Business Monthly, v. XIII, n. 1, p. 88-93.

Caldwell-Ewart, C., 2003, Reading vines to manage wine quality: Practical Winery and Vineyard, v. 78, July/August, 2003, p. 16-24.

Chen, J., Hubbard, S., and Rubin, Y., 2001, Estimating Hydraulic Conductivity at the South Oyster Site from Geophysical Tomographic Data using Bayesian Techniques based on the Normal Linear Regression Model: Water Resources Research, v. 37, n. 6, p. 1603-1613.

Chone, X., Van Leeuwen, C., Dubourdieu, D., and Gaudilliere, J.P., 2001, Stem water potential - Sensitive indicator of grapevine water status: Annals of Botany, v. 87, p. 447-483.

Davis, J.L., and Annan, A.P., 1989, Ground-penetrating radar for high resolution mapping of soil and rock stratigraphy: Geophysical Prospecting, v. 37, n. 5, p. 531-551.

Dry, P.R., Loveys, B.R., Stoll, M., Stewart, D., and McCarthy, M.G., 2000, Partial rootzone drying - an update: Australian Grapegrower and Winemaker, v. 438, p. 35-39.

Fisher, E., McMechan, G.A., and Annan, A.P., 1992, Acquisition and processing of wide aperture ground-penetrating radar data: Geophysics, v. 57, n. 3, p. 495-504.

Freeman, B.M., and Kliewer, W.M., 1983, Effect of irrigation, crop level and potassium fertilization on Carignane vines: 2. Grape and wine quality: American Journal of Enology and Viticulture, v. 34, n. 3, p. 97-207.

Friedman, S.P., 1998, A saturation degree-dependent composite spheres model for describing the effective dielectric constant of unsaturated porous media: Water Resources Research, v. 34, p. 2949-2961.

Grattan, S.R., Bowers, W., Dong, A., Snyder, R.L., Carroll, J.J., and George, W., 1998, New crop coefficients estimate water use of vegetables, row crops: California Agriculture, v. 52, n. 1, p. 16-21.

Grote, K., Hubbard, S., and Rubin, Y., 2003, Field-Scale Estimation of Volumetric Water Content using GPR Groundwave Techniques: Water Resources Research, v. 39, n. 11, 1321, doi 10.1029/2003WR002045.

Hardie, W.J., and Considine, J.A., 1976, Response of grapes to water-deficit stress in particular stages of development: American Journal of Enology and Viticulture, v. 27, p. 55-61.

Harstock, N.J, Mueller, T.G., Thomas, G.W., Barnhisel, R.I., and Wells, K.L., 2000, Soil electrical conductivity variability: *in* Robert, P.C. *et al.* (eds.), Proceedings of the Fifth International Conference on Precision Agriculture and Other Resource Management, Bloomington, MN, July 16-19, 2000, ASA, CSSA, and SSSA, Madison, WI, published as CD-ROM.

Hendrickx, J.M.H., Wierenga, P.J., and Nash, M.S., 1990, Variability of soil water tension and soil water content: Agricultural Water Management, v. 18, p. 135-149.

Hubbard, S. S., Chen, J., Peterson, J., Mayer, E.L, Williams, K.H., Swift, D.J, Mailloux, B., and Rubin, Y., 2001, Hydrogeological characterization of the South Oyster Bacterial Transport Site using geophysical data: Water Resources Research, v. 37, n. 10, p. 2431-2456.

Hubbard, S.S., Grote, K. and Rubin, Y., 2002, Estimation of near-subsurface water content using high frequency GPR ground wave data: Leading Edge of Exploration, Society of Exploration Geophysics, v. 21, n. 6, p. 552-559.

Hubbard, S.S., Grote, K., Lunt, I., Pierce, L., Peterson, J., and Rubin, Y., 2003, Spatial and temporal variability of soil moisture content obtained using GPR data: Accuracy, resolution and implications for Precision Viticulture: Abstracts with Programs, Geological Society of America Annual Meeting, November 2–5, 2003, v. 35, n. 6, p. 187.

Huisman, S., Hubbard, S., Redman, D., and Annan, P., 2003, Monitoring Soil Water Content with Ground-Penetrating Radar: A Review: Vadose Zone Journal, v. 2, p. 476-491.

Johnson, C.K., Mortensen, D.A., Weinhold, B.J., Shanahan, J.F., and Doran, J.W., 2003, Site specific management zones based on soil electrical conductivity in a semiarid cropping system: Agronomy Journal, v. 95, p. 303-315.

Johnson, L., Bosch, D., Williams, D., and Lobitz, B., 2001, Remote sensing of vineyard management zones: implications for wine quality: Applied Engineering in Agriculture, v. 17, p. 557-560.

Johnson, L.D., Roczen, D., Youkhana, S., Nemani, R., and Bosch, D., 2003, Mapping vineyard leaf area with multispectral satellite imagery: Computers and Electronics in Agriculture, v. 38, n. 1, p. 37-48.

Johnson, L.F., 2003, Temporal stability of an NDVI-LAI relationship in a Napa Valley vineyard: Australian Journal of Grape and Wine Research, v. 9, p. 96-101.

Kennedy, J., 2002, Understanding grape berry development: Practical Winery and Vineyard, July/August 2002, p. 14-23.

Lamb, D.W., and Bramley, R.G.V., 2001, Innovations and technology - managing and monitoring spatial variability in vineyard productivity: Australian Association of Natural Resource Management, v. 4, n. 1, p. 25-30.

Loague, K.L., and Gardner, G., 1990, R-5 revisited: 1. Spatial variability of infiltration on a small rangeland catchment: Water Resources Research., v. 26, n. 5, p. 957-971.

Loveys, B., Stoll, M., Dry, P., and McCarthy, M., 1998, Partial rootzone drying stimulates stress responses in grapevine to improve water use efficiency while maintaining crop yield and quality: Australian Grapegrower and Winemaker, v. 414, p. 108-113.

Lund, E.D., Christy, C.D., and Drummond, P.E., 2001, Using yield and soil electrical conductivity (EC) maps to derive crop production performance information: *in* Robert, P.C. *et al.* (eds.), Proceedings of the Fifth International Conference on Precision Agriculture and Other Resource Management, Bloomington, MN, July 16-19, 2000, ASA, CSSA, and SSSA, Madison, WI, published as CD-ROM.

Lunt, I., Hubbard, S.S., and Rubin, Y., 2005, Soil moisture content estimation using ground-penetrating radar reflection data: Journal of Hydrology, doi:10.1016/j.jhydrol.2004.10.014, v. 307/1-4, p. 254-269.

Martinez, A., and Byrnes, A. P., 2001, Modeling Dielectric-constant Values of Geologic Materials - An Aid to Ground-penetrating Radar Data Collection and Interpretation: Current Research in Earth Sciences, Kansas Geological Survey Bulletin 247, part 1, http://www.kgs.ku.edu/Current/2001/martinez/martinez1.html

Matthews, M.A., Anderson, M.M., and Schultz, R.H., 1987, Phenologic and growth responses to early and late season water deficits in Cabernet Franc: Vitis, v. 26, p. 147-160.

Michot, D., Benderitter, Y., Dorigny, A., Nicoullaud, B., King, D., and Tabbagh, A., 2003, Spatial and temporal monitoring of soil water content with an irrigation corn crop cover using surface electrical resistivity tomography: Water Resources Research, v. 39, n. 5, doi:10.1029/2002WR001636.

Miller, V.C, 1972, Soil Survey of Sonoma County, California. USDA Soil Survey,

United States Government Printing Office, Washington D.C., 188 p.

Naor, A., 1998, Relations between leaf and stem water potentials and stomatal conductance in three field-grown woody species: Journal of Horticultural Science and Biotechnology, v. 73, p. 431-436.

Or, D., and Rubin, Y., 1993, Stochastic Modeling of Unsaturated Flow in Heterogeneous Media with Water Uptake by Plant Roots: Tests of the Parallel Columns Model Under Two Dimensional Flow Conditions: Water Resources Research, v. 29, n. 12, p. 4109-4119.

Reedy, R.C., and Scanlon, B.R., 2003, Soil water content monitoring using electromagnetic induction: Journal of geotechnical and geoenvironmental engineering, v. 129, n. 11, p. 1028-1039.

Rhoades, J.D., Raats, P.A.C., and Prather, R.J., 1976, Effects of liquid-phase electrical conductivity, water content, and surface conductivity on bulk soil electrical conductivity: Soil Science Society of America Journal, v. 40, p. 651-655.

Rice, T.J., 2002, Importance of Soil Texture to Vineyard Management: Practical Winery and Vineyard, March/April 2002, p. 22-29.

Roth, K., Schulin, R., Fluhler, H., and Attinger, W., 1990, Calibration of time domain reflectometry for water content measurement using a composite dielectric approach: Water Resources Research, v. 26, p. 2267-2273.

Saddiq, M.H., Wierenga, P.J., Hendrickx, J.M.H., and Hussain, M.Y., 1985, Spatial variability of soil water tension in an irrigation soil: Soil Science, v. 140, p. 126-132.

Sheets, K.R., and Hendrickx, J.M.H., 1995, Non-invasive soil water content measurement using electromagnetic induction: Water Resources Research, v. 31, p. 2401-2409.

Snyder, R.L., Lanini, B.J., Shaw, D.A., and Pruitt, W.O., 1989, Using reference evapotranspiration (ETo) and crop coefficients to estimate crop evapotranspiration (ETc) for trees and vines: Cooperative Extension, University of California, Berkeley, CA, Leaflet No. 21428, 8 p.

Stamp, J., 2003, Partial Rootzone Drying: Wine Business Monthly, January, 2003, p. 44-50.

Topp, G.C., Davis, J.L., and Annan, A.P, 1980, Electromagnetic determination of soil water content: measurements in coaxial transmission lines: Water Resources Research, v. 16, p. 574-582.

van Overmeeren, R.A., Sariowan, S.V., and Gelrels, J.C., 1997, Ground penetrating radar for determining volumetric soil water content; results of comparative measurements at two test sites: Journal of Hydrology, v. 197, p. 316-338.

Van Zyl, J.L., 1986, Canopy temperature as a water stress indicator in vines, South Africa: Journal of Enology and Viticulture, v. 7, 53-60.

Wample, R. L., Mills, L., and Davenport, J.R., 1999, Use of precision farming practices in grape production: *in* Robert, P., Rust, R.H., and Larson, W.E., (eds.). Proceedings of the. 4th International Conference on Precision Agriculture, Minneapolis-St. Paul, MN, 19-22 July, 1998. ASA/CSSA/SSSA Press, Madison, WI. p. 897-905.

Western, A.W., and Grayson, R.B., 1998, The Tarrawarra data set: Soil moisture patterns, soil characteristics, and hydrological flux measurements: Water Resources Research, v. 34, n. 10, p. 2765-2768.

White, R.E., 2003, Soils for fine wines: Oxford University Press, New York, 279 p.

Williams, L.E., 2001, Irrigation of Winegrapes in California: Practical Winery and Vineyard, November/December 2001, p. 42-55.

Wilson, J.E., Terroir, 1998, The role of Geology, Climate and Culture in the Making of French Wines: Mitchell Beazley, London, United Kingdom, 336 p.

Wright, Terry, 2001, Diverse geology/soils impact wine quality: Practical Winery and Vineyard, v. XXIII, n. 2, September-October, 2001, p. 40-49.

PAPER 4 FROM THE "TERROIR, GEOLOGY AND WINE: A TRIBUTE TO SIMON J. HAYNES" SESSION HELD AT THE GEOLOGICAL SOCIETY OF AMERICAN ANNUAL MEETING, SEATTLE, WASHINGTON, NOVEMBER 3, 2003

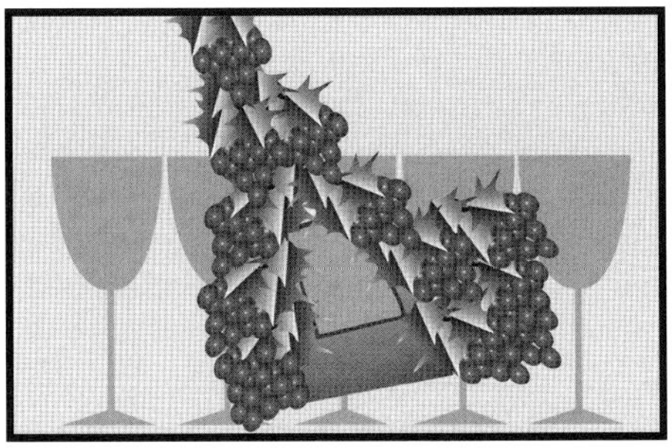

Climate and Terroir: Impacts of Climate Variability and Change on Wine

Gregory V. Jones
Department of Geography, Southern Oregon University, 1250 Siskiyou Blvd, Ashland, Oregon 97520, USA, gjones@sou.edu

SUMMARY

As part of the concept of terroir, climate is a factor that influences both the general viability of a region to ripen a specific variety of grapes and the resulting wine style. Climate variability and change affect every form of agriculture and are seldom more evident than with the production of high quality wines where narrow climate zones for optimum quality place them particularly at risk. While climates have changed dramatically during the history of the cultivation of grapes for wine, the changes in the recent past and those projected in the near future have received a great deal of attention for their potential impacts on today's wine industry. Research has shown that both recent short and long-term climate changes have resulted in generally warmer and longer growing seasons with less frost risk in many of the world's best wine regions. While these changes have been related to greater production and quality, future climate projections indicate the potential for threshold issues in which too much warming will potentially alter traditional wine styles and/or varieties planted, and likely bring about spatial shifts in viticultural viability.

RÉSUMÉ

Cet élément du concept de terroir qu'est le climat détermine la viabilité d'une région à permettre le mûrissement d'une variété définie de raisin ainsi que l'élaboration du style du vin qu'on en tire. La variabilité et le changement du climat affectent toute forme d'agriculture, mais les effets sont rarement plus évidents que sur la production de grands vins où l'optimisation de la qualité nécessite une viticulture en des zones climatiques aux conditions très circonscrites, d'où les risques. Bien que les climats aient changé dramatiquement dans l'histoire de la culture des vignes vinicoles, les changements climatiques récents ainsi que les changements projetés dans un avenir rapproché ont été beaucoup étudiés en fonction de leurs impacts sur l'actuelle industrie vinicole. Les compte rendus des recherches montrent que les changements climatiques, tant à courts et qu'à longs termes, ont amené des saisons viticoles plus chaudes et plus longues comportant moins de risques de gel, et cela dans nombres des meilleures régions vinicoles du monde. Bien que ces changements aient été associés à une amélioration de la production et de la qualité, les projections climatiques montrent qu'il est possible qu'un seuil soit atteint, au-delà duquel ce réchauffement pourrait altérer le style les vins traditionnels et/ou les variétés utilisées, entraînant des changements spatiaux probables de la viabilité viticole.

INTRODUCTION

Climate is a very complex, highly variable, and pervasive factor in our natural Earth- and human-based systems. From controlling vegetation patterns and geological weathering characteristics, to influencing water resources and agricultural productivity, climate is at the heart of the delicate equilibrium that exists on Earth. Climate is a wide-ranging factor in virtually all forms of agriculture, from influencing spatial variations of crop viability to largely determining year-to-year yield variability; climate is an important issue in determining where and how crops are grown. Climate-related impacts on agriculture are generally related to climate variability over the short term (i.e., intra-annual and inter-annual) and climate change over the long term (i.e., decades to centuries or longer). Understanding such relationships is seldom more important than today, as variations and/or changes in climate can greatly impact the ability to produce a given crop.

The Climate Component of Terroir

Grapevines are some of the oldest cultivated plants and have been historically associated with Mediterranean climates (e.g., Italy). Today, however, grapevines for wine production (note that all occurrences of grapes or grapevines refer to grapes for wine production, unless stated otherwise) are grown in many types of climates throughout the mid latitudes: Mediterranean, marine west coast (e.g., Oregon), humid subtropical (e.g., eastern Australia), and semi-arid continental climates (e.g., eastern Washington state) (Fig. 1). The climates of these regions are an integral part of the notion of terroir, the French concept in which a complex interplay of physical factors (Wilson, 1998; Haynes, 1999; Meinert and Busacca, 2000) and cultural influences (Vaudour, 2000) interact to define the wine styles and quality that come from any site or region. Climate is one part of the continuum that includes the physical landscape influences of soil and terrain, which in combination largely determine the grape variety that can be grown in a given region. Once grape variety-site characteristics are considered, the remaining viticulture (the science of the cultivation of grapevines) and enology (the science of the making of wines) aspects, which include regional associations and cultural traditions, result in the defining wine style a region produces.

Of all of the site factors, climate arguably exerts the most profound effect on the ability of a region or site to produce quality grapes. Worldwide, the average climatic conditions of a wine region determine to a large degree the grape varieties that can be grown there, whereas wine production and quality are chiefly influenced by site-specific factors, husbandry decisions, and short-term climate variability (Jones and Hellman, 2003). While the average climate structure is very important for grape growth and wine quality, weather and climate factors on daily and hourly time scales are critical, and include: solar radiation, heat accumulation, temperature extremes (including high temperatures during the summer, winter freezes and spring and fall frosts), diurnal temperature ranges, precipitation (especially during flowering and ripening stages), wind, and extreme weather events such as hail. Owing to inadequate weather station locations, poor data sharing between countries, and disagreement on what weather and climate factors are most important, the climate component of terroir has been studied largely through spatial and temporal averaging, and by comparing new wine region climates to analogous climates in historical regions.

Temperature and Growing Season

Although the climate-based aspects of terroir are related to various factors that operate from the macroscale to the microscale, measures of temperature on daily, weekly, monthly, and growing season time scales have most often been employed to define the spatial differences between regions. In general, growing season length and temperatures are critical aspects because of their major influence on grape ripening and fruit quality, and therefore varietal adaptation to a specific terroir. It is in its ideal climate that a given grape variety can achieve optimum ripening profiles of sugar, acid, and flavour to maximize a given style of wine and the vintage quality.

The growing season necessary for wine grapes varies from region to region but averages approximately 170-190 days (Mullins *et al.*, 1992). The

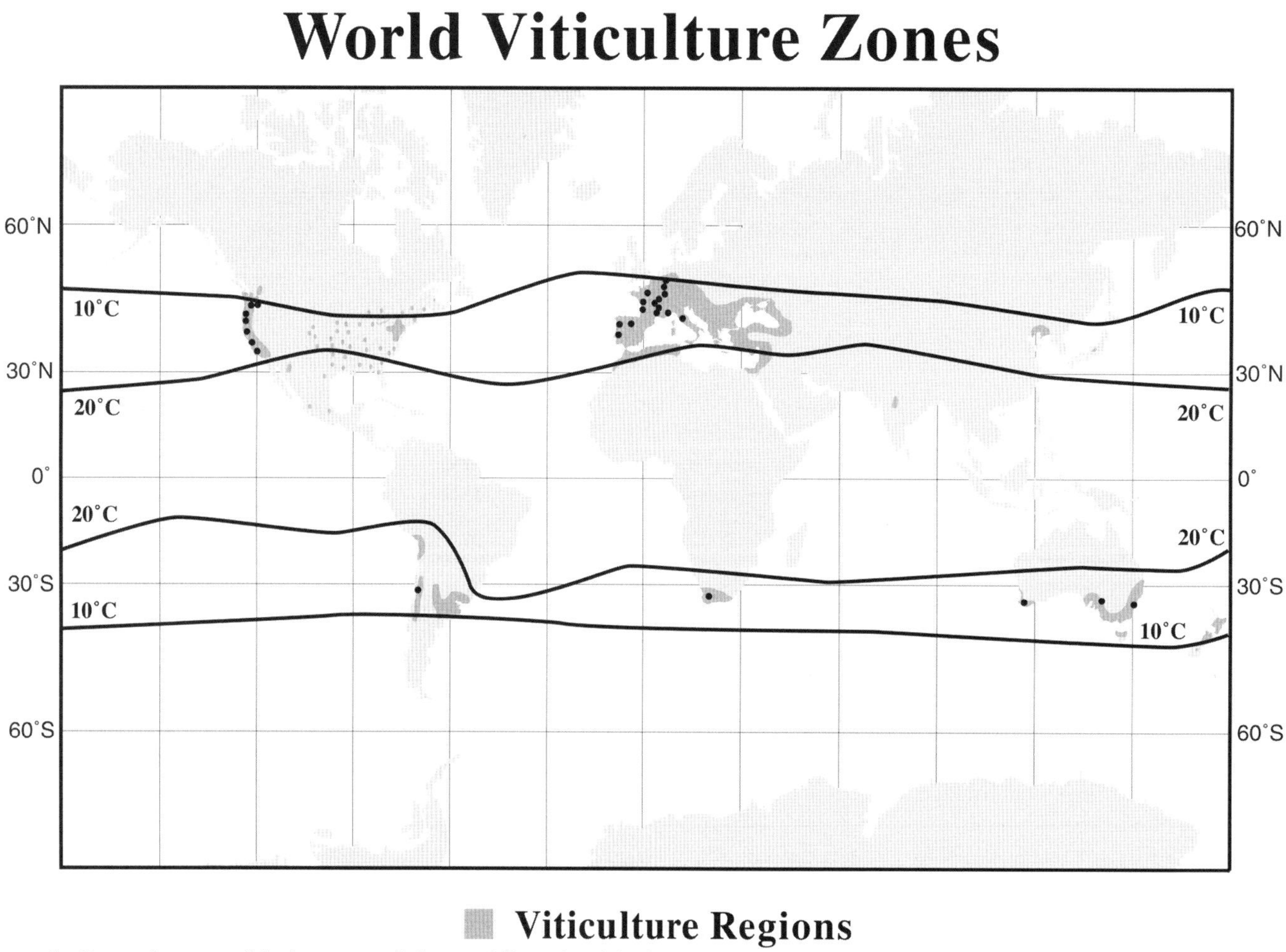

Figure 1. General geographical extent of the world's main viticulture regions (adapted from de Blij, 1983). Contours represent the mean annual 10°C and 20°C isotherms as a proxy for the latitudinal limits of the majority of the world's grape growing areas. Neither latitude nor isotherms fully accounts for the location or quality of individual vineyard regions, however. The solid dots represent the wine regions studied by Jones et al. (2005).

length of the frost-free season is important to the onset of bud break, flowering, and the timing of harvest. Prescott (1965) noted that in most areas suitable for grape production the mean temperatures of the warmest and coldest months are more than 18.9°C and -1.1°C, respectively. Winter and spring temperatures are also important in that frost and low temperatures can cause injury to grapevines. Research has also shown that there is a minimum winter temperature that the grapevines can withstand. This minimum ranges from -5°C to -20°C, and is chiefly controlled by microscale climate variations which in turn reflect location and topography (Amerine *et al.*, 1980; Winkler *et al.*, 1974). Temperatures below these thresholds will damage plant tissue by the rupturing of cells, the denaturing of enzymes by dehydration, and the disruption of membrane function (Mullins *et al.*, 1992). Another important influence on winter damage is vine health prior to winter dormancy. Diseases such as moisture-induced downy mildew that ultimately cause defoliation and reduced sugar accumulation, can affect winter hardiness (Amerine *et al.*, 1980).

Heat Summation Index

In France during the mid nineteenth century, A.P. de Candolle observed that vine growth started when the mean daily temperature reached 10°C. This led to the notion of a heat summation above a base temperature that could define vine growth and grape maturation. Amerine and Winkler (1944) followed this concept and developed a heat summation index for California that is widely used as a guide for selecting appropriate grape varieties and for determining a given area's suitability to produce quality wine grapes. In the Northern Hemisphere the heat summation index is calculated for the period of April 1 through October 31 by summing each day's average temperature above a base of 10°C, the minimum temperature at which vine growth occurs. Using the index, five climatic regions were defined, Winkler regions I-V, from which a minimum and maximum threshold of 950 and 2900 degree-days, based on °C respectively, was calculated for the cultivation of grapes. Although it varies by grape variety and many different terroir factors, the best quality wine typically is produced from regions that experience between 1400-2000 degree-days. However, this index has been found to be less appropriate for determining viticultural viability outside of California (Gladstones, 1992; Spellman, 1999; Jones and Davis, 2000).

Bioclimatic Indices

Other bioclimatic indices have been used to characterize a region's potential for viticulture and are mostly developed on the basis of temperature. Various forms of a heliothermal index (Branas, 1974; Huglin, 1978) and a latitude-temperature index (Jackson and Cherry 1988; Kenny and Shao, 1992) have been used to help define the suitability of a region to the planting of certain grape varietals. Smart and Dry (1980) developed a simple classification of viticultural climates that uses five dimensions of mean temperatures, continentality (defined as the difference between the average mean temperature of the warmest and coolest months), sunlight hours, aridity (based upon the difference between rainfall and evaporation), and relative humidity. Gladstones (1992) developed a classification similar to Amerine and Winkler (1944) but refined it by imposing an upper limit on mean temperatures, a correction factor for latitude, and a correction for each month's temperature range. In a maturity classification, Jones *et al.* (2005) related climate and ripening potential for different vari-

Table 1: Wine region average growing season temperatures as analyzed by Jones et al. (2005) sorted into their respective climate maturity groupings as depicted in Figure 2.

Region	Growing Season[a] Tavg (°C)	Climate Maturity Grouping[b]
Mosel Valley	13.0	Cool
Alsace	13.1	
Champagne	14.5	
Rhine Valley	14.9	
N. Oregon	15.2	Intermediate
Loire Valley	15.3	
Burgundy-Côte	15.3	
Burgundy-Beaujolais	15.8	
Chile	16.3	
E. Washington	16.5	
Bordeaux	16.5	
C. Washington	16.6	
Rioja	16.7	
S. Oregon	16.9	
C. California	17.0	Warm
South Africa	17.1	
N. California	17.4	
N. Rhône Valley	17.6	
N. Portugal	17.7	
Barolo	17.8	
S. Rhône Valley	18.2	
Margaret River	18.6	
Chianti	18.8	
Hunter Valley	19.8	Hot
Barossa Valley	19.9	
S. Portugal	20.3	
S. California	20.4	

Note that the growing season average temperatures depicted here are derived from a 0.5° x 0.5° grid and not from any one station [see Jones et al. (2005) for details].

[a] The growing season is Apr-Oct in the Northern Hemisphere and Oct-Apr in the Southern Hemisphere.

[b] The climate maturity groupings are based upon the average growing season temperatures and the ability to ripen a given variety [see Figure 2 and Jones et al. (2005) for details].

eties based on average growing season temperatures – cool, intermediate, warm, and hot climates – that were derived from the benchmark climates of the world's premium quality wine producing regions (Table 1; Fig. 2).

Importance of Climate: the Bordeaux Example

While no one study can ever capture the sum of all influences in the terroir debate, Van Leeuwen *et al.* (2004) may have come the closest to date. Studying numerous plots in the St. Émilion and Pomeral regions of Bordeaux, France, the authors examined the effect of vintage (climate), soil, and grape variety (Cabernet Sauvignon, Cabernet Franc, and Merlot) on wine quality parameters. The parameters included plant effects that influence quality (i.e., phenology, shoot growth cessation, vine water status, etc.) and yield components such as fruit weight, sugar and acid levels, and phenolics (e.g., anthocyanin). Their results showed that climate was the dominant factor, accounting for over 50% of the variation when averaged across all quality parameters. Soil type and structure was the next most important factor, accounting for a quarter of the resulting wine quality. Varietal differences, while not as important as climate or soil, still accounted for 10% of the variation in quality parameters. While the study controlled for many cultural practices (e.g., pruning type and timing, trellising, and harvest timing), it would appear in this case that the cultural component of terroir roughly accounts for only 15-20% of the variation in important wine quality parameters. If results from this study are applicable elsewhere, it would appear that climate plays the most significant role in wine quality.

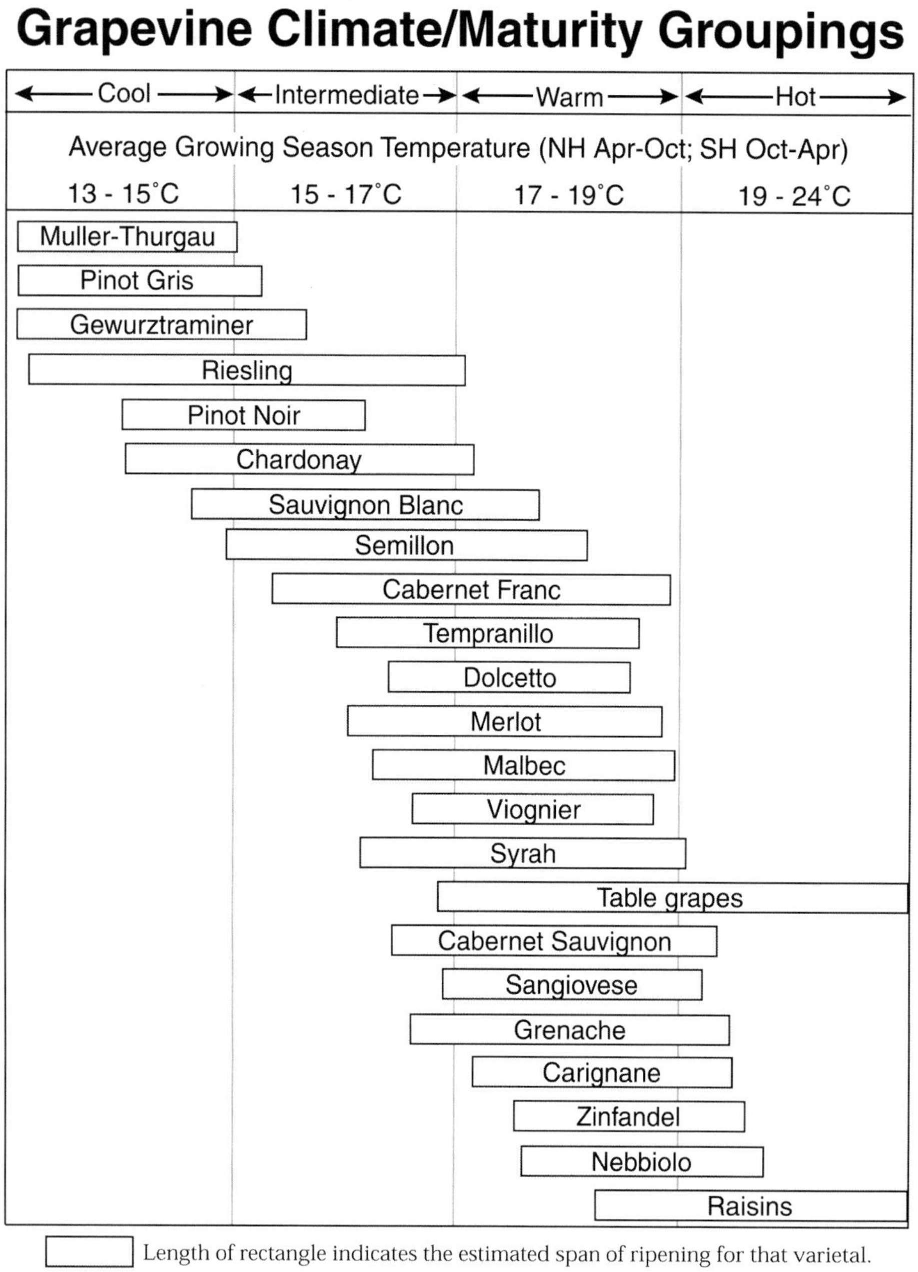

Figure 2. Climate maturity groupings based on average growing season temperatures and the estimated span of varietal ripening potential that occurs within and across the groups. While V. vinifera varieties are not typically used for table grapes and raisins, they are included for comparison of warm to hot climate wine grape production. Note that the average growing season temperatures are depicted in Table 1 and are derived from grids, not station data; therefore the values given may deviate slightly from any one station in a given region (Jones *et al.*, 2005).

Climate Variability, Change, and Agricultural Responses

It is clear from the observational record that global climates have changed over both short and long time scales. Global observations from the 20th century indicate that the mean surface temperature of the planet has increased by 0.6 ± 0.2°C with land areas warming more than oceans (Houghton *et al.*, 2001). In most cases observed atmospheric warming trends have been found to be asymmetric with respect to seasonal and diurnal cycles, with greatest warming occurring during the winter and spring and at night (Karl *et al.*, 1993). Precipitation changes have been much more spatially variable, but show 5-10% increases over broad areas of the Northern Hemisphere while other regions have seen declines of 5-15% (IPCC, 2001). Some regions have also seen an increased frequency of heavy precipitation events, while other regions have seen moderate changes in drought frequency and severity. In addition, evidence from paleoclimatological studies has revealed large fluctuations in global and regional climate over the past million years. Furthermore, there are indications that over the last 2000-3000 years climates have shifted from cooler to warmer periods, with the most recent century potentially being the warmest in the last millennium (Mann *et al.*, 1999). Our understanding of the potential range of climate's natural variation is still somewhat limited, however.

Contemporary global climate change due to increased levels of atmospheric CO_2 has received considerable attention (Climate Research Board, 1979; CO_2/Climate Review Panel, 1982; Carbon Dioxide Assessment Committee, 1983; IPCC 2001; and others). The effects have been observed in the general climate record (e.g., Jones *et al.*, 1986) and simulated through a variety of general circulation models (GCMs; see Cess *et al.*, 1990; and Gates *et al.*, 1998 for good reviews). The observed trends in temperatures have been related to agricultural production viability through their impact on winter hardening potential, frost occurrence, and growing season lengths (Menzel and Fabian, 1999; Carter *et al.*, 1991; Easterling *et al.*, 2000; Nemani *et al.*, 2001; Moonen *et al.*, 2002; Jones, 2005). On the other hand, GCM simulations – although far from perfect representations of climate reality – all indicate that climate change will continue to occur in some manner owing to anthropogenic influences. While the initial magnitude of climate change, as simulated by GCMs, has been reduced due to improved parameterization (IPCC, 2001), climate change at any magnitude or in any direction will still have an effect on agricultural productivity. The overall impacts of climate change on agriculture ultimately will depend on plant physiological requirements and the spatial variations, seasonality, and magnitude of the warming (McCarthy *et al.*, 2001; Butterfield *et al.*, 2000).

Although the forecast of temperature increases for individual regions is under debate owing to the scale problem associated with GCM projections (Giorgi and Mearns, 1991; Jones *et al.*, 1995), there is one area of no debate - the atmospheric concentration of carbon dioxide will continue to rise for at least a few decades to come (Climate Research Board, 1979; CO_2/Climate Review Panel, 1982; Carbon Dioxide Assessment Committee, 1983; IPCC 2001; and others). Under optimum growing conditions, the atmospheric concentration of CO_2 is the limiting factor in photosynthesis, and increasing CO_2 concentrations will increase the water use efficiency in plants in the absence of other climate-related changes (Rosenberg, 1981a,b). Therefore, more carbon dioxide can potentially lead to greater production. However, many observational and modeling studies on the effects of increases in CO_2 on various plant systems (Reekie and Bazzaz, 1991; Hänninen, 1995; Maytin *et al.*, 1995) indicate that increased carbon uptake is partitioned, with more entering vegetative tissues and less entering reproductive organs, which could spell a reduction in grain and fruit quality. More complete research into changes in the quality of plant tissue and fruit along with better understanding of how other impacts (e.g., concomitant changes in temperature) from CO_2–induced climate change is needed.

Agricultural productivity varies with climate variations both between and within years. The observed human activities that drive changes in atmospheric composition and climate could therefore be expected to alter, perhaps significantly, the relative levels of agricultural productivity in different regions of the globe. Agriculture systems typically are chosen and developed based on the mean climatic conditions of a region and its variability, but it has been suggested that variations in the probability and duration of extreme events are more important than changes in the mean (Waggoner, 1989; Katz and Brown, 1992). Extreme events including floods, high temperatures, frost, drought, hail, rainfall, and wind all affect agriculture in some way. If extreme events increase in frequency, plant systems would tend to fail to recover over the short term and/or adjust over the long term without some form of stress and damage occurring. In addition, as an agricultural system becomes less resilient, it is more susceptible to increased outbreaks in disease and pests. Therefore, climatic variability and its relation to climate change will play an important role in the long and short-term structure and yields of agriculture systems, including the growing of wine grapes.

Past Climate Variability, Change, and Viticulture: the European Example

Viticulture and enology combine to produce a very geographically defined and therefore climatically distinct agricultural pursuit. Experience over thousands of years has resulted in grapes being grown and wine being produced in mostly mid-latitude regions that are roughly defined by their ability to ripen fruit to produce certain wine styles (Fig. 1). Although many new world wine regions are relatively young, long-term historical records of European viticulture have been maintained for nearly a thousand years. Records of harvest dates and yield were initially maintained by monks during the Middle Ages and later by merchants and the prominent châteaux. These records indicate that the region has experienced wide fluctuations of climate and viticultural productivity in the past (e.g. Penning-Roswell, 1989). Gladstones (1992) depicted climatic variability in Europe from historical records and details that, during the medieval 'Little Optimum,' temperatures were about 1°C above the present day and the warmth lasted from about the eighth century to the early fourteenth century. During this time, vineyards were planted over most of southern England and along the coasts of the North and Baltic Seas. Subsequently the 'Little Ice Age' brought an abrupt fall in temperatures and a prolonged cold period that lasted from the early fourteenth century through the early 1800s. While there were some large fluctuations in weather during this period, the overall climate regime was extremely rough on all aspects of agriculture in Europe with failed crops and famine for many consecutive years. Vineyards throughout the British Isles and northern Europe died out and harvests did not occur for many years in southern Europe and the Mediterranean. Temperatures during the coldest decades of the Little Ice Age were about 2°C lower than the warm periods of the Middle Ages, and at least 1°C lower than today (Le Roy Ladurie, 1971).

Pfister (1988) used the recorded dates of harvests and other vine developmental stages to study the direct effects of climate variability on viticulture in Europe from the Middle Ages to 1860. He inferred that temperatures during the growing season in the High Middle Ages must have averaged 1.7°C warmer than today, and that harvest dates began around the first of September compared to early to mid

October today. These grape harvest records correlated well with other long-term records of climate change as evidenced in glacial advances and retreats, ice core analyses, palynological studies, varve chronologies, and dendrochronologies. Carrying the viticulture record even further into the twentieth century, Ray (1985) compiled the dates of the start of vintage harvest for a single château in Bordeaux (Lafite). While he noted that some of the variation is due to biotic influences (mildew, fungus, pests, diseases, etc.), the dates correspond very well with land-based climatologies of temperature (Jones *et al.*, 1986). In addition, recent research by Chuine *et al.* (2004) and colleagues used contemporary grape harvest dates from Burgundy to reconstruct spring-summer temperatures from 1370 to 2003 and, while the results indicate that temperatures as high as those reached in the warm 1990s have occurred several times in the region since 1370, the extremely warm summer of 2003 appears to have been higher than in any other year since 1370.

Climate Variability and Climate Change Impacts on Wine

Recent research on the impacts of climate on wine factors has followed two lines of analysis: 1) how short-term climate variability and trends have impacted grapevine phenology, production, and quality (e.g., Braslavska, 2000; Jones and Davis, 2000; Esteves and Orgaz, 2001), and 2) how future climate change may potentially impact the spatial viability of wine growing regions, vine and grape growth potential, and overall wine quality and styles (Kenny and Harrison, 1992; Bindi *et al.*, 1996; Jones *et al.*, 2005). Owing to the fact that many viticulture climates are dry during the growing season, most of the research conducted to date has focused on temperature-related impacts.

Climate Variability Impact: Small Scale

Climate variability impacts agriculture and, in particular, grape and wine production are typically related to site-specific extreme events such as frost, hail, and heavy rain, or to broader scale interannual cycles of warmer, colder, wetter, or drier conditions brought about by recurring modes of atmospheric circulation and/or sea surface temperatures. Analyzing a nearly 50-year record of grape and wine data for Bordeaux (1949-1997), Jones and Davis (2000) found trends towards earlier phenological events (bud break, flowering, véraison, and harvest), shorter intervals between events, and a lengthening of the growing season. This study also found that the two main varieties grown in Bordeaux, Merlot and Cabernet Sauvignon, exhibited larger berry weights and higher sugar to total acid ratios, which resulted in higher vintage ratings and less year-to-year variability in quality. Warmer temperatures during the growing season and a reduction of ripening period rainfall were found to be the most significant climate factors in Bordeaux phenology, composition, and wine quality. Wine production trends in Bordeaux were found to be more variable than phenology, composition, and quality; however, the study revealed that rainfall during the physiologically important stages of flowering and maturation tended to decrease production (Jones and Davis, 2000). Similar results were found for California's Napa and Sonoma Valleys where, since the 1950s, grape growers have seen dramatic increases in premium wine quality (along with decreased year to year variability in quality), grape yield, and crop value (Nemani *et al.*, 2001). This research found that the climate of the region has experienced greater warming at night and during spring. As a consequence of the asymmetric warming, the diurnal temperature range – the difference between daily maximum and minimum temperatures – declined by 1.9°C over 47 years. Although the average annual warming trend was a modest 1.1°C over the 47 years, there was a 71% decline in frost frequency (20-day reduction) and a 25% increase in the length of the frost-free growing season (65-day increase). The decline in frost frequency was significantly correlated with the increase in vintage ratings. A possible explanation for such a relation could be that frosts may damage the primary buds on the vine, leaving secondary or tertiary buds that are less fruitful, along with delaying subsequent plant physiology and ultimately leading to uneven maturity and poor wine quality. Additionally, for the same time period, grape yields grew 34% and were significantly influenced by minimum temperatures in spring and decreases in summer vapor pressure deficits. If the current trends in frost frequency continue, Napa/Sonoma could become a frost-free climate in another decade or two. In a larger study covering the entire west coast of the United States, Jones (2005) documented climate variability during 1948-2002 for the principal grape growing regions in the states of California, Oregon, and Washington. This research showed that on average, most regions have experienced a decline in frost frequency, earlier last spring frosts, later first fall frosts, longer frost-free periods, and warmer growing seasons with greater heat accumulation.

Climate Variability Impact: Large Scale

Examining larger scale influences of climate variability, Rodó and Comín (2000) detailed how the North Atlantic Oscillation (NAO; a north-south seesaw in the pressure fields of the North Atlantic) and the El Niño-Southern Oscillation (ENSO; a combined east-west seesaw in sea surface temperatures and pressure fields in the tropical Pacific) impacted grape production and wine quality in Spain. Acting as a climate "teleconnection", where the impact of an atmospheric variation in one region is conveyed to other regions; El Niño events (warmer than average sea surface temperatures in the eastern tropical Pacific) were found to have significant impact on wine quality, providing more rain during the growing season in Spain over the last thirty years. The NAO, on the other hand, did not impact wine quality even though the condition is closer to the wine producing regions of Spain. Esteves and Orgaz (2001) found similar results for the Viseu wine region of Portugal. The authors conducted a spectral analysis between wine quality and teleconnection indices (ENSO and NAO) and determined that wine quality in Portugal follows a 3 to 7 year cycle that is similar to that of the ENSO cycle. Whereas the NAO had some impact on Portuguese wine quality, its effect was much less than that of ENSO. Jones (1997) also found that the NAO was insignificant for Bordeaux phenology, grape composition, yields, and wine quality. The authors of each of the papers noted above speculate that the best explanation for these observations is due to the fact that the NAO is a dominant winter mode of circulation variability, whereas ENSO can influence moisture levels throughout the year depending on its strength and length of activity. For the wine regions of the west coast of the United States, Jones (2005) analyzed short-term climate variability in relation to the Pacific Decadal Oscillation (PDO, a measure of the dominant variability in sea surface temperatures in the North Pacific) and ENSO, finding that the PDO was the most dominant influence on important viticulture climate parameters. However, neither the PDO nor ENSO was significantly related to vintage ratings for Napa (Jones, 2005). The geographical location of Australian and Chilean wine producing regions places each near one of the main centers of action of ENSO. For Australia, ENSO is the main contributor of rainfall variability and El Niño years often result in moderate to severe drought. A similar but opposite situation is experienced in Chile where greater than average rainfall occurs during El Niño years. Although direct studies of the impacts of ENSO on wine production and quality in Chile and Australia are incomplete, climate modeling indicates that ENSO conditions are expected to increase in frequency and severity with climate change. Finally, the authors in the studies noted above state that since many of the circulation modes can be assessed months ahead of a given growing season, there may be some potential for seasonal prediction that would allow a greater latitude of planning for the impacts noted above. Clearly, more research needs to be done in the area of climate variability modes and their impact on viticulture and wine quality for other areas of the world.

Climate Change Impacts

Lough *et al.* (1983) examined scenarios of a warmer world in Europe using the noted 20th century warming as an analog. They compared thermal, moisture, and pressure changes over the two coolest and warmest twenty year periods during the 1900s, and used the results to construct scenarios of the impacts of future climate change on agriculture and energy usage. While noting the difficulty in assessing the consequences of technological advances and the "fertilizer effect" of increasing CO_2 concentration, these authors showed that the length of the growing season should expand over all of Europe with precipitation increasing in the north and decreasing in the south. Using mean climate values and Broadbent's (1981) compilation of vintage ratings for Bordeaux and Champagne, the authors noted that the climate variables explained 58 percent and 63 percent of the vintage ratings, respectively. Their warmer world scenarios suggested that conditions would lead to improvements in wine quality, particularly in Bordeaux. From recent analyses by Jones *et al.* (2005), where vintage ratings for Bordeaux and Champagne have trended higher with less year-to-year variability, the work of Lough *et al.* (1983) appears to have been generally correct for these regions.

Tate (2001) discussed the general impact of future climate change on grape production and wine quality. Although no data were analyzed, this author described the potential consequences of climate change as affecting where grapes can be grown ("ideal" locations today ceasing to be so in the future and more poleward locations becoming viable), changing the distribution and intensity of pests and diseases, causing sea level changes that could alter coastal climates, and increasing CO_2 such that it might impact grape components or the texture of oak wood used in barrels. In another overview of potential climate change impacts, Schultz (2000) discussed how shifts in precipitation would greatly affect the dry summer regimes of most high quality wine regions. In regions where irrigation practices are controlled or completely forbidden (i.e., many parts of France) the tendency to drier summers could be detrimental to yields and flavor profiles. Schultz (2000) also detailed how changes in CO_2, temperature, and solar radiation likely would have direct impacts on yields, grape composition, and flavor development.

Addressing the climate change effects for a specific area and grape varieties, Bindi *et al.* (1996) studied the effects of increased CO_2 levels and the associated changes in climate on Cabernet Sauvignon and Sangiovese grapes in Northern Italy. Using field data from 1992-1994 and a model of grapevine growth and yield, they found that projected temperature increases resulted in a composite 23 day reduction in the interval from bud break to harvest, and that doubled concentrations of CO_2 in the atmosphere resulted in a 36% increase in yield. However, the combined effects of increases in temperature and CO_2 resulted in an increase in yield variability with the potential to create greater economic risk. Pincus (2003) provided insight into the connections between wine, place, and identity – the heart of terroir – by examining how climate change may make the "associations between wine and place difficult or impossible to maintain." Examining climate projections to 2025 and discussing the results with grape growers and wine makers in various regions, Pincus (2003) found that some form of adaptation – including changing varieties, vineyard management, and regional identities – must occur for the industry to survive. Examining projected climate change scenarios for 2030 and 2070 for Australian viticulture, McInnes *et al.* (2003) found that temperatures are predicted to warm by 1.0-6.0°C by 2070, thus increasing the number of hot days and decreasing frost risk, while precipitation changes are predicted to be more variable, potentially resulting in greater growing season stress on irrigation. These authors indicate that the challenges facing the Australian wine industry include more rapid phenological development, changes in suitable locations for some varieties, a reduction in the optimum harvest window for high quality wines, excessive vegetative growth due to increased CO_2 levels, and greater management of already scarce water resources.

In a multi-region analysis of the impacts of climate change on wine quality, Jones *et al.* (2005) analyzed growing season temperatures in 27 of arguably the best wine producing regions in the world, in terms of being highly recognized for quality (Fig. 1). The authors used average growing season temperatures as these values typically define the climate-maturity ripening potential for varieties grown in cool, intermediate, warm, and hot climates (Fig. 2; Table 1). For example, Cabernet Sauvignon is grown in regions that span intermediate to hot climates with growing season averages that range from roughly 16.5-19.5°C (e.g., Bordeaux or Napa). Results from the analysis of 1950-1999 revealed that 17 of the 27 wine regions experienced statistically significant warming during their respective growing seasons. Figure 3 provides an example of the warming where both the Bordeaux and Napa regions warmed by 1.8°C and 1.2°C from 1950-1999, respectively. Also note that Napa has a warmer average growing season and that Bordeaux has greater growing season temperature variability. In the study, a large majority of the U.S. and European wine regions saw significant temperature increases whereas the majority of the Southern Hemisphere locations changes were not significant. Averaged across all wine regions with significant trends, the warming was 1.3°C. The most dramatic of these changes, confirmed by another observation-based climatology (Moisselin *et al.*, 2002), occurred in the northern Rhone Valley of France where the growing season warmed by 4.1°C.

Jones *et al.* (2005) also examined the relationships between average growing season temperatures and vintage ratings as given by Sotheby's and the Wine Enthusiast (Stevenson, 2002; Mazur, 2002). Ratings are commonly used to compare vintages and have a strong influence on the economic success of a wine producing region (de Blij, 1983). Although vintage ratings are inherently subjective, the correlations between different rating systems typically are very high ($r > 0.9$; Jones, 1997) indicating their usefulness as a wine quality metric. Furthermore, in the absence of a worldwide systematic, consistent, and readily available set of compositional data, vintage ratings currently provide the best metric for global scale analyses of wine quality. Vintage ratings do have limitations, however, as they are often averaged over large regions, effectively masking inter-region variations, and not all ratings systems cover all regions, nor have all regions received the publicity necessary to be more thoroughly rated. The Sotheby's vintage ratings data analyzed by Jones *et al.* (2005) represent one of the most complete rating systems and, when combined with the Wine Enthusiast ratings for South Africa and Chile (Mazur, 220), cover 27 regions and 30 categories of wine over varying time periods during 1967-2000 (Table 2; Fig. 1). Some regions are divided into subregions or wine styles with separate ratings, whereas others are simply divided into ratings for red and white wines. For 25 of the 30 wine regions or categories of wine, vintage ratings have shown trends of increasing overall quality with less vintage-to-vintage variation. As an example, Figure 4 depicts the vintage ratings for 1963-2000 for red wines from the Médoc and Graves region of

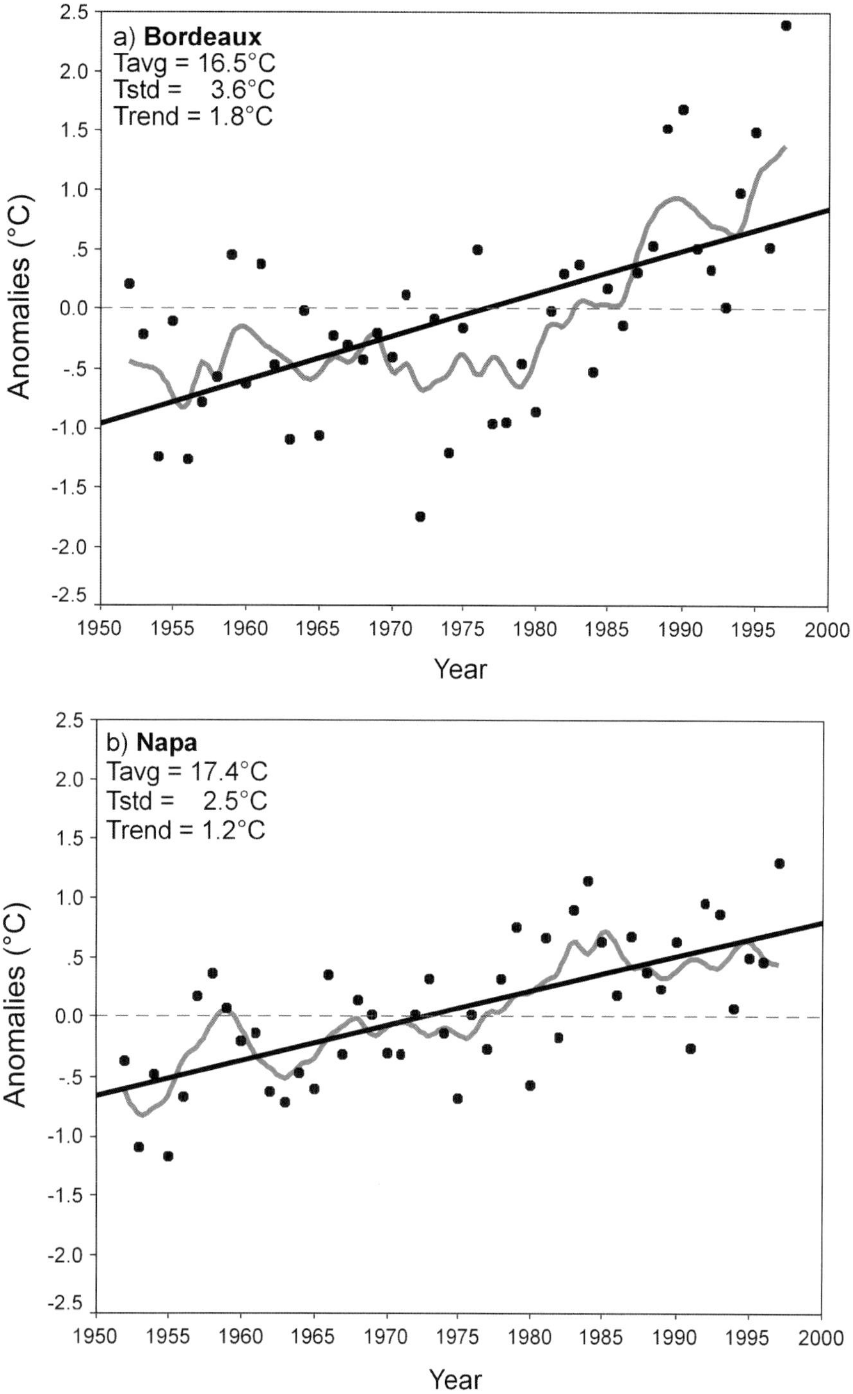

Figure 3. Observed growing season average temperature anomalies for a) Bordeaux, France, and b) Napa, California, as analyzed by Jones et al. (2005). The temperature data are monthly values extracted from a 0.5° x 0.5° grid centered over the wine producing regions for 1950-1999. Tavg is the average growing season temperature (April-October in the Northern Hemisphere and October-April in the Southern Hemisphere). Tstd is the standard deviation of monthly temperatures during the growing season, and the Trend is over the 50-year period.

Bordeaux, and red wines from California: a general trend over time to better quality, with less variability, can be seen. Also note the greater relative variability in Bordeaux vintages as compared to California, indicating the impact of greater growing season climate variability as seen in Figure 3. Examining the relationships between climate and wine quality found that the majority of the trends in vintage ratings were significantly related to average growing season temperatures. Whereas the effect varied by region, vintage ratings on average rose by 13.3 points (on a 100 point scale) for every 1°C warmer the growing season. From 10-62% of the variation in vintage ratings (32% average) was related to growing season temperature variations, with the most significant results being found in the cooler climate regions (e.g., the Mosel and Rhine Valley of Germany).

To examine future climate change, Jones *et al.* (2005) used output from the HadCM3 general circulation model (GCM; Gordon *et al.*, 2000; Pope *et al.*, 2000) from 1950-2049 for 25 grid cells encompassing the same wine regions as described above. A comparison of the two periods, 1950-1999 and 2000-2049, suggested that mean growing season temperatures would warm by an average 1.2°C over the 27 wine regions studied with the differences for Bordeaux and Napa 1.2°C and 1.7°C, respectively (Fig. 5). The projected changes are greater for the Northern Hemisphere (1.3°C) than the Southern Hemisphere (0.9°C). Examining the rate of change projected for 2000-2049 revealed significant changes in each wine region with trends ranging from 0.2°C to 0.6°C per decade. Overall changes during the 2000-2049 period averaged 2.0°C across all regions with the smallest warming in South Africa (0.9°C/50 years) and greatest warming in Portugal (2.9°C/50 years). For the Napa and Bordeaux regions, decadal trends are 0.4 and 0.5°C while the overall change is 2.2 and 2.3°C, respectively (Fig. 5). In addition, the HadCM3 model predicted significant increases in growing season temperature variability and warming during the dormant season across most regions.

While the observed warming of the last fifty years appears to have mostly benefited the quality of wine grown worldwide, the predicted regional warming rates and magnitudes detailed by Jones *et al.* (2005) for the next half century have numerous potential impacts on grapevine growth, grape production, and wine production in the future. The impacts are not likely to be uniform across all varieties and regions, but are more likely to be related to a climatic threshold whereby any continued warming would push a region outside the ability to ripen similar varieties. Note that a wine region, on average, can be positioned within the range of the climate maturity types based on its average growing season average temperature (Fig. 2). For example, if a region has an average growing season temperature of 15°C and the climate warms by 1°C, that region is climatically more conducive to

Table 2: Wine regions and categories of wine analyzed by Jones et al. (2005). The wine regions correspond to the locations shown in Figure 1.

Region	Categories of Wines in Sotheby's Vintage Ratings
C. Washington E. Washington N. Oregon S. Oregon	US - Pacific Northwest Red US - Pacific Northwest White
N. California C. California S. California	US - California Red US - California White
N. Portugal	Vintage Port
S. Portugal	No Specific Rating Provided
Rioja	Rioja Red
Barolo	Barolo Red
Chianti	Chianti Red
Rhine Valley	Rhine Valley White
Mosel Valley	Mosel-Saar-Ruwer Valley White
N. Rhône Valley	N. Rhône Valley Red
S. Rhône Valley	S. Rhône Valley Red
Loire Valley	Loire Valley Red Loire Valley Sweet White
Alsace	Alsace White
Champagne	Vintage Champagne
Burgundy-Côte Burgundy-Beaujolais	Burgundy - Côte D'Or Red Burgundy - Côte D'Or White Burgundy - Beaujolais Red
Bordeaux	Bordeaux - Médoc and Graves Bordeaux - St. Émilion and Pomeral Bordeaux - Sauternes and Barsac
Hunter Valley	Hunter Valley Red Hunter Valley White
Margaret River	Margaret River Red Margaret River White
Barossa Valley	Barossa Valley Red Barossa Valley White
South Africa[a] Chile[a]	Overall Vintage Overall Vintage

[a] Rating data for South Africa and Chile are from a different source than the other locations (see text for details).

ripening some varieties, while potentially less conducive to ripening for others. If the magnitude of the warming is 2°C or larger, a region may potentially shift into another climate maturity type (e.g., from intermediate to warm, using the climate scale of Jones *et al.* 2005 – cool, intermediate, warm, and hot – as noted above). While the range of potential varieties that a region can ripen will expand in many cases, if a region is a hot climate maturity type and warms beyond what is considered viable, grape growing becomes challenging and may even be impossible. The wine quality issues related to climate change and shifts in climate maturity potential are evidenced mostly through more rapid plant growth and out-of-balance ripening profiles. If a region currently experiences a maturation period (véraison to harvest) that allows sugars to accumulate, maintains acid levels, and produces the optimum flavor profile for that variety, balanced wines result. In a warmer than ideal environment, the grapevine will go through its phenological events more rapidly, resulting in earlier sugar ripeness and, while the grower or winemaker is waiting for flavors to develop, the acidity is lost through respiration resulting in "flabbly" wines (high alcohol with little acidity for freshness). In addition, harvests that occur earlier in the summer, in a warmer part of the growing season (e.g., September instead of October in the Northern Hemisphere), will result in hot and potentially desiccated fruit without greater irrigation inputs.

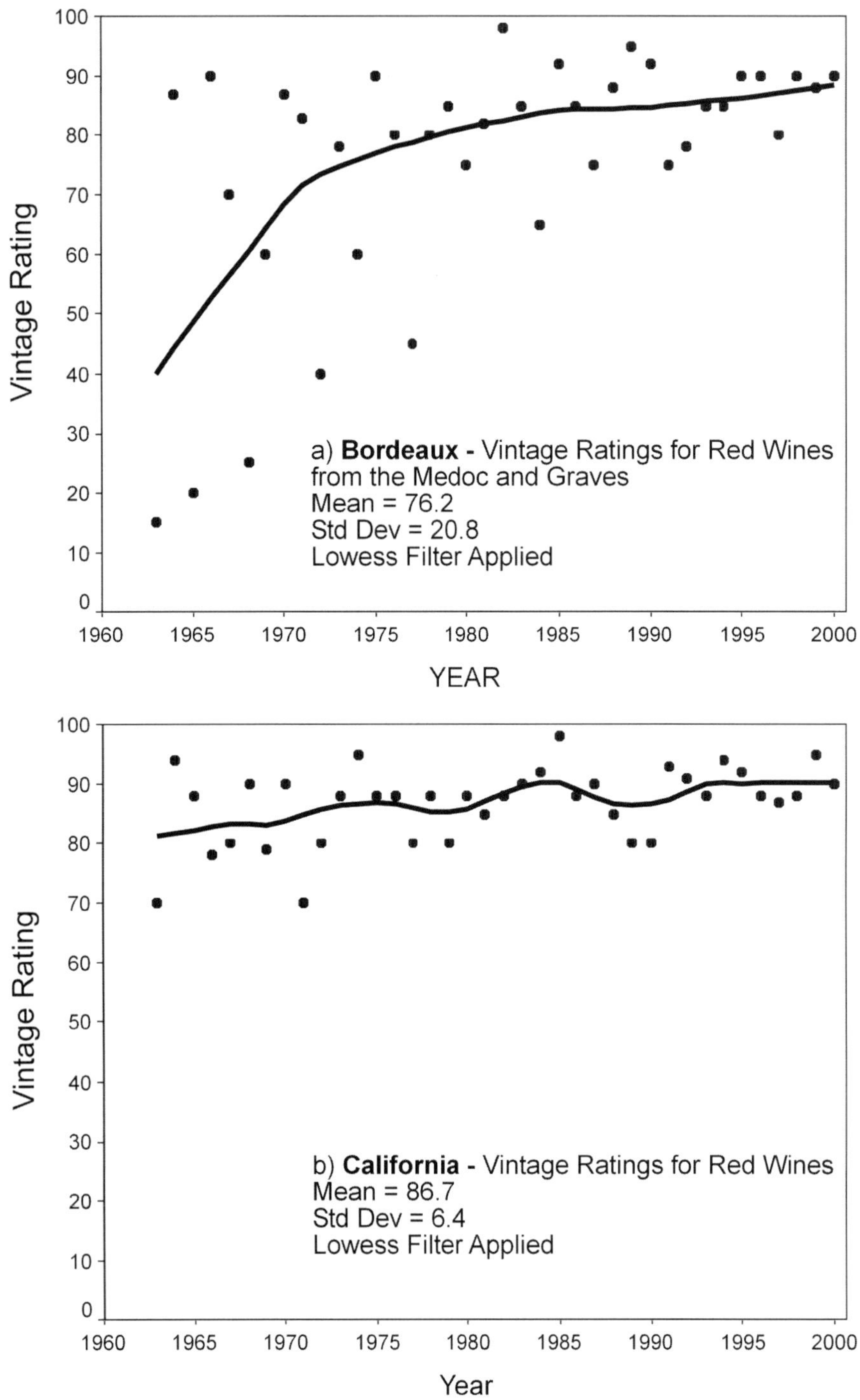

Figure 4. Vintage ratings for (a) red wines from the Médoc and Graves regions of Bordeaux and (b) red wines from California as analyzed by Jones et al. (2005). The ratings are from Sotheby's (Stevenson, 2001) and are based on a 0-100 scale. A LOWESS filter is applied to indicate the underlying pattern in the ratings.

CONCLUSIONS

Cultivation of grapevines for wine production has a rich geographical and cultural history of development. This history has helped shape the notion of terroir, which describes the interrelationships among climate, soil, landscape, and the people who cultivate the grapevine for the production of wine. While there is much debate as to what is the most important aspect of terroir (e.g., Seguin, 1986; and others), environmental change has the potential to impact the balance that exists in any region. Today's viticultural regions for premium quality wine production are located in narrow climatic zones that put them at particular risk from both short term climate variability and long term climate change. This was clearly evident during the Little Ice Age when viticultural viability was threatened throughout much of Europe. The warming of the last century or more appears to have largely benefited grape growing and wine production through the expansion of viable growing regions, providing longer growing seasons, earlier phenological development, and more optimum ripening leading to better overall quality. It is clear that advances in viticultural practices such as irrigation, nutrition, trellising, and pest/disease control, and more knowledge and experience in winemaking techniques, certainly have contributed to larger yields and better quality. In spite of such advances, however, grape growers and winemakers generally believe climate plays the most significant role in determining the overall quality and style of wine from a given region, and that year-to-year variations in the quantity and quality of vintages are controlled by climate variability.

Future climate change issues for the wine industry are mostly related to changes in temperature, precipitation, and CO_2 concentrations and their interactive effects on the spatial viability of wine producing regions, plant growth timing, production, and quality. The climate change scenarios predicted for grape growing regions point to the potential for regional changes in viticultural viability owing to changes in growing season temperatures and precipitation seasonality and distribution (Pincus, 2003; McInnes *et al.*, 2003; Jones *et al.*, 2005). Regions with cool growing season temperatures today (e.g., the Mosel and Rhine Valleys of Germany – more poleward locations) theoretically would have reduced year-to-year vintage quality variations and potentially could ripen warmer climate varieties (Schultz, 2000). Other regions, currently with warmer growing seasons (e.g., the Iberian Peninsula and Chianti – more subtropical regions) may become too warm for the existing varieties grown there, while the hottest regions may lose production viability altogether. While growing season changes clearly are important, projected dormant period temperature changes would also affect viticulture by reducing winter freeze damage in some regions (e.g., eastern Washington state), while other regions (e.g., parts of California and Australia) would have very mild winters where the hardening of latent buds may not occur and pests limited by winter minimum temperatures may increase in number or severity. Although grape growing typically requires less water demand than many other crop systems, changes in seasonally dependent snowmelt or rainfall could also place added stress on vines in water-limited regions. Given the observed and modeled acceleration of vegetative and reproductive growth of grapevines in a warmer climate, a general trend of increased yields and higher sugar contents is predicted for several growing regions and varieties. However, thresholds may be reached by which grape quality could be jeopardized when varieties come to maturity too early in the season and the typical balance between sugar and acids cannot be achieved. This suggests increasing potential economic risks for grape growers and winemakers (Bindi and Fibbi, 2000). Changes in growth and quality due to increases in CO_2 are more complex owing to the interactive effects with changes in temperature and moisture availability. Observations and modeling indicate that photosynthesis and water-use efficiency (ratio of photosynthesis to water consumption) is stimulated by increased CO_2 and that grapevine pro-

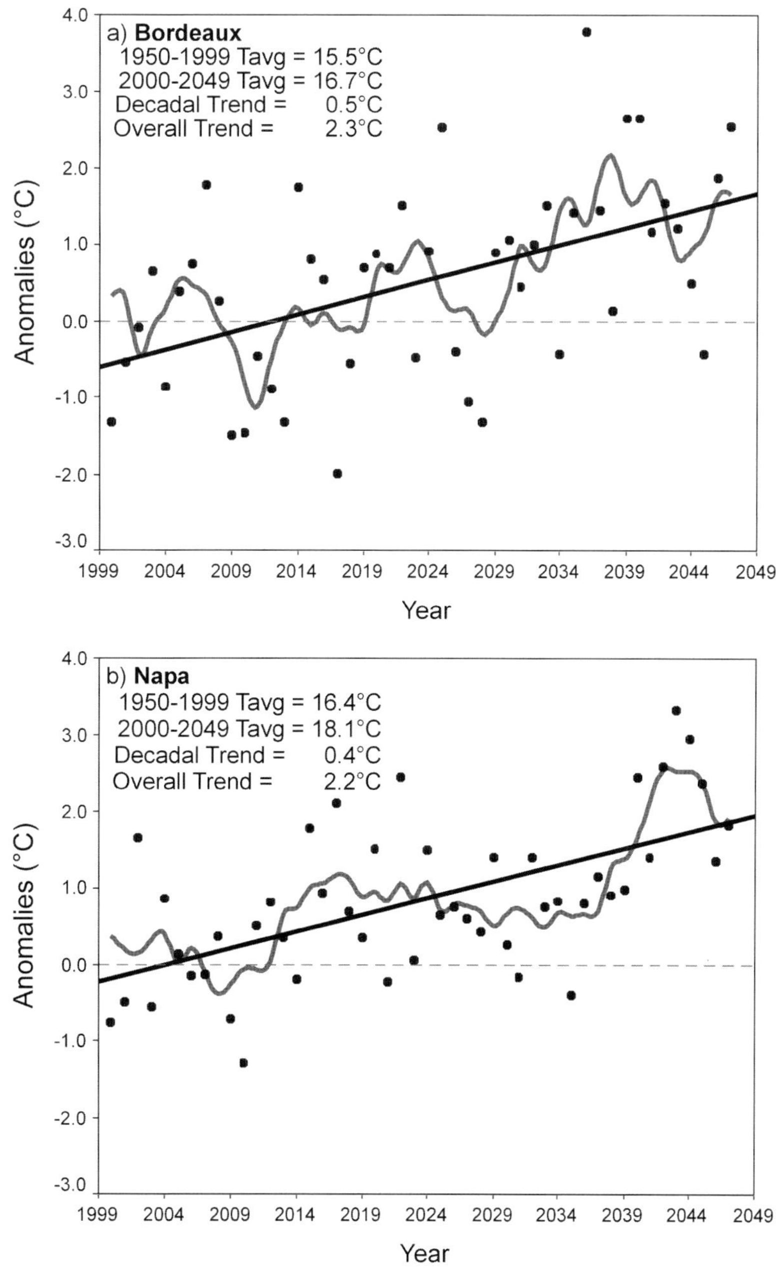

Figure 5. Modeled growing season average temperature anomalies for a) Bordeaux and b) Napa as analyzed by Jones et al. (2005). The modeled temperature data are from the HadCM3 climate model on a monthly time scale extracted from a 2.5° x 3.75° grid centered over the wine producing regions for 2000-2049. The anomalies are referenced to the 1950-1999 base period from the HadCM3 model. Note that the difference between the 1950-1999 growing season average temperature in Figure 3 and the 1950-1999 growing season average temperature shown here are due to the larger size of the grid square used in the HadCM3 model. Trend values are given as an average decadal change and the total change over the 50-year period.

duction should increase without causing negative influences on the quality of grapes and wine (Bindi *et al.*, 2001).

In most of the wine producing regions of the world, especially those with the longest history, both physical and cultural landscapes and local economies are shaped by wine production. In these regions the wine industry drives regional development and dominates many economic sectors from production to trade to tourism. While much uncertainty still exists in the magnitude and rate of climate change, any change is likely to bring about cultural change where regional identities may shift with the varieties and wine styles that can be produced there. To prepare for the future, the industry will most certainly need to integrate planning and adaptation strategies to adjust accordingly to the predicted changes in climate.

ACKNOWLEDGEMENTS

I would like to thank the many researchers from various fields of study who have delved into the notion of terroir and those that have studied climate-related impacts on the wine industry. The data in the Jones *et al.* (2005) analysis were provided by the Climate Impacts LINK Project (DEFRA Contract EPG 1/1/124) on behalf of the Hadley Centre and U.K. Meteorological Office; and the Center for Climatic Research, Department of Geography, University of Delaware. Tom Stevenson, author of the New Sotheby's Wine Encyclopedia: A Comprehensive Reference Guide to the Wines of the World, is thanked for his insights and discussion on wine ratings. This paper was critically read by L.D. Meinert and an anonymous reviewer: comments from these individuals and from R.W. Macqueen improved the presentation and are gratefully acknowledged. Finally, thanks are due Simon Haynes, whose interest in terroir and research provided the arena for Geoscience Canada's Geology and Wine series and the Geological Society of America's 2003 annual meeting tribute session on terroir.

REFERENCES

Amerine, M. A. and Winkler, A. J., 1944, Composition and quality of musts and wines of California grapes: Hilgardia, v. 15, p. 493-675.

Amerine, M. A., Berg, H. W., Kunkee, R. E., Ough, C. S., Singleton, V. L. and Webb, A. D., 1980, The Technology of Wine Making, (4th ed.): AVI Publishing Company, Inc., Westport, Connecticut. 795 p.

Bindi, M., Fibbi, L., Gozzini, B., Orlandini, S., and Miglietta, F., 1996, Modeling the Impact of Future Climate Scenarios on Yield and Variability of Grapevine: Climate Research, v. 7, p. 213-224.

Bindi, M. and Fibbi, L., 2000, Modeling climate change impacts at the site scale on grapevine: *in* Downing, T.E, and Harrison, P.A., Butterfield, R.A., and Longsdale, K.G., (eds.): Climate Change, Climatic variability and Agriculture in Europe: An Integrated Assessment. Final Report: Environmental Change Institute, University of Oxford, p. 117-134.

Bindi, M., Fibbi, L., and Miglietta, F., 2001, Free Air CO_2 Enrichment (FACE) of grapevine (Vitis vinifera L.): II. Growth and quality of grape and wine in response to elevated CO_2 concentrations: European Journal of Agronomy, v. 14, n. 2, p. 145-155.

Branas, J., 1974, Viticulture: Dehan, Montpellier, 990 p.

Braslavska, O., 2000, Trends in the grapevine phenology - Slovakia, Dolne Plachtince, period 1971- 2000: Slovak Hydrometeorological Institute, Národny Klimaticky Program SR, v. 2000, n. 8, p. 69-77.

Broadbent, M., 1981, The Great Vintage Wine Book: Sotheby's, London, 432 p.

Butterfield, R. E., Gawith, M. J., Harrison, P. A., Lonsdale, K. J. and Orr, J., 2000, Modelling climate change impacts on wheat, potato and grapevine in Great Britain: *in* Downing, T. E., Harrison, P. A., Butterfield, R. E. and Lonsdale, K. G., (eds.): Climate Change, Climatic Variability and Agriculture in Europe: An Integrated Assessment. Final Report: Environmental Change Institute, University of Oxford, 411 p.

Carbon Dioxide Assessment Committee, 1983, Changing Climate (U.S. National Research Council): National Academy Press, Washington, D.C., 496 p.

Carter, T. R., Parry, M. L. and Porter, J. H., 1991, Climatic change and future agroclimatic potential in Europe: International Journal of Climatology. v. 11, p. 251-269.

Cess, R. D. *et al.*, 1990, Intercomparison and interpretation of climate feedback processes in 19 atmospheric general circulation models: Journal of Geophysical Research, v. 95, p. 16601-16615.

Chuine, I., Yiou, P., Viovy, N., Seguin, B., Daux, V., and Le Roy Ladurie, E., 2004, Grape Ripening as a Past Climate Indicator: Nature, v. 432, p. 289-290.

Climate Research Board, 1979, Carbon Dioxide and Climate: A Scientific Assessment (U.S. National Research Council): National Academy Press, Washington, D.C. 22 p.

CO_2/Climate Review Panel, 1982, Carbon Dioxide and Climate: A Second Assessment (U.S. National Research Council): National Academy Press, Washington, D.C. 72 p.

de Blij, H. J., 1983, Geography of viticulture: rationale and resource: Journal of Geography, v. 82, n. 3 p. 112-121.

Easterling, D. R., Evans, J. L., Groisman, P., Ya., P., Karl, T. R., Kunkel, K. E., and Ambenje, P., 2000, Observed variability and trends in extreme climate events: A brief review: Bulletin of the American Meteorological Society. v. 81, p. 417-425.

Esteves, M. A. and Orgaz, M. D., 2001, The influence of climatic variability on the quality of wine: International Journal of Biometeorology, v. 45, n. 1, p. 13-21.

Gates, W. L., *et al.*, 1998, An Overview of the Results of the Atmospheric Model Intercomparison Project (AMIP I): Bulletin of the American Meteorological Society, v. 73, p. 1962-1970.

Giorgi, F. and Mearns, L. O., 1991, Approaches to the simulation of regional climate change: A review: Reviews of Geophysics. v. 29, p. 191-192.

Gladstones, J., 1992, Viticulture and Environment: Winetitles, Adelaide, Australia. 310 p.

Gordon, C., Cooper, C., Senior, C. A., Banks, H., Gregory, J. M., Johns, T. C., Mitchell, J. F. B., and Wood, R. A., 2000, The simulation of SST, sea ice extents and ocean heat transports in a version of the Hadley Centre coupled model without flux adjustments: Climate Dynamics, v. 16, p. 147-168.

Hänninen, H., 1995, Effects of climatic change on trees from cool and temperate regions: an ecophysiological approach to modeling of bud burst phenology: Canadian Journal of Botany, v. 73, p. 183-199.

Haynes, S.J., 1999, Geology and Wine 1. Concept of terroir and the role of geology: Geoscience Canada, v. 26, p. 190-194.

Houghton J.T. *et al.*, 2001, Climate Change 2001: The Scientific Basis, Contribution of the Working Group 1 to the Third Assessment of the Intergovernmental Panel on Climate Change: Cambridge University Press, Cambridge, U.K., 944 p.

Huglin, P., 1978, Nouveau mode d'evaluation des possibilities heliothermiques d'un milieu viticole: C.R. Academy of Agriculture in France (n. 111726).

Jackson, D. I. and Cherry, N. J., 1988,

Prediction of a district's grape-ripening capacity using a latitude-temperature index: American Journal of Enology and Viticulture. v. 39, n, 1, p. 19-28.

Jones, G. V. and Davis, R. E., 2000, Climate influences on grapevine phenology, grape composition, and wine production and quality for Bordeaux, France: American Journal of Enology and Viticulture, v. 51, n. 3, p. 249-261.

Jones, G. V. and Hellman, E., 2003, Site Assessment: *in* Hellman, E. (ed.), "Oregon Viticulture" 5th Edition: Oregon State University Press, Corvallis, Oregon, p. 44-50.

Jones, G. V., 1997, A Synoptic Climatological Assessment of Viticultural Phenology: Dissertation, University of Virginia, Department of Environmental Sciences, 394 p.

Jones, G.V., White, M.A., Cooper, O.R., and Storchmann, K., 2005, Climate Change and Global Wine Quality: Climatic Change, v. 73, n. 3, p. 319-343.

Jones, G.V., 2005, Climate Change in the Western United States Grape Growing Regions; Acta Horticulturae (ISHS), v. 689, p. 41-60.

Jones, P. D., Bradley, R. S., Diaz, H. F., Kelly, P. M. and Wigley, T. M. L., 1986, Northern Hemisphere surface air temperature variations: Journal of Climate and Applied Meteorology. v. 25, p. 161-179.

Jones, R. G., Murphy, J. M., and Noguer, M., 1995, Simulation of climate change over Europe using a nested regional climate model. Part 1: Assessment of control climate including sensitivity to location of lateral boundaries: Quarterly Journal of the Royal Meteorological Society. v. 121B, p. 1413-1449.

Karl, T. R., Jones, P.D., Knight, R.W., Kukla, G., Plummer, N., Razuvayev, V., Gallo, K.P., Lindseay, J., Charlson, R.J., and Peterson, T.C., 1993, A new perspective on global warming: asymmetric trends of daily maximum and minimum temperature: Bulletin of the American Meteorological Society, v. 74, p. 1007–1023.

Katz, R. W. and Brown, B. G., 1992, Extreme events in changing climate: variability is more important than averages: Climatic Change, v. 21, p. 289-302.

Kenny, G. J. and Harrison, P. A., 1992, The effects of climate variability and change on grape suitability in Europe: Journal of Wine Research. v. 3, p. 163-183.

Kenny, G. J. and Shao, J., 1992, An Assessment of a latitude-temperature index for predicting climate suitability for grapes in Europe: Journal of Horticultural Science, v. 67, n. 2, p. 239-246.

Le Roy Ladurie, E., 1971, Times of Feast, Times of Famine: A History of Climate Since the Year 1000: Doubleday & Company, Inc., Garden City, New York. 426 p.

Lough, J. M., Wigley, T. M. L. and Palutikof, J. P., 1983, Climate and climate impact scenarios for Europe in a warmer world: Journal of Climate and Applied Meteorology, v. 22, p. 1673-1684.

M. Mazur, Wine Enthusiast's 2002 Vintage Chart. The Wine Enthusiast Magazine (http://www.winemag.com/vintage.cfm).

Mann, M. E., Bradley, R. S. and Hughes, M. K., 1999, Northern Hemisphere temperatures during the past millennium: Inferences, uncertainties, and limitation: Geophysical Research Letters, v. 26, p. 759-762.

Maytin, C. E., Acevedo, M. F., Jaimez, R., Andressen, R., Harwell, M. A., Robock, A. and Azocar, A., 1995, Potential effects of global climate change on the phenology and yield of maize in Venezuela: Climatic Change, v. 29, p. 189-211.

McCarthy, J.J. *et al.*, 2001, Climate Change 2001: Impacts, Adaptation, and Vulnerability, Contribution of the Working Group 1 to the Third Assessment of the Intergovernmental Panel on Climate Change: Cambridge University Press, Cambridge, 1032 p.

McInnes, K. L., Whetton, P. H., Webb, L., and Hennessy, K. J., 2003, Climate change projections for Australian viticultural regions: The Australian and New Zealand Grapegrower and Winemaker, February 2003, p. 40-47.

Meinert, L.D., and Busacca A.J., 2000, Geology and Wine 3: Terroirs of the Walla Walla Valley appellation, southeastern Washington, USA: Geoscience Canada, v. 27, p. 149-172.

Menzel, A. and Fabian, P., 1999, Growing season extended in Europe: Nature, v. 397, p. 659.

Moisselin, J-M., Schneider, M., Canellas, C. and Mestre, O., 2002, Climate change over France during the 20th century: A study of long-term homogenized data of temperature and rainfall: La Météorlogie v. 38, p. 45-56.

Moonen, A. C., Ercoli, L., Mariotti, M. and Masoni, A., 2002, Climate change in Italy indicated by agrometeorological indices over 122 years: Agricultural and Forest Meteorology, v. 111, p. 13-27.

Mullins, M. G., Bouquet, A. and Williams, L. E., 1992, Biology of the Grapevine: Cambridge University Press, Great Britain. 239 p.

Nemani, R. R., White, M. A., Cayan, D. R., Jones, G. V., Running, S. W., and Coughlan, J.C., 2001, Asymmetric climatic warming improves California vintages: Climate Research, v. 19, n. 1, p. 25-34.

Penning-Roswell, E., 1989, Wines of Bordeaux, (6th ed.): Penguin Books, London/New York. 610 p.

Pfister, C., 1988, Variations in the spring-summer climate of central Europe from the High Middle Ages to 1850, *in* Wanner, H., and Siegenthaler, U., (eds.), Long and Short Term Variability of Climate: Springer-Verlag, Berlin, p. 57-82.

Pincus, R., 2003, Wine, Place, and Identity in a Changing Climate: Gastronomica, v. 3, n. 2, p. 87-93.

Pope, V. D., Gallani, M. L., Rowntree, P. R. and Stratton, R. A., 2000, The impact of new physical parametrizations in the Hadley Centre climate model — HadAM3: Climate Dynamics, v. 16, p. 123-147.

Prescott, J. A., 1965, The climatology of the vine (Vitis Vinifera L.) the cool limits of cultivation: Transcriptions of the Royal Society of Southern Australia, v. 89, p. 5-23.

Ray, C., 1985, Lafite: The Story of Château Lafite-Rothschild: Christie's Wine Publications, London, 135 p.

Reekie, E. G. and Bazzaz, F. A., 1991, Phenology and growth of four annual species grown in ambient and elevated CO_2: Canadian Journal of Botany, v. 69, p. 2475-2481.

Rodó, X. and Comín, F. A., 2000, Links between large-scale anomalies, rainfall and wine quality in the Iberian Peninsula during the last three decades: Global Change Biology, v. 6, n. 3, p. 267-279.

Rosenberg, N. J., 1981a, The increasing CO_2 concentration in the atmosphere and its implication on agricultural productivity: effects on photosynthesis, transpiration and water use efficiency: Climatic Change, v. 3, p. 265-279.

Rosenberg, N. J., 1981b, The increasing CO_2 concentration in the atmosphere and its implication on agricultural productivity: effects through CO_2 induced climatic change: Climatic Change, v. 3, p. 239-254.

Schultz, H. R., 2000, Climate change and viticulture: a European perspective on climatology, carbon dioxide, and UV-B effects: Australian Journal of Grape and Wine Research v. 6, p. 2-12.

Seguin, G., 1986, 'Terroirs' and pedology of wine growing: Experentia, v. 42, p. 861-872.

Smart, R. E. and Dry, P.R., 1980, A climatic classification for Australian viticultural regions: Australian Grapegrower and Winemaker, v. 17, p. 8-16.

Spellman, G., 1999, 'Wine, weather and Climate': *Weather*, v. 54, p. 230-239.

Stevenson, T., 2001, New Sotheby's Wine Encyclopedia: A Comprehensive

Reference Guide to the Wines of the World: Dorling Kindersley, London, 3rd Edition, 601 p.

Tate, A. B., 2001, Global warming's impact on wine: Journal of Wine Research, v. 12, n. 2, p. 95-109.

Van Leeuwen C., Friant P., Chone X., Tregoat O., Koundouras S. and Dubourdieu D., 2004, The influence of climate, soil and cultivar on terroir: American Journal of Enology and Viticulture, v. 55, n. 3, p. 207-217.

Vaudour, E., 2002, The quality of grapes and wine in relation to geography: Notions of terroir at various scales: Journal of Wine Research, v. 13, n. 2, p. 117-141.

Waggoner, P. E., 1989, Anticipating the frequency distribution of precipitation if climate change alters its mean: Agricultural and Forest Meteorology, v. 47, p. 321-337.

Wilson, J.E., 1998, Terroir: the Role of Geology, Climate, and Culture in the Making of French Wines: Mitchell Beazley, London, U.K., 336 p.

Winkler, A. J., Cook, J. A., Kliewere, W. M. and Lider, L. A., 1974, General Viticulture, (4th ed.): University of California Press, Berkeley, 740 p.

PAPER 5 FROM THE "TERROIR, GEOLOGY AND WINE: A TRIBUTE TO SIMON J. HAYNES" SESSION HELD AT THE GEOLOGICAL SOCIETY OF AMERICAN ANNUAL MEETING, SEATTLE, WASHINGTON, NOVEMBER 3, 2003

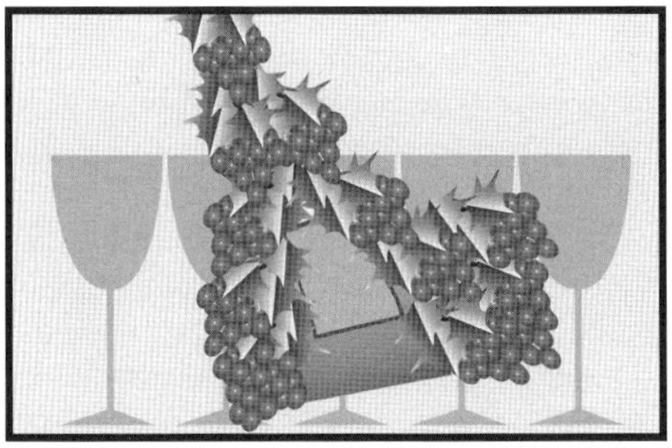

A Review of Napa Valley, California, Terroir

Jonathan Swinchatt[1], David G. Howell[2] and Michael Silacci[3]

[1]*EarthVision, Inc., 52 Cook Hill Road, Cheshire, Connecticut, USA, 06410.swin36@cox.net*

[2]*United States Geological Survey, 345 Middlefield Road, Menlo Park, CA 94025, USA, dhowell@usgs.gov*

[3]*Opus One, P.O. Box 6, Oakville, CA 94562, USA*

SUMMARY

The evolution of the Napa region, California, has created a geologic architecture and topographic complexity that provides a diversity of microenvironments in which to grow grapes. Sedimentary and volcanic bedrock and reworked material eroded from it characterizes vineyard substrates. These geologic factors are primary determinants of grape quality and character, controlling the substrate texture and structure, which in turn modulates the availability of water. Thus a geologic view of substrate provides a level of understanding, including some degree of predictive capability, which is absent from conventional soil descriptions and maps. In Napa, we group all vineyard substrates within three broad categories - residual, alluvial, and fluvial materials. Each of these categories can be mapped and analyzed in terms of the mix of geologic processes that have influenced any particular piece of ground. Some vineyards are developed entirely on one or another of these types of materials, whereas others may include all three categories. Understanding the substrate from a geologic perspective can contribute to the process of making fine wine: an example from Stag's Leap Wine Cellars evinces a convergence between geologic understanding and a winemaker's sensory evaluation.

RÉSUMÉ

L'évolution géologique de la région de Napa en Californie a créé une configuration géologique et topographique qui offrent une diversité de micro-environnements propices à la viticulture. Le sous-sol de ces vignobles est constitué de roches sédimentaires, volcaniques ainsi que d'unités remaniées de celles-ci. Les facteurs géologiques sont des déterminants de première importance de la qualité du raisin et de son caractère, et ils déterminent la structure et la texture du substratum, lesquels modulent l'offre hydrique. Il s'ensuit qu'une perspective géologique du substratum permet un niveau de compréhension auquel correspond une certaine capacité de prédiction, que ne permettent pas la cartographie et la description conventionnelles des sols. Nous regroupons tous les substratums des vignobles de la Napa en trois grandes catégories de matériaux : résiduels, alluvionnaires et fluviatiles. Chacune de ces catégories peut être cartographiée et analysée dans le contexte de la suite des processus géologiques qui ont modelé la zone considérée. Alors que certains vignobles sont limités à l'un ou l'autre de ces types de matériaux, d'autres peuvent s'étendre indifféremment sur toutes ces catégories. Il en ressort la compréhension du substratum d'un point de vue géologique peut bonifier le processus de création de vins de bonne qualité. L'exemple de Stag's Leap Wine Cellars montre à l'évidence la relation profitable entre compréhension géologique et l'évaluation sensorielle du vigneron.

INTRODUCTION

Historically, agricultural soil analysis, including conventional soil descriptions and chemical analyses, has been the primary source of vineyard information available to winegrowers. A typical soil description provided to a winegrower might state that the soil is a "sandy loam with 80 percent cobbles and gravel." Here the focus is on the finer fraction of the deposit, with the dominant component - cobbles and gravel - assigned an accessory role. To a geologist, however, the cobbles and gravel are an integral part of the substrate and provide vital information concerning the conditions under which the substrate formed as well as an indication of the source from which the material was derived. The soil description quoted above might refer to river channel deposits, alluvial fan sediments, or colluvium, here identified as residual materials. These deposits govern the texture and structure of the substrate, lateral and vertical variation of that substrate, and the affect of this variation on a variety of factors, particularly on drainage and the availability of moisture. In addition, soil descriptions tend to generalize and lump together materials that are geologically distinct, either in processes of origin or in terms of the bedrock on which they lie. Soil maps in Napa and Sonoma, for example, typically show little correlation with bedrock or other geologically-defined entities such as alluvial fans or fluvial deposits.

Owing to valley-wide denudation during sea-level low stands, the sediments in the Napa Valley are generally less than 15,000 years old. Through much of the area, soils have not had long to form. Under these circumstances, geologic attributes are more influential than in wine-growing areas where soils are older and more fully developed. Many winemakers and winegrowers have observed that drainage, and the ways in which water is made available to vine roots, are important factors in growing high-quality grapes that make great wine. The relationship between soil moisture and vine roots is influenced by a variety of substrate conditions including grain size distribution, pore space size and shape, clay mineral content, and discontinuities such as marked bedding planes that tend to inhibit root growth. These substrate conditions are controlled predominantly by geologic processes and the products thereof, the particulate deposits - residual, alluvial, and fluvial - in which grapes grow.

NAPA VALLEY AVA AND ITS GEOLOGIC SETTING

The Napa Valley AVA (American Viticultural Area) lies in the Coast Ranges of California in Napa County just north of San Francisco Bay (Fig. 1). It contains within its boundaries 14 smaller AVAs (Fig. 2), most of which lie within the Napa Valley per se or in the adjacent mountains—the Vaca Mountains on the east and the Mayacamas Mountains on the west (see Fig. 5). The viticultural heart of Napa lies here, in the valley and its bordering highlands. Within the Napa Valley AVA the mountains rise to heights of slightly over 2500 feet (~760 m). Compressional forces associated with the San Andreas Fault System formed the mountains and the valley that lies in their midst. This was the most recent major event in a history that stretches back 145 million years.

A Brief Geologic History of California

Prior to 145 MA, the land we now call California did not exist. At that time, the coast of North America lay along what is now the Sierra Nevada, then a linear range of active volcanoes. Offshore, in the Pacific Ocean, a volcanic island arc was creeping toward the continent; soon, it would slam into North America, a formative event in the birth and evolution of present-day California. This accreted arc is preserved as bits and pieces of ocean crust in the western foothills of the Sierra Nevada. Soon after this collision event, the subduction zone jumped to the west, trapping a chunk of ocean crust and adding it to the North American plate. This material, known as the Coast Range Ophiolite, forms the foundation for much of the material that later came together to form California (Bailey, Irwin and Jones, 1964; Ernst, 1981).

For the next 80 million years - from about 140 to 60 million years ago - the position of subduction throughout California remained relatively stable. As the volcanoes of the Sierra Nevada grew, they weighted the crust, forming a trough along the length of the range on its western edge. Seaward of the trough lay a crustal bulge that roughly defined the new, western, geologic boundary of continental North America. The range remained an active volcanic arc for the entire 80 million years of relative crustal stability. During this time, the granular products of weathering and erosion were carried to the shore of the adjacent trough by streams and rivers, and then into its depths by turbidity currents. Eventually, 50,000 feet (15,240 m) of these sediments accumulated, forming the sandstones and shales of the Great Valley succession. The sandstones are mostly arkose - they changed slightly through time, reflecting compositional changes as the volcanic arc evolved. The style of sedimentation reflects broad sea level changes combined with on-going volcanic, climatic, and tectonic events in the Sierra Nevada. The Great Valley succession is notably well understood. (Ingersoll, 1979, 2004).

Not so the Franciscan Formation, which was accumulating during this time on the other side of the trough, seaward of the crustal bulge where subduction continued at the continental edge. Some 7,000 kilometers of ocean crust slid under North America: seamounts, island arcs, deep sea and continental margin sediments - all were scraped from the surface of the oceanic plate and plastered onto the North American continent, forming the Franciscan Formation which today underlies much of coastal California. The Franciscan is a diverse, heterogeneous, and puzzling set of rocks. Individual components - a seamount, a deep sea fan, a chunk of ocean crust - are recognizable in individual outcrops, but how this outcrop relates to the one around the corner is considerably less clear. The history of the Franciscan and the mechanics of what brought together the individual components remain elusive in the details. High pressure—low temperature metamorphism and extensive zones of tectonic disruption at all scales clearly suggest an array of accretionary tectonic processes rumbling in the depths of the subduction zone (Ernst, 1981; Wakabayashi, 2004).

This long period of spatially stable subduction at the continental edge ended about 60 million years ago when the angle of the descending ocean plate flattened and centers of volcanism associated with the subduction mechanism moved progressively eastward as far as Colorado. The decrease in the angle of subduction appears to reflect younger, more buoyant, in-coming crust that sank more slowly into the mantle.

The San Andreas Fault System and Its Effects

The seminal event in the history of California, the creation of the San Andreas Fault System, occurred about 25 million years ago (Atwater, 1970; Wallace, 1990). The mechanics are complex but the pattern is easy to discern. The movement of three plates - the Farallon, Pacific, and North American - brought them together at a triple junction in what is now southern California (Fig. 3). The geometry created a transition from subduction to translation—the motion of the San Andreas fault and its brethren, the Hayward, Rogers Creek, and the various others that splay off the main San Andreas fault and affect much of the California coast. As these three plates continued to move, the triple junction ran up the coast like the toggle of a zipper. The translational faults following along behind, separating California into two parts, east and west of the complex San Andreas fault system.

The transition from subduction to translation also induced a moving wave of volcanic activity associated with the triple junction, a process involving

Figure 1. Satellite view of the San Francisco Bay area with Napa County outlined.

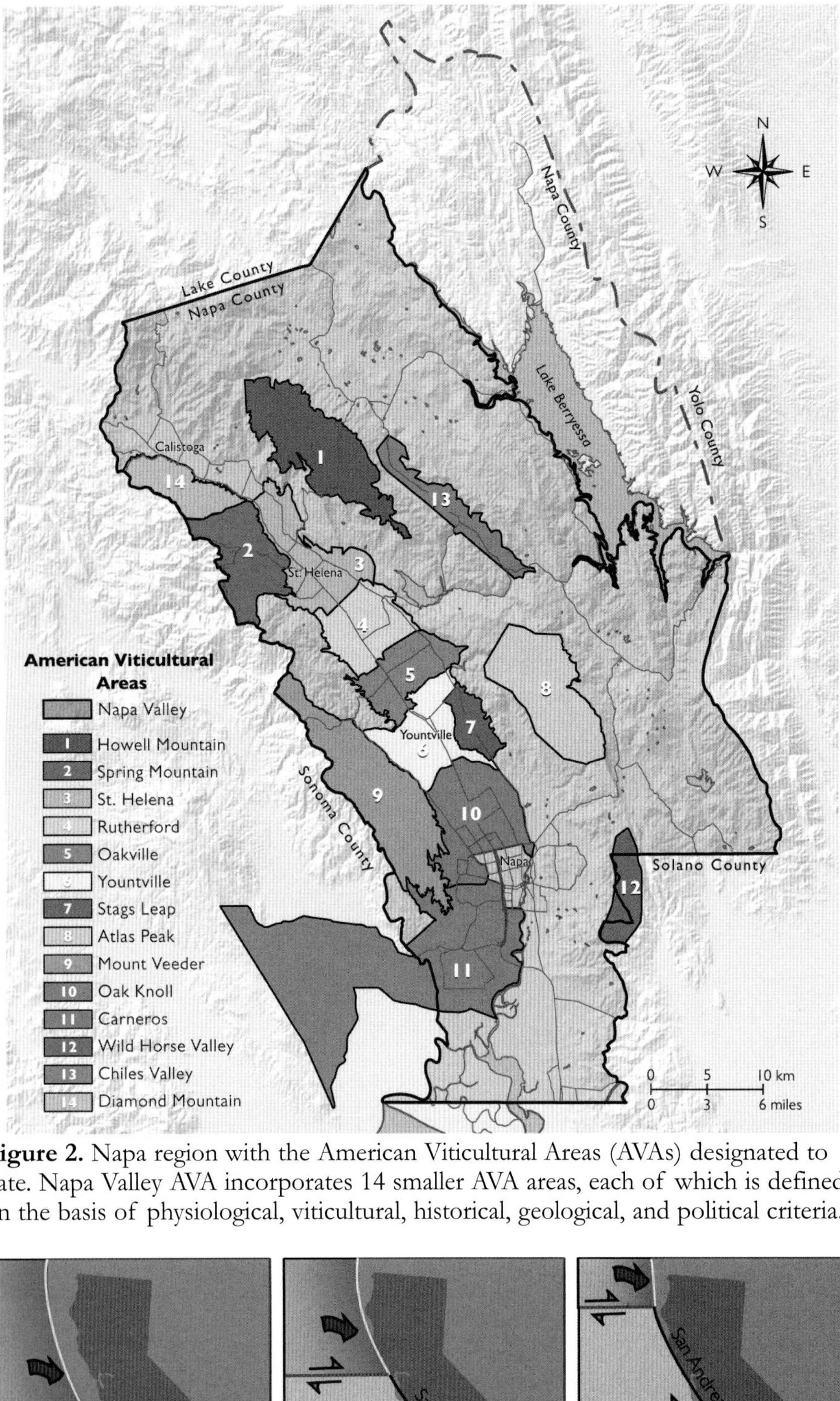

Figure 2. Napa region with the American Viticultural Areas (AVAs) designated to date. Napa Valley AVA incorporates 14 smaller AVA areas, each of which is defined on the basis of physiological, viticultural, historical, geological, and political criteria.

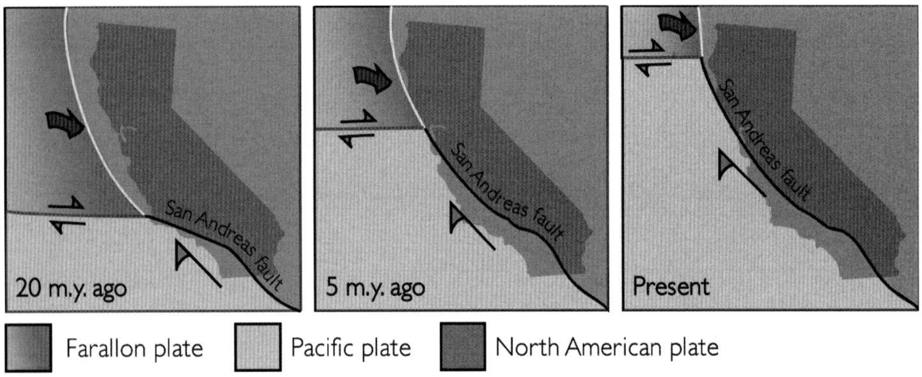

Figure 3. Creation of the San Andreas fault system by a triple junction moving up the west coast of North America. The triple junction is the location, a point on a simplified diagram such as this, at which the Farallon, Pacific, and North American plates are in contact. Spatial and temporal relationships are considerably more complex than shown here.

the creation in the crust of a seemingly empty space or "slab-window" (Dickinson and Snyder, 1979; Severinghaus and Atwater, 1990). Magmatic activity occurred as the space rapidly filled with buoyantly rising hot mantle material from the now naked aesthenosphere. The volcanism began in southern California about 24 million years ago and skipped along as the triple junction "zipped up" the coast, but with no distinctive pattern other than association with the triple junction. At each volcanic location, a thick covering of lava and ash would blanket the land for a few tens or hundreds of square miles. Older accumulations were sliced up and offset by the growing faults, making the bedrock geology ever more complex. About seven million years ago the volcanism reached Napa, then an area of low topographic relief. Over the following five million years, volcanism would cover some hundreds of square miles, with the Sonoma Volcanics forming a thick veneer over the earlier Franciscan Formation and Great Valley sequence.

Forming the Napa Valley

The present form of the Napa Valley is a result of a compression component of motion along the San Andreas fault. At its inception, movement along the fault contained a small component of tension - the Pacific plate was moving northwest at a slight angle to the line of the fault, creating a pull on the crust that initiated a group of basins that contain some of the richest oil accumulations, in terms of volume of oil per cubic meter of rock, in the world. The southern San Joaquin, Santa Barbara, Santa Maria, Los Angeles and other basins of the California continental borderlands, and basins along the coastal region of central California including the ancestral San Francisco Basin, formed at this time (Blake and others, 1978; Argus and Gordon, 2001). The cold waters of the Pacific margin that filled these deep, oxygen-deprived basins were rich with marine algae. With little circulation and no oxygen in the bottom waters, the depths of these basins were filled with organic "gunk" that was later transformed into prolific volumes of oil and natural gas.

About three million years ago, the motion of plates along the Pacific margin changed direction very slightly, enough to create a small component of compression, perhaps ten percent of the total movement along the fault (Page and Brocher, 1993; Jones and others, 1994; DeMets and others, 1994; Atwater and Stock, 1998). This component of east-west compression was enough to deform the Napa region into its present configuration. In the east, thrust slices of Great Valley sequence and Sonoma Volcanics were stacked up to form the Vaca Mountains, while simple folds formed in the west to create the Mayacamas Mountains, with the Napa Valley a syncline that separates the two ranges (Figs. 4, 5). This activity is still occurring: In 2001, an earthquake shook the Mayacamas Mountains, rattling crockery and breaking chimneys. The Napa region still evolves. Those interested in a broader view of California geology should see Harden, 1997; Hyndman and Alt, 2000; and McPhee, 1994.

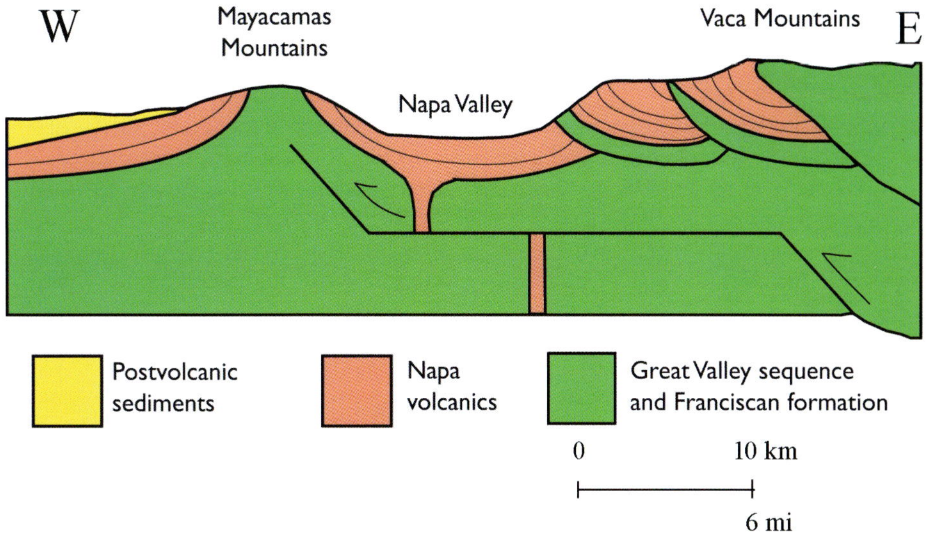

Figure 4. Simplified, hypothetical bedrock cross-section of the Napa Valley. Note that this cross-section does not show the veneer of residual, alluvial, and fluvial sediments that lie above bedrock, or the soil that has formed on these geological materials.

GEOLOGY AND TERROIR IN THE NAPA VALLEY AVA

Bedrock

Sonoma Volcanics

The bedrock geology of the Napa Valley AVA is structurally complex but straightforward in terms of the distribution of rock units. Throughout their length, the Vaca Mountains are formed of the Sonoma Volcanics (= Tertiary Napa volcanics of Fig. 5). These rocks, which include lava flows, pyroclastic flows, volcanic ash, lahars, and stream deposited sediments, also cover the Mayacamas Mountains north of St. Helena (Graymer, Jones and Brabb, 2002; and Graymer, Jones and Brabb, in press). They form a band of varying thickness along the base of these mountains from St. Helena to near the City of Napa (Fig. 5). They also appear as a thin band in the Carneros District. Sonoma Volcanics also make up most of the hills and knobs in the valley. Overall, they are the dominant rock type in the area bounded by the Vaca and Mayacamas Mountains, the heart of the Napa Valley AVA.

Wineries that produce from grapes grown on the Sonoma Volcanics often imply that this imbues the grapes and wine with some special quality and/or character. We think the dominance of volcanic rocks has no major significance, at least in terms of wine quality. Throughout the world, great wines are made from grapes grown on rocks as disparate as limestone and granite; volcanic rocks are just part of the mix. It is important to note that the Sonoma Volcanics include a wide range of rock types, each of which provides a different texture and structure to the substrate, as well as differences in mineralogy and chemistry. To lump all these volcanic lithologies together masks differences that might well prove to influence grapes and wine.

Franciscan Formation

Franciscan Formation and Great Valley succession rocks may be less volumetrically important than the Sonoma volcanics, but they have influenced some hallowed Napa Valley ground. Franciscan rocks cover the central Mayacamas Mountains; while few wineries are located directly on these rocks, they have fed debris to a stretch of the western edge of the Napa Valley that has supported the growth of great wines throughout Napa's winemaking history. Streams that drain Franciscan rocks feed the "Rutherford Bench," an alluvial fan that has supported vineyards of George de la Tour, BV, and Inglenook (now Niebaum-Coppola).

Vineyards located on rocks of the Franciscan Formation are faced with some difficulties not seen elsewhere, in particular a high content of magnesium and nickel; magnesium limits the uptake of potassium, an element vital to healthy plants, and nickel is toxic to vines (Sherlock and others, 1993). An additional problem is the presence of serpentinite, which occurs in small masses throughout the Franciscan, although only in one significant concentration within the Napa Valley, at the foot of the Mayacamas Mountains south of St. Helena (Fig. 5). The high concentrations of magnesium, nickel, and boron in serpentinite preclude it from being productive for grapes, at least without the application of extreme agricultural methods.

Great Valley Succession

From Oakville AVA (Fig. 2) south, Great Valley succession rocks form the surface of the Mayacamas Mountains, except for a fringe of Sonoma Volcanics near their base. The Mount Veeder AVA (9, Fig. 2) is located on Great Valley rocks, which have also fed some historically rich ground in the valley, particularly the Oakville Bench. This "bench" is made up of two alluvial fans that are home to the To Kalon vineyard (owned mostly by Robert Mondavi Wines), Martha's Vineyard, and John Daniels' property Napanook, now Dominus. Great Valley succession rocks also appear at the surface through much of the northeastern part of the Napa Valley AVA, and in small areas elsewhere.

Vineyard Substrate

In Napa, three types of vineyard substrate are based on geologic origins—

residual, alluvial, and fluvial (Swinchatt and Howell, 2004).

Residual Materials

Residual materials are those that have developed directly on bedrock and have moved little if at all. In geologic terminology, this material is commonly known as colluvium. Stagecoach Vineyards provides the most extreme example. At Stagecoach, deep weathering of Sonoma Volcanics has produced residual materials in places reportedly 16 feet (~5 m) thick. Core stones are present here and in many Napa vineyards, reflecting the still competent part of what is otherwise fully decayed (= weathered in place) bedrock. Some 500,000 tons of core stones up to 12 feet (~3 ½ m) in diameter were ripped from this vineyard during preparation for planting (Fig. 6). Residual materials cover most of the hilly topography in the Napa Valley AVA (Fig. 7). Residual materials are most commonly very thin - less than two feet thick but in some places only a few inches - and very well drained. Some winegrowers have had to crack bedrock with dynamite in order to plant vines.

Alluvial Materials

These occur as fans that line the western edge of the Napa Valley and appear sporadically in the east (Fig. 7). The west side fans south of St. Helena (Fig. 2) are broad with almost imperceptible slopes; those north of the town are narrow and steep. East side fans are small and widely separated. Even the deeply incised

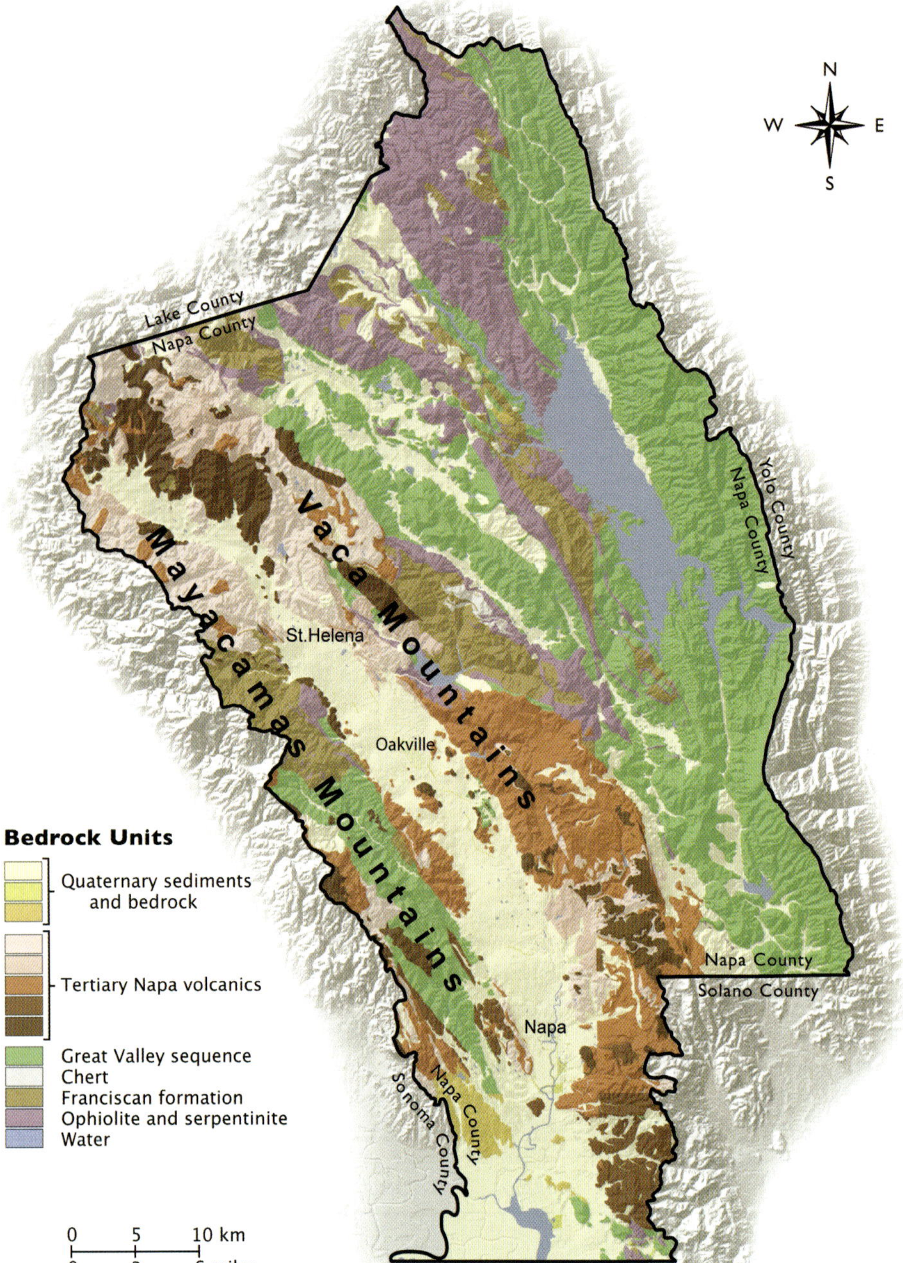

Figure 5. Geologic Map of the Napa region. Napa Valley is bounded by the Vaca Mountains on the northeast and the Mayacamas Mountains on the southwest. The valley occupies the area underlain by Quaternary sediments between these two ranges. "Napa Volcanics" is an informal name we have used for the Sonoma Volcanics in the Napa region.

Figure 6A. Large core stones that were raked from Stagecoach Vineyards. **6B.** Core stones are formed when water penetrates fractured volcanic flow rock, rounding and transforming the edges of blocks that, over time, are further decayed (weathered) until they appear to be isolated boulders separated by soft granular material. Note concentric layers around the solid cores, all that remains of what was a solid layer of rock, once a lava flow. These altered materials - the core stones and the soft matrix between - are an example of residual materials (see text for discussion).

Rector and the very large Henessey drainages have small, low-relief fans at their mouths. The materials eroded from these drainages were swept from the valley at a time of lower sea level: the present configuration of alluvial fans represents deposition over only the past 15,000 years (Swinchatt and Howell, 2004). Wines produced from the west side fans include, among many others, BV George de la Tour, Opus One, Mondavi To Kalon Reserve, Spottswoode, and Dominus. Wines from the east side fans, where the grapes are exposed to greater heat in the afternoon, include Stag's Leap Wine Cellars Fay and SLV, Araujo Estate Eisele Vineyard, Clos du Val, Gemstone, and Screaming Eagle. Rudd Estate is grown on an interesting mix of alluvial fan and residual materials.

Fluvial Materials

Deposits of the Napa River (the fluvial materials of Fig. 7) form a thin meandering band pushed to the east side of the valley by the growth of the west side alluvial fans. The fluvial deposits are the most fine-grained and agriculturally vigorous of Napa substrates: deep, soft, and easily penetrated by vine roots. Grapes grown on this substrate tend to exhibit herbal and vegetal notes that can be unpleasant if too intense.

THE DIVERSITY OF NAPA TERROIR

This overview is, at best, a first approximation of what the Napa Valley provides in terms of terroir. One of the Napa Valley's singular attributes is a diversity of micro-environments that reflects variation of substrate, topography, sun exposure, and climate (Swinchatt, 2002; Swinchatt and Howell, 2004). Detailed vineyard studies reveal a world in which a zone a few tens of feet wide might contain three separate physical terroirs. An example from the Stags Leap District is particularly illuminating.

Stag's Leap Wine Cellars: The Detail of Terroir

Stag's Leap Wine Cellars (SLWC) is thought by some to be one of the greatest vineyards in California, perhaps in the world. SLWC produced the Cabernet Sauvignon wine that won a 1976 Paris tasting organized as a USA bicentennial celebration. This event brought world focus on California and attention to SLWC and its owner/winemaker, Warren Winiarski.

Stag's Leap Wine Cellars estate property consists of two vineyards, Fay and SLV, each containing several separate vineyard blocks (Fig. 8). It was wine from Winiarski's first vintage from the SLV vineyard that surprised the French at the 1976 Paris tasting. These two vineyards lie at the south end of a small valley (Stags Leap vale) that is home also to Shafer, Stags Leap Winery, Steltzner, and vineyards owned by Joseph Phelps and other wineries. Clos du Val and Chimney Rock lie just to the south. Fay and SLV vineyards comprise about 140 acres, lying between a volcanic knoll on the west and the slopes of the Vaca Mountains on the east. On summer afternoons, strong breezes from the

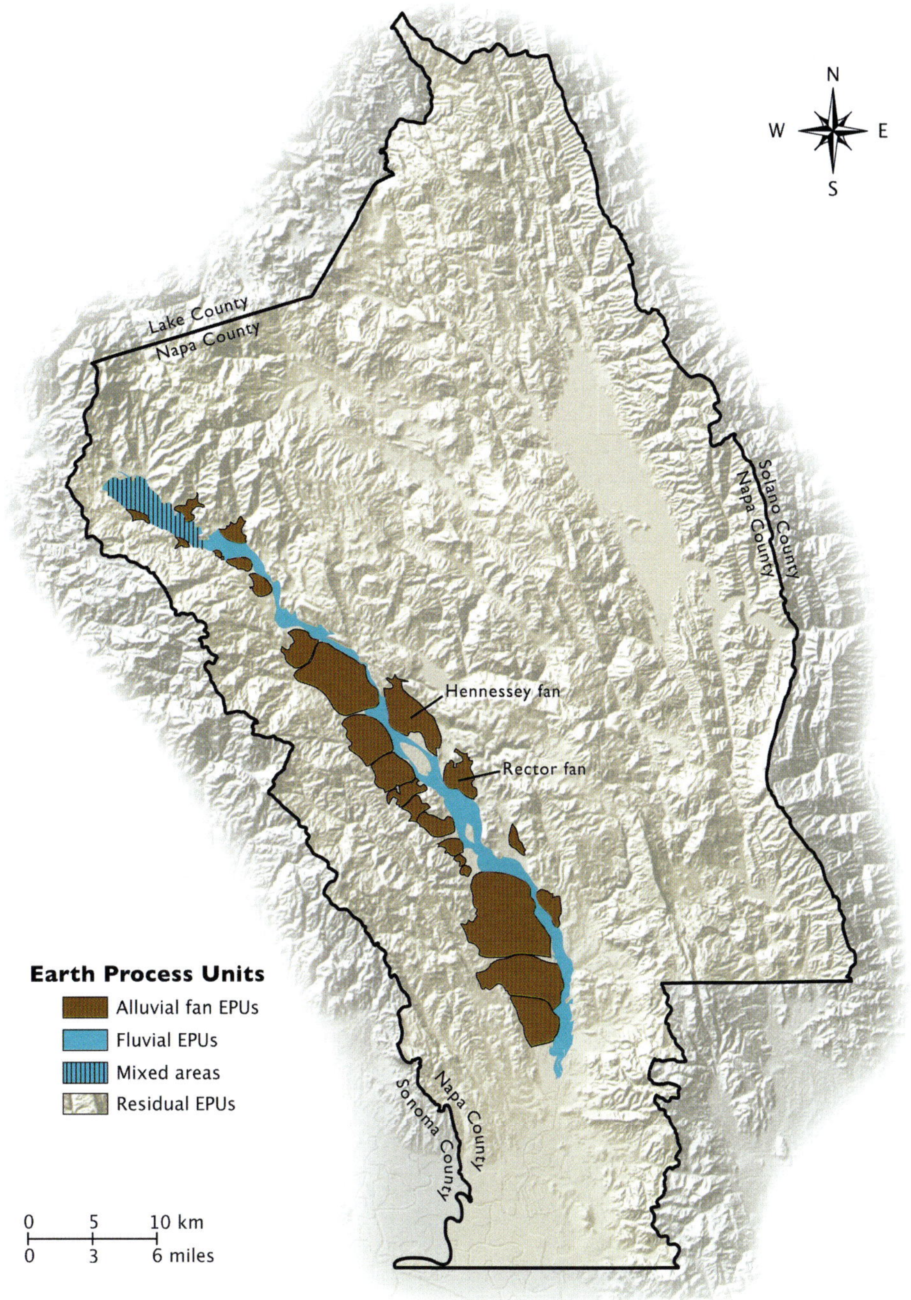

Figure 7 In the Napa region, we identify three types of vineyard substrate related to geologic processes of origin. Residual materials cover most of the upland topography. Alluvial fans line the west side of the valley and occur more sporadically on the east. Fluvial materials border the Napa River. An area to the northwest contains variable amounts of these three materials, and is thus identified as a mixed area.

Figure 8. Air photo of vineyards of Stag's Leap Wine Cellars. Fay vineyard consists of the two rows of vineyard blocks on the north side (left side of figure) of the property, SLV vineyard lies to the south (right) of Fay vineyard. Vineyard blocks are numbered; many are divided into sub-blocks, not indicated here, that are planted with different root stocks and grape clones. The distance from the eastern boundary of block 10 to the western boundary of block 2 is about 2500 feet (~760 m).

south are funneled through the vale, cooling the grapes and aiding in the preservation of acids and volatile flavor and aroma compounds that are often lost in the heat of the afternoon sun.

Fay and SLV vineyards occupy the slight, westward facing, slopes of a small alluvial fan that formed at the mouth of a draw in the Vaca Mountains that feeds an intermittent stream. The stream was relocated to the boundary between the two vineyards after Winiarski bought SLV in the late 1960s. In the northwest corner of Fay, the SLWC fan intersects with a larger fan that heads in the upper reaches of the Stags Leap vale (Fig. 9). In the summer of 2003, we examined 91 backhoe pits and trenches in Fay and SLV vineyards. Thirty-nine of these were in Fay block 8, 28 in Fay block 5, and the remaining 24 were distributed along avenues throughout both vineyards (Fig. 8). The holes ranged in depth from two to a little over seven feet (~ ½ - 2 m) and reveal distinct geologic units, providing a tantalizing connection between geology and the taste buds of winemaker Michael Silacci.

The backhoe pits revealed that the SLWC alluvial fan in Fay vineyard consists of two distinct lobes formed by debris flows, and a more diffuse array of channel deposits dominated by overbank material in SLV (Fig. 10). The block 8 debris flow materials contain an array of highly weathered volcanic boulders and cobbles up to two feet in diameter. In the two blocks to its west (6 and 4, Figs. 8, 10), this coarse layer drops in the section, large clasts decrease in size, and the percent of fine matrix increases (Fig. 11). At the eastern edge of block 8, boulders - some larger than 20 inches (~50 cm) in diameter - lie at the surface. The clasts are tightly packed with perhaps 10 percent sand and gravel matrix. On the southern edge of block 8, the coarse material lies 20 inches below the surface (Fig. 12c). Here, the clasts average 6 inches in diameter, are less abundant and are mixed with gravel. The matrix, which comprises perhaps 25 percent of the volume, is a mix of sand, silt, and clay. These observations suggest mass-flow deposition with some later winnowing of the coarsest sections. The full size of such a debris flow deposit is difficult to determine, but it extends progressively farther below the surface,

Figure 9. Most of Fay vineyard and all of SLV vineyard lie on what we identify as the Stag's Leap alluvial fan. In the northwest corner of Fay vineyard, the Stag's Leap alluvial fan intersects a larger alluvial fan that arises in the upper regions of the Stag's Leap vale. Dashed line marks the approximate intersection of these two fans. Aerial photography @ 2003 AirphotoUsa LLC.

Figure 10. (Top right) Fay vineyard is underlain by the deposits of four debris flows, one revealed in the block 8 backhoe pits (Block 8 debris flow), two (one atop the other) in block 5 (Block 5C/D debris flow), and a small surface fan (Small debris flow) in block 7 (see Fig. 8 for block locations). Toward the west (down in the figure), these coarse sediments interfinger with overbank deposits and sediments delivered from the northern part of the Stag's Leap vale. Substrate in SLV vineyard on the right side of this figure is a combination of stream channel and overbank deposits.

westward across block 6 and into at least the eastern edge of block 4 (Figs. 8, 10).

The boulders and cobbles in the block 8 flow are highly weathered and easily cut with a chisel pick. Similarly weathered and decayed rocks occur in place at or near the surface in the Vaca Mountains, as we have seen at Stagecoach Vineyards (Fig. 6). The debris flow appears to have been derived from such weathered volcanics, with further decay taking place after deposition. Throughout most of block 8, a surface layer up to 20 inches thick, composed of gray-brown silty sand with variable amounts of volcanic cobbles and gravel, overlies the debris flow. At the eastern edge of the block this layer contains only a modest amount of clay, but the clay fraction increases away from the axis of the flow and westerly through the block. The surface layer is a mix of organic debris, fine material washed from the slopes above, and material from the debris flow brought to the surface by agricultural practices.

Although the pits on the northern and eastern edges of block 5 (Fig. 8) reveal two well-defined coarse deposits,

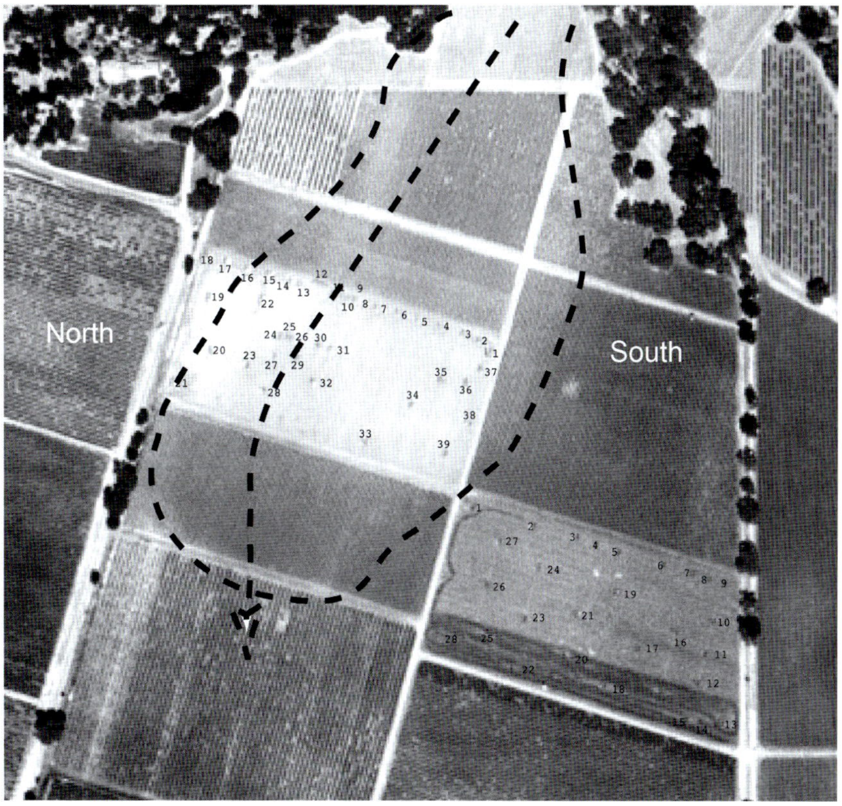

Figure 11 (Bottom right) Axis of the block 8 debris flow of the Fay vineyard is marked by the dashed line in the middle of the figure. Outer dashed line encloses an area in which the debris flow lies within 5 feet (~1.5 m) or less of the surface. The debris flow drops in the section and becomes finer-grained away from the axis. East-west distance across the vineyard is approximately 2500 feet (~760 m).

pits toward the south and west contain progressively more complex sedimentary sections with beds of sand, gravel, and clay interspersed with the debris flows (Fig. 13). It appears that two distinct flows divided into a number of smaller distributary channels down the fan, with fine-grained overbank deposits separating coarser channelized materials.

The boulders and cobbles of the block 5 debris flows are fresh and hard, unlike the highly weathered and decayed boulders and cobbles of block 8. The block 5 boulders have thin weathering rinds and are primarily of a single volcanic type.

SLV vineyard, while part of the same SLWC alluvial fan, contains distinctly different sediments. Fay is dominated by diversity, SLV by uniformity; Fay sediments are mostly the coarse deposits of debris flows, SLV the finer sediments of fluvial processes. Nine trenches in the vineyard avenues of SLV

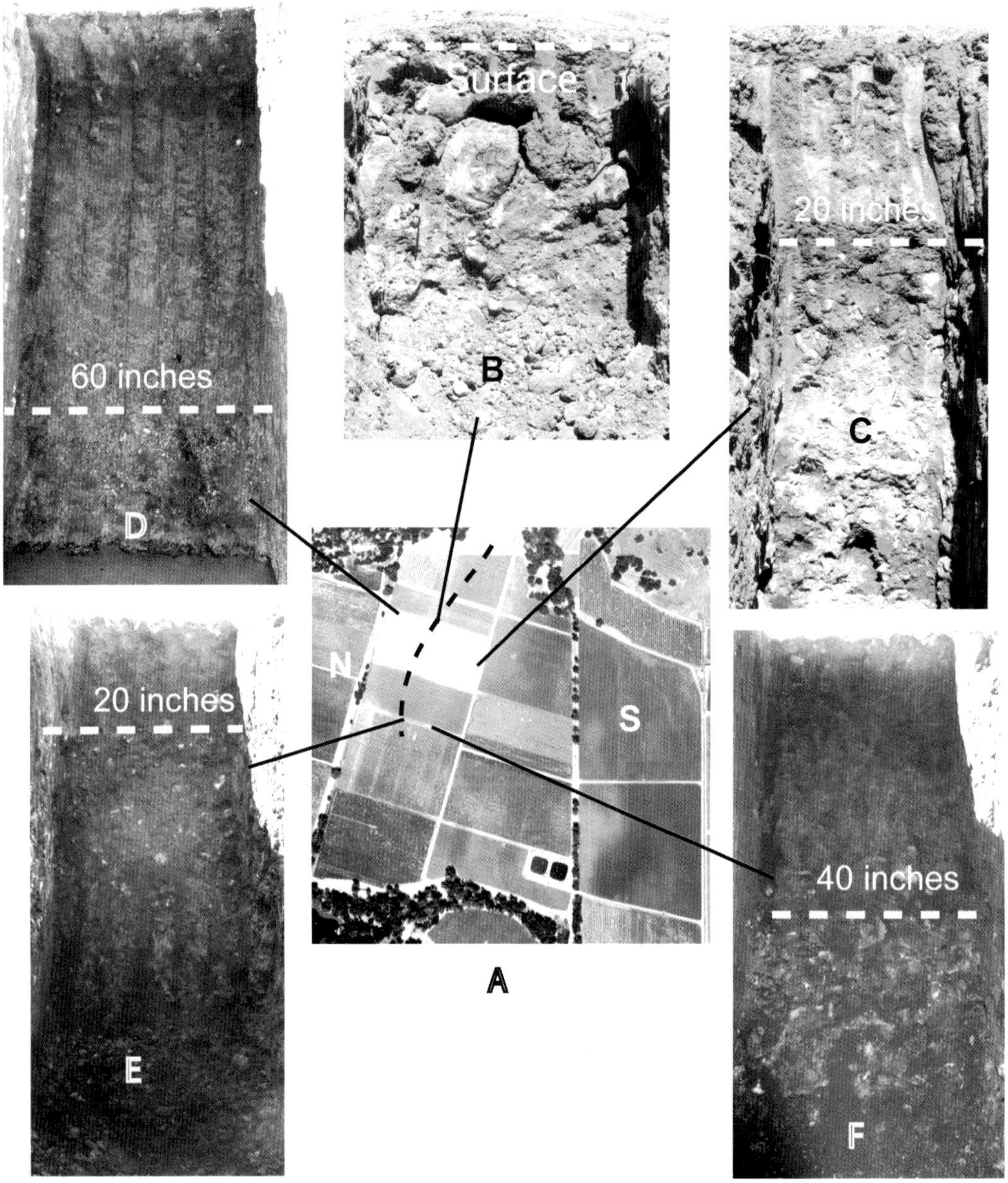

Figure 12 Backhoe pit photographs of the Fay vineyard block 8 debris flow illustrating how it varies away from the axis of the flow (black dashed line in 12A, centre of Figure). Dashed white lines indicate depth to the top of the debris flow. 12A shows the location of the pits. Large boulders that extend to the surface dominate pit shown in 12B, on the axis of the flow. Toward the edges of the debris flow deposit, the coarse sediment drops in the section, to 20 inches (~51 cm) in 12C, and more abruptly on the north to 60 inches (~1.5 m) in 12D. To the west, on the axis of the debris flow deposit, the coarse material still lies near the surface (12E) but drops abruptly to the north and south, for example to 40 inches (~1 m) in 12F. The east-west distance across the vineyard is approximately 2500 feet (~760 m). Pit photos are not all to the same scale. Aerial photograph © 2003 AirphotoUSA LLC.

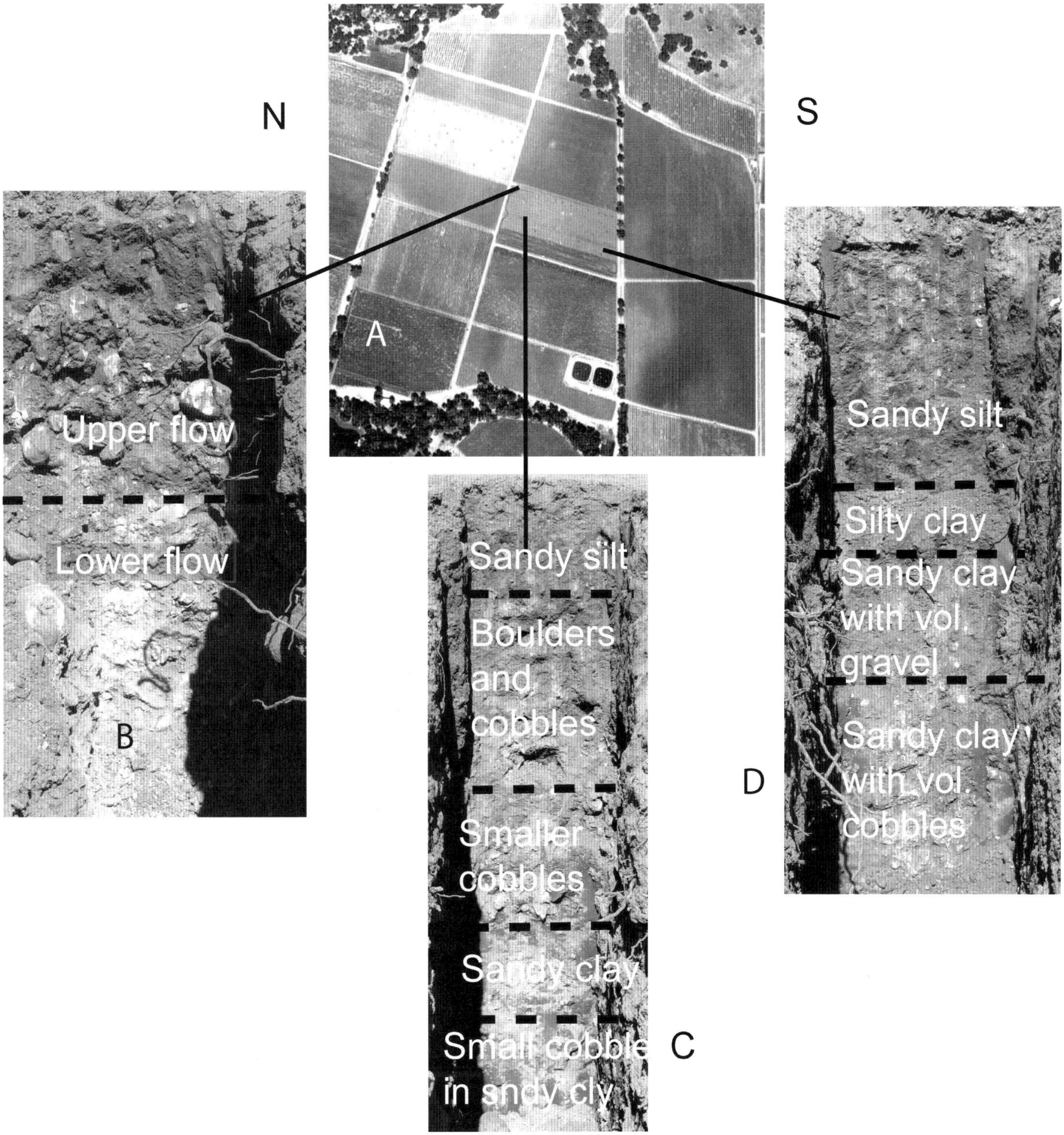

Figure 13 Backhoe pit photos from Fay vineyard block 5 debris flows; locations shown in 13A. The axes of the flows trend close to the north side of the block where they are stacked one directly atop the other (13B). To the south, the debris flow deposits are progressively more complexly interbedded with overbank deposits. In 13C, an upper boulder and cobble layer suggests the upper debris flow, but beneath this lies a mix of near-channel and overbank deposits. At the southern end of the block, 13D, the coarse fraction of the debris flows is present only as cobbles and gravel dispersed in sandy clays. Pit photos are not all to the same scale. East-west distance across the vineyard is approximately 2500 feet (~760 m). Aerial photograph © 2003 AirphotoUSA LLC.

Figure 14 Backhoe pit photos showing variation down the fan through SLV vineyard. Sediments generally become finer grained to the west with less interbedded sand and gravel and few cobbles. 14A shows aerial photo of vineyard. At the eastern end of the vineyard, 14B, two feet (~0.6 m) of boulders and cobbles overlie volcanic bedrock. Farther west, 14D, the section is quite uniform, composed of clay-rich sand with scattered volcanic pebbles and cobbles. At the western edge of the vineyard, 14C, the section is dominated by silty clay. Pit photos are not all to the same scale. East-west distance across the vineyard is approximately 2500 feet (~760 m). Aerial photograph © 2003 AirphotoUSA LLC.

showed a general pattern of channel and overbank stream sediments that range from clay-rich sands with scattered pebbles and cobbles with some gravel lenses in the eastern part of the block to predominantly sandy clays in the west (Fig. 14). Coarse surface deposits in the eastern part of the SLV vineyard indicate the most recent stream course, now diverted to the manmade channel that separates SLV and Fay.

Depositional History

The sediments of Fay and SLV vineyards record a series of events in the evolution of the SLWC fan complex. The block 8 flow moved directly west out of the mountain drainage rather than following the southerly course of the modern natural stream. The apex of the block 8 debris flow is topographically higher than the apex of the block 5 flows; this debris flow blocked an existing drainage and forced subsequent flows to follow a more southerly course. The second (upper) of these flows again blocked the drainage, forcing still younger streams progressively south to the position marked on USGS topographic maps. The youngest streams built up the deposits that now underlie SLV-gravels, sands, silts, and clays
The highly decayed boulders and cobbles in the Block 8 debris flow indicate deep weathering in the source, extensive weathering after the deposit formed, or a combination of the two. Deeply weathered material near the surface in the Vaca Mountains would be the first to be delivered to the Stags Leap vale during torrential storms, exposing less weathered rocks such as the boulders and cobbles that make up the block 5 debris flows.

In the western part of Fay vineyard, SLWC fan deposits mix and interfinger with sediments delivered from the northern reaches of the Stags Leap vale. These two sources join along a zone that crosses the southeast corner of Fay block 4 and continues across the northwest corner of Fay block 3. Between this line and the hill that rises to the west, the finer fractions of the SLWC alluvial fan deposits, delivered from the east, mixed with sediments from the north.

Geology and Grape Character

The way water is made available to vines

appears to be a primary control on grape quality and character. For years prior to the recognition of this fan architecture in the Fay vineyard, grapes were picked selectively on the basis of taste. Inside a section of rows that had the shape of a "Hershey kiss" (Fig. 15), the grapes were distinctly different from those on the margin of this subjectively defined area, and provided superb wines. As it turns out the "Hershey kiss" configuration lies atop the Block 8 fan.

The specific geologic elements of the debris flow that influence the character and quality of the grapes are not known, but we can make a few observations. Inside the "Hershey kiss," a surface layer 0 - 12 inches (0 - ~30 cm) thick lies above a layer of tightly packed boulders and cobbles with a small amount of sand and gravel matrix (Fig. 15). The surface layer is a mixture of topsoil, fine sediment from the slopes above, and material mixed into the surface layer from below by agricultural practices. Outside the "kiss", the surface layer thickens from 12 inches to 20 inches (~30 - 50 cm) or more, and the boulders and cobbles of the debris flow are more loosely packed, separated by a matrix of gravel, sand, silt, and clay that increases in volume and decreases in grain size away from the "kiss." While these facts reveal little in themselves, they do suggest two variables that might influence the development of the vines and grapes: drainage and clay mineral content.

Effects of Drainage

Drainage in the core of the block - the area of the "Hershey kiss" outlined in Figure 15 - is greatest where we found cobbles and boulders with little or no clay. Outward from this area, the percentage of matrix increases and clay becomes a significant component. The finer matrix impedes drainage and some clay minerals absorb water in their crystal structure. In addition, the pore volume available for water storage, during periods of heavy rain in the winter and spring for example, increases outward from the core. Within the area of the "kiss," most of the sediment volume is accounted for by rock - the boulders and cobbles that make up the debris flow in this area. Vines outside the "kiss" area potentially have more access to water than do vines within it. Conventionally, this possibility would be interpreted in terms of stress - that vines within the "kiss" area, with less moisture available, are more highly stressed than those outside the "kiss" area. An alternative view is that the vines outside the "kiss" area are over-watered and thus perhaps a bit too vigorous and somewhat unbalanced (see Swinchatt and Howell, 2004, for a discussion of balanced vines). Variation in the thickness of the surface layer might be an additional influence to the degree that it affects drainage and moisture in the near-surface zone.

CONCLUSIONS

Figure 15 Block 8 "Hershey kiss" shown in 15A. In the area within the black dashed line, which outlines the "Hershey kiss", the top of the debris flow lies within 12 inches (~30 cm) of the surface. The shape of this "Hershey kiss"area closely matches the grape picking pattern for this block, as determined by Michael Silacci's taste buds. In 15B on the debris flow axis (blue dashed line), the debris flow sediments are very coarse, with little matrix, and extend to the surface. Away from the flow axis, the debris flow sediment drops in the section to ~11 inches (~28 cm) near the edge of the "kiss", 15C, with boulders and cobbles more widely dispersed in a finer matrix. These changes are accentuated outside the "kiss" area, 15D, E, and F, where the surface layer above the debris flow increases in thickness, the cobbles become progressively smaller away from the "kiss" area, and the matrix increases in percentage and decreases in average grain size. Note: Focus here is entirely on block 8. To the east (up on the figure; north is to the left and south to the right) the top of the debris flow also lies close to the surface. Aerial photograph © 2003 AirphotoUSA LLC.

The Napa Valley displays an array of microenvironments available for growing grapes. With growing interest in the connection between wine and terroir, the availability of diverse niches and the potentially unique wines produced from them can provide significant consumer interest and marketing potential. In such an environment, however, terroir needs to be viewed on a small scale. In Napa, to speak of the terroir of an AVA, for example, provides little useful information. Several of the AVAs contain a variety of geologic, topographic, and climatic zones, negating any meaning they might have in the context of terroir. Indeed, the example of Stag's Leap Wine Cellars indicates that the geologic component of terroir can vary considerably even within a single vineyard block. It is not surprising, therefore, that every vineyard we have examined has proven to have a distinct geologic character, and thus a unique terroir.

The type of information described herein is useful to winegrowers and winemakers in planning new vineyards, replanting vineyard blocks, and managing existing vines. Vineyard managers are often seeking greater uniformity in order to achieve an equally high standard of quality throughout a vineyard and to make vineyard management less complex and more precise. Note, however, that the search is for uniform quality, not uniform character.

Clear and unembellished information that relates a wine to its terroir is also useful to wine marketers and consumers alike. We compare this association to that of provenance in the art world. Knowing the origin and history of a painting or sculpture is vital to establishing its authenticity, but it also adds interest and even a hint of mystery. Knowing something of the terroir—in all its aspects—from which a wine arises adds a vibrant dimension to the sensory experience.

In France, with its extended history of winemaking, terroir is associated with vineyards of longstanding reputation. Even so, the average wine drinker knows little of the land from which the wine came or how it differs from its neighbors. Such information is difficult to access, and is generally available in less detail than our studies suggest is warranted. In North America, where the winemaking history is short, even less information is available. Vineyard designated wines develop a reputation, but few wine enthusiasts know much about the actual character of the terroir. Thus there is a real opportunity to delineate all aspects of terroir and make this information more generally available to the public-at-large.

ACKNOWLEDGEMENTS

Parts of this paper summarize information from two publications; a report to the Napa Valley Vintners Association (Swinchatt, 2002), and a book published by the University of California Press (Swinchatt and Howell, 2004). We would like to thank all our friends and colleagues at these organizations, particularly Jeff Adams of the Napa Valley Vintners Association and Blake Edgar of the University of California Press, for their encouragement and support of our work and its publication. We also thank the many winegrowers and winemakers in the Napa Valley who generously shared with us their experiences and opinions concerning these thorny issues of grapes and ground. We also thank the reviewers of this paper, particularly Rob MacNaughton and also Roger Macqueen and Larry Meinert, for useful suggestions for revision of the initial manuscript. Above all, we thank Warren Winiarski, whose curiosity, insight, and thirst for knowledge led him to commission this study and whose generosity allowed it to be shared.

REFERENCES

Argus, D. F., and Gordon, R. G., 2001, Present tectonic motion across the Coast Ranges and San Andreas fault system in central California: Bulletin of the Geological Society of America, v. 113, p. 1580-1592.

Atwater, Tanya, 1970, Implications of plate tectonics for the Cenozoic tectonic evolution of western North America: Bulletin of the Geological Society of America, v. 81, p. 3513-3536.

Atwater, T. M., and Stock, J., 1998, Pacific-North America plate tectonics of the Neogene southwestern United States: An update: International Geology Review, v. 40, p. 375-402.

Bailey, E.H., Irwin, W.P., and Jones, D.L., 1964, Franciscan and related rocks and their significance in the geology of western California: California Division of Mines and Geology Bulletin 183, 177 p.

Blake, M. C. Jr., Campbell, R.H., Dibblee, T.W. Jr., Howell, D.G., Nilsen, T.H., Normark, W.R., Silver, E.A., and Vedder, J.G., 1978, Neogene basin formation and hydrocarbon accumulation in relation to the plate-tectonic evolution of the San Andreas fault system: American Association of Petroleum Geologists Bulletin, v. 62, n. 3, p. 344-372.

DeMets, D. C., Gordon, R. G., and Stein, S., 1994, Effects of recent revisions to the geomagnetic-reversal time scale on estimates of current plate motions: Geophysical Research Letters, v. 21, p. 2191-2194.

Dickinson, W. R., and Snyder, W. S., 1979, Geometry of subducted slabs related to San Andreas transform: Journal of Geology, v. 87, p. 609-627.

Ernst, W.G. (ed.), 1981, The geotectonic development of California
(Rubey volume 1): Prentice-Hall, Englewood Cliffs, N.J., 706 p.

Graymer, R.W., Jones, D.L., and Brabb, E.E., 2002, Geologic map and map database of Northeastern San Francisco Bay Region, California: United States Geological Survey Miscellaneous Field Studies MF 2403.

Graymer, R.W., Jones, D.L., Brabb E.E., in press, Geologic map and map database of eastern Sonoma and western Napa Counties, California: United States Geological Survey Miscellaneous Field Studies.

Harden, Deborah, 1997, California Geology: Prentice-Hall, Englewood Cliffs, N.J., 479 p.

Haynes, S. J., 1999, Geology and Wine 1. Concept of Terroir and the Role of Geology: Geoscience Canada, v. 26, p. 190-194.

Haynes, S. J., 2000, Geology and Wine 2. A geological foundation for *terroirs* and potential sub-*appellations* of Niagara Peninsula wines, Ontario, Canada: Geoscience Canada, v. 27, p. 67-87.

Howell, D. G., and Swinchatt, J., 2000, A discussion of geology, soils, wine, and history of the Napa Valley region: California Geology, v. 53, n. 3, p. 4-12.

Hyndman, Donald W, and Alt, David D., 2000, Roadside Geology of Northern and Central California: Mountain Press Publishing Company, 384 p.

Ingersoll, R. V., 1979, Evolution of the Late Cretaceous forearc basin of northern and central California: Geological Society of America Bulletin, Cordilleran Section, 73 p.

Ingersoll, R. V., 2004, The Great Valley Group: The arc-trench gap: Geological Society of America Special Paper 338: Classic Cordilleran Concepts: A View from California: p. 135-153.

Jones, D.L., Graymer, R., Wang, C., McEvilly,

T.M., and Lomax, A., 1994, Neogene transpressive evolution of the California Coast Range: Tectonics, v. 13, n. 2, p. 561-574.

McPhee, John, 1994, Assembling California: Noonday Press, 224 p.

Meinert, L. D. and Busacca, A.J., 2000, Geology and Wine 3: Terroirs of the Walla Walla Valley appellation, southeastern Washington State, U.S.A.: Geoscience Canada, v. 27, p. 149-170.

Meinert, L. D. and Busacca, A.J., 2002, Geology and Wine 6. Terroir of the Red Mountain Appellation, Central Washington State, U.S.A.: Geoscience Canada, v. 29, p. 149-168.

Page, B.M. and Brocher, T.M., 1993, Thrusting of the central California margin over the edge of the pacific plate during the transform regime: Geology, v. 21, p. 635-638.

Pomerol, Charles (ed.), 1989, The Wines and the Winelands of France: Robertson McCarta Limited, London, U.K., 370 p.

Severinghaus, J., and Atwater, T., 1990, Cenozoic geometry and thermal state of the subducting slabs beneath western North America, *in* Wernicke, B.P. (ed.), Basin and Range extensional tectonics near the latitude of Las Vegas Nevada: Geological Society of America Memoir 176, p. 1-22.

Sherlock, R. L., Amelia, M., Logan, V., and Jowett, E. C., 1993, Silica carbonate alteration of serpentinite, implications for the association of precious metal and mercury mineralization in the Coast Ranges: *in* Rytuba, James J. (ed.), Active geothermal systems and gold-mercury deposits in the Sonoma-Clear Lake volcanic fields, California: Society of Economic Geologists Guidebook Series, v .16, p. 90-116.

Swinchatt, Jonathan, 2002, The Foundations of Terroir in the Napa Valley: Geology, Landscape, and Climate: unpublished Report for the Napa Valley Vintners Association, 106 pages text, 116 illustrations; EarthVision, Inc. Cheshire, CT.

Swinchatt, Jonathan, and Howell, David G., 1990, EarthNectar: Napa Valley Wines and the Earth from which they Arise (52 minute video); EarthVision, Inc. Cheshire CT.

Swinchatt, Jonathan, and Howell, David G., 2004, The Winemaker's Dance: Exploring Terroir in the Napa Valley: University of California Press, Berkeley, California, 229 p.

Wallace, R. E. (ed.), 1990, The San Andreas Fault System: United States Geological Survey Professional Paper 1515, 283 p.

Wilson, James E., 1998, Terroir: The Role of Geology, Climate, and Culture in the Making of French Wines: University of California Press, Berkeley, California, 336 p.

PAPER 6 FROM THE "TERROIR, GEOLOGY AND WINE: A TRIBUTE TO SIMON J. HAYNES" SESSION HELD AT THE GEOLOGICAL SOCIETY OF AMERICAN ANNUAL MEETING, SEATTLE, WASHINGTON, NOVEMBER 3, 2003

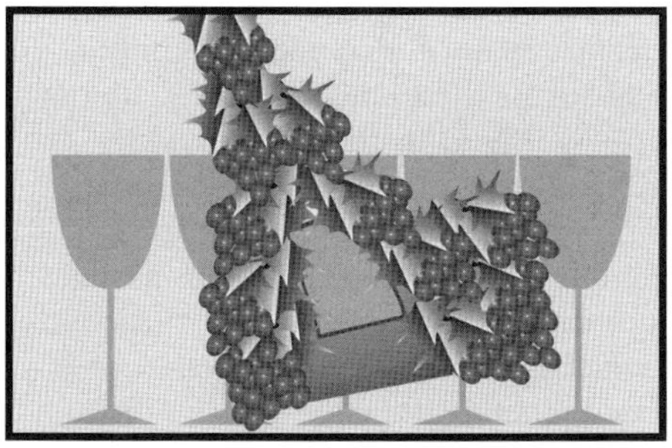

Terroir, wine and geoscience fieldwork education in Languedoc-Roussillon, southern France

Roger J. Suthren[1], Michael B. Fowler[2], and Paul D. Guion[3]

[1]*Geography, Earth and Environmental Sciences, University of Derby, Kedleston Road, Derby DE22 1GB, U.K., r.suthren@derby.ac.uk*

[2]*Earth and Environmental Sciences, University of Portsmouth, Burnaby Road, Portsmouth PO1 3QL, U.K.*

[3]*10 Green Lane, Belper, Derbyshire, DE56 1BY, U.K.*

SUMMARY

Geological training of undergraduate students in the Languedoc-Roussillon region of southern France involves the use of wine as a theme to link aspects of regional and applied geology, to introduce new geological field skills, and to increase student engagement on fieldwork. Here, in the largest wine-producing region in the world, varied geology and geological structure provide a great range of terroirs and microclimates for vine growing, resulting in a wide variety of wine styles and qualities. Geologically, two east-west trending mountain belts – the Pyrenees in the south, and the Montagne Noire-Cévennes in the north - are separated by a large sedimentary basin, the Languedoc Foreland Basin, filled mainly by continental sediments from Late Cretaceous to Eocene time and now forming lower ground in the centre of the region. The Languedoc basin passes south into the foothills of the Pyrenees, where basin sediments were involved in Pyrenean deformation. The resultant east-west ridges and valleys are exploited for viticulture. Post-Pyrenean sedimentary basins and Quaternary volcanoes provide added interest, and important applied aspects include quarrying, mining, hydrogeology and engineering geology. Student exercises that we have found of particular value include geological mapping and structural, stratigraphic and sedimentological studies at several scales; close comparison of geological contacts with vineyard or terroir boundaries; and soil nutrient analyses as related to vine hardiness and growth rates. Superb outcrop exposures and scenery, and excellent regional museums, as well as fine wines, make for instructive and memorable student field experiences.

RÉSUMÉ

La formation géologique d'étudiants du premier degré dans la région de Languedoc-Roussillon dans le sud de la France, permet de recourir au vin comme thème unificateur de certains aspects de la géologie régionale et de la géologie appliquée, de présenter de nouvelles techniques de géologie de terrain, et de renforcer la motivation des étudiants pour la géologie de terrain. Ici, dans la plus grande région vinicole au monde, la variété du contexte et des structures géologiques explique la présence d'une large gamme de terroirs et de microclimats propices à la viticulture, d'où la production d'une grande variété de styles et de qualités de vin. Du point de vue géologique, il existe deux chaînes de montagnes de direction est-ouest, soit les Pyrénées au sud, et la Montagne Noire-Cévennes au nord, de part et d'autre d'un grand bassin sédimentaire, le bassin d'avant-pays du Languedoc, renfermant principalement des sédiments continentaux du Crétacé supérieur à l'Éocène inférieur et qui constituent des basses terres au centre de la région. Vers le sud, le bassin du Languedoc devient le piedmont des Pyrénées, où les sédiments du bassin ont subit une déformation pyrénéenne. Les crêtes et vallées orientées est-ouest qui en résultent portent des vignes vinicoles. Des bassins sédimentaires post-pyrénéens ainsi que des volcans quaternaires ajoutent à l'intérêt géologique, et on y trouve des exemples de géologie appliquée comme des carrières, des mines ainsi que des applications de l'hydrogéologie et du génie géologique. Parmi les travaux pratiques offerts aux étudiants, il y a la cartographie géologique, les études structurales, stratigraphiques et sédimentologiques à diverses échelles; l'étude comparée des contacts géologiques et des limites des vignobles ou des terroirs; et les analyses des éléments nutritifs des sols en fonction de la robustesse et des taux de croissance des vignes. La beauté des affleurements et des paysages, l'excellence des musées ainsi que les grands vins de la région font de l'exercice une expérience formatrice et mémorable.

INTRODUCTION

This article aims to introduce the geology, scenery and terroirs of the world's largest vineyard area (Johnson and Robinson, 2001), the Languedoc-Roussillon region (Cavaillé, *in* Pomerol, 1989) of southern France (Fig. 1), and to emphasize the value of this complex and beautiful area for geoscience fieldwork education. The wide range of geological and landscape features provides great variation in bedrock geology, soil types, aspects, microclimates and terroirs.

Terroir is defined by Wilson (1998) as the totality of the elements of the vineyard habitat. Terroir and wine provide an attractive theme to link aspects of regional and applied geology, to introduce new field skills, and to increase student engagement on field courses, study tours for amateur groups and independent fieldwork projects. "In the Languedoc, geology and topography are an open book" (Marcus, 2000): the high percentage of outcrop, the variety of geology, structure and landforms in a relatively small area, and a pleasant climate, make this an ideal area for geological fieldwork.

Languedoc-Roussillon is a broad arcuate region, extending inland from that part of the Mediterranean coast, which is bounded to the south by the frontier with Spain, and to the east by the River Rhône (Fig. 1). Its southern physical boundary is the Pyrenees Mountains, from which the topography declines gradually northwards to a central plain drained by the rivers Aude and Hérault and their tributaries. This low ground continues east through the coastal plain to the lower Rhône valley. North of the coastal plain, the land rises to the hills of the Montagne Noire and the Cévennes, the southernmost part of the Massif Central. North again are the bare limestone plateaux of Les Causses.

The modern administrative region of Languedoc-Roussillon, and its component *départements* (Fig. 2) of Aude, Gard, Hérault, Lozère (not shown) and Pyrénées-Orientales, encompasses only part of the ancient provinces of Languedoc and Roussillon. Roussillon corresponds to the modern département of Pyrénées-Orientales, and forms the French part of Catalonia.

The climate of Languedoc-Roussillon is typically Mediterranean. The wine-growing area has been described by Cavaillé (1989) as a huge, sun-drenched arc protected from continental and Atlantic influences. Consequently there are few 'bad' vintages (Strang, 2002) compared with areas where the climate is less certain. It is hot in summer, cooler and more humid in winter. Mean temperatures range between a maximum of over 30° C, and a minimum of 3° C; extremes are greater, reaching 45° C in the exceptional summer of 2003. Sunny days are the norm, with around 2700 hours of sunshine annually in Montpellier (Fig. 1). Mean annual precipitation is on the order of 550 to 700 mm on the lower ground, with maximum aridity close to the Mediterranean coast, and up to 1400 mm in the mountains (average figures from 1971 to 2000: www.meteofrance.com/FR/climat/france.jsp). Rainfall is very unevenly distributed through the year, with a dry summer, and a rainy autumn and winter. Much rain falls during intense thunderstorms, often generating flash floods and producing hail, a serious hazard to the vines. Frost and snow rarely occur below 600m a.s.l., allowing vines to be grown up to this altitude or higher (Cavaillé, 1989). Winds are often strong, with dominant dry winds (the *Cers* and *Tramontane*) from the west, competing with a moist easterly wind (the *Marin*) from the Mediterranean. The western parts of the Languedoc-Roussillon region are more influenced by weather systems from the Atlantic, and tend to be cooler and more moist.

Soil cover on the higher ground, where the highest quality grapes are

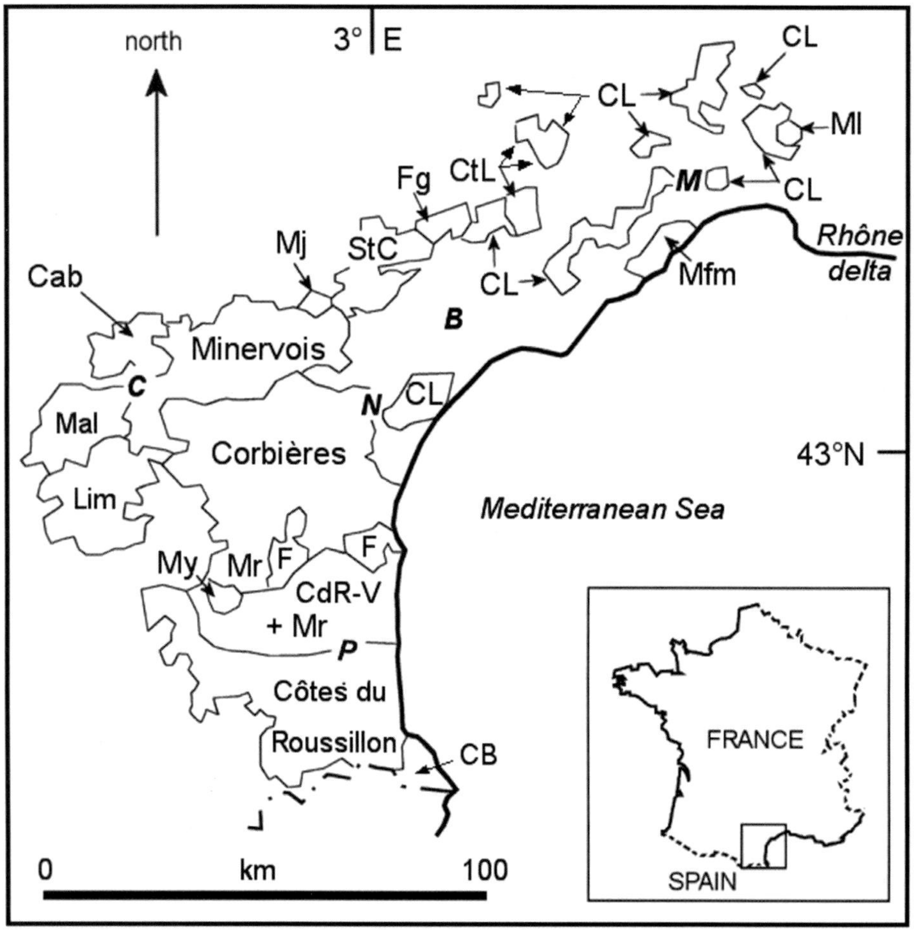

Figure 1. The wine appellations of Languedoc-Roussillon: all of the areas shown have Appellation d'Origine Contrôlée (= AC or AOC) status, except Côtes de la Malepère (VDQS; see text). Some smaller appellations are not shown, and there is often overlap between vineyard areas. Location of the region within France is shown on the inset map (bottom right). **Cities:** *B* - Béziers; *C* - Carcassonne; *M* - Montpellier; *N* - Narbonne; *P* – Perpignan. **Appellations**: **Cab** - Cabardès; **CB** - Collioure and Banyuls; **CdR-V** - Côtes du Roussillon Villages; **CL** - Côteaux du Languedoc (numerous separate vineyard areas); **CtL** - Clairette du Languedoc; **F** - Fitou; **Fg** - Faugères; **Lim** - Limoux and Blanquette/Crémant de Limoux; **Mal** - Côtes de la Malepère; **Mfm** - Muscat de Frontignan and Muscat de Mireval; **Mj** - Muscat de St Jean de Minervois; **Ml** - Muscat de Lunel; **Mr** - Rivesaltes and Muscat de Rivesaltes; **My** - Maury; **StC** - St-Chinian.

grown, typically is thin, and supports a natural flora dominated by drought-resistant plants, including herbs such as thyme, rosemary and lavender, and shrubs or small trees such as box, evergreen oaks and conifers. The typical flora - *garrigue* - developed on limestone soils differs somewhat from that developed on acid soils, and known as *maquis*. Thicker, more fertile soils are developed on the Tertiary to Quaternary sediments of the coastal plain, where minor areas of annual crops such as maize, sunflowers and vegetables relieve the monoculture of vineyards producing large quantities of mainly lower quality wine grapes.

This simplistic subdivision of the wine areas into slopes and plains belies the enormous range of geology, topography, soil types and terroir to be found across the region. Traditionally, terroir has been a more important influence on wine production than grape varieties, and most traditional wines of the region still are a blend of wines from several grape varieties. Over the last 20 years or more, named grape varietals have played a more important role, and the region now displays an exciting mix of traditional and 'New World' viticulture and winemaking.

Published literature on the relationship between geology, terroir and wine in Languedoc-Roussillon is limited. An important contribution is Albert Cavaillé's chapter in Pomerol's (1989, p. 216-241) geological field guide to the wine regions of France. Its usefulness is increased by frequent cross-references to the regional geological guides series (for this region, Gèze, 1979; Jaffrezo, 1977). James Wilson's (1998) book on French terroir has been widely acclaimed. The short chapter on Languedoc-Roussillon gives a useful overview, but contains a number of errors of geography and geology. Strang (2002) covers individual appellations in greater depth, concentrating on the wines and individual growers, with less detail on soils and bedrock. Several standard works on wine contain short but useful sections on the region's vineyards (e.g. Anglade, 1990; Johnson and Robinson, 2001). Other publications cover the geology of the region at the level of the general interest reader (Bilotte et al., 1989; Garrone, 1991; Bousquet, 1997). A page of Internet links on the region's geology and scenery (www.virtual-geology.info/languedoc/terroir.html) includes colour versions of the illustrations in this article. The Internet also provides further information on the Appellation d'Origine Contrôlée (= AC or AOC) wines of Languedoc (www.languedoc-wines.com), and the regional Vins de Pays (www.vins-de-pays.info).

WINES OF LANGUEDOC-ROUSSILLON

History

This region has a 2500-year history of vine cultivation. After about 600 B.C., the Greeks introduced the first vines, but there is no definite evidence for winemaking until the region became part of the Roman Empire around 122 B.C. Narbo Martius (now Narbonne, Fig. 1) became provincial capital of one of the Romans' most important wine-producing regions. After the fall of the Roman Empire in the 5th century A.D., viticulture went into a long decline. Replanting began seriously in the 9th century A.D., and wine production was important from the Middle Ages onwards, with the strong influence of the Church, who were major landowners (Simms and Simms, 1991). The Canal du Midi, linking the Mediterranean and the Languedoc vineyards to the Atlantic at Bordeaux, was an important transport link for the wine industry from its completion in 1681 (www.canalmidi.com). Until the early 19th century, Languedoc wines were produced in the upland areas, whilst the fertile plains were used mainly for growing grain. From the mid 19th century on, the plains became the site of industrial-scale planting of inferior vines to produce enormous volumes of low-quality wine (Cavaillé, 1989; Anglade, 1990).

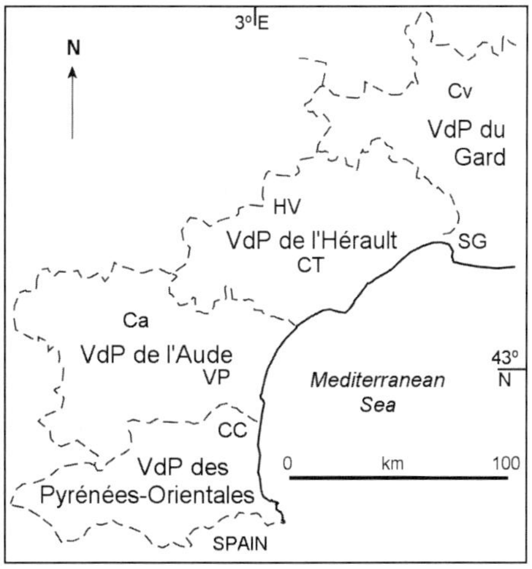

Figure 2. The Vins de Pays (VdP) areas of Languedoc-Roussillon. The map covers most of the region included in the broad VdP d'Oc. The main départemental VdPs are shown, along with a few examples of smaller VdP areas: **Ca** - VdP de la Cité de Carcassonne; **CC** - VdP des Côtes Catalanes; **CT** - VdP des Côtes de Thongue; **Cv** - VdP des Cévennes; **HV** - VdP de la Haute Vallée de l'Orb; **SG** - VdP des Sables du Golfe du Lion; **VP** - Vdp de la Vallée du Paradis.

Grapes and Wines

Wine Production and Producers

Languedoc-Roussillon is the largest wine-producing area in the world, with some 298,000 ha (736,000 acres) of vineyards, producing more than 30 per cent of all French wine: around 1,800,000,000 litres per year, more than 80% of total production being red. Some 18% of the volume produced is of AC status, and another 62% is Vin de Pays (regional wines): both designations include controls of quality and of geographic origin. The remaining 20% of regional wine production goes to Vins de Table, the lowest rung on the quality ladder. The latter are mainly for domestic consumption, sold in bottle or in bulk, and only rarely reach the wine consumer outside France.

Some 70% of regional production is from 376 co-operative wineries (*caves co-operatives*), supplied by tens of thousands of growers. Growers own their own vineyards, but the wine is made at a central winery for the village or group of villages. The rise of the *cave co-operative* movement in the 20th century was an important force for change, including closer control over hygiene and conditions of fermentation. There are around 2600 large and small private grape and wine producers (*vignerons* and *vigneronnes*). Recent years have seen an influx of large businesses and 'flying winemakers' from Australia and New Zealand. Some co-operatives and private producers have now grouped into larger organizations for marketing purposes. Unlike some regions of the world, most growers of grapes here are also involved in wine-making. (Data from: Anglade, 1990; Johnson and Robinson, 2001; www.vins-de-pays.info; www.onivins.fr).

Until the 1980s, this region was viewed overwhelmingly as a 'wine lake', accommodating much of the European Community's surplus of inferior wine. Since then there has been a concerted effort to grub out inferior vines and replace them with superior varieties or with other crops. Almost all of the infamous Aramon and many Carignan vines, the source of much of the cheaper red wine, have disappeared. Thus the quantity of wine produced has decreased markedly, whilst quality and prices have increased, but there are still many good quality wines available at moderate prices. Of the progress made in the recent past, Johnson and Robinson (2001, p. 138) observed: "...by the end of the 1990s, the Languedoc-Roussillon had established itself as France's best-value and most exciting wine region by far".

Present Situation

Whilst the region's output is dominated by red wines, there is a substantial production of still rosé and white wines, sparkling white wines, and fortified wines from both red (Grenache noir) and white (chiefly Muscat) grapes. There is a broad spectrum of wine styles. Some traditional reds, a minority of them aged in oak, can be laid down for many years. By contrast, there is an increasing production of many named varietal 'New World' style wines made for drinking young. There is, however, a great difference in vineyard practice between the New World of southern France and the New World of the western and southern hemispheres. New World techniques such as computer-controlled drip irrigation and intensive canopy control are rarely seen in Languedoc-Roussillon, where the grape-ripening process is left largely to nature, the traditional approach in France.

Generally, the highest quality wines are produced from grapes grown on stony slopes with poorer soils and slightly cooler climates. Some 67% of AC wines from Languedoc-Roussillon are red, with a further 8% of rosé wines from the same grape varieties. In the vineyard, grape varieties and the percentage of land which each occupies, planting density, cultivation and pruning of vines are controlled by regulations which vary from one AC area to another (Johnson and Robinson, 2001; Strang, 2002) and are overseen by the Institut National des Appellations d'Origine (INAO) (www.inao.gouv.fr). Yields are usually restricted to 40 hectolitres per hectare or less. In the small hilly vineyards, grapes can only be harvested manually, and indeed this is compulsory for certain appellations. In the winery, regulations control vinification methods and minimum alcohol content, and wines are subject to tasting and analysis before the award of AC status. With the exception of AC Fitou, AC Clairette du Languedoc and some fortified wines in 1948, Languedoc-Roussillon wines were not accorded AC status until the 1970s or later. In the 1950s to 1970s some wines bore the lower status of VDQS (Vins Délimités de Qualité Supérieur; Cavaillé, 1989), and have since been promoted to AC. Only one VDQS remains, in Côtes de la Malepère (Fig. 1).

AC wines use traditional Mediterranean grape varieties, also found in adjacent regions such as the southern Rhône and northern Spain. For red and rosé wines, the most important traditional grapes are Carignan, Grenache noir, Syrah (= Shiraz), Cinsaut (or Cinsault), and Mourvèdre. Whilst Carignan has been regarded as a 'work-horse' variety, old vines may produce low yields of high quality wines in good areas. Most traditional red wines are a blend of separately vinified grape varieties, but there is an increasing frequency of single variety AC wines (e.g. Syrah, or old vines Carignan) as, for example, in AC Minervois and AC St-Chinian. Bordeaux varietals are not traditionally used, except in the far west of the region where the increasing Atlantic influence on climate is complemented by a westerly influence on grape variety. Merlot, Cabernet Sauvignon, Cabernet Franc and Cot are blended with Mediterranean varieties in AC Cabardès and VDQS Côtes de la Malepère (Fig. 1).

Although only 6% of the AC wine volume is white, the list of approved grape varieties is longer than for reds (Strang, 2002). Traditional white grapes include Grenache Blanc, Maccabeu, Clairette, Bourboulenc (= Malvoisie), Picpoul, Muscat, Vermentino and Mauzac (used almost exclusively for the sparkling champagne-method Limoux wines). Unusually, some of the white AC wines are made from a single grape variety, notably AC Clairette du Languedoc, and Picpoul de Pinet (in AC Côteaux du Languedoc).

Sweet fortified wines (Vins Doux Naturels, or VDN), produced by arresting fermentation by the addition of alcohol, are a speciality of Roussillon and some smaller areas in Languedoc. There are essentially two styles: delicate, peachy Muscats, and port- or Madeira-like reds, mainly from the Grenache noir grape.

Wine Quality Levels

In the last few years the AC system in Languedoc-Roussillon has started to define quality at a more detailed level. The AC Côteaux du Languedoc grouping (Fig. 1) has granted independence to some vineyards, such as those of AC St-Chinian and AC Faugères. Within larger appellations, individual villages or small groups of villages have begun to be recognised, by inclusion of their names in the appellations. This resembles the system in Bordeaux, for example, where individual villages (e.g. AC Pauillac) have a higher status than the regional appellation (e.g. AC Médoc). In Languedoc-Roussillon, both the region and the village are included in the AC name. The first were AC Côtes du Roussillon-Villages Caramany, AC Côtes du Roussillon-Villages La Tour de France, and AC Minervois La Livinière. More recently this status has been awarded to AC Côtes du Roussillon-Villages Tautavel, AC Côtes du Roussillon-Villages Lesquerde, and AC St-Chinian Berlou (2003). AC Corbières Boutenac (2004) is the first named village in Corbières.

Lower quality (at least in theory) Vins de Pays and Vins de Table vines flourish on the coastal plain and valley alluvium, where machine harvesting is widespread. Languedoc-Roussillon produces 70% of all French Vins de Pays wines (www.vins-de-pays.info). These must be from a specific area; produced from grape varieties approved for that area; and must observe maximum crop yields and minimum alcohol contents. This is a relatively new category, introduced in 1983, and the wines are often good value.

There is a single classification, Vin de Pays d'Oc, covering wines from the entire region, and a category for each of its four wine-producing départements (Fig. 2). These broad designations allow blending of moderately priced wines to produce a consistent product. More interesting are Vins de Pays from 55 smaller specified areas, ranging in size from vineyards surrounding single villages to much larger areas: examples are shown in Figure 2. The smaller Vin de Pays areas have more stringent regulations for grape varieties, yields, analysis and tasting (Johnson and Robinson, 2001).

There are two principal groups of Vins de Pays wines: blends from traditional grape varieties in areas not given AC status - *vins de terroir*, and wines from mainly non-traditional grape varieties. Bordeaux, Burgundy and Rhône grape varieties dominate the latter, especially Cabernet Sauvignon, Merlot and Syrah for red wines, and Chardonnay, Sauvignon Blanc and Viognier for whites. Many such wines are made from single varietals, and are known as *vins de cépage*. Most Vin de Pays wine production is aimed at the moderately priced supermarket shelf, but some individual producers are experimenting with high quality (and high price) Vins de Pays, avoiding the more restrictive AC regulations, and sometimes producing great, unconventional wines in the manner of some 'super Tuscan' producers in Italy.

To summarize, there is a concentration on terroir in traditional Languedoc-Roussillon wines, and on grape variety in more modern wine styles. The greatest success is likely to be achieved by careful matching of grape variety with terroir, a process which is far from complete when compared with other French wine areas (e.g. Wilson, 1998). Geology and terroir also play an important role in marketing, with mountains, rocks and fossils making frequent appearances on wine bottles, both as evocative images and in the often flowery language of the back label.

REGIONAL GEOLOGY AND TERROIRS OF LANGUEDOC-ROUSSILLON

The principal geological elements of the region (Fig. 3) are two east-west trending mountain belts - the Pyrenees in the south, and the Montagne Noire-Cévennes in the north. They are separated by a large sedimentary basin, the Languedoc Foreland Basin, forming lower ground in the centre of the region. Still further north are the vast limestone plateaux of Les Causses. The mainly east-west trending ridge and valley topography, defined by the geological structure, has a strong influence on several elements of terroir. On a smaller scale, tectonic fracturing enhanced by weathering is important, locally enabling vines to grow in rock without a well-developed soil. The region is bounded to the east by an undulating coastal plain, underlain by cover sediments of Oligocene to Recent age, and Pliocene to Pleistocene basaltic volcanoes.

Les Causses: the Limestone Plateaux

Northeast of the Montagne Noire, large expanses of more or less flat-lying

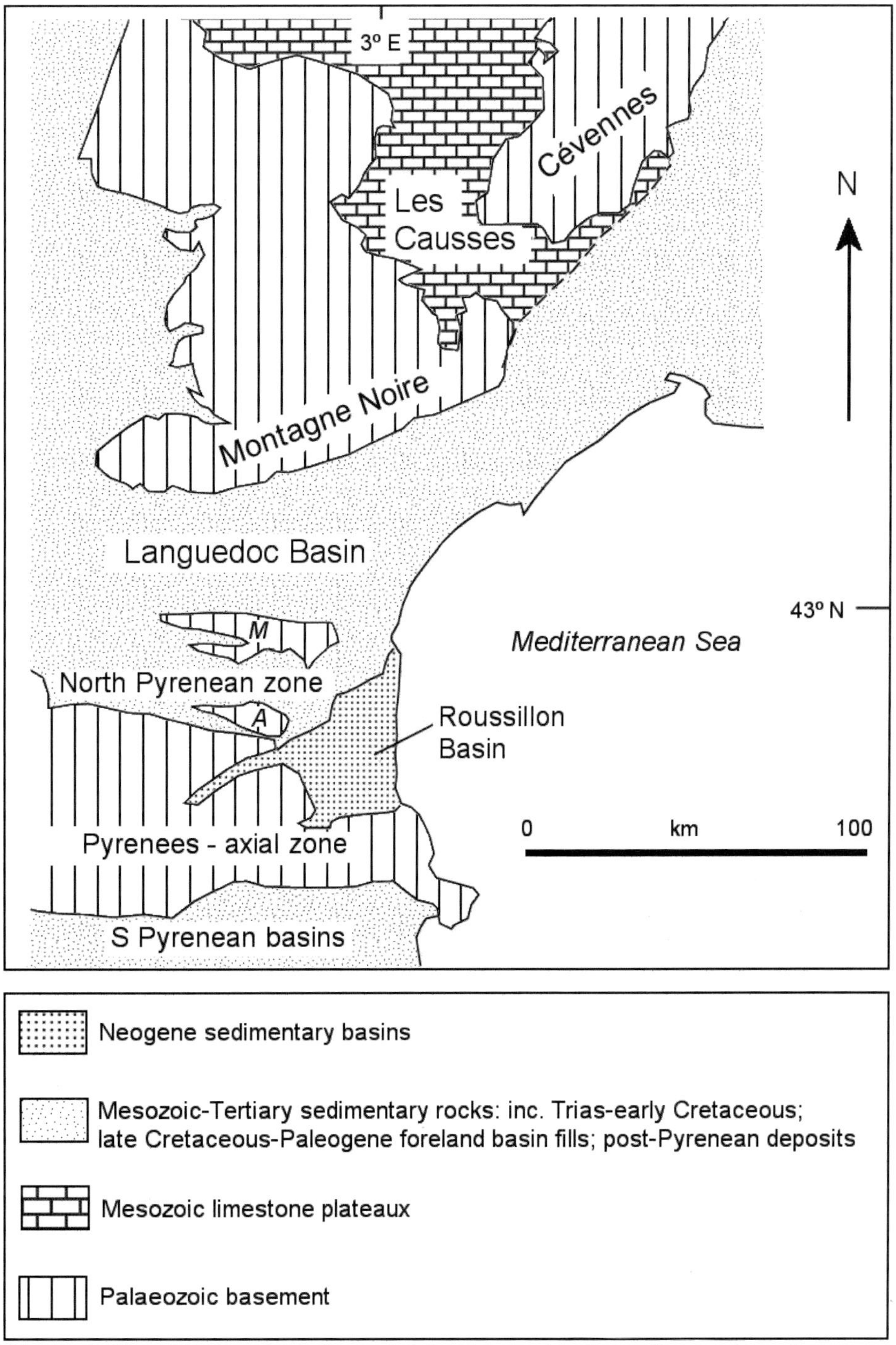

Figure 3. Simplified geological map of the Languedoc-Roussillon region of Southern France. In the North Pyrenean Zone, sedimentary rocks of the Languedoc Foreland Basin, and underlying Mesozoic and Paleozoic rocks, are involved in fold-thrust tectonics. Northern satellite massifs of the Eastern Pyrenees include Massif d'Agly **(A)** and Massif de Mouthoumet **(M).**

Mesozoic limestones form barren, thin-soiled, sparsely vegetated plateaux of Les Causses (Fig. 3). The thick limestone succession lies with profound unconformity on Paleozoic sedimentary, igneous and metamorphic rocks of the Cévennes and Montagne Noire terrain. Triassic red beds are developed locally above the unconformity and below the limestones. Long periods of attack by acid waters have allowed major cave and karst systems to develop within the limestones. Whilst the plateaux are too high topographically to support viticulture, they nevertheless form an important terroir, that of the Lacaune breed of sheep. Their milk, curdled and matured in limestone caves, is the basis of the blue cheeses of Roquefort, which have their own AC status, and provide an excellent accompaniment to red and fortified wines from further south.

Montagne Noire

South of Les Causses lies the Montagne Noire terrain - the 'Black Mountain' (Fig. 3). This highland area, up to 1210 m high, is the southernmost part of a much larger late Paleozoic mountain belt, the French Massif Central. The 'axial zone' of the Montagne Noire shows high grade metamorphic and granitic igneous rocks, interpreted as gneiss domes, developed in either an extensional or a strike-slip regime (van den Driessche and Brun, 1992; Franke, 2001). The ground underlain by these axial zone rocks is generally too high to support viticulture (Fig. 4), except at its western end in the AC Cabardès vineyards, whose steep, stony soils are developed on granite, gneiss and schist (Strang, 2002). Elsewhere, the west-southwest-east-northeast trending mountain range protects the vineyard areas to the south from cold northerly winds (Cavaillé, 1989).

The axial zone of the Montagne Noire is structurally overlain to both north and south by Paleozoic sedimentary successions in large fold nappes. In the southern nappes (Fig. 4), the sedimentary succession includes Cambrian shallow marine limestones and sandstones, Ordovician continental slope and deep marine mudstones and sandstones, Devonian deep marine and shelf carbonates, and Carboniferous deep marine sediments. These last were involved in continental margin collapse during synsedimentary Variscan tectonics (Engel et al., 1978), creating spectacular olistostromes of blocks tens to hundreds of metres across, mainly of Devonian and Carboniferous limestone. These olistoliths are contained in a 'matrix' of Carboniferous shale on which vines, such as those of Cabrières (AC Côteaux du Languedoc) and AC Faugères are grown. Post-tectonic Upper Carboniferous (Stephanian) sediments lie with profound angular unconformity on earlier, deformed sedimentary rocks.

Figure 4. View of Montagne Noire from the south. Beyond the village of Cessenon in the foreground, vines grow on Holocene alluvium of the River Orb, rising to gentle slopes underlain by Eocene-aged sediments. The steeper wooded hills in the distance beyond are developed on Paleozoic nappes, and support AC St- Chinian vines in places. The distant ridge forming the skyline is the axial zone gneisses and granites of the Montagne Noire at elevations up to ~1090 m. above sea level.

Wine-Making, Southern Montagne Noire peripheral zone

Of particular interest in wine-making on the southern margin of the Montagne Noire are Ordovician marine sandstones and shales and, to a lesser extent, similar lithologies of the lower Carboniferous, prized for growing sturdy vines in their well-drained, acid soils consisting largely of rock fragments on top of weathered bedrock (Fig. 5). These soils are the basis for robust and often long-lived red grape vines, especially in AC St-Chinian, AC Faugères and AC Minervois, and the Cabrières terroir of AC Côteaux du Languedoc.

In Berlou, the first 'named village' (AC St-Chinian Berlou, 2003) in the St-Chinian appellation, vines grow exclusively on such Ordovician *schistes* (Fig. 5). "Schist" is a common but erroneous translation for this rock type, correctly translated as shale, but at Berlou more accurately described as interbedded fine sandstone and shale. Although these acid soils are dry, AC St-Chinian Berlou forbids irrigation, and grapes must be manually harvested. Berlou is a fine illustration of the contention of Christophe Baron of the Cailloux Vineyard, Walla Walla Valley of Washington and Oregon, that "the best wines in France grow in rocks" (pers. comm., 2003). Old Carignan vines (sometimes 100 years in age) produce high quality grapes here.

Additional economic interest of these Paleozoic units includes the mining of copper and gold (and by-product arsenic, used for many years as a fungicide in the vineyards). Since antiquity, red 'marbles' from the Devonian limestones have been exploited for use in public buildings such as the palaces of Versailles, and the White House in Washington D.C.

Some vineyard areas cross the unconformity separating the Montagne Noire from the Languedoc Basin to the south (Fig. 3). The large appellations of AC St-Chinian (Fig. 4) and AC

Minervois straddle this geological divide, and wines from the same grape varietals grown on the different terroirs have distinctly different characters. The AC Côteaux du Languedoc (Fig. 1), the second largest Languedoc AC area by volume of wine produced (Strang, 2002), is a loose grouping of 17 named and contrasting sub-appellations within a larger area given only the general AC appellation. They lie partly on Paleozoic rocks of the Montagne Noire, and partly on younger cover rocks of the Languedoc Basin to the south of the Montagne Noire (Fig. 3).

Languedoc Foreland Basin

The east-west trending Languedoc Foreland Basin (Fig. 3) is an eastward continuation of the Aquitaine Foreland Basin, and together these basins form a broad corridor of low ground between the Atlantic and the Mediterranean. Foreland basin development was intimately linked to the Pyrenees Mountain Belt of southern France and northern Spain. The Pyrenees are a compressional Alpine-type mountain belt formed during late Cretaceous to early Tertiary times. The Montagne Noire foreland continues southwards as the largely concealed basement beneath the Languedoc Foreland Basin, and has a sharp contact, usually an angular unconformity, sometimes a steep fault, with the overlying Languedoc Basin sediments. The Languedoc Foreland Basin has an asymmetrical fill, thickest in the centre and south, with a thinner, less complete succession onlapping the Montagne Noire to the north. In some thrust sheets within the basin, foreland basin sediments lie unconformably on older Triassic to early Cretaceous sedimentary rocks, predominantly carbonates, with tropical bauxitic soils sporadically developed at the contact.

The stratigraphy of the Languedoc Foreland Basin fill was described by Bilotte et al. (1989), and aspects of the sedimentology by Freytet and Plaziat (1982) and Plaziat et al. (1987). The basin has a complex history of late Cretaceous and Tertiary fill. Deposition was controlled by paleogeographic changes associated with convergence between the European and Iberian plates, and eventual collision and partial subduction of the continent of Iberia. Foreland basin development began in early late Cretaceous time with shallow marine deposits, including rudist reefs (Fig. 6). These are succeeded by a thick succession of continental siltstones, sandstones and conglomerates deposited in fluvial channels and floodplains, lacustrine siltstones and limestones, and a wide variety of palaeosols (Fig. 7). These continental sediments range in age from Late Cretaceous (Campanian) to Late Paleocene. A marine transgression from the west in late Paleocene to early Eocene time deposited thick fossiliferous sandstones and marls (the *Ilerdian*). A return to continental conditions resulted in further fluvial and lacustrine deposition in mid to late Eocene time, including sandstones of the Molasse de Carcassonne.

The Languedoc Foreland Basin sediments and underlying rocks were involved in Pyrenean deformation. South-to-north compression resulted mainly in east-west ridges and valleys, giving a strong structural grain to the landscape, and controlling the slope and aspect of vineyards (Fig. 8). The degree of folding and faulting varies. At the simplest level, tilting of early Tertiary rocks in the southern Minervois resulted in gentle south-facing dip slopes and steeper north-facing scarps, both exploited for growing vines. The effects of deformation generally become more intense towards the south, in the direction of the overthrusting Pyrenean mountain belt. In the St-Chinian area, however, strong compressional tectonics reach to the northern margin of the Languedoc Foreland Basin (Fig. 8).

Figure 5. Vines growing in rubble from Ordovician marine sandstones and shales in the AC St-Chinian Berlou vineyards

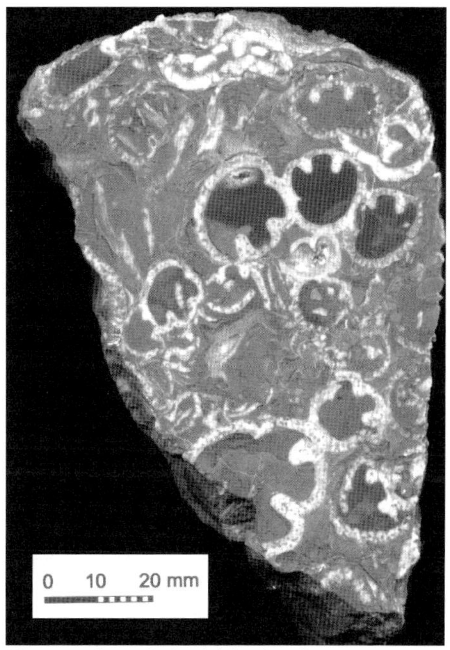

Figure 6. Polished block of Upper Cretaceous rudist reef rock, showing rudist bivalves in cross-section; Fontfroide, AC Corbières.

Here, and further south in Corbières, complex thrusting, folding and repetition and inversion of the stratigraphy are seen (Fig. 9a). An important factor is the presence of thick evaporite units of late Triassic age: these acted in ductile fashion, with major thrusting developed along them (Fig. 9b).

Tectonic control of topography may be as important an element of terroir as bedrock geology in these vineyard areas. Whilst vines are grown predominantly on the softer lithologies such as late Cretaceous flood plain sediments (Fig. 10a), it is not uncommon to find vines on deep-ploughed limestone rubble on the more resistant limestone ridges, where thin *terra rossa* type soils have developed (Fig. 10b).

In the northeastern part of the foreland basin, deformed sediments underlie several of the AC Côteaux du Languedoc vineyard areas, including that of the eponymous white grape, Picpoul de Pinet. This terroir lies on an area of important late Cretaceous (Maastrichtian) dinosaur nest sites, displayed at the Plaine des Dinosaures open-air museum near Mèze, between Béziers and Montpellier (Fig. 1).

Terroirs and Wines

Much of the southern portion of the basin and of the Pyrenean foothills is covered by the AC Corbières and AC Fitou vineyards (Fig.1). This area of more than 90 villages is, in terms of volume, the most important AC in Languedoc-Roussillon, and the fourth largest in France (Strang, 2002). AC Corbières was subdivided into 11 named terroirs, possibly based more on marketing and on local political factors than on geological and soil elements of terroir. Nine villages in the south of AC Corbières also produce AC Fitou (Fig. 1), in two separate areas (Cavaillé, 1989). From the rugged landscapes of the Hautes Corbières in the south, the land descends northwards to the gentler slopes of the plain, providing a great variety of terroirs and microclimates. Red wine grapes are dominant, and grow on a wide range of lithologies including Paleozoic shales and sandstones of the Massif de Mouthoumet (Fig. 3), Jurassic and lower Cretaceous limestones, and folded, thrust foreland basin sediments (Figs. 7, 9). High limestone ridges in the east are part of a large thrust sheet (the

Figure 7. Continental sedimentary rocks of the Languedoc Foreland Basin. **A.** Aerial view of Albas, AC Corbières. The succession dips and youngs to north (left). View shows alluvial and lacustrine rocks of Upper Cretaceous to Paleocene age; Eocene marine marls are capped by reef limestones on far left skyline. Vineyards to left of the village are located on valley colluvium. **B.** Deep, conglomerate-filled channel eroded into palaeosol, Upper Cretaceous, Arques, AC Blanquette de Limoux. **C.** Paleosol with vertical root traces, Lower Paleocene, Arques. **D.** Fluvial channel conglomerates and sandstones, Lower Paleocene, Albas.

eastern Corbières 'nappe') emplaced northwards during the Pyrenean orogeny; the thrust surface, now partly exhumed, has several hundred metres of relief. One ridge, geologically an eastern outlier of the Corbières, forms the southernmost Côteaux du Languedoc terroir of La Clape (east of Narbonne, Fig. 1), reputed to be the site of the best Roman vineyards (Strang, 2002).

At the western end of the wine-producing part of the basin, along the cooler upper Aude River valley, vines flourish on Late Cretaceous to Eocene foreland basin sediments, and on underlying Paleozoic rocks of the Massif de Mouthoumet (Fig. 3). Of great interest here are the finds of dinosaur bones (e.g. Buffetaut and Le Lœuff, 1997), and more recently of an articulated skeleton ("Eva": www.dinosauria.org) in Maastrichtian fluvial sediments (Fig. 7). The most common remains are of a group of herbivorous sauropods, the titanosaurs. A new genus, *Ampelosaurus* (dinosaur of the vines) was first discovered here (Le Lœuff, 1995). In the same stratigraphic units nearby are important egg and nest sites (Cousin et al., 1994). Fossils and reconstructions form part of the excellent displays at the Dinosauria museum in Espéraza (www.dinosauria.org), and the excavation site may be visited.

The dinosaur-bearing sediments form some of the soils of the Limoux vineyards (Fig. 1). AC Blanquette de Limoux, a champagne-method sparkling wine, is produced largely from the Mauzac grape, with some Chardonnay and Chenin Blanc. This is the oldest sparkling wine in France, dating back at least to 1531 (Strang, 2002), and predating Champagne by at least 150 years. A more up-market version, AC Crémant de Limoux, contains higher proportions of Chardonnay and Chenin Blanc. The same area was awarded a new appellation, AC Limoux (1993) for still white wines dominated by barrel-fermented Chardonnay. Red varietal Vins de Pays include Pinot Noir, Merlot and Cabernet Franc.

North of Limoux, the furthest west of the Languedoc appellations, VDQS Côteaux de la Malepère (Fig. 1) is developed on late Eocene fluvial conglomerates and sandstones shed from the rising Pyrenees into the foreland basin. The Eocene sandstones here and further north were the major source of building stone for the construction and restoration of the medieval fortified city of Carcassonne (Fig. 1). The interesting red wines are a mix of Bordeaux and Mediterranean grape varieties, as found in AC Cabardès to the north (Fig. 1).

North Pyrenean Foothills

The Languedoc Foreland Basin passes gradually southward into the foothills of the Pyrenees. The foreland basin sediments were themselves involved in the compressive deformation which produced the Pyrenees Tertiary fold-mountain belt. Major structures are dominated by southward dipping, northward verging thrusts. Allochthonous units consist largely of Mesozoic and Paleozoic rocks transported some 20 km northwards (Fischer, 1984). Thus, vineyard areas such as southern AC Corbières and AC Limoux (Fig. 1) can be said to belong to both the foreland basin and the fold-thrust belt of the North Pyrenean zone. Immediately south of the high Corbières lie the best vineyards of Roussillon, the AC Côtes du Roussillon-Villages, consisting of 28 villages including 4 named villages. Most of their red grape vines

Figure 8. Aerial view of mainly late Cretaceous limestone ridges repeated by folding and thrusting. The ridges provide the topographic framework for AC St-Chinian vineyards. Note the patchwork of small vineyards, mainly on softer lithologies between limestone ridges. South is to the top, and the town of St-Chinian is at upper far left.

grow on distinctive dark grey-brown soils on hummocky topography developed on deepwater calcareous mudstones and impure limestones of Early Cretaceous (Albian) age. These vineyards are also the source of the largest production in Languedoc-Roussillon of sweet, fortified wines termed Vins Doux Naturels, from both red and white grapes, especially from AC Rivesaltes, AC Muscat de Rivesaltes and AC Maury. Some of the best vines, including those of AC Côtes du Roussillon-Villages Caramany, grow on the Agly massif (Fig. 3), a northern satellite of the Pyrenees, in harsh conditions on fractured gneisses and schists with little true soil. This massif is bounded to the south by the North Pyrenean fault zone, a fundamental crustal fracture which may represent the suture between the Iberian and European plates. The fault zone is roughly followed by the boundary between AC Côtes du Roussillon-Villages and AC Côtes du Roussillon to the south.

The Pyrenees

In the south, the mountain belt of the Pyrenees stretches from the Atlantic Ocean in the west to the Mediterranean in the east, forming a natural topographic border between Spain and France. The 'axial zone' forming the high Pyrenees, and its eastward continuation, the Chaîne des Albères, are part of a Late Paleozoic (Variscan) mountain belt with deformed sedimentary rocks, low to high-grade metamorphism, and granitic intrusions.

The Early Tertiary orogen is superimposed on the Paleozoic one, with major fold nappes and foreland basins on both sides of the chain. The deep structure of the Pyrenees as revealed by the ECORS (Etude Continentale et Océanique par Réflection et Réfraction Sismique) seismic reflection profile some 100 km to the west (Choukroune, 1989) indicates limited subduction of the Iberian continental lithosphere beneath the southern edge of the European plate (Roure et al., 1989). Oblique convergence, involving anti-clockwise rotation of the Iberian plate relative to the European plate, began in Late Cretaceous (Late Campanian) time with eventual continental collision in mid-late Eocene time (Engeser and Schwentke, 1986). Total shortening is estimated at greater than 100 km (Williams and Fischer, 1984). Uplift ceased during the Miocene, when the Pyrenees were broadly at sea level. Then up to 3 km of uplift occurred in Pliocene to Recent times (McCaig and Wickham, 1986). Thus, the present relief may be a result not so much of the major compressive events of the Eocene as of much more recent uplift owing to fault reactivation (Gourinard and Bandet, 1980).

Figure 9. Pyrenean deformation. **A.** View to the southeast of the Ripaud fold from St. Victor Ermitage, AC Corbières. This is a footwall syncline below a major north-directed thrust, marked by arrows. **B.** Complex deformation in ductile Upper Triassic evaporites, near Durban, AC Corbières; scale in centre of photograph.

Geology and Wines

AC Banyuls, AC Collioure and some of the southernmost AC Côtes du Roussillon vineyards (Fig. 1) are located geologically on the axial zone of the Pyrenees, much reduced in height as the mountains descend towards the Mediterranean, forming the rugged coastline of the Côte Vermeille, which continues south into Spain as the Costa Brava. AC Banyuls - sweet, fortified red wines from the Grenache noir grape with considerable potential for bottle ageing - is produced in the same small, steep area as the robust AC Collioure dry reds, also dominated by Grenache noir. The soils are mainly fragmented, weathered Cambrian phyllites. Steep, terraced vineyards rising to 750m above sea level have numerous stone walls and elaborate drainage channels to protect against soil erosion (Cavaillé, 1989).

Post-Pyrenean Sedimentary Basins and Volcanoes
Geology, Terroirs and Wines

The lower lying areas of the broad Mediterranean coastal plain supply much of the bulk wine industry, as noted above. Quality wines are produced from a number of areas, however, particularly on 'islands' of slightly higher, steeper ground. Younger sediments and sedimentary rocks unconformably overlie the Paleozoic to Eocene units deformed by the Variscan and Pyrenean orogenies. In latest Oligocene time, faulting and subsidence formed the Narbonne-Sigean basin (Bilotte et al., 1989; Szulc et al., 1991). Sub-horizontal carbonate, clastic and evaporite sediments of late Oligocene and younger age fill the basin. Extensive gypsum deposits were mined until 1991.

A large part of the Mediterranean coastal plain is underlain by Miocene marine and continental sediments, deposited during a highstand when the Mediterranean was more than 100m higher than its present level, and the shoreline lay tens of kilometres west of its present position. These post-tectonic sedimentary rocks lie unconformably on deformed Mesozoic and Tertiary rocks. In southern AC St-Chinian, the Miocene shoreline may be traced as a steep cliff with boulder talus at its foot, passing offshore into coarse bioclastic gravels rich in quartz, and oyster banks. The Miocene forms characteristic yellow, fossiliferous soils occupied by vineyards of parts of AC St-Chinian, AC Corbières and AC Côteaux du Languedoc.

Huge volumes of Pliocene to Quaternary clastic sediments were deposited by high-energy river systems as a result of erosion during glacial and interglacial periods with fluctuating sea levels. The largest are those forming the plains and terraces of the Roussillon embayment (Fig. 3): the vast outwash plains of the three main rivers draining the eastern Pyrenees, the Agly, the Têt and the Tech, are now left as high terraces incised by the present courses of these rivers. The terraces provide a distinctive and extensive terroir of the AC Côtes du Roussillon (Fig. 1), in which the 'soils' consist largely of water-rounded cobbles. Most of the 120 villages of AC Côtes du Roussillon, and some belonging to AC Rivesaltes immediately to the north, grow their vines on such ground. On the lower parts of this coastal plain, and on the alluvial valley floors of the main rivers, fruit orchards and vegetables have supplanted the vine.

Similar terroirs are seen in the far northeast of the region, where AC Costières de Nîmes vines grow on rounded cobbles, the outwash of an earlier Quaternary incarnation of the Rhône River and its delta, now incised by the modern river.

Another distinctive Quaternary lithology is calcareous tufa or travertine, deposited from springs emerging from limestones of various ages. In places this has built up extensive terraces used for vineyards, as at l'Estabel in Cabrières (AC Côteaux du Languedoc, Fig. 1). In others, calcareous tufa has made an indirect contribution to the wine industry, quarried as a light, strong and easily worked construction material, formerly used for constructing wine presses, barrel supports and window and door frames.

Finally, a group of Pliocene to Pleistocene volcanoes extends northwards from the coast at Cap d'Agde. These lie on a more or less north-south trend, a southward continuation of the much more extensive Neogene to Recent volcanism of the Massif Central, related to young extensional tectonics. The AC Clairette du Languedoc (Fig. 1) vineyards are underlain partly by alkali basalt lava flows and scoria cones, containing mantle peridotite nodules. This Chasselas grape also forms the basis for the important table grape trade around Clermont l'Hérault (Cavaillé, 1989).

All of these younger rocks are affected by Tertiary to Recent extensional faulting, such as the major Cévennes fault system. This fault, and the faults bounding the Roussillon basin, are still active. The region experiences frequent moderate earth tremors, with major earthquakes occurring every few centuries.

GEOLOGICAL FIELD EDUCATION IN LANGUEDOC-ROUSSILLON

Languedoc-Roussillon is an excellent area for students to examine different aspects of regional geology and to link them to interpret the geological and tectonic evolution of a complex region. On recent final-year undergraduate field courses at Oxford Brookes University, the geology of wine has been used as a theme to link the different aspects of regional geology, an approach that significantly increases student engagement with fieldwork.

Among the factors which make this region so valuable for fieldwork education at different academic levels is the highly varied geology, including geological settings and rock types not normally encountered by students in Britain and northern Europe. Examples, some noted above, include well-exposed sedimentary successions and geological structures in a deformed foreland basin and young mountain belt; marine carbonate facies including Devonian carbonate mud mounds quarried for 'marble'; pelagic limestones with iron-manganese nodules; evaporites of Triassic and Tertiary age; clear evidence of Tertiary sea-level change; and volcanoes

Figure 10. Vineyards of AC St-Chinian, on Upper Cretaceous sediments. **A.** Boundary between top surface of lacustrine limestone (left) and overlying red siltstones runs along the edge of the vineyard, and is easily mapped. Vines grow on the softer siltstone lithology. **B.** Vines growing in deep-ploughed lacustrine limestone.

of Quaternary age. The region also provides a wealth of applied geology and evidence of the use of geological materials through nearly half a million years of human history, as well as examples of modern geological hazards, especially flash flooding.

Several stratigraphic sections of global importance occur in Languedoc, especially in the Upper Paleozoic deep marine limestones of the Montagne Noire. The Global Boundary Stratotype and Point (GSSP) for the Devonian-Carboniferous boundary (www.stratigraphy.org/devcar.htm) occurs, very conveniently, in the Cabrières vineyards of AC Côteaux du Languedoc. In addition, two of the Upper Devonian stage boundaries have their GSSPs here, in AC St-Chinian: the base of the Frasnian stage, which is also the base of the Upper Devonian (www.stratigraphy.org/frasn.htm); and the base of the Famennian stage (www.stratigraphy.org/famen.htm), immediately above a mass extinction level of global importance, the Upper Kellwasser Horizon. Visiting these sections allows students to appreciate the importance of a global stratigraphic framework and the methods used to construct it.

Numerous well-exposed sections of Paleozoic, Mesozoic and Cenozoic sedimentary rocks occur that are suitable for recording and interpreting sedimentological information, at various scales and from a wide range of depositional settings. Examples range from Devonian-Carboniferous pelagic limestones and cherts, to sections in which Tertiary shallow marine sediments pass upward into fluvial and lacustrine deposits. Some of these sedimentary sections are abundantly fossiliferous, and provide excellent opportunities for detailed paleontological, paleoecological and biostratigraphical studies. Examples include late Cretaceous rudist reefs, and early Eocene rocks rich in benthic foraminifera and molluscs. The fauna and sedimentary settings in these deposits of the Tethyan realm contrast markedly with their age equivalents in northern Europe. Field study of the fill of the Languedoc foreland basin linked well to topics and methods covered in a basin analysis course given immediately before the field course.

Modern sedimentary environments are also well represented, and provide useful comparisons for the interpretation of the ancient successions. Students gain experience of depositional settings usually unfamiliar to them, including evaporites in lagoons of the Mediterranean margin, the deposits of ephemeral rivers, and an active fault-bounded basin.

Student Exercises

Geological Mapping; Regional Geology

As in many undergraduate geoscience courses, over the years there has been a move away from the 'Cook's tour' type of field trips, toward fieldwork based on a variety of supervised exercises carried out by students to develop specific skills and areas of knowledge. Examples of such student project work, carried out individually or in small teams, include structural analysis and mapping of small-scale structures in areas of compressional tectonics, leading to the interpretation of regional structure in the Montagne Noire (Variscan orogeny: late Paleozoic) and AC St-Chinian (Pyrenean orogeny: early Tertiary). The good exposure enables students to see and understand folds and thrusts in three dimensions. In the southern Montagne Noire, a geological traverse allows students to observe increasing metamorphic grade and deformation from the margin to the axial zone of the metamorphic belt. Assessment of the field courses required students to produce a synthesis of the geological evolution of the region, based on field evidence, and presented as posters. The high quality and originality of many of these posters clearly demonstrates the enthusiasm generated by fieldwork in this region.

For more than 20 years the authors have supervised undergraduates who have chosen Languedoc-Roussillon as the location for their independent mapping projects, an important element of most British geology degree courses. Students spent several weeks of the summer before their final year making a detailed geological map at 1:10,000 scale, and researching a report on the geology of an area of several square kilometres. Requirements for such areas include good exposure and access, varied lithologies, sufficient structural complexity, and the likelihood of matching the student's ability to the difficulty of the area. More resistant lithologies, especially limestones, are well-exposed, and often highlight tectonic structures (Fig. 8). Terraced vineyards provide outcrops and float in softer, less well-exposed formations suited to vine growing. Vineyard and terroir boundaries commonly correspond to geological contacts (Fig. 10a), and thus students learn to appreciate connections between geology, landforms, and land use. Project work of this type imparts both subject-specific and transferable skills. On seeking employment after graduation, students have found that employers value highly their experience of project work and logistics in a foreign country.

Soil and Nutrient Analysis; Mine Visits

One example of skill development is an exercise designed to investigate links between soil chemistry and grape vine growth, using standard Rapide soil nutrient analysis kits. In a selected vineyard, small teams of students take soil samples and use colorimetric tests to determine pH and content of potassium, magnesium, phosphate and nitrate (Fig. 11). The advantage of this method is that it gives almost instant results. In a field of young Syrah vines, students were able to determine low levels of most soil nutrients, possibly explaining why the vines were getting off to a slow start. Analytical results from other terroirs of the same property are provided for comparison, and students are able to correlate nutrient levels from soils developed on different lithologies. The field analyses are followed by a visit to the winery, a talk from the vignerons about their terroirs, traditional wine-making methods and their wines, and a tasting. The exercise is a good example of developing skills with a clear scientific purpose. For students interested in geochemistry, there is opportunity for further laboratory work to determine soil trace element levels using ICPAES analysis. This study of modern soils complements the study of semi-arid paleosols in the late Cretaceous and early Tertiary, and tropical, humid bauxitic and lateritic paleosols in the early Cretaceous. The results of the chemical analyses of modern vineyard soils are applied in discussion of the composition of ancient soils.

Other wine-related localities

include a visit to a large, disused underground gypsum mine now used as a wine ageing cellar, providing an opportunity to study fresh exposures of Tertiary evaporites. Metalliferous mining also has been important, particularly in the Montagne Noire, where copper mines exploited in Neolithic and Roman times are visited. The Salsigne gold mine in the AC Cabardès vineyards, at one time the largest gold mine in Europe, closed recently. The large open pit clearly demonstrates the complex, overturned fold structure of the Lower Paleozoic rocks of the Montagne Noire. The mine's closure presents interesting environmental problems in terms of landscape restoration and minimization of arsenic and cyanide release to the environment.

Virtual field trips

During a final-year elective course (www.brookes.ac.uk/geology/8361/8361welc.html) designed to develop both scientific and communication skills, students had the opportunity to produce virtual field trips, and several chose to focus on aspects of the geology of Languedoc-Roussillon, based on their experiences on field courses or mapping projects. These may be viewed online (www.virtual-geology.info/lstudentpages/).

Other Attributes of Interest for Student Studies

Sea-cliff and artificial exposures through late Neogene basaltic volcanoes demonstrate complex relationships and rapid lateral and vertical changes, allowing able students to interpret a sequence of events. Student project work has included petrography and SEM-EDS analysis of basalts and mantle peridotite nodules from these volcanoes.

In addition to the links between geology and wine, other aspects of the applied geosciences are evident in Languedoc-Roussillon. These emphasize the broader appeal of the region to students in geography, environmental sciences, ecology and related disciplines. Hydrology and water supply are of utmost importance in this partly semi-arid region, and are linked inextricably to the classic surface and subterranean karst features of the limestone areas, of various ages from Cambrian to Tertiary. Another aspect of hydrology is the devastating flash floods that accompany sporadic thunderstorms, such as those of 1999 in which 35 people lost their lives across the region (www.corbieresweb.com/english/inondation.htm). The effects of these disasters on watersheds, settlements, vineyards, infrastructure and people are still clearly visible, and provide students with an insight into the human effects of natural processes, and an introduction to the construction and use of hazard maps. Landslides and earthquakes are further examples of natural hazards whose past and future effects may be studied in this region.

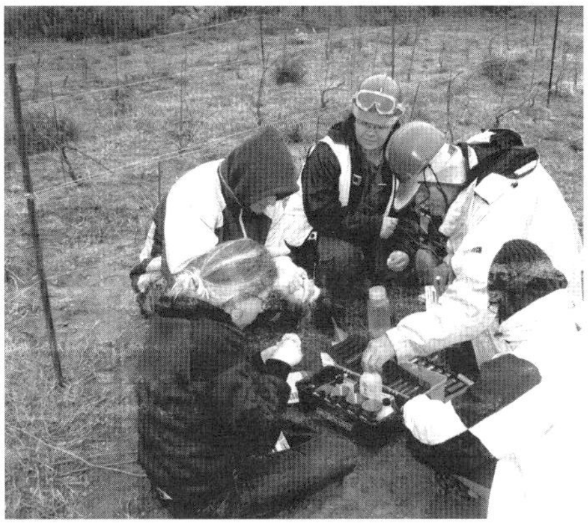

Figure 11. Students involved in soil nutrient analysis in a vineyard of young Syrah vines, Albas, AC Corbières.

Aspects of civil engineering and engineering geology are readily studied in the roads and bridges which cross the upland parts of Languedoc-Roussillon. The most spectacular example in recent years has been the construction of the A75 viaduct over the Tarn gorges at Millau, completed in January 2005, and arguably the tallest bridge in the world. This bridge is one of the last links in the completion of a continuous autoroute from Paris over the Massif Central to the Languedoc coast and into northeastern Spain.

Student Feedback

Undergraduate feedback on these field courses, through questionnaires and informal evaluation, has been strongly positive. In 2001, with responses from 28 out of 38 students (74%), almost all students agreed that the fieldwork had enabled them to acquire an integrated view of geology by gaining experience of rocks and geological relationships. They felt able to unravel complex geological histories, satisfying one of the main learning outcomes. The applied aspects of the course also were satisfied, with a majority of students agreeing that the fieldwork had enabled them to become more aware of the role of geology in society. There were no specific questions about wine-related geology, but only one of 28 students expressed the view that the limited time spent visiting wineries would have been better spent elsewhere! Most other comments were positive: students found the field course interesting, enjoyable and intellectually challenging, and felt that experiencing the geology of a different country was valuable.

CONTINUING EDUCATION IN LANGUEDOC-ROUSSILLON

Several groups and individuals, including the first author, have conducted study tours or "geological vacations" in the region, for groups of interested amateurs, who are often the most enthusiastic of students. Such tours allow a broader view of the links between geology, landscape, natural history, history, archaeology and use of geological materials. Many of the locations visited on undergraduate field courses as outlined above are also suitable for study tours, often with emphasis on different aspects. Sites visited also include museums such as that located over the site of a Gallo-Roman pottery where amphorae for storing wine were manufactured from local clay, and a quarry where quartz-rich Eocene limestones were formerly worked for millstones. At the Tautavel museum and cave site, hominid and other fossil finds illustrate 450,000 years of human history. Show caves and karst scenery include those of Roquefort, and production of cheese, olive oil and other foods can be shown to have geological relevance. A vital part of the scientific process on such study tours is the rigorous evaluation of the products of the terroirs!

The varied geology and the potential to practice a wide range of fieldwork skills development all suggest that the region would provide a good location for a North American-style field camp of several weeks duration.

CONCLUSIONS

Languedoc-Roussillon is very much a region in evolution in terms of its dynamic geology, and of the industries deriving from the rocks and soils, notably wine-making and tourism. The region will continue to be an attractive location for geoscience fieldwork education because of its broad range of well-exposed geology and landforms, and the opportunity for students from northern Europe and further afield to visit and study an unfamiliar region and culture.

ACKNOWLEDGEMENTS

We wish to acknowledge the co-operation of vignerons, vigneronnes and friends in Languedoc-Roussillon. We thank Nick and Clare Bradford of Domaine des Pensées Sauvages at Albas (AC Corbières) for visits to their caveau and vineyards, and provision of soil samples. Members of the Harrow and Hillingdon Geological Society were willing 'guinea pigs' on a study tour to the region in September 2003. Through the eyes of many past and present project and research students at Oxford Polytechnic and Oxford Brookes University, we have come to know the region, its geology and its vineyards in much more detail than would otherwise have been possible. We thank Roger Macqueen, Larry Meinert and an anonymous reviewer for significant suggestions in improving the manuscript.

REFERENCES

Anglade, P., (ed.), 1990, Larousse Wines and Vineyards of France: Ebury Press, London, United Kingdom, 639 p.

Bilotte, M., Capera, J.C., Debat, P., Debroas, E.J. Duranthon, F., Marre, J. Pulou, R., Rey, J. and Vizlainu, D., 1989, Géologie du Département de l'Aude: Mémoire de la Société d'Etudes Scientifiques de l'Aude, 80 p.

Buffetaut, E., and Le Lœuff, J., 1997, Late Cretaceous dinosaurs from the foothills of the Pyrenees: Geology Today, March-April, p. 60-68.

Bousquet, J.-C., 1997, Géologie du Languedoc-Roussillon: Les Presses du Roussillon/Editions du B.R.G.M., Montpellier, France, 142 p.

Cavaillé, A., 1989, Languedoc-Roussillon, in Pomerol, C., (ed.), The Wines and Winelands of France: Robertson McCarta, London, United Kingdom, p. 216-241.

Choukroune, P., 1989, The ECORS Pyrenean deep seismic profile reflection data and the overall structure of an orogenic belt: Tectonics, v. 8, p. 23-39.

Cousin, R., Breton, G., Fournier, R., and Watte, J.-P. 1994. Dinosaur egg-laying and nesting in France, *in* Carpenter, K., Hirsch, K.F. and Horner, J.R., eds., Dinosaur Eggs and Babies: Cambridge University Press, New York, p. 56-74.

Engel, W., Feist, R. and Franke, W., 1978, Synorogenic gravitational transport in the Carboniferous of the Montagne Noire (S-France): Zeitschrift der Deutschen Geologischen Gesellschaft, v. 129, p. 461-472.

Engeser, T. and Schwentke, W., 1986, Towards a new concept of the tectonogenesis of the Pyrenees: Tectonophysics v. 129, p. 233 - 242.

Fischer, M. W., 1984, Thrust tectonics in the North Pyrenees: Journal of Structural Geology, v. 6, p. 721-726.

Franke, W., 2001 (abstr.), Exhumation of LP gneisses in the Montagne Noire (S-France): the collapse has collapsed: European Union of Geosciences XI, Abstracts, Symposium FMF1, p. 235.

Freytet, P. and Plaziat, J.-C., 1982, Continental carbonate sedimentation and pedogenesis - Late Cretaceous and Early Tertiary of southern France: Contributions to Sedimentology, no. 12, E. Schweitzerbart'sche Verlagsbuchhandlung, Stuttgart, Germany, 213 p.

Garrone, B., 1991, La Géologie de l'Hérault: Les Ecologistes de l'Euzière, Cahiers de Cuculles no. 3, Prades-le-Lez, France, 92 p.

Gèze, B., 1979, Languedoc Méditerranéen - Montagne Noire: Guides Géologiques Régionaux, Masson, Paris, France, 191 p.

Gourinard, Y., and Bandet, Y., 1980, La morphogenèse des Pyrénées Orientales, in Solé, L., and Souquet, P., (eds.), La Chaîne Alpine des Pyrénées, 26th International Geological Congress Bulletin, Bulletin des Centres de Recherches Exploration-Production Elf Aquitaine, Mémoire no. 3, p. 172-173.

Jaffrezo, M., ed., 1977. Pyrénées orientales, Corbières: Guides Géologiques Régionaux. Masson, Paris, France, 191 p.

Johnson, H., and Robinson, J., 2001, World Atlas of Wine, 5th Edition: Mitchell Beazley, United Kingdom, 352 p.

Le Lœuff, J., 1995. *Ampelosaurus atacis* (nov. gen., nov. sp.), un nouveau Titanosauridae (Dinosauria, Sauropoda) du Crétacé supérieur de la Haute Vallée de l'Aude: Les Comptes rendus de l'Académie des Sciences, Paris, series IIa, v. 321, p. 693-699.

Marcus, K., 2000, Languedoc wakes up: Wine Spectator Online. www.winespectator.com/Wine/Archives/Show_Article/0,1275,2580,00.html

McCaig, A.M., and Wickham, S.M., 1986, The tectonic evolution of the Pyrenees: a workshop: Journal of the Geological Society of London, v. 141, p. 379 - 381.

Plaziat, J.-C., Freytet, P. and Marec, P., 1987, Sédimentation molassique et paléopedogenèse en Languedoc: Les Editions de l'Association des Sédimentologistes Français, Livre no. 3, 127 p.

Pomerol, C., (ed.), 1989, The Wines and Winelands of France; Geological Journeys (English Language Edition): Robertson McCarta, London, United Kingdom, 370 p.

Roure, F., Choukroune, P., Berastegui, X., Munoz, J. A., Villien, A., Matheron, P., Bareyt, M., Seguret, M., Camara, P., Deramond, J., 1989, ECORS deep seismic data and balanced cross sections: geometric constraints to trace the evolution of the Pyrenees: Tectonics, v. 8, p. 41-50.

Simms, P. and Simms, S., 1991,The Wines of Corbières and Fitou: Cristal/Privat, Toulouse, France, 128 p.

Strang, P., 2002, Languedoc-Roussillon: The Wines and Winemakers: Mitchell Beazley, London, United Kingdom, 192 p.

Szulc, J., Roger, P., Mouline, M. P. and Lenguin, N., 1991, Evolution of lacustrine systems in the Tertiary Narbonne Basin, northern Pyrenean foreland, southeast France: Special Publication, International Association of Sedimentologists, v. 13, p. 279-290.

van den Driessche, J. and Brun, J.P., 1992, Tectonic evolution of the Montagne Noire (French Massif Central): a model of extensional gneiss dome: Geodinamica Acta, v. 5, p. 85-99.

Williams, G.D. and Fischer, M.W., 1984, A balanced cross section across the Pyrenean orogenic belt: Tectonics, v. 3, p. 773-780.

Wilson, J., 1998, Terroir: The role of Geology, Climate and Culture in the Making of French Wines: Mitchell Beazley, London, United Kingdom, 336 p.